The IMA Volumes in Mathematics and its Applications

Volume 63

Series Editors
Avner Friedman   Willard Miller, Jr.

Institute for Mathematics and
its Applications
IMA

The **Institute for Mathematics and its Applications** was established by a grant from the National Science Foundation to the University of Minnesota in 1982. The IMA seeks to encourage the development and study of fresh mathematical concepts and questions of concern to the other sciences by bringing together mathematicians and scientists from diverse fields in an atmosphere that will stimulate discussion and collaboration.

The IMA Volumes are intended to involve the broader scientific community in this process.

Avner Friedman, Director
Willard Miller, Jr., Associate Director

* * * * * * * * * *

IMA ANNUAL PROGRAMS

1982–1983	Statistical and Continuum Approaches to Phase Transition
1983–1984	Mathematical Models for the Economics of Decentralized Resource Allocation
1984–1985	Continuum Physics and Partial Differential Equations
1985–1986	Stochastic Differential Equations and Their Applications
1986–1987	Scientific Computation
1987–1988	Applied Combinatorics
1988–1989	Nonlinear Waves
1989–1990	Dynamical Systems and Their Applications
1990–1991	Phase Transitions and Free Boundaries
1991–1992	Applied Linear Algebra
1992–1993	Control Theory and its Applications
1993–1994	Emerging Applications of Probability
1994–1995	Waves and Scattering
1995–1996	Mathematical Methods in Material Science

IMA SUMMER PROGRAMS

1987	Robotics
1988	Signal Processing
1989	Robustness, Diagnostics, Computing and Graphics in Statistics
1990	Radar and Sonar (June 18 - June 29)
	New Directions in Time Series Analysis (July 2 - July 27)
1991	Semiconductors
1992	Environmental Studies: Mathematical, Computational, and Statistical Analysis
1993	Modeling, Mesh Generation, and Adaptive Numerical Methods for Partial Differential Equations
1994	Molecular Biology

* * * * * * * * * *

SPRINGER LECTURE NOTES FROM THE IMA:

The Mathematics and Physics of Disordered Media

Editors: Barry Hughes and Barry Ninham
(Lecture Notes in Math., Volume 1035, 1983)

Orienting Polymers

Editor: J.L. Ericksen
(Lecture Notes in Math., Volume 1063, 1984)

New Perspectives in Thermodynamics

Editor: James Serrin
(Springer-Verlag, 1986)

Models of Economic Dynamics

Editor: Hugo Sonnenschein
(Lecture Notes in Econ., Volume 264, 1986)

H.S. Dumas   K.R. Meyer   D.S. Schmidt
Editors

Hamiltonian Dynamical Systems
History, Theory, and Applications

With 52 Illustrations

Springer-Verlag
New York Berlin Heidelberg London Paris
Tokyo Hong Kong Barcelona Budapest

H.S. Dumas
Department of Mathematical Sciences
University of Cincinnati
Cincinnati, OH 45221 USA

K.R. Meyer
Institute for Dynamics
Department of Mathematical Sciences
University of Cincinnati
Cincinnati, OH 45221 USA

D.S. Schmidt
Department of Computer Science
University of Cincinnati
Cincinnati, OH 45221 USA

Series Editors:
Avner Friedman
Willard Miller, Jr.
Institute for Mathematics and its
  Applications
University of Minnesota
Minneapolis, MN 55455 USA

Mathematics Subject Classifications (1991): 70-06, 58F99, 01A99, 58F05, 70H99, 70G99

Library of Congress Cataloging-in-Publication Data
Hamiltonian dynamical systems : history, theory, and applications / H.
  S. Dumas, K. R. Meyer, D. S. Schmidt, editors.
      p.   cm. — (The IMA volumes in mathematics and its
  applications ; v. 63)
      Proceedings of the International Conference on Hamiltonian
  Dynamical Systems, held at the University of Cincinnati, Mar. 1992.
      Includes bibliographical references.
      ISBN-13:978-1-4613-8450-2   e-ISBN-13:978-1-4613-8448-9
      DOI: 10.1007/978-1-4613-8448-9
      1. Hamiltonian systems — Congresses. I. Dumas, H. Scott,
  II. Meyer, Kenneth R. (Kenneth Ray), 1937-   . III. Schmidt, Dieter
  S. IV. International Conference on Hamiltonian Dynamical Systems
  (1992:University of Cincinnati) V. Series.
  QA614.83.H335   1995
  514′.74 — dc20   94-44179

Printed on acid-free paper.

Production managed by Hal Henglein; manufacturing supervised by Jeffrey Taub.
Camera-ready copy prepared by the IMA.

9 8 7 6 5 4 3 2 1

ISBN-13:978-1-4613-8450-2

The IMA Volumes
in Mathematics and its Applications

Current Volumes:

FOREWORD

This IMA Volume in Mathematics and its Applications

HAMILTONIAN DYNAMICAL SYSTEMS: HISTORY, THEORY, AND APPLICATIONS

is based on the proceedings of an IMA Participating Institutions (PI) International Conference held at the University of Cincinnati in March 1992. Each year the 29 Participating Institutions select, through a competitive process, several conferences proposals from the PIs, for partial funding. This conference brought together a whole spectrum of researchers in mathematics and physics as well as researchers in the history of science. We thank H. Scott Dumas, Kenneth R. Meyer, and Dieter S. Schmidt for organizing the meeting and for editing the proceedings.

Avner Friedman

Willard Miller, Jr.

PREFACE

From its origins nearly two centuries ago, Hamiltonian dynamics has grown to embrace the physics of nearly all systems that evolve without dissipation, as well as a number of branches of mathematics, some of which were literally created along the way. The power and elegance of Hamiltonian methods were seized upon in the mid-nineteenth century by theoretical astronomers, who reduced some of their problems to finding enough integrals of Hamiltonian systems. Their enthusiasm was soon dampened however, by the difficulty of finding these integrals in all but the simplest cases. Still, though their success was not fully understood, formal calculations based on Hamiltonian methods were extremely useful in predicting celestial motions. Efforts focused on this paradox near the end of last century by H. Poincaré showed that the emphasis on solving the equations for physical systems was largely futile, and that the important features of a system's behavior could be found by other means. While astronomers sought to improve the accuracy of their planetary orbit calculations, Poincaré instead asked— and formulated methods for understanding—whether the solar system in its present configuration is ultimately stable. These new methods altered the course of mathematics itself, as the fields of topology and dynamical systems were born out of the qualitative, geometric view of problems in celestial mechanics and Hamiltonian systems.

The qualitative approach to dynamical systems has remained an active research area, especially now that computers provide a kind of mathematical laboratory in which to view systems of increasing complexity. Driven partly by the discovery in these experiments of strange new phenomena, and partly by the maturation of the mathematical disciplines arising out of Poincaré's work, the results and methods of Hamiltonian systems have grown and penetrated neighboring fields of research. As a result, old questions are resolved and new questions arise at a faster and faster pace.

The contents of this volume reflect the wide scope and increasing influence of Hamiltonian methods, in both theory and applications, and are the result of much writing and editing following the International Conference on Hamiltonian Dynamical Systems held at the University of Cincinnati in March of 1992. We were fortunate to have present at the conference a whole spectrum of researchers in mathematics and physics from more than half a dozen countries, as well as several researchers in the history of science. Their perspective is displayed in the historical articles here, so that these proceedings are a slice not only of state-of-the-art methodology in Hamiltonian dynamics, but also of the bigger picture in which that methodology is embedded.

We are pleased to acknowledge the National Science Foundation, the Institute for Mathematics and its Applications (IMA) at the University of Minnesota, and the Charles Phelps Taft Foundation at the University of

Cincinnati for the generous support that made the conference possible. We also thank Patricia V. Brick, Stephan J. Skogerboe, and Kaye Smith of the IMA for their expert help in preparing this volume.

H.S. Dumas
K.R. Meyer
D.S. Schmidt
University of Cincinnati

Conference on Hamiltonian Dynamical Systems
Institute for Dynamics
University of Cincinnati
March 25–28, 1992

1. Jacques Laskar
2. Carles Simo
3. Pierre Lochak
4. Ernesto Lacomba
5. Judith Arms
6. Curtis Wilson
7. Craig Fraser
8. Bruce Pourciau
9. N.P.K. Swami
10. Martin Kummer
11. Roger Broucke
12. Ernesto Perez-Chevala
13. Scott Dumas
14. Antonio Giorgilli
15. Jeff Xia
16. Don Saari
17. Clark Robinson
18. Esmeralda Sousa Dias

19. Jim Murdock
20. Liam Healy
21. Vince Coppola
22. Bruce Miller
23. Ed Wright
24. Rick Wicklin
25. Jurgen Poschel
26. Dan Offin
27. Debra Lewis
28. Ken Meyer
29. Joaquin Delgado
30. Wolfgang Scherer
31. Jaume Llibre
32. Luis Martinez
33. Tasso Kapper
34. Rick Moeckel
35. Leonid Faybusovich
36. Patrick McSwiggen

37. L.C. Wang
38. John Maddocks
39. Jürgen Scheurle
40. Eugene Gutkin
41. Victor Donnay
42. Chris Golé
43. Dick Hall
44. Peter Rejto
45. Andre Deprit
46. Dave Hart
47. Dieter Schmidt
48. Sergio Alvarez
49. Antonio Elipe
50. Joe Gruendler
51. Edoh Y. Amiran
52. Boris Kupershmidt

CONTENTS

THE CONCEPT OF ELASTIC STRESS IN EIGHTEENTH-CENTURY MECHANICS: SOME EXAMPLES FROM EULER*

CRAIG G. FRASER†

Introduction. In the historical evolution of any highly mathematized physical theory there occurs an interesting interplay between the development of physical concepts and ideas, on the one hand, and the application and extension of mathematical techniques on the other. The subject of continuum mechanics offers in this respect a very clear example. Combining mathematical elements taken from calculus, analytic geometry and linear algebra, it provides a general and powerful approach to the investigation of physical phenomena, one based nevertheless on a specific and highly idealized conception of the way in which matter interacts.

The subject as it emerged in the writings of Navier and Cauchy in the 1820s had roots in several different branches of pure and applied science.[1] Considerable knowledge was available about engineering design and the strength of material structures, a field of research that had been investigated in the eighteenth-century and was familiar to French-trained engineers in the post-Revolutionary period. Theoretical mechanics, the science that became known in nineteenth-century France as "rational" mechanics, had reached a very high level of development in the classic treatises of Euler, Lagrange and Laplace. Finally, the advent of the wave theory of light led to an interest in analyzing wave phenomena based on an elastic-solid model of the ether.

In surveying the eighteenth-century background to Cauchy's work one it struck by how little of the modern theory was in place in the period before Coulomb. A wide range of problems was considered, numerous special results were reached, and sophisticated mathematical techniques were frequently employed; overall, however, the subject remained in a very rudimentary state of development. Even the most basic results concerning the distribution of shear stress in beams, for example, were beyond the scope of the theory, such as it was then.

The present paper seeks to examine aspects of the early history of the concept of stress, focusing on examples from the work of Euler, and to indicate some of the difficulties that were involved in the original formulation of the stress principle. It is intended as a case study in concept formation

---

* Paper presented at International Conference on Hamiltonian Dynamical Systems, March 1992, University of Cincinnati.

† Institute for the History and Philosophy of Science and Technology, Victoria College, University of Toronto, Toronto, Ontario M5S 1K7, Canada, Tel: 416-9778-5135; Fax: 416-978-3003; E-Mail: cfraser@epas.utoronto.ca.

[1] Belhoste [1991, Chapter 6] provides a discussion of Cauchy's early work on elasticity.

1

within the domain of mathematical physics. In reference to physical science in the eighteenth-century the historian Thomas Hankins [1985, 30] has observed: ". . . then as now physical concepts developed much more slowly than did new mathematical techniques. Physical concepts are extraordinarily difficult to create. Mathematical methods may not be any simpler, but when they are needed they seem to be found much more quickly."[2]

The stress principle. The stress principle is used to analyze the static and dynamic behaviour of a continuous body when it is acted upon by applied forces. There are two distinct ideas involved in this principle. The first, which is logically independent of the concept of stress *per se*, is that a body may for the purposes of analysis be imagined to be divided into two parts: one of these parts is isolated for study and the effect of the second part upon the first is replaced by a specified set of forces acting at the boundary. The second idea is that the given separation into two parts may be regarded as being accomplished by a plane of division, with the effect of the second part on the first being understood in terms of forces per unit area or stresses distributed over this plane and acting across it.

It is this second idea that is taken up and developed in the full theory of stress. One distinguishes between normal and shear stress and analyzes the mechanical behaviour of the body in terms of the relations between the various stress components acting on an infinitesimal cube of material contained in the body.

Background to Euler's early research. In the early eighteenth century there was considerable work on determining the shape assumed by a perfectly flexible body under various loadings. The basic idea, developed by the brothers Johann and Jakob Bernoulli and published by Jakob Hermann, was to analyze the tensions acting on a segment of a hanging cable and to obtain a differential equation that described the resulting curve. Since the cable was regarded as inextensible and perfectly flexible no assumptions were necessary concerning the elastic behaviour of the material comprising the cable.

Assumptions about elasticity did enter into research on two distinct and important problems, the problem of fracture and the problem of the elastica. In the study of fracture one attempted to determine the maximum load that a beam of given material and dimensions can sustain without breaking. In the problem of bending one was concerned with determining the shape assumed by a rod or lamina in equilibrium when subject to given

---

[2] Hankins is referring to the history of physical science where the mathematical techniques in question were already largely in place. It should be noted that within mathematics itself the process of concept formation is far from straightforward. The establishment of such subjects as Galois theory, differential geometry or partial differential equations illustrates the slowness and difficulty with which the relevant techniques and concepts were identified and developed. On a more philosophical level one could argue that the absence of an empirical referent in pure mathematics places a special premium on the formation and development of concepts there.

external forces. In the case of the elastica these forces were assumed to act at the ends of the rod causing it to bend into a curve.

In the problem of fracture researchers such as Leibniz, Varignon and Parent obtained results that can be readily interpreted in terms of modern formulas and theory. Typically they assumed that the beam was joined transversally to a wall and that rupture occurred at the joining with the wall. Here the physical situation directly concentrated attention on the plane of fracture. The conception then current of the loaded beam as comprised of longitudinal fibres in tension is readily understood today in terms of stresses acting across this plane.

In the problem of elastic bending by contrast researchers were much slower to develop an analysis that connected the phenomenon in question to the internal structure of the beam. Here there was nothing in the physical situation that identified for immediate study any particular cross-sectional plane. In all of Jakob Bernoulli's seminal writings on the elastica the central idea of stress fails to receive clear identification and development.

Two papers by Euler. The formative period in the development of Euler's early research in mechanics occurred during the ten years following his move at age twenty to St. Petersburg in 1728. A colleague of Hermann's (until 1731) and Daniel Bernoulli's, he benefited from the favourable intellectual atmosphere that existed at this time for research in the Russian academy. If one surveys his early work on mechanics it shows a familiarity with Johann and Jakob Bernoulli's writings in the *Acta Eruditorum*, Newton's *Principia Mathematica*, Hermann's *Phoronomia*, Taylor's *Methodus incrementorum* and Varignon's various papers on particle dynamics in the Paris Academy's memoirs.

(a) Euler's "De oscillationibus annulorum elasticorum"

An example illustrating how Euler understood elastic deformation during the period is provided by an unpublished paper, dating (we believe) from sometime the 1730s.[3] Euler considered an "annulus" (washer-like ring) that is disturbed from its equilibrium position and set as a consequence into motion. Figure 1 shows a part $abBA$ of the ring in its normal and stressed configuration. The segment $AaeE$ is regarded as being composed of concentric filaments. The inner line ae remains constant under deformation while the outer line AE is stretched to $A\varepsilon$. The triangle $eE\varepsilon$ shows the stretching of the filaments as one proceeds outward from a to A. Euler sets $c = Aa$, $ds = ab$ and $dt = E\varepsilon$.

---

[3] The essay was eventually published in 1862. Dating of the paper is discussed by Fraser [1992, p. 244, n.6], who suggests the middle 1730s as the most plausible time of composition.

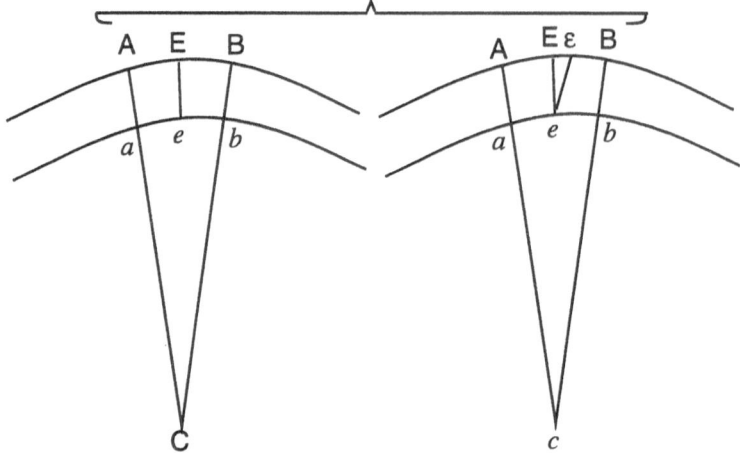

Figure 1 (from Euler [1862, Fig. 3])

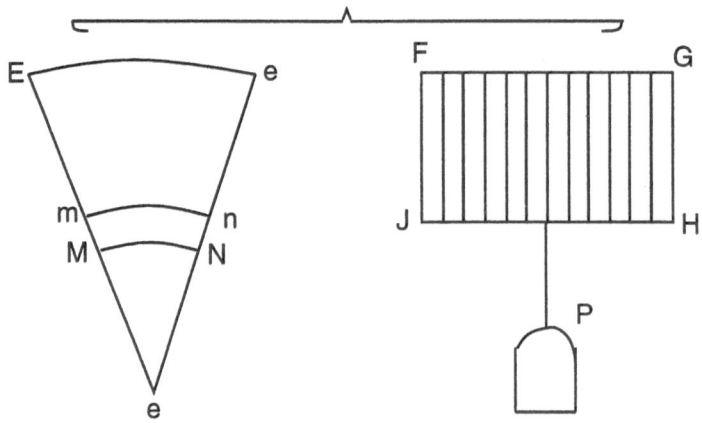

Figure 2 (from Euler [1862, Fig. 4])

    To obtain a measure of the elastic force Euler considered the material
membrane $FGHJ$ (Figure 2) consisting of the extended part of a series of
stretched filaments. $FJ = g$ is the magnitude of the extension and $FG = f$
is the width of the membrane. He supposed that the weight P is sufficient
to sustain this stretching so that P/fg is a measure for the given material
of the elastic force per unit of extension and per unit of width.
    The stretched part of the ring segment $AaeE$ is comprised of the trian-
gle $eE\varepsilon$ (Figure 2). Consider the portion $MNnm$ of $eE\varepsilon$ located a radial
distance $eN = x$ from e. $MNnm$ is composed of a series of concentric
extended filaments bounded by MN and mn. Since $MN = (x/c)\,(E\varepsilon) =
(x/c)dt$ the area of $MNnm$ is $(xdxdt)/c$. The elastic force that gives rise to

$MNnm$ is therefore equal to $(Pxdxdt/cfg)$. Euler calculated the moment of this force about the point e to be $(Px^2dxdt)/c^2fg$. (Rather curiously he took x/c instead of x as a measure of the moment arm eM.) By integrating this expression from 0 to c he obtained a value of the total "force of cohesion", $(Pcdt)/(3fg)$.[4] By relating this formula to the radii of curvature of ab in its normal and deformed states he arrived at an expression that he was able to use to investigate the vibratory motion of the ring.[5]

What is striking in Euler's treatment of this problem is the absence of anything that could be interpreted from a later perspective as an application of the stress principle. The elastic forces that arise are regarded as being distributed over the plane in which they act, not over a transverse cross-section. This plane is not regarded logically as something that divides the body into two parts; rather it is viewed as a component element (rather like a membrane) of the body upon which it is a necessary to calculate the forces acting. These forces are themselves viewed as an absolute function of the displacement dt. Thus Euler lacked the concept of elastic strain. The formula $(Pcdt)/(3fg)$ itself fails to relate in a satisfactory manner the bending moment to the cross-sectional structure of the ring.[6]

(b) Euler's "Solutio problematis de inveniendi curva quam format lamina utcunque elastica in singulus punctis a potentiis quibuscunque sollicitata"

Although unsuccessful Euler's paper "De oscillationibus" is of interest because of the detailed picture it presents of his understanding at this time of elastic phenomena. In this published writings he would abandon any attempt at analyzing the elastic behaviour of a body in terms of its internal constitution. The approach that he would follow publicly throughout his career was established in the 1732 paper "Solution of the problem of finding the curve formed by an elastic lamina acted upon at each point by arbitrary soliciting powers". Its purpose was to provide a coordinated treatment of the various results then available concerning mechanical lines, the catenary, parabola, elastica, velarium, lintearia and so on.

---

[4] The text that is reprinted in the *Opera* has $Pc^2dt/3fg$ rather than $Pcdt/3fg$. (The latter is what actually follows from the preceding step of the derivation.) That this is a misprint is evident from the fact that Euler immediately sets $dt = (a-b)cds/ab$ and obtains $Pcc(a-b)ds/3abfg$, the formula he actually works with in the paper. (That the c^2 is a typographical error is also clear from the fact that the Euler always denotes the square of a single-letter variable a by aa rather than a^2.)

[5] A detailed critical account of Euler's analysis is presented by Cannon and Dostrovsky [1981, 37–43].

[6] Through a series of "corrections" it is possible to obtain the modern formula for the bending moment of a prismatic beam from Euler's procedure. In calculating the moment of the elastic force we use x rather than x/c as the moment arm; we replace dt by the strain dt/ds; we incorporate the thickness h of the ring into the final formula; finally, we interpret $E = P/fg$ as "Young's modulus". With these changes Euler's formula becomes $E(c^2/3)(dt/ds)$, which is the flexure formula for a prismatic beam in which the neutral axis is assumed to lie on an outer surface. (It is on the basis of an argument something like this that Truesdell [1960, 145] arrives at a high evaluation of Euler's paper.)

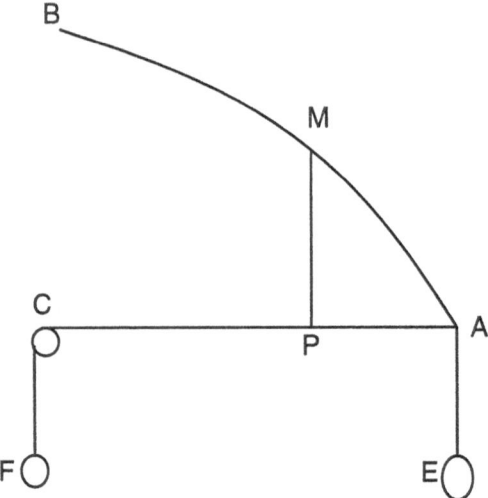

Figure 3 (from Euler [1732, Fig. 4])

In the "general problem" considered by Euler we are given a heavy lamina BMA that is subjected at its end to vertical and horizontal forces E and F (Figure 3). It is necessary to determine the shape of the curve subject to this loading. Consider any point M on the lamina. Euler took the resisting moment at M to be Av/r, where v is an elastic constant, A is a constant of proportionality and r is the radius of curvature of the curve at M. The resisting moment will balance the moment of all forces acting on the section of the system to the right of M. Let A be the origin of a Cartesian coordinate system, $AP = x$, $PM = y$. The moment about M of the forces E and F is $Ex + Fy$. Euler calculated the moment of the body forces acting on the lamina MA to be

$$\int_0^x P\,dx + \int_0^y Q\,dy,$$

where P and Q are equal to

$$P = \int_0^x F_V\,dx, \qquad Q = \int_0^y F_H\,dy,$$

F_v being the vertical force per unit distance along the x-axis and F_H being the horizontal force per unit distance along the y-axis. (Euler obtained these last expressions in a preliminary lemma. The result is demonstrated geometrically in terms of curves and graphs. The argument reduces to the following. If $M(x)$ is the moment due to the vertically-acting body forces then we have $dM = \int_0^x (F_V(t)dx)dt$. Hence it is clear that $P = \int_0^x F_V(t)dt$.)

The total moment of all forces must balance the resisting moment, leading to the general equation

(1) $$\frac{Av}{r} = Ex + Fy + \int_0^x Pdx + \int_0^y Qdy.$$

In the case of perfectly flexible bodies the elastic force $v = 0$ and (1) reduces to

(2) $$Ex + Fy + \int_0^x Pdx + \int_0^y Qdy = 0.$$

If furthermore we suppose $Q = 0$ we have the case of the hanging cable given by

(3) $$Ex + Fy + \int_0^x Pdx = 0.$$

Differentiation yields this equation in the form

(4) $$Edx + Fdy + Pdx = 0.$$

Euler further differentiated (4) with respect to s, where s is the distance along the curve:

(5) $$dP + \frac{Fdxddy - Fdyddx}{dx^2} = 0.$$

Because ds is constant we have $d(dx^2 + dy^2) = 0$ or $dxddx + dyddy = 0$. (5) may therefore re-expressed in the form

(6) $$dPdx^2dy = Fds^2ddx.$$

An alternate form of (6) was obtained by Euler by setting the radius of curvature r equal to $dsdy/ddx$:

(6') $$rdPdx^2 = Fds^3.$$

Euler observed that all the various types of the catenary are expressed by (6').

Euler illustrated (6) with several examples, one of which will be described here. Suppose the curve BMA is a segment of a light flexible hanging cable fixed at B from which a uniform heavy horizontal load is suspended (Figure 4). We have $E = 0$ and $dP = adx$ where a is a constant. F may here be regarded as the tension supporting the cable at A. Equation (4) becomes $adx^3dy = Fds^2ddx$, which integrated yields $ay = -(Fds^2)/(2dx^2)$. Because the applied force at A acts to the right we replace F by $-F$ and (using $ds^2 = dx^2 + dy^2$) arrive at $dx = (dy\sqrt{F})/\sqrt{(2ay - F)}$. This is the equation, Euler noted, of the Apollonian parabola.

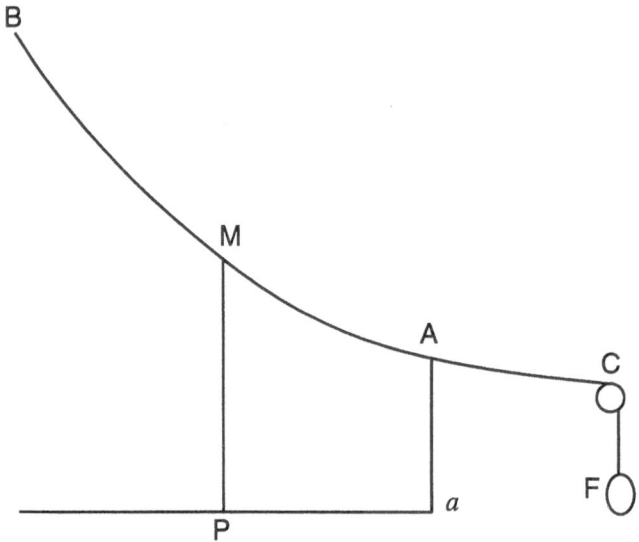

Figure 4 (from Euler [1732, Fig. 5])

Returning later in the paper to the more general equation (1) Euler considered examples in which the elastic constant v is non-zero, deriving differential equations to describe the resulting curve. His research here was preparatory to his famous essay of 1744 on the elastica in which he provided a full and systematic analysis of the different solutions obtained from (1) when F_H and F_V are zero.[7]

Let us turn now to a critical evaluation of Euler's 1732 paper. Earlier researchers in mechanics – Jacob Bernoulli and Hermann foremost among them – had analyzed the catenary and elastica separately. The equation of the catenary was obtained by examining tensions along a segment of the hanging cable following the approach that is more or less customary today. The equation of the elastica by contrast was obtained by examining the elastic behaviour of the fibres that make up the lamina and arriving at an estimate for the resisting moment at each point. Using the fact that the resisting moment of a perfectly flexible body is everywhere zero, Euler was

---

[7] Euler's essay of 1744 is the subject of [Fraser 1991].

able by means of equation (1) to provide a single method that covered both flexible and elastic lines.

Euler's condition for equilibrium is that the moment of all forces acting on the body to the right of a given point M must balance the resisting moment at M. In effect he is dividing the line into two parts, isolating for study the part MA, and examining at the boundary point M the effect of the second part BM on MA. This principle had been used implicitly by Jakob Bernoulli and Hermann when they equated the tension acting at a given point of a hanging cable to the sum of all forces acting on the cable to the right (or to the left) of this point. Euler himself came to the realization by 1750 that the principle could be used in balancing either the force or the moment acting at a point.[8] In a memoir of 1771, "Genuine principles of the doctrine of the equilibrium and motion of flexible and elastic bodies", he provided a more explicit statement of this condition, supplementing it with a derivation of the general differential equation relating the tension, normal shear and moment acting at an arbitrary point of a line. This paper, as well as one published five years later in 1776, contained Euler's finished theory of the static equilibrium of mechanical lines.

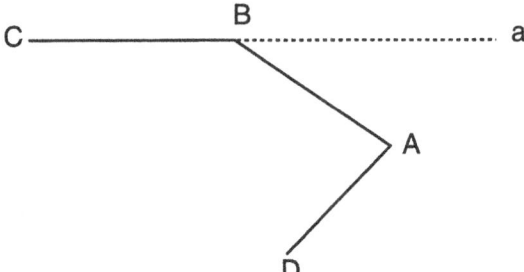

Figure 5 (from Euler [1732, Fig. 1])

Central to Euler's conception of an elastic line is the assertion that the resisting moment at a given point is proportional to the radius of curvature of the line there. This proposition is a consequence of the "hypothesis" presented at the beginning of the 1732 paper. Euler considered two rods CB and Ba joined initially at B in a straight line (Figure 5). The action of the force or power AD at A causes the segment Ba to assume the position BA. Euler stated that the moment of this power at B will be proportional conjointly to the "elastic force" at B and the angle aBA. This hypothesis, he wrote [1732, 71], "is commonly assumed and could probably be demonstrate physically if the angle aBA is extremely small."

---

[8] See Euler's unpublished notebook EH 5, pp. 268–269, cited in [Truesdell 1961, p. 395, note 1, example 3].

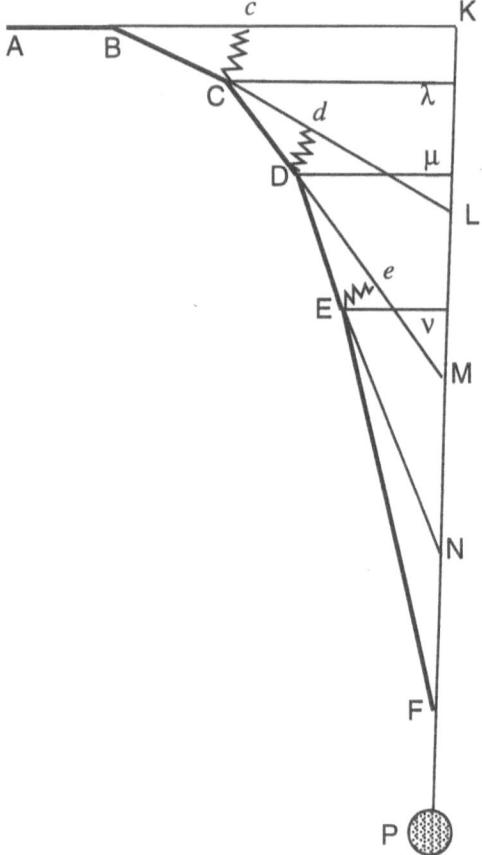

Figure 6 (from Lagrange [1771, Fig. 1])

Note that in this conception physical objects such as rods and beams are idealized as one-dimensional lines, possessing a certain degree of elasticity but lacking any internal constitution that can be further analyzed. This point of view seems to have become standard in eighteenth-century theoretical mechanics. A paper published by Lagrange in 1771 opened with a discussion of the elastic curve. The author wished to show that the moment of an applied weight P acting at the end of the curve would at each point equal P times the perpendicular distance from the point to the vertical through P. The validity of this result for rigid bodies was known and it was necessary to demonstrate it for elastic bodies. To do so Lagrange considered a curve made up of polygonal segments; at each vertex a force represented as sort of an elastic hinge was assumed to exert a resisting moment (Figure 6). He devised a certain argument on the basis of this model to establish the result, which was then extended to continuous curves by assuming that the number of sides of the polygonal line increases indefinitely.

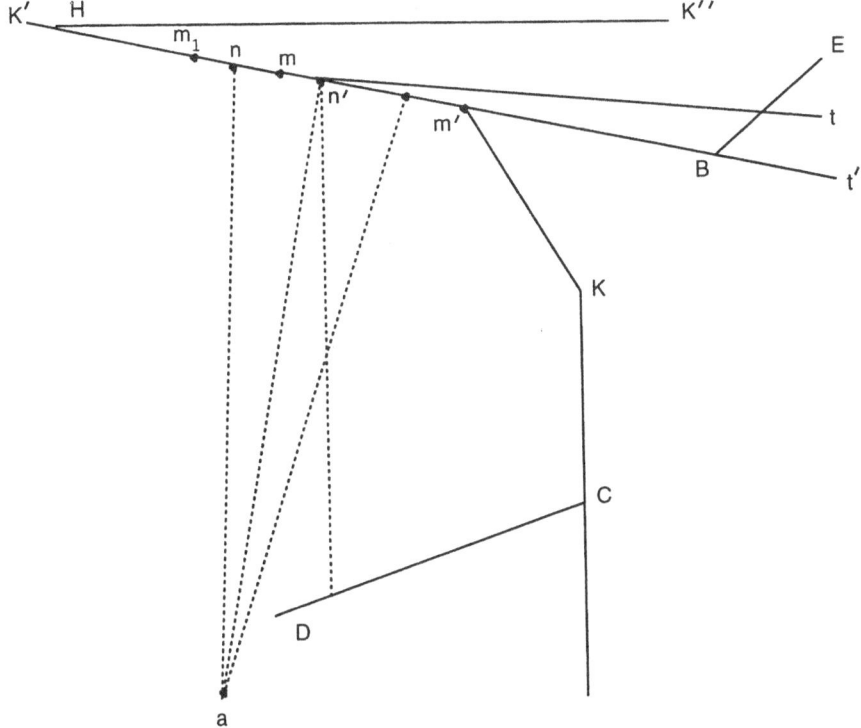

Figure 7 (from Poisson [1811, Fig. 40])

In 1811 in the first volume of his *Traite de Mécanique* Siméon Poisson adopted Euler's conception of the elastic line. He justified treating the elastic lamina as a geometrical line by noting that its thickness would be assumed constant along its length. In an extended discussion (pp. 212–230) he analyzed an elastic line as an infinite-sided polygonal line in which an elastic moment is exerted at each vertex (Figure 7). He assumed as hypothesis the proportionality of the moment to the angle between two successive polygonal segments.

In the second edition of the *Traité*, published in 1833, Poisson discarded this model, replacing it by an analysis in terms of stresses acting across cross-sectional planes.[9] He provided no discussion of this change, although

---

[9] Poisson's analysis of the static equilibrium of an elastic lamina in the 1833 edition of the *Traité* is presented on pp. 600–607 of volume 1. On pp. 617–620 he presents an analysis of elastic bending for various assumptions concerning the shape of the transverse cross-section of the lamina.

it is clear that the appearance of stress analysis in the treatises of Navier and Cauchy influenced his treatment. By the 1830s the older Eulerian conception of an elastic line had become something of an historical curiosity.

The employment of the elastic line as a conceptual entity was indicative of the distinctive style of theoretical mechanics as it was practiced in the eighteenth-century. It was at this period not simply a case of applying mathematics to physical reality. In the characteristic processes of abstraction and conceptualization the entities under study emerged as objects possessing both a mathematical and physical identity. The elastic line was a kind of mathematico-physico hybrid. Although treated formally as a geometrical object in a quasi-Archimedian sense, its ultimate ontological status (paradoxically) was that of a physical object. There was an understanding of the relationship of mathematics to empirical reality in which the autonomous mathematical character of conceptual entities was acknowledged to a greater degree than it would be in later physics.[10]

  Concluding remarks. If one surveys eighteenth-century work for anticipations of the concept of stress they are found in researches that focus on concrete physical problem. Problems such as the fracture of beams or the buckling of columns led researchers to analyze the internal forces acting across a given cross-sectional plane, be it the plane of fracture or the plane perpendicular to the long axis of the column. In each case the object under study was a definite engineering structure possessing a well defined physical identity.

When scientists attempted to develop a general mathematical theory of elasticity they turned to models that were physically very restrictive. Euler's treatment of the statics of elastic lines is a case in point. His investigation was limited at the outset by the adoption of an idealization of physical structures that excluded any analysis of their properties in terms of the concept of internal stress.[11] He was himself conscious of limitations of his approach, writing [1771, 381] at the beginning of his 1771 paper that "we are still far removed from a complete theory which is capable of determining the figure of elastic surfaces as well as bodies", in consequence of which he would restrict his study to "simple strings whether perfectly flexible or elastic, as they have been treated till now by geometers."

It was Cauchy's achievement to develop a theory that was mathemat-

---

[10] Further discussion of the general philosophical point at issue here is provided by Grosholz [1990].

[11] The papers of 1771 and 1772 are excellent representatives of exact science as it was practiced by Euler. A given point of view is worked out in a systematic and methodical manner with the careful derivation and study of the requisite differential equations. What results is an orderly and satisfactory analysis whose overall character and scope are substantially restricted by the limitation of the initial point of view. (Miller's [1916, 238] general comments are relevant here: 'The great volume of Euler's writings is partly due to the fact that he went into great details, and presented even the simpler matters with considerable completeness.")

ically sophisticated and at the same time grounded in a physical analysis generally applicable to continuous structures and media. Consideration of the eighteenth-century background enables us to appreciate better his originality by indicating some of the difficulties of mathematical technique, physical conception and underlying methodology that were involved in the early evolution of the subject.

REFERENCES

BELHOSTE, B., 1991: *Augustin–Louis Cauchy: A Biography*, 1991, Springer–Verlag.

BERNOULLI, JAKOB I., 1694: *Curvatura laminae elasticae. Ejus identitas cum curvatura lintei a pondere inclusi fluidi expansi. Radii circulorum osculantium in terminis simplicissimis exhibiti; Una cum novis quibusdam theorematis huc pertinentibus, &c.*, Acta eruditorum. Also in Jakob Bernoulli's *Opera* V.1 (1744), 576–600.

———————— 1695: *Explicationes, annotationes et additiones ad ea quae in actis superiorum annorum de curva elastica, isochrona paracentrica, & velaria, hinc inde memorata, & partim controversa leguntur; ubi de linea mediarum directionum, aliisque novis*, Acta eruditorum, 537–553. In *Opera* V. 1 (1744), 639–663.

———————— 1706: *Veritable hypothèses de la résistance des solides, avec la démonstration de la courbure des corps qui font ressort*, Mémoires de l'académie des sciences de Paris 1705, 176–186. In *Opera* V. 1, 976–989.

CANNON, J.T. AND DOSTROVSKY, S., 1981: *The evolution of dynamics: Vibration theory from 1687 to 1742*, Springer–Verlag, New York.

EULER, L., 1732: *Solutio problematis de inveniendi curva quam format lamina utcunque elastica in singulis punctus a potentiis quibuscunque sollicitata*, Commentarii academiae scientiarum Petropolitanae 1728 V. 3, 70–78; in *Opera omnia* Ser. 2 V. 10 (1948), 1–16.

———————— 1744: *Methodus inveniendi lineas curvas maximi minimive proprietate gaudentes, sive solutio problematis isoperimetrici latissimo sensu accepti*. Additamentum I, "De curvis elasticis", occupies pp. 245–320. The work is reprinted as volume 24 of Euler's *Opera omnia*, Ser. 1 (1952). Additamentum I (pp. 231–308) is supplied by C. Carathéodory with notes containing references to Euler's other works.

———————— 1771: *Genuina principa doctrinae de statu aequilibrii et motu cororum tam perfecte flexibilium quam elasticorum*, Novi commentarii academiae scientiarum Petropolitanae 15 (1970), 381–413. In *Opera omnia* (1957) Ser. 2 V. 11, 37–61.

———————— 1776: *De gemina methodo tam aequilibrium quam motum corporum flexibilium determinandi et utriusque egregio consensu*, Novi commentarii academiae scientiarum Petropolitanae 20 (1775), 286–303. In *Opera omnia* Ser. 2 V. 11 (1957), 180–193.

———————— 1862: *De oscillationibus annulorum elasticorum*, Opera postuma V. 2, pp. 129–131, reprinted in *Opera omnia* Ser. 2 V. 11 (1948), 378–382.

FRASER, C.G., 1991: *Mathematical technique and physical conception in Euler's investigation of the elastica*, Centaurus 34, 211–246.

GEORGE, S.G., RETTGER, E.W. AND HOWELL, E.V., 1943: *Mechanics of Materials*, Second edition, McGraw–Hill, New York and London.

GROSHOLZ, E.R., 1990: *Problematic objects between mathematics and mechanics*, paper presented at a meeting at the Philosophy of Science Association.

HANKINS, T., 1985: *Science and the Enlightenment*, Cambridge University Press.

HERMANN, J., 1716: *Phornomia, sive de viribus et motibus corporum solidorum et fluidorum*, Amsterdam.

HEYMAN, J., 1972: *Coulomb's memoir on statics*, University Press, Cambridge.

LAGRANGE, J.L., 1771: *Sur la force des ressorts pliés*, Mémoires de l'académie de Berlin 1770 V. 25, 167–203. Reprinted in Lagrange's Oeuvres V. 3, 77–110.

MILLER, G.A., 1916: *Historical introduction to mathematical literature*, Macmillan, New York.

POISSON, S.D., 1811: *Traité de mécanique, two volumes.* Paris. Second edition "considérablement augmentée", 1833.

SOUTHWELL, R.V., 1941: *An introduction to the theory of elasticity for engineers and physicists,* second edition, Oxford University Press.

TIMOSHENKO, S.P., 1953: *History of strength of materials,* McGraw–Hill, New York.

TODHUNTER, I. AND PEARSON, K., 1886: *A history of the theory of elasticity and of the strength of materials,* Volume 1, University Press, Cambridge.

TRUESDELL, C., 1960: *The rational mechanics of flexible or elastic bodies 1638–1788,* published as V. 11 part 2 of Euler's *Opera omnia Series 2.*

_____ 1968: *The creation and unfolding of the concept of stress,* Chapter 4 of Truesdell's *Essays in the history of mechanics* (Springer–Verlag), pp. 184–238.

BOOK TWO OF RADICAL PRINCIPIA

BRUCE POURCIAU*

Introduction. By 1694 NEWTON had devised some quite radical plans for revising his 1687 *Principia*. "Newton came in the early 1690's to conceive a grand scheme of revision of the published *Principia*," writes WHITESIDE,

> in which not only its particular verbal and mathematical errors were to be corrected but, much more radically, the redundant in its logical and expository framework was to be cut out and the flimsier portions of the remaining structure were to be strengthened and supported and (where necessary) completely rebuilt. To David Gregory, during a visit by him to Cambridge in early May 1694, he proved unprecedentedly expansive regarding his intentions, elaborating for him a detailed overview of his plans for revision, and also showing him the manuscript papers in which he had already himself gone some way towards implementing them.♠

Two months after this meeting, GREGORY summarized many of the intended revisions in a retrospective memorandum. Of the plans for the early parts of the first edition, he wrote that

> many corrections are made near the beginning: some corollaries are added; the order of the propositions is changed and some of them are omitted and deleted. He deduces the computation of the centripetal force of a body tending to the focus of a conic section from that of a centripetal force tending to the centre . . . ; moreover the proofs given in propositions 7 to 13 inclusive now follow from it just like corollaries. Sections IV and V are removed and become parts of a separate treatise . . . .†

Later in the same memorandum, GREGORY records the intention to

> . . . attach two treatises, one about the geometry of the ancients . . . on the problem posed by Pappus and repeated by Descartes, on plane loci . . . . Part of this treatise will be on the discovery of orbits with a given focus or even when neither focus is given, and all this is the subject-matter of Sections IV and V. The second

---

* Department of Mathematics, Lawrence University, Appleton, Wisconsin 54912, U.S.A.
♠ WHITESIDE (1967–1981: VI, 568).
† TURNBULL (1961: 384–386).

treatise will contain his Method of Quadratures. . . . In the
author's manuscript where these things are recorded one may see
the methods of infinite series, of tangents, of maxima and minima,
of curvature and of the rectification of curves.§

NEWTON apparently had various ideas for attaching mathematical tracts
to the *Principia*, from appending a set of sixteen calculus propositions to
collecting in a separate book (to be called "Liber Geometriae") all the
purely mathematical parts of the first edition.¶
    In the end, as we know, not one of these grand schemes ever appeared in
print,◊ but behind the details of these radical plans for the second edition,
NEWTON's fundamental agenda remains very clear:

- to bring out, in general, the underlying mathematical content
- to make explicit, in particular, the crucial reliance on curvature
  and the fluxionic character of the analysis
- to stress the separation between geometry on the one hand and
  dynamics on the other
- to involve new, more general theorems in a tighter, more elegant
  framework

Our plan for the present work is simple: to carry out NEWTON's agenda in a
radical revision of the early portions of the *Principia*. The individual pieces
of this radical revision—the passages, propositions, lemmas, axioms, and
definitions—derive from Newtonian sources, but vary in their authenticity.
Some are NEWTON word for word, others are NEWTON whittled to fit our
scheme, and many are NEWTON only implicitly, for they derive not from
NEWTON's words, but from what seems implicit in his work. As we move
from lemma to lemma in this "Radical Principia," we offer commentaries
to identify sources and degrees of authenticity, to note the consequences
of certain placements or wordings, and to point out contrasts with the
published *Principia*. In this project, we have been constrained by a desire
to "keep the faith" with the Newtonian style and spirit, and, along these
lines, we have strived to hold the gross anachronisms—both in notion and
notation—to a countable number. Our hope for this work is that it may
lend some insights into the logical, mathematical, and dynamical structure
of the *Principia*.
    "Radical Principia" comes in two parts: Book One (Preliminary Math-
ematics) and Book Two (Centripetal Motions). To satisfy space constraints, *in these Proceedings we present only Book Two*. The original,

§ *Ibid.*, p. 385.
    ¶ COHEN (1971: 345–346) and WHITESIDE (1967–1981: VI, 600–609).
    ◊ Here and there, however, bits and pieces of these schemes were tucked into nooks
and crannies of the second (1713) and third (1726) editions.

uncut "Radical Principia" appears in *Archive for History of Exact Sciences*. See POURCIAU (1992b).

We employ a touch of artifice in our presentation. *On the following page, in the Editor's Note, you will read the story of how I came to edit a recently discovered manuscript of* NEWTON'S. *Do not believe a word of it.* It is all pretense, but it is pretense with a purpose, for this artifice, among other pleasantries, suggests a simple and coherent scheme for distinguishing the text of the radical revision from our comments on the text: the comments reside in the editor's footnoted commentaries.

Editor's Note#

During the spring of 1991, I received a curious package from a friend. The package contained a manuscript, yellowed with age, and an accompanying note. In Greenwich, England to handle the estate of his aunt Edith Flamsteed Pond, my friend, it seems, had come across the manuscript while clearing out his aunt's attic and had sent it to me because, as he put it, "You're the only mathematician I know, and besides, you have some interest in the history of this sort of thing, don't you?" Looking at the Latin manuscript, strewn with lemmas, formulas, and diagrams, I thought at first it was the work of the Rev. John FLAMSTEED, the Astronomer Royal in NEWTON's time, but a closer inspection revealed the manuscript to be a revision—incomplete and radical, yet still a revision—of *Philosophiae Naturalis Principia Mathematica*, at least the early portions, and it was then I became convinced that its author was Isaac NEWTON himself.

Today, many months later, despite the unanimous and firm opinion of every consulted NEWTON scholar that the manuscript is an obvious fraud, an amateurish hoax littered with anachronisms and inconsistencies of style and substance, I still prefer to believe in its authenticity. As far as a date for the manuscript and how it may have come to rest among FLAMSTEED's papers, I have only a guess. On November 17, 1694, in a letter to FLAMSTEED requesting his observations of the moon's position, NEWTON wrote,

> I desire that you send the right ascensions and meridional altitudes of the moon, in your observations of the last six months. . . And for the trouble you are at in this business, besides the pains you will save of calculating . . . the observations you communicate to me, and the satisfaction you will have to see the theory you have ushered into the world brought (as I hope) to competent perfection, and received by astronomers, I do intend to gratify you to your satisfaction: though at present I return you only thanks@ and the enclosed which I began but soon gave over for lack of time during summer past.

If the present manuscript is in fact the enclosure referred to in this letter, as I believe it is, then the manuscript would have been composed by NEWTON in the summer of 1694.

\# A fable.

@ There is fact in the fable: Rev. John FLAMSTEED *was* the Astronomer Royal, he and NEWTON *did* correspond, this *is* part of a letter from NEWTON to FLAMSTEED dated 17 November 1694, but the *actual* letter ends with the word 'thanks'. See COHEN (1971: 176, Note 21). The insertion, "and the enclosed . . . summer past," is an invention to help our fable along.

Finding no NEWTON scholar willing to undertake the translation and editorial duties, I have undertaken these burdens myself. My commentaries reside in the footnotes.£

BRUCE POURCIAU
Appleton, Wisconsin

£ We are thus presenting the text of this radical revision as if it were a recently discovered, possibly Newtonian, manuscript. As the supposed editor of this manuscript, my comments on the text will appear in the footnotes, but to separate fact from fable in these notes, we shall refer to the author of the manuscript as "the author" and cite NEWTON only for works that are undeniably his.

Contents

---

(1) The manuscript begins with this title, Principia, and follows with a table of contents. For reference, we record here a table of contents for the early portions of the published *Principia*:

Comparing the organization of the published versions with the manuscript's radical revision, we find some obvious contrasts. In the manuscript: (a) Much of the *Principia*'s purely mathematical or geometrical material has been placed in a separate leading book. This division would seem consistent with the well-documented intentions NEWTON had in the early 1690's to revise the 1687 *Principia* in ways which emphasized the mathematical content. See, for example, WHITESIDE (1967–1981: VI, 568–609). (b) Fluxions appear very early, combined with ultimate ratios, in a section *preceding* the definitions of speed and acceleration, an order which permits more precision in these definitions. (c) An entire section is devoted to the study of curvature. (d) Force and mass are nowhere to be seen. (e) The celebrated 'Laws of Motion' from the published *Principia* have apparently disappeared.

(2) The manuscript contains no ode and no preface, despite the listing of these items in the table of contents. Presumably, the author had not bothered to copy out Edmund HALLEY's ode from the first edition of the *Principia*, nor yet composed a preface for this radical revision.

Book Two[3]

CENTRIPETAL MOTIONS[4]

SECTION I[5]

Definitions of motion, speed, and acceleration

DEFINITION I

I say we have a motion if for each number t there is a corresponding point $Q=Q_t$ in space whose position varies smoothly with t. If $P = Q_0$, we will say the motion starts at P. As t increases the motion passes through various

---

(3) In these Proceedings, we present only Book Two of the radical revision. The original, uncut manuscript appears, with commentaries, in the article "Radical Principia," *Archive for History of Exact Sciences*, 44 (1992), pages 331–363.

(4) Note the strict separation achieved in this manuscript between the intrinsic *geometry* in Book I (Sections I and III) and the *kinematics* beginning here. Curvature belongs to the geometry of a curve, for it is independent of the specific parametric representation (or *motion*, as NEWTON puts it). Notions such as speed and acceleration *do* of course depend on the motion on the curve. To see many of the definitions and propositions of this manuscript, and of this book on centripetal motions in particular, in more modern dress, consult POURCIAU (1992a).

(5) In this radical revision of the early portions of the *Principia*, (inertial) mass and force are never defined and play no part in the proceedings. We may be startled by this at first, but it is consistent with NEWTON's intention, in his unpublished revisions of the 1687 edition, to stress the underlying mathematics. Moreover, the published *Principia* does not actually require, in any logical sense, the notions of mass or force until the treatment of the *two*-body problem in Section XI(!) of Book I. Had the author completed the present revision, he would, presumably, have presented definitions of mass and force at some later point in the exposition. The two books he *did* complete of this radical revision treat the one-body problem only, and the solution of this problem involves no physics at all, only geometry and kinematics, that is, only mathematics.

TRUESDELL (1967: 238–258, especially 255) has written that "whatever his theology and philosophy, this is NEWTON's grand lesson for us: Disregard the 'causes'—whatever that strange word may mean—and cleave to relations among the phenomena." In the present manuscript, force, perhaps the ultimate cause, has been ignored. Read COHEN (1970: 143–177) and WESTFALL (1971) for illumination on the notion of force in NEWTON's dynamics. Of course the notion of force cannot be so easily ignored in any adequate treatment of the deeper problems of mechanics, such as the n-body problem for n greater than one. To see a precise, axiomatic theory of forces, look at NOLL (1974), pp. 269–272 and 311–313. A simplified version of NOLL's theory can be found in TRUESDELL (1992), pp. 19–29 and 156–165.

points which form a curve, a curve called the trajectory of the motion.[6]
When the trajectory is part of a conic section or spiral, for example, we
shall say we have a conic or spiral motion.

If it pleases us, we may think of t as measuring the uniform flow of
time. [7]

DEFINITION II

If the trajectory of a motion starting at P is a line, and if by $s = s_t$ we
mean the length of the segment PQ_t, then the speed of the motion is the
fluxion $\dot{s}$, and the acceleration of this linear motion is given by the second
fluxion $\ddot{s}$.[8]

DEFINITION III

For nonlinear motions, we let $s = s_t$ measure the length of the arc from P
to Q_t. We call $\dot{s}$ the speed of the motion, as before. The second fluxion $\ddot{s}$
gives the arc-length acceleration.

In the nonlinear case, the second fluxion $\ddot{s}$ does not give the full acceleration, for it measures only the variation in speed and ignores acceleration
effects due to the *shape* of the trajectory. For arbitrary nonlinear motions,

---

(6) The subscript *notation* may seem anachronistic, but the *notion* it records is
surely not, for NEWTON understood very well and in his work made clear the distinction
between a motion on the one hand and a trajectory or orbit on the other. For example,
in Book I of the 1726 *Principia*, Section VI bears the title, "How the motions are to be
found in given orbits," and this same section begins with the problem, "To find at any
assigned time the place of a body moving in a given parabolic trajectory." One could
hardly be more clear.

(7) On referring to a variable as 'time,' NEWTON makes the following comment in
his 1671 tract on series and fluxions: ". . . I shall, in what follows, have no regard to
time, formally so considered, but from quantities propounded which are of the same
kind shall suppose some one to increase with the equable flow: to this all the others
may be referred as though it were time, and so by analogy the name of 'time' may not
improperly be conferred upon it." WHITESIDE (1967–1981: III, 73).

(8) 'Fluxion' is Newton's word for derivative. Refer to Lemma II on 'moments of
genitum' in Section I, Book II, of the 1726 *Principia* or to the tract on series and fluxions
in WHITESIDE (1967–1981: III, 72–75). The published *Principia* contains earlier but less
explicit references to fluxions. For example, in Lemma IX (Section I of Book I) one can
see a hidden definition of derivative, phrased geometrically: "If a right line AE and a
curve line ABC, both given by position, cut each other in a given angle A; and to that
right line, in another given angle, BD, CE are ordinately applied, meeting the curve in
B, C; and the points B and C together, approach towards, and meet in, the point A: I
say that the areas of the triangles ABD, ACE, will ultimately be one to the other in the
duplicate ratio of the sides." NEWTON (1726: 48). See Section II of Book I in "Radical
Principia," POURCIAU (1992b), for the treatment of fluxions in the radical revision.

it may be difficult to devise a single measure of the full acceleration,[9] but for a special class of motions, those we shall call centripetal, it is a simple matter.

Consider any motion starting at P with speed ν. The trajectory has a certain tangent line at P. On this tangent line, choose the point L_t so that $PL_t = \nu t$. (A linear motion on the tangent beginning at P and continuing with constant speed ν for a time t would end up at the point L_t.)

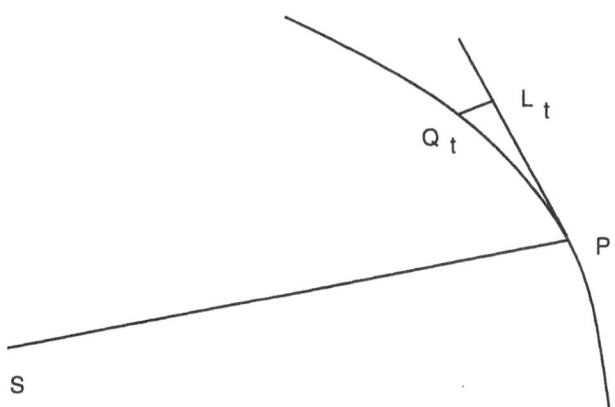

DEFINITION IV

If there exists a fixed point S such that at each point P the line segment Q_tL_t becomes ultimately parallel[10] to the radius SP, we shall say we have a

---

(9) Without the precise notion of directional variation provided by vector calculus, this *would* be difficult. The published editions of the *Principia* tell us how to measure the acceleration in a *centripetal* motion—read the definition for the 'accelerative quantity of a centripetal force,' NEWTON (1726: 6)—but they remain mute on measuring the acceleration in an *arbitrary* motion.

(10) Although the manuscript provides no definition for the phrase "becomes ultimately parallel," the intent appears intuitively clear, and any remaining ambiguities do not cause serious problems. (However, see Note 17, regarding the proof of the Theorem on Areas.) In the published *Principia* (Definition V, Book I), Newton defines a centripetal force as ". . . that by which bodies are drawn or impelled, or any way tend, towards a point as to a centre." To unearth the intended meaning buried in this definition, one can open the *Principia*, stop at each proposition which assumes a centripetal force, and study the given proof. In all such proofs, Newton takes the centripetal force assumption to mean the "ultimately parallel" condition we see stressed in this manuscript. (See especially the geometric construction in Newton's proof of Propositions I and II in Book I.) A translation of "Q_tL_t becomes ultimately parallel to SP" into the language

centripetal motion or, equivalently, a motion about S.[11] *For a centripetal motion, the acceleration is the second fluxion* $\ddot{d}$ *(when t vanishes) of the length* $d_t = Q_t L_t$[12]

We imagine a motion about S as being at each point P instantaneously in free fall toward S, and, this being the case, the deviation $d_t = Q_t L_t$ as a measure of the distance fallen toward S in time t. In this light, our definition of acceleration for a motion about S becomes a natural outgrowth of Definition II, our definition of acceleration for linear motions.

Corollary.[13] *In any centripetal motion, the acceleration may be resolved along oblique lines according to the parallelogram law.* For all displacements may be so resolved, and hence also any fluxion or second fluxion of a displacement.

———————

of vector calculus might be: the vector product

$$\frac{\overrightarrow{Q_t L_t}}{t^2} \times \overrightarrow{SP}$$

vanishes as $t \to 0$.

(11) The language introduced here can shorten certain common statements. For instance, both "A body revolves in any orbit about an immovable center S" and "A body is urged by a centripetal force continually directed toward the point S" become, simply, "Given a motion about S."

(12) Those who might see this fluxionic definition of centripetal acceleration as an obvious anachronism might do well to open the published *Principia* to Definition VII— "The accelerative quantity of a centripetal force is the measure of the same, proportional to the velocity which it generates in a given time"—and ask themselves to reset this definition using fluxions. The only reasonable version for fluxions turns out to be exactly the version recorded here in this manuscript.

In a modern verification that $\ddot{d}$ (when t is zero) measures the (scalar) acceleration at P, we observe that

$$\overrightarrow{Q_t} = \overrightarrow{P} + \overrightarrow{v}\, t + \frac{1}{2}\, \overrightarrow{a}\, t^2 + \cdots$$

$$\overrightarrow{L_t} = \overrightarrow{P} + \overrightarrow{v}\, t$$

where $\overrightarrow{v}$ and $\overrightarrow{a}$ are the (vector) velocity and acceleration at P. Therefore,

$$2\frac{\overrightarrow{Q_t} - \overrightarrow{L_t}}{t^2} \to \overrightarrow{a}$$

so that, in particular,

$$2\frac{d_t}{t^2} \to a.$$

But, by Taylor's Theorem,

$$2\frac{d_t}{t^2} \to \ddot{d}$$

(13) Replaces Corollary I and II of the Laws in the published editions.

SECTION II[(14)]
Centripetal motions in general orbits

THEOREM ON AREAS[(15)]

A motion in a fixed plane whose radius SP from a fixed point S sweeps out area at a uniform rate must be a motion about S, and conversely.

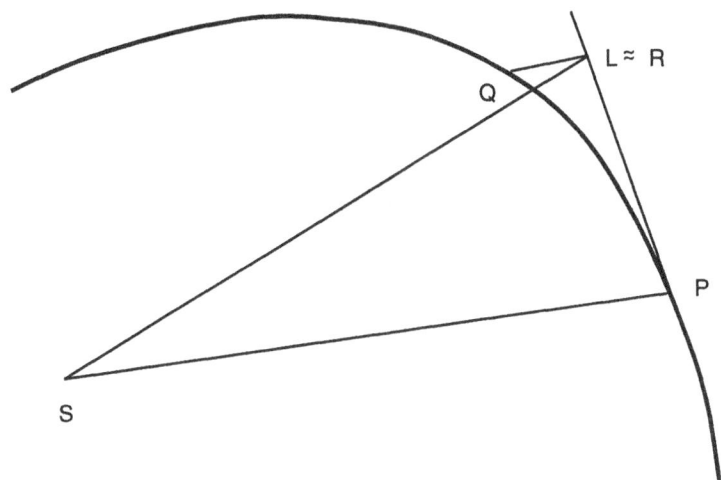

Suppose we have a motion $Q = Q_t$ in a fixed plane which starts at P with speed ν. On the tangent line at P Let L_t, as before, satisfy $PL_t = \nu t$,

---

(14) It may seem odd that the manuscript could now begin to analyze the acceleration in centripetal motions without having yet recorded the celebrated three 'Laws of Motion.' We *will* see Law I—"Every body perseveres in its state of rest, or uniform motion in a right line, unless it is compelled to change that state by forces impressed thereon," NEWTON (1726: 19)—but not as a law. It surfaces, in *force*-less form, as Corollary III of the fundamental Theorem on Speeds and Curvatures in the present section. Laws II and III, however, have completely disappeared in this revision, essentially because (inertial) mass and force have no role in the one-body problem. One might say that NEWTON's three laws of motion have been replaced in this manuscript by a single hypothesis, the Axiom of Vanishing Fluxions (Section II, Book I, of "Radical Principia"): *If the fluxion of the ordinate vanishes everywhere, then the ordinate remains constant.* See POURCIAU (1992b).

(15) The manuscript combines Propositions I and II (Book I, Section II) of the published *Principia* into a single theorem and presents a new demonstration resting on fluxions in general and the Axiom of Vanishing Fluxions—see Note 14— in particular. Although carrying its own problems—see Note 17 —the revised proof shores up certain weaknesses in the published argument. Refer to WHITESIDE (1967–1981:VI, 35, Note 19) for a critique of the corresponding argument in *De Motu*.

and choose a point $R = R_t$ on this line making QR parallel to SP. By $A = A_t$ we mean the area of the "sector" SPQ. For very small t, notice that A is nearly the area of triangle SPR, for the triangles SPR and SPQ have equal area. This being the case, we see that QL is essentially parallel to SP if and only if L is close to R, which happens if and only if R moves at almost constant speed on the tangent line, which occurs when and only when the area of triangle SPR increases at an almost constant rate, and this takes place when and only when the area A of the "sector" increases at nearly a constant rate.

Thus, on one hand, if we have a motion about S, so that at each point P of the orbit QL is ultimately parallel to SP, then A increases at essentially a uniform rate near every fixed time t, or, equivalently, the first fluxion $\dot{A}$ is essentially constant near every t. We infer that the second fluxion $\ddot{A}$ vanishes at each t. But then, by the Axiom of Vanishing Fluxions[16] (Book I, Section II), the first fluxion $\dot{A}$ is a uniform constant over the entire motion, as required. On the other hand, if the area A increases uniformly, then at every point P, QL must be ultimately parallel to SP, as we saw above, and this means we have a motion about S.

To complete the proof of the Theorem on Areas, we must show that any motion about S always takes place in a fixed plane. Suppose, then, that we are given a motion about S which starts at a point P. The motion begins in the plane which contains S and the tangent line at P. Let assume that through some later time T the motion has remained in this fixed plane, but that, if possible, just after the time T, the motion and hence the point Q begins to pass out of this plane. Then the orbit is not essentially planar near the point $p = Q_T$. Yet the ultimately parallel condition must hold at the point p, which means, in particular, that the orbit *is* essentially planar near p.[17] This contradiction indicates that such a time T cannot exist, and it follows that the motion does indeed remain in a fixed plane throughout.

Corollary.[18] *Given any motion about a point S, the speed varies inversely as SY, the perpendicular let fall from S onto the tangent to the orbit.* For then the "areal speed" $\dot{A}$ remains a constant, say h, and if t is sufficiently small, h_t is nearly the area of the "sector" SPQ, which is almost the area of the triangle SPQ, which in turn equals half of $SY \cdot PQ$.

---

(16) Read the end of Note 14.

(17) Exactly what is meant by "essentially planar" remains unclear, as does precisely how this property follows from the 'ultimately parallel condition.' Of course, we should not be surprised at the failure to be completely convincing when the argument begins to center on the *direction* of the motion, for it is just then that the advantages of vector calculus are most sorely missed.

(18) In the published editions of the *Principia*, this is Corollary I of Proposition I (Section II, Book I).

Dividing by t, we find, ultimately[19], $h = \frac{1}{2}SY \cdot \nu$.

THEOREM ON SPEEDS AND CURVATURES[20]

For any motion about S, the acceleration varies as ν^2/PV, where PV is the chord of curvature through S[21] and ν is the speed of the motion.

By Definition IV (Book II, Section I), we measure at P the (centripetal) acceleration of a motion $Q = Q_t$ about S by computing the second fluxion $\ddot{d}$ at P. Here $d = d_t$ stands for the deviation QL, where $L = L_t$, positioned on the tangent at P, satisfies $PL_t = \nu t$. From the Lemma on Second Fluxions (Book I, Section II), $2d/t^2$ has $\ddot{d}$ at P as its ultimate ratio.[22] Clearly,

$$\frac{d}{t^2} = \frac{QL}{(PL/\nu)^2} = \frac{\nu^2}{PL^2/QL},$$

but as the motion is a motion about S, for small t we know L becomes very nearly R, where QR is parallel to SP, and therefore d/t^2 is essentially

$$\frac{\nu^2}{PR^2/QR},$$

---

(19) The radical revision and the *Principia* both write 'ultimately,' 'ultimate value,' or 'ultimate ratio' to mean a passage to the limit. In the uncut version of "Radical Principia," POURCIAU (1992b), we find the following definition leading off the section on ultimate ratios and fluxions: *In what follows, if we should happen to mention quantities as evanescent, we shall always mean a quantity which comes closer to zero than any given number. We call a number the ultimate value of certain quantities or ratios of quantities when that number is approached more closely than any assignable number; that is, when the difference is evanescent. The ultimate value of a ratio (generally a ratio of evanescent quantities) is often referred to as the ultimate ratio. When we say one evanescent quantity varies ultimately as another, we mean that the ultimate ratio of these evanescent quantities is a number of neither zero nor infinite.*

(20) The author has apparently opted to stress the reliance on curvature by taking Corollary 4 of NEWTON's basic Proposition VI (Section II, Book I, 1726 *Principia*) as the fundamental Theorem on Speeds and Curvatures. Certain extant manuscripts, where NEWTON wrote out intended revisions of the 1687 *Principia*, show him considering v^2/PV as the primary measure of centripetal acceleration. Read WHITESIDE (1967–1981: VI, 548, Note 25) for an entertaining and enlightening footnote on the history of the v^2/PV measure. See also AITON (1964: 89).

(21) Let $\mathcal{C}$ be the trajectory of the motion. At P draw the circle of curvature to $\mathcal{C}$. In this circle, the chord PV passing through a given point S is called the chord of curvature through S. Refer to Section III of Book I in the uncut manuscript, POURCIAU (1992b).

(22) For the Lemma on Second Fluxions, read POURCIAU (1992b). Using an approximation by circular arcs, this lemma argues that $\ddot{d}(0) = \lim_{t \to 0} 2d(t)/t^2$. The Lemma on Second Fluxions replaces, in the published *Principia*, Lemma X of Book I: "The spaces which a body describes by any finite force urging it, whether the force is determined and immutable, or is continually diminished, are in the very beginning of the motion one to the other in the duplicate ratio of the times," NEWTON (1726: 49). See also Note 12.

whose ultimate value is ν^2/PV by the Curvature Formula[23] (Book I, Section III).[24]

Corollary I.[25] *Given a motion about S, the acceleration varies inversely as $SY^2 \cdot PV$.* For by the corollary to the Theorem on Areas, the speed ν varies inversely as the perpendicular SY on the tangent.

Corollary II.[26] *In a circular motion about the center, the acceleration varies directly as ν^2/r, where r is the radius.*

Corollary III.[27] *When the acceleration of a centripetal motion vanishes everywhere, the orbit must be a (uniformly traversed) line.* This follows directly from the Proposition on Vanishing Curvature[28] (Book I, Section III).

---

(23) The curvature Formula, Section III of Book I in the uncut manuscript, tells us that the quotient PR^2/QR has the chord of curvature PV as its ultimate value. Compare the tract on series and fluxions, WHITESIDE (1967–1981: III, 155), and Lemma XI (Section I, Book I) of the 1726 *Principia*.

(24) Compare the unconvincing argument for Proposition VI in the third edition, NEWTON (1726: 68). Here, in the present revision, the demonstration is more cogent. For discussions of the published proof of Proposition VI, consult BRACKENRIDGE (1988: 458–463), HERIVEL (1964: 351–354), HERIVEL (1970: 120–135), and WHITESIDE (1970: 116–138).

(25) Listed as Corollary 3 of Proposition VI, NEWTON (1726: 69). According to BRACKENRIDGE (1988: 451–476), this measure of centripetal acceleration occupies a central place in NEWTON's mature dynamics. It certainly occupies a central place in the present manuscript, as the author uses it exclusively in the following section.

(26) Appears as Corollary 1 of Proposition IV, NEWTON (1726: 64).

(27) Here we see NEWTON's first law of motion, stated in the language of acceleration rather than force, surfacing as a corollary of the Theorem on Speeds and Curvatures and the Proposition on Vanishing Curvature, and not as an axiom or law. The development in the present radical revision rests on one axiom only—the Axiom of Vanishing Fluxions (Section II, Book I, of the uncut manuscript): *If the fluxion of the ordinate vanishes everywhere, then the ordinate remains constant.*

(28) According to the Proposition on Vanishing Curvature, *if the curvature vanishes everywhere, then the curve must be a line.* See POURCIAU (1992b), Section III, Book I. One can think of this proposition as capturing the underlying geometric content of NEWTON's first law: "Every body perseveres in its state of rest, or of uniform motion in a right line, unless it is compelled to change that state by forces impressed thereon." NEWTON (1726: 19).

LEMMA ON THE EXISTENCE
OF CENTRIPETAL MOTIONS[29]

Suppose given any planar curve C and any point S off the curve. Assume only that no tangent to C ever passes through S. Then on C there exists a motion about S which starts at a given point with a given speed.

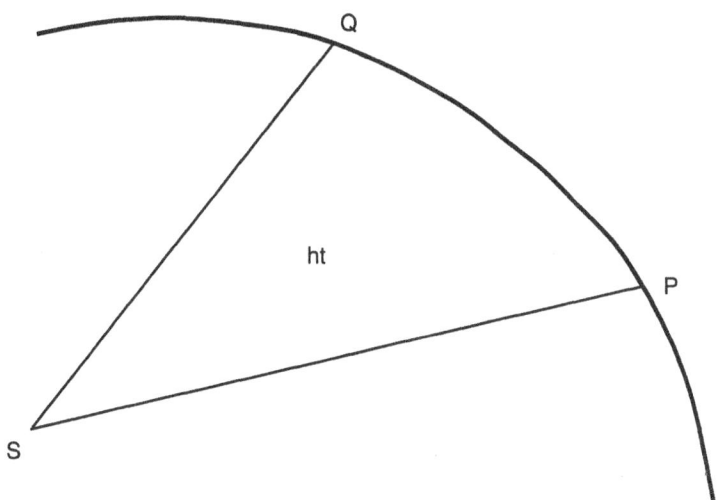

Let P be the given point on the curve and let ν be the given speed. Put $2h$ equal to $SY \cdot \nu$, where SY is the perpendicular let fall from S onto the tangent line to the curve at P. For each t, place Q_t on the curve C so that

---

(29) In the published editions of the *Principia*, when NEWTON has a given curve and a given point off the curve, he tacitly assumes that a motion about the point—see Definition IV (Section I, Book II) of this manuscript—must exist on the curve. For example, the demonstration of Proposition XVII, NEWTON (1726: 91–92), begins: "Let the centripetal force tending to the point S ... be such, as will make the body p revolve in any given orbit pq." FROST (1883: 224) noted the same hidden assumption in NEWTON's brief sketch of a proof that centripetal motions having inverse-square acceleration must enjoy conic orbits. See Corollary 1 of Propositions XI–XIII, NEWTON (1726: 85). For more on this little gap and the recent consternation it has caused, consult ERLICHSON (1990), POURCIAU (1991: 159–172), POURCIAU (1992a), WEINSTOCK (1982: 610–617), and WEINSTOCK (1989: 846–849). Apparently, in the present radical revision, the author has decided to bring this little gap out of hiding by giving it an explicit statement and proof. Even in the published editions, this Lemma on the Existence of Centripetal Motions was already established implicitly by NEWTON's argument for Propositions I and II (Section II, Book I)—see ERLICHSON (1992)—and it was also a simple corollary (at least for conic curves) of his work in Section VI (Book I), where given any conic section, he constructs a motion on the conic about a given point.

A_t, the area of the "sector" SPQ_t, equals ht.[30] Clearly then the fluxion $\dot{A}$ is the constant h. Therefore, by the Theorem on Areas, the motion $Q = Q_t$ must be a motion about S. Further, by the corollary to the Theorem on Areas, the speed of this motion at P is ν, as required.

THEOREM ON THE UNIQUENESS
OF CENTRIPETAL MOTIONS[31]

No two distinct motions about a given point, which leave from a given point in a given direction with a given speed, can share the same acceleration law throughout.

---

(30) Since the area swept out increases continuously as Q moves on the given curve away from P, it is clear that such a point Q_t must exist.

(31) In the published *Principia*, this result appears as the uniqueness portion of the fundamental existence-uniqueness theorems, Propositions XLI and XLII, in Book I, Section VIII (!). Why, in this radical revision, does the author choose to insert the Theorem on the Uniqueness of Centripetal Motions so much earlier? Because he must invoke it in his solutions of the Inverse Problem: from the (centripetal) acceleration and the center, to determine the orbit. See Corollary 1 of Propositions XI–XIII in the second or third edition of the *Principia* and KEPLER's Theorem in Section III following in the present revision. The manuscript contains no proof of the Theorem on the Uniqueness of Centripetal Motions, but a note in the margin suggests that the author was hoping to simplify the published *Principia*'s demonstration of Proposition XLI for this special case.

SECTION III[32]

Centripetal motions in particular orbits

PROPOSITION ON SPIRAL MOTIONS[33]

In any (constant angle) spiral motion about the pole, the acceleration varies inversely as the cube of the radius.

By Corollary I to the Theorem on Speeds and Curvatures (Book II, Section II), the acceleration varies inversely as $SY^2 \cdot PV$, but according to the Proposition on Spirals[34] (Book I, Section III), PV, the chord of curvature through the pole S, varies as the radius SP, and on a constant angle spiral SY also varies as SP. It follows that $SY^2 \cdot PV$ varies as SP^3.

THEOREM ON CONIC MOTIONS[35]

In the figure below, assume DC to be parallel and both SY and PN to be perpendicular to the tangent line at P. Moreover, suppose the ordinate PM to be perpendicular to the focal axis.

---

(32) After the hard work on curvature (Section III of Book I, POURCIAU (1992b)), the results in this section appear almost as trivial observations.

(33) Proposition IX, in Section II of Book I of the 1726 *Principia*.

(34) Read POURCIAU (1992b), Section III, Book I.

(35) Leaving aside the present revision, if we study NEWTON's extant manuscripts which describe his plans for radically revising the early portions of the 1687 *Principia*, we find two grand schemes for arriving at results equivalent to the Theorem on Conic Motions recorded here. Refer to WHITESIDE (1967–1981: VI, scheme 1 on 572–581, scheme 2 on 581–589) for the actual manuscripts as well as insightful commentary, and consult BRACKENRIDGE (1989: 24–31) for an excellent overview of these schemes. If we concentrate only on how NEWTON deduces the force through a focus of a central conic, then the logical flow for each scheme becomes simple enough to record here:

SCHEME 1
[force to center + force comparison] ⇒ force to any point ⇒ force to focus

SCHEME 2
[curvature to center + buried comparison] ⇒ curvature to any point ⇒ force to any point ⇒ force to focus

Apart from the stress on curvature and the more hidden reliance on comparison in the second plan, the two schemes exhibit a strikingly similar structure. The flow of the present manuscript is of course most like that of the second scheme.

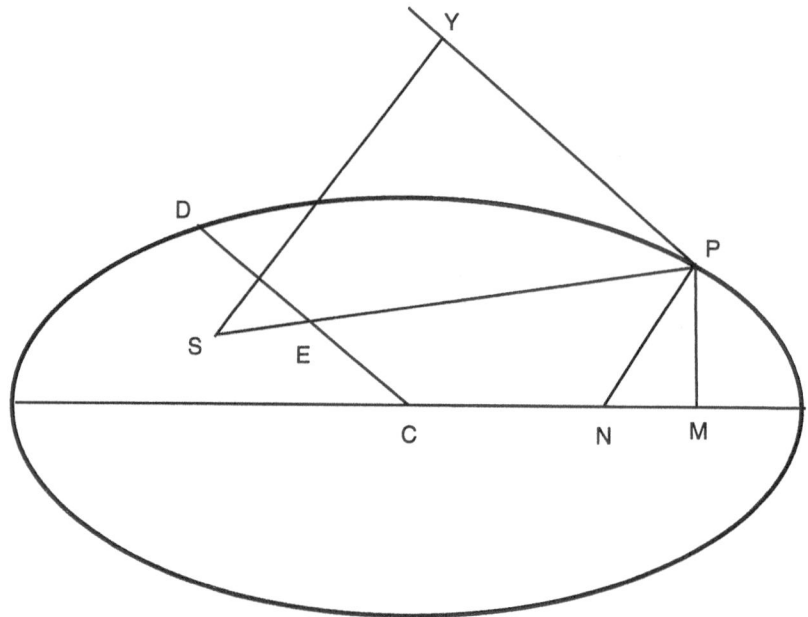

Then:[36]

(1)   *In a central conic motion about any point S, the acceleration varies directly as* PE^3/SP^2.[37] For the acceleration varies inversely as $SY^2 \cdot PV$, and by Corollary I of the Proposition on Central Conics[38] (Book I, Section III), PV varies as

$$\frac{1}{PE^3}\left(\frac{SP}{SY}\right)^2 .$$

(2)   *In a central conic motion about the center C, the acceleration varies as the distance* CP.[39] For then PE is CP.

(3)   *In a central conic motion about a point S infinitely remote, the acceleration will vary directly as the cube of* PE.[40] For then SP is given.

---

(36) Parts (1) through (5) appear in the extant unpublished revisions. See WHITE-SIDE (1967–1981: VI, 579–581 and 585–589).

(37) In essence this statement appears as a final scholium to Section III of Book I in the second and third editions. See NEWTON (1726: 93) and read also WHITESIDE (1967–1981: VI, 588 and 589, Notes 46 and 50).

(38) From POURCIAU (1992b), Section II of Book I, we import the Proposition on Central Conics: *For a central conic, the chord of curvature through the center is the* latus rectum. This elegant result, phrased with less stark simplicity, forms the centerpiece of Newton's second extant grand scheme for revising the 1687 *Principia*. Read POURCIAU (1992b), Note 34, for more detail.

(39) Proposition X in the published *Principia*.

(40) A similar result—incompletely stated: see Note 44 below and WHITESIDE (1967–1981: VI, Note 99 on 137 and Note 46 on 588)—can be found in a scholium following Proposition VIII (Section II of Book I), NEWTON (1726: 74).

(4) *In a circular motion about a point S on the circumference, the acceleration varies inversely as the fifth power of the distance SP,*[41] For in this case, PE will be reciprocally as SP.

(5) *Given any conic motion (the parabola included) about a focus S, the acceleration varies inversely as the square of the distance SP.*[42] For by Corollary I of the Proposition on Conic Curvature[43] (Book I, Section III), the chord of curvature through the focus, on any conic section, varies as the square of SP/SY.

(6) *Suppose S to be so remote and in such a direction that all the lines SP may be taken for parallels which are perpendicular to the focal axis. Then given any conic motion (the parabola included) about S, the acceleration will vary inversely as the cube of the ordinate PM.*[44] For, on the one hand, by Corollary II of the Proposition on Conic Curvature,[45] the chord of curvature perpendicular to the focal axis varies as PN^2PM. But, on the other hand, SY varies as PM/PN, as SP is given. Thus $SY^2 \cdot PV$ varies as the cube of PM.

---

(41) Appears as Corollary 1 to Proposition VII in the second and third editions.

(42) The manuscript records here NEWTON's celebrated solution, for conic motions, of the Direct Problem for conic motions: from the orbit shape and the center, to determine the (centripetal) acceleration. The solution in the published *Principia*, where the proof looks quite different, can be found in Propositions XI–XIII. For modern and Newtonian style treatments of the Direct and Inverse Problems, read POURCIAU (1992a).

(43) From POURCIAU (1992b), Section III of Book I, we record the statement of the Proposition on Conic Curvature: *In any conic section let S be a focus, let SY be the perpendicular from S onto the tangent line at P, and let the diameter of curvature at P meet the focal axis at N. Then the diameter of curvature at P varies as the cube of SP/SY or, equivalently, as the cube of PN.*

(44) Proposition VIII (Section II of Book I in the 1726 *Principia*) reads as follows: "If a body moves in the semi-circumference PQA; it is proposed to find the law of the centripetal force tending to a point S, so remote, that all the lines PS, RS drawn thereto, may be taken for parallels." After the proof of this proposition, NEWTON appends a generalizing scholium: "And by a like reasoning, a body will be moved in an ellipse, or even in an hyperbola, or parabola, by a centripetal force which is reciprocally as the cube of the ordinate directed to an infinitely remote centre of force." WHITESIDE (1967–1981: VI, Note 99 on 137) has noted, quite correctly, that the scholium, as stated, is false. However, the scholium becomes true provided we assume that the lines of force are not merely parallel, but also perpendicular to the focal axis—as in part (6) of the Theorem on Conic Motions, above, in the present radical revision—and it may well be that NEWTON had this additional assumption in mind but not in quill when he penned this scholium.

(45) Refer to Book I, Section III, of the original, uncut manuscript, in POURCIAU (1992b).

KEPLER'S THEOREM[46]

Supposing the acceleration to vary inversely as the square of the distance from the center, any centripetal motion must be a conic motion about the focus.

For simplicity in the statement of Kepler's Theorem, we have excepted the case of motion along a fixed line through the center S, which occurs if the motion leaves any given point heading precisely toward or away from S. To prove the theorem, let us choose any point P on the orbit of the given inverse square motion. At P the given motion has a given speed ν and a given acceleration and therefore, by the Theorem on Speeds and Curvatures, a given chord of curvature PV through S. At P the given orbit has a tangent line PT. Using the Proposition on Conic Curvature[47] (Book I, Section III), we can construct a conic section $\mathcal{C}$ having focus at S and passing through P with tangent PT and with focal chord of curvature PV[48]. On $\mathcal{C}$, courtesy of the Lemma on the Existence of Centripetal Motions (Book II, Section II), there exists a motion about the focus S which leaves P in the direction PT with speed ν. Yet this conic motion about the focus must have an inverse-square acceleration by part (5) of the Theorem on Conic Motions above. Thus the given motion and the constructed conic motion both have inverse-square accelerations, but in fact they share precisely the *same* inverse square acceleration (that is, they share the same proportionality constant), because they have the same speed and curvature at P. Invoking now the Theorem on the Uniqueness of Centripetal Motions (Book II, Section II), the given and constructed conic motions cannot be distinct, and it follows that the given inverse-square motion about S is, in fact, a conic motion with S as focus, as desired.[49]

---

(46) Here the radical revision sets out NEWTON's solution, with an expanded argument, to the Inverse Problem for inverse-square acceleration. Corollary 1 of Propositions XI–XIII records this wonderful result in the published editions of the *Principia*.

In *De Motu*, NEWTON cites KEPLER in a scholium: "Therefore the major Planets revolve in ellipses having a focus in the center of the sun; and by radii drawn to the sun describe areas proportional to the times, altogether as Kepler supposed." See HALL and HALL (1962: 253 and 277) and COHEN (1971: Note 3 on 28). But NEWTON is surely less generous in the published *Principia*. We celebrate the appearance, in the present manuscript, of KEPLER's name at this most appropriate place. We find it appropriate to name this theorem after KEPLER, not because he was the first to suggest a proof—this was NEWTON—nor even because he was the first to state this theorem as conjecture—this was HOOKE—but rather because he was the first to know that the planets actually do traverse ellipses with the sun at one focus.

(47) Section III, Book I, POURCIAU (1992b).

(48) Read POURCIAU (1992a) for details on this conic construction. See also Proposition XVII in the published *Principia* for an equivalent conic construction described by NEWTON.

(49) In the 1687 edition, NEWTON left KEPLER's Theorem—Corollary 1 of Propositions XI–XIII in the *Principia*—as a bald claim, with no proof. More than twenty years later, in October of 1710, Johann BERNOULLI wrote to Jakob HERMANN, complaining

ACKNOWLEDGEMENT

For their many informative and insightful comments, I would like to thank J. Bruce Brackenridge, C. Truesdell, D.T. Whiteside, and C. Wilson. The responsibility for the remaining problems in accuracy or elegance should be placed at my door. I would like to thank K. Meyer for the invitation to speak at the International Conference on Hamiltonian Dynamical Systems in the spring of 1992.

Bruce Pourciau
Appleton, Wisconsin

---

about this state of affairs, but one year earlier, NEWTON had already decided to fill the hole by inserting an argument for Corollary 1. Here is part of a letter, dated 11 October 1709, from NEWTON to Roger COTES, the editor of the second (1713) edition: "I forgot also to add the following Note to the end of Corol. 1 pag. 55 1in. 6. Nam datis umbilico et puncto contactus & positione tangentis, describi potest Sectio conica quæ curvaturam datam ad punctum illud habebit. Datur autem curvatura ex data vi centripeta: et Orbes duo se mutuo tangentes eadem vi describi non possunt." See HALL and TILLING (1975: 5–6). The same argument passed also into the third(1726) edition with one small but important addition: NEWTON inserts, in two places, the *speed* of the body, pairing it with the force. Thus we have the final third edition argument for Corollary 1 of Propositions XI–XIII: "For the focus, the point of contact, and the position of the tangent being given, a conic section may be described, which at that point shall have a given curvature. But the curvature is given from the centripetal force and the bodies velocity given: and two orbits mutually touching one the other, cannot be described by the same centripetal force and the same velocity." Read WHITESIDE (1967–1981: VI, 146–149, Note 124) and WHITESIDE (1992?) for more of the historical context surrounding this brief argument for Corollary 1 (= KEPLER's Theorem). WHITESIDE (in a private communication) points out that NEWTON in later years appealed only to Proposition XVII to justify his claim in Corollary 1. Proposition XVII *can* be made the basis of a proof of Corollary 1, provided we make additional appeals to the Lemma on the Existence of Centripetal Motions and the Theorem on the Uniqueness of Centripetal Motions at appropriate states of the argument.

One would never characterize the few lines NEWTON appended to Corollary 1 as giving a complete proof of the corollary, but rather perhaps the sketch of an outline of a proof. Most scholars seem to believe that this sketch does indeed form the outline of a correct, cogent demonstration, but there have been dissenters. Recently, WEINSTOCK (1982 and 1988) has claimed that the sketch contains an intrinsic flaw, a logical circularity in fact, which prevents it from being the basis for any proof. This claim has been countered by others, including, for example, ERLICHSON (1990 and 1992), POURCIAU (1991 and 1992a), and WHITESIDE (1992?).

In the present radical revision, Corollary 1 of Propositions XI–XIII has been elevated to the status of a theorem—in the published editions of the *Principia*, it should never have appeared as a mere corollary, because of its importance and because it fails to follow, in any *simple* way, from the three propositions that precede it—and NEWTON's very brief argument for Corollary 1 has been expanded into the clear outline of a correct demonstration, including an explicit reference to the Lemma on the Existence of Centripetal Motions, which in the published argument seems to be accepted as an unspoken obvious fact. (Read Note 29.) To see a complete proof of KEPLER's Theorem, a proof guided by NEWTON's sketch, and all the mathematical details, both Newtonian and modern, examine POURCIAU (1992a).

REFERENCES

AITON, E.J., (1964). *The inverse problem of central forces*, Annals of Science 20, 81–99.

BRACKENRIDGE, J. BRUCE, (1988). *Newton's mature dynamics: revolutionary or reactionary?*, Annals of Science 45, 451–476.

BRACKENRIDGE, J. BRUCE, (1989). *Newton's unpublished dynamical principles*, Annals of Science 47, 3–31.

CLARKE, JOHN, (1730). *A demonstration of Some of the Principal Sections of Sir Isaac Newton's Principles of Natural Philosophy. In Which His Peculiar Method of Treating That Useful Subject is Explained, and Applied to Some of the Chief Phenomena of the System of the World*, London: printed for James and John KNAPTON.

COHEN, I. BERNARD, (1970). *Newton's second law and the concept of force in the Principia*, In *The Annus Mirabilis of Sir Isaac Newton: 1666-1966*, edited by Robert PALTER, pages 143–185. Cambridge (Massachusetts): The M.I.T. Press.

COHEN, I. BERNARD, (1971). *Introduction to Isaac Newton's Principa*, Cambridge (Massachusetts): Harvard University Press.

ERLICHSON, HERMAN, (1990). *Comment on 'Long buried dismantling of a centuries-old myth: Newton's Principia and inverse-square orbits'.*", American Journal of Physics 58, 882–884.

ERLICHSON, HERMAN, (1992). *The instantaneous impulse construction as a formula for central force motion on an arbitrary plane curve with respect to an arbitrary force centre in the plane of that curve*, Annals of Science 49, 369–375.

FIELD, J.V. AND GRAY, J.J., (1987). *The Geometrical Work of Girard Desargues*, New York: Springer-Verlag.

FROST, PERCIVAL, (1883). *Newton's Principia, First Book, Sections I, II, III, With Notes and Illustrations, and a Collection of Problems Principally Intended as Examples of Newton's Methods*, London: Macmillan and Co.

HALL, A.R. AND HALL, MARIE BOAS, eds. (1962). *Unpublished Scientific Papers of Isaac Newton: A Selection From the Portsmouth Collection in the University Library Cambridge*, Cambridge (England): Cambridge University Press.

HALL, A.R. AND TILLING, L., eds. (1975). *The Correspondence of Isaac Newton, Volume V: 1709-1713*, New York: Cambridge University Press.

HERIVEL, JOHN W., (1964). *Newton's first solution to the problem of Kepler motion*, British Journal for the History of Science 2, 351–354.

HERIVEL, JOHN W., (1970). *Newton's achievement in dynamics*, In *The Annus Mirabilis of Sir Isaac Newton: 1666–1966*, edited by Robert PALTER, pages 120–135, Cambridge (Massachusetts): The M.I.T. Press.

NEWTON, ISAAC, (1726). *The Mathematical Principles of Natural Philosophy*, English translation in two volumes by Andrew MOTTE, 1729. All page references are to the facsimile reprint, with introduction by I. Bernard COHEN, London: Dawsons of Pall Mall, 1968.

NOLL, WALTER, (1974). *The Foundations of Mechanics*, New York: Springer-Verlag.

POURCIAU, BRUCE, (1991). *On Newton's proof that inverse-square orbits must be conics*, Annals of Science 48, 159–172.

POURCIAU, BRUCE, (1992a). *Newton's solution of the one-body problem*, Archive for History of Exact Sciences 44, 125–146.

POURCIAU, BRUCE, (1992b). *Radical Principia*, Archive for History of Exact Sciences 44, 331–363.

SALMON, GEORGE, (1879). *A Treatise on Conic Sections: Containing an Account of Some of the Most Important Modern Algebraic and Geometric Methods*, London: Longmans, Green, and Co.

TRUESDELL, C., (1967). *Reactions of late baroque mechanics to success, conjecture, error, and failure in Newton's Principia*, The Texas Quarterly, 238–258. Revised in (1968) for Essays in the History of Mechanics. New York: Springer-Verlag.

TRUESDELL, C., (1992). *A First Course in Rational Continuum Mechanics*, Second Edition. New York: Academic Press.

TURNBULL, H.W., (1961). *The Correspondence of Isaac Newton*, Volume 3: 1688–1694. Cambridge (England): Cambridge University Press, for the Royal Society.

WESTFALL, RICHARD S., (1971). *Force in Newton's Physics: The Science of Dynamics in the Seventeeth Century*, New York: American Elsevier.

WEINSTOCK, ROBERT, (1982). *Dismantling a centuries-old myth: Newton's Principia and inverse-square orbits*, American Journal of Physics 50, 610–617.

WEINSTOCK, ROBERT, (1989). *Long-buried dismantling of a centuries-old myth: Newton's Principa and inverse square orbits*, American Journal of Physics 57, 846–849.

WHITESIDE, D.T., ed. (1967–1981). *The Mathematical Papers of Isaac Newton*, Volumes I–VIII. Cambridge (England): Cambridge University Press.

WHITESIDE, D.T., (1970). *Sources and strengths of Newton's early mathematical thought*, In *The Annus Mirabilis of Sir Isaac Newton: 1666–1966*, edited by Robert PALTER, pages 69–85. Cambridge (Massachusetts): The M.I.T. Press.

WHITESIDE, D.T., (1991). *The prehistory of the Principia from 1664 to 1686*, Notes and Records of the Royal Society of London 45, 11–61.

WHITESIDE, D.T., (1992?). *Corol. 1 to Propositions 11–13 of Book I of Newton's Principia*, Preliminary version.

FACTORING THE LUNAR PROBLEM: GEOMETRY, DYNAMICS, AND ALGEBRA IN THE LUNAR THEORY FROM KEPLER TO CLAIRAUT

CURTIS WILSON*

As everyone knows, mathematical astronomy from Hipparchus up to but not including Kepler proceeded by the compounding of circular motions, and the apparatus could get top-heavy. Tycho Brahe, for one was bothered by geometrical complication. Here is his theory (Fig. 1.1), developed during the 1590s, for the two main inequalities in the Moon's longitude.

The so-called *first inequality* had been accounted for by Hipparchus, in the second century B.C.; the *second inequality* was first dealt with by Ptolemy, in the second century A.D. [1]. The first inequality was determined from lunar eclipses. Lunar eclipses were the data to start with, because they are independent of lunar parallax, the apparent displacement of the Moon due to the observer's not being at the Earth's center. This can go as high as 1°. Until you know pretty well where the Moon is, relative to the Earth's center, you can't get a measure of parallax and so subtract it out of the observations.

Hipparchus found that uniform motion of the Moon on an epicycle, combined with uniform motion of the epicycle on a circle concentric with the Earth, would predict the Moon's longitude in lunar eclipses (Fig. 1.2). The epicycle got once round in 27.3 days but the Moon had to go about 5.6 hours more to return to the high point on the epicycle; in other words, the apse advanced about 3° a month. In Tycho's diagram (Fig. 1.1), the large deferent circle plus the *two* epicycles at the top are the substitute for this Hipparchian theory. Why *two* epicycles? I'll come to that in a moment. Let me first describe the second inequality, known since the 17th century as the *evection*.

The second inequality showed itself when the Moon was in quadrature to the Sun. For these positions the Hipparchian theory failed. Ptolemy found he could account for the deviations if he put Hipparchus's circle-plus-epicycle on a crank, pulling the whole apparatus in toward the Earth when the Moon was in first or last quarter, then pushing it out again. The apparent departures from uniform motion were thus enlarged in the quadratures, as the observations required. Unfortunately, the mechanism also implied an enlargement in the Moon's apparent diameter, up to about twice its least size; which doesn't happen.

This discrepancy between theory and observation seems to have gone unmentioned in Europe till Regiomontanus remarked on it in the 15th century. Copernicus undertook to eliminate it. He discarded the crank mech-

---

* St. John's College, P.O. Box 2800, Annapolis, MD 21404.

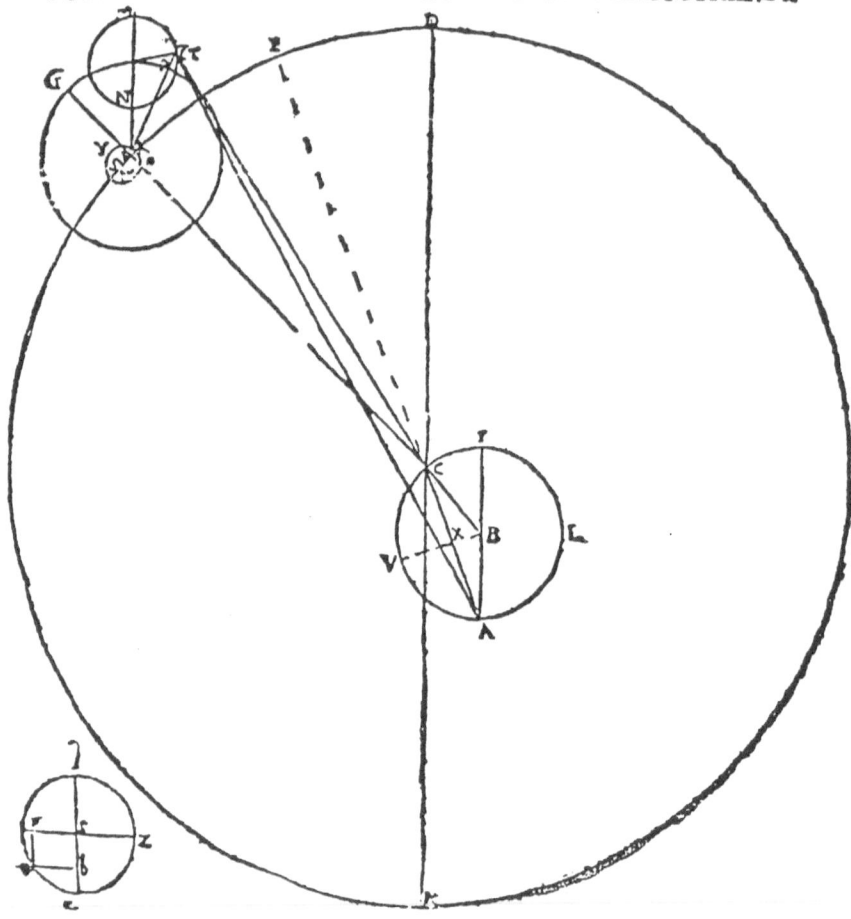

FIG. 1.1. *Tycho's Lunar Theory.*
The diagram is from Tycho's Astronomiae Instauratae Progymnasmata (Prague, 1602),
where it appears on p. 24 of a 28-page section on the lunar theory which is inserted
following p. 112 of the main text. (Each page number in this section is preceded by the
letter "O".)

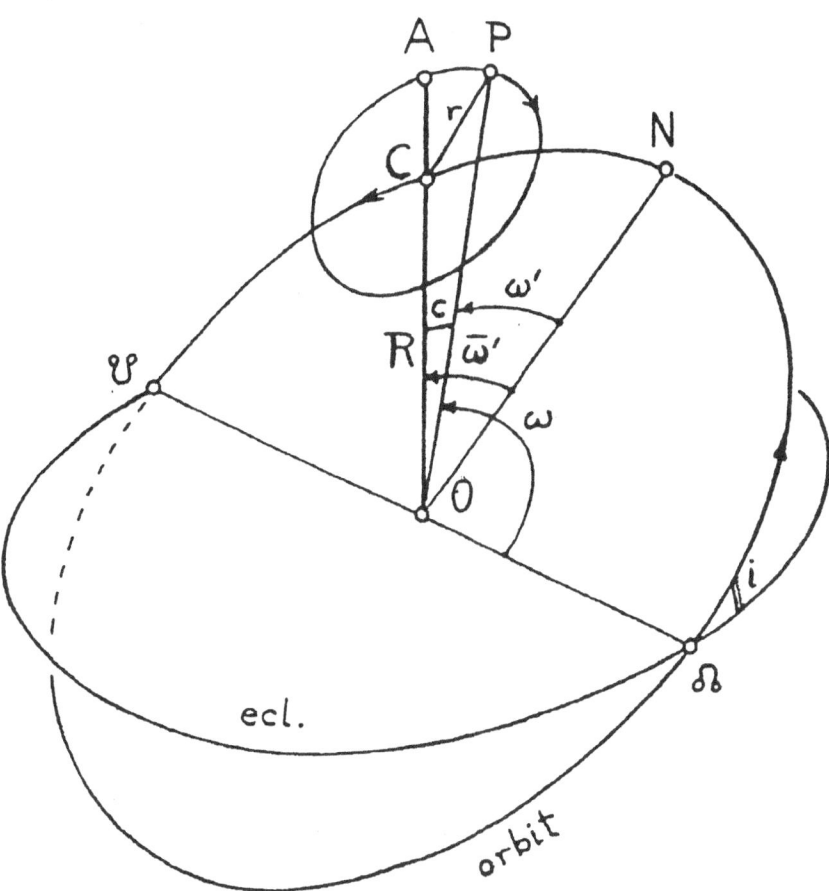

FIG. 1.2. *The Hipparchian lunar theory. The figure is taken from Neugebauer, A History of Ancient Mathematical Astronomy, p. 1228, Figure 72.*

anism, replacing it by a small epicycle mounted atop the single epicycle of the Hipparchian theory. The Moon went round this small epicycle twice a synodic month, in such a way as to enlarge the Moon's distance from the center of the larger epicycle whenever the Moon was in quadrature, and so give larger equations of center, or departures from uniform motion. The limits of variation in the Moon's apparent diameter were reduced from Ptolemy's 2:1 ratio to about 13:10 in Copernicus's theory [2].

Tycho retained this small Copernican epicycle for the evection. In our diagram (Fig. 1.1), it is *not* the small epicycle atop the big epicycle; rather it is the hypocycle ALCV in the middle. A is the Earth's center. C, the center of the big, deferent circle, goes around the little circle ALCV in such a way that it coincides with A when the Moon is in the line through Earth and Sun, in the syzygies as we say, and is at the top of the little circle whenever the Moon is in quadrature. The effect is to enlarge the equations of center for the latter positions. Whether the little circle is a hypocycle or an epicycle doesn't matter: the two arrangements are equivalent because of the commutativity of vector addition. Why did Tycho choose hypocycle in place of epicycle? He already had one epicycle atop another in his account of the first inequality, and apparently he was loath to pile up more epicycles.

And why did Tycho use *two* epicycles for the first inequality? An important reason, I believe, was this. Copernicus, by introducing his small epicycle for the evection, reduced the ratio of greatest to least lunar parallax to about 13:10. That was still too large. Tycho could not measure lunar diameters very precisely - the screw micrometer was not invented yet - but a ratio of 13:10 would imply a variation in the Moon's diameter from about 35' to about 27'; the observed variation was less than half as much.

An obvious way to reduce the variation was to introduce a Ptolemaic-style equant point, such as Ptolemy had used for the planets (Fig. 1.3). This was a point (here Z) other than the center (D) of the deferent or large circle, about which the motion of the epicycle's center (point B) was angularly uniform. In Ptolemy's theories for Venus and the outer planets the equant point (Z) was twice as far from the Earth (E) as the center (D). This arrangement gave you the same maximum equation of center as if you put the center of the deferent at the double distance EZ, with uniform motion about this center. In the lunar theory, the equant would reduce errors in lunar parallax, and in longitude as well.

Unfortunately, it was a matter of principle with Tycho not to use an equant. To him, as to Copernicus, the Ptolemaic equant was anathema. It makes the motion on the circle non-uniform, and so violates what Copernicus called the first principle of the astronomical art. Copernicus had shown how to get almost the same effect, with one epicycle mounted atop another, and the two epicycles turning in opposite directions. Tycho regarded this invention of a substitute for the equant as Copernicus's greatest achievement. And it is this Copernican device that we have here in the double epicycle at the top of the Tychonic diagram.

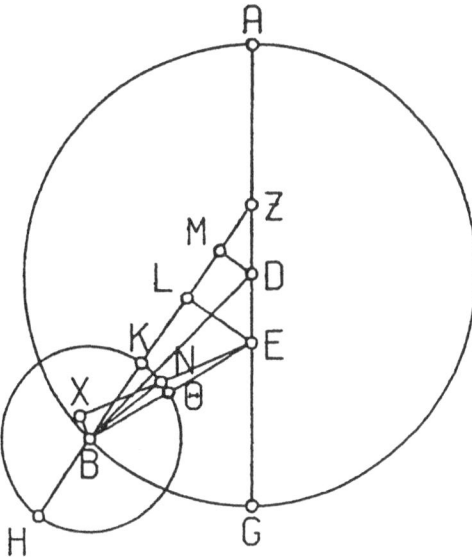

FIG. 1.3. *Ptolemy's Equant*
The figure is taken from G.J Toomer, Ptolemy's Almagest (Springer-Verlag, 1984), p.
500.

So Tycho was the first to introduce what amounts to an equant into the lunar theory, but it appears as a double epicycle. And then, rather than mounting the Copernican epicycle for the evection on top of the double epicycle, he made it into a hypocycle. There are other signs that Tycho was loath to pile up epicycles. The annual equation, which he and Kepler discovered independently, is not in his diagram. He tried to account for it by incorporating it in the equation of time; a ploy that makes the diurnal motion of the heavens non-uniform. The hypothesis was not refuted till Huygens in mid-century showed that, to within the errors of his pendulum clock, the Earth rotated uniformly. Another inequality discovered by Tycho was the Variation, an oscillation of about 40′ amplitude reaching its maximum in the octants of the syzygies. Tycho did not incorporate this inequality in the geometry of his diagram, but computed it separately as a harmonic oscillation along the diameter of a circle (see the separate circle at the lower left of Fig. 1.1). Evidently geometrical complication was something to shun.

I turn now to Kepler. From the word Go, Kepler was moved by the aesthetic simplicity of the heliocentric arrangement. Very early he adopted a dynamic point of view, with each planet moving in an *orbit* subject to *forces*. To Kepler the equant point was an index to the dynamics of planetary motion, for it implied that the planet was moving more slowly when farther from the Sun. He was also impressed by the fact that planets farther from the Sun move more slowly, not just in angle about the Sun, but in *linear* speed. Both facts meant to him that the Sun was the *cause* of motion, and that its force varied inversely with distance. Note that *force* here does not have the Newtonian sense, but is proportional to velocity. Kepler invented the tern *inertia*, but to him it meant merely the tendency of a body to stay put. Any motion betokened the presence of a force, proportional to speed.

The assumption that an immaterial force issues from the Sun, its strength varying inversely with distance, led him to his area law, the constancy of areal velocity; indeed, an inverse variation of the transverse component of the velocity with distance from the Sun strictly *implies* the constancy of areal description. With the aid of the area law, he was able to arrive at the elliptical orbit; he deluded himself into believing he had excluded all empirically acceptable alternatives [3]. The area law and the elliptical orbit are thus not mere empirical laws; the truth is less simple. Note that Kepler's dynamics is such as to force him to introduce a separate mechanism to vary the planet's distance from the Sun [4].

After Kepler had arrived at the first two planetary laws, he turned, in the late teens of the 17th century, to the lunar problem. He accepted Tycho's theory as a correct account of the appearances, but sought to transform it in accordance with his new ideas about the dynamics of planetary motion. The Moon's basic orbit, he assumed, was an ellipse, traversed in accordance with the area law. Thus, according to Kepler, the Moon was moved round by a *motive* force issuing from the Earth, and diminishing in strength with distance from the Earth's center. Another force caused it to vary its distance from the Earth.

Let me describe the way the elliptical orbit and areal rule work for Kepler, in the determination of equations of center (Fig. 1.4). The Moon *L* moves about *E* the Earth in such a way as to sweep out equal areas in equal times. *O* is the center of the ellipse. The equation of center is the angle by which the Moon as seen from the Earth departs from uniform motion. It is composed of two parts. First, the area of the little triangle OEL, compared with the whole area of the ellipse taken as 360°; Kepler calls this the *physical equation*, meaning that it corresponds to a real inequality of motion on the orbit. Second, the angle OLE, which Kepler calls the *optical equation*, and which is due simply to the displacement of *E* from the center of the ellipse. For small eccentricities these two parts are nearly equal.

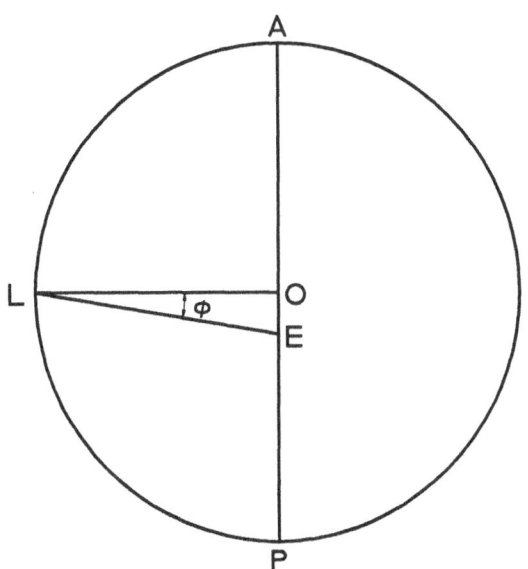

FIG. 1.4. *Kepler's Lunar Theory:*
First Inequality Eccentricity = OE/OA = 0.04362
Optical Equation   = ϕ = arcsin 0.04362 = 2°30′
Physical Equation   = Area LEO/Area Ellipse ×360° = 2°30′
Total Equation of Center = 5°0′

What now about the second inequality, the evection? This inequality, like all the inequalities other than the first, was evidently the work of the Sun. How did the Sun add its influence to the Earth's motive force? Not by acting directly on the Moon, for the Moon sometimes moved *with* the turning rays issuing from the Sun, and sometimes *against* them, and the effects were independent of this difference. Kepler concluded that what the Sun's rays did was to enhance the Earth's power to move the Moon, both in pushing the Moon round the Earth and in varying its distance from the Earth. In instantiating these dynamical ideas, Kepler in 1618 arrived at a theory in which an elliptical orbit oscillated in position and shape, to account for the first and second inequalities. This was identical with a lunar theory put forward a little later by one Jeremiah Horrocks, of whom more in a moment. [5]

Let me use Horrocks's diagram (Fig. 1.5) to illustrate the theory. In position 1 the line of apsides, AP, in the Moon's orbit is at right angles to the line of syzygies, through the Earth E and the Sun S at the center; in this position the equation of center and the eccentricity are at their smallest values. In the course of some three and a half months, the Earth, carrying the lunar orbit, moves into position 3, where line of apsides AP and line of syzygies ES coincide. Now the eccentricity and equation of center are at their maxima. As we go on to position 5, they have returned to their minima again, and at 7 they are once more maximal. Meanwhile, in the intermediate positions, 2, 4, 6, and 8, the line of apsides departs from its mean position, oscillating to one side and the other. In a moment I shall describe how Horrocks came to this theory. Kepler arrived at it by a piece of free geometrical invention, uniting areas and angles in a diagram so as to obtain the simplest representation of the total equation of center. The result was in accord with his idea about the dynamics involved, and he knew it to be very nearly right - very nearly in agreement with Tycho's theory. But he soon abandoned it, because it did not agree strictly with that theory.

For observe: the period of this inequality is about seven months; whereas the second inequality, from Ptolemy to Tycho, was always conceived as completing itself twice each synodic month. Moreover, in the Horrocksian theory, when the Earth is in position 3 or 7, the eccentricity is a maximum, so that the Moon in its apogee is farther from the Earth, and in its perigee closer to the Earth, than anywhere else. Kepler feared to introduce such variations in Earth-Moon distance, contrary to Tycho's theory.

So we find Kepler backing and filling. Here is his final lunar theory (Fig. 1.6). The Earth is at T, the Moon at L, and C is the center of the Moon's elliptical orbit, so that TC is the eccentricity of the Moon's orbit, which stays constant as C moves round the zodiac. The equation of center of the first inequality is calculated in the usual way. As for the second inequality, in this theory it is altogether a physical equation, given by the

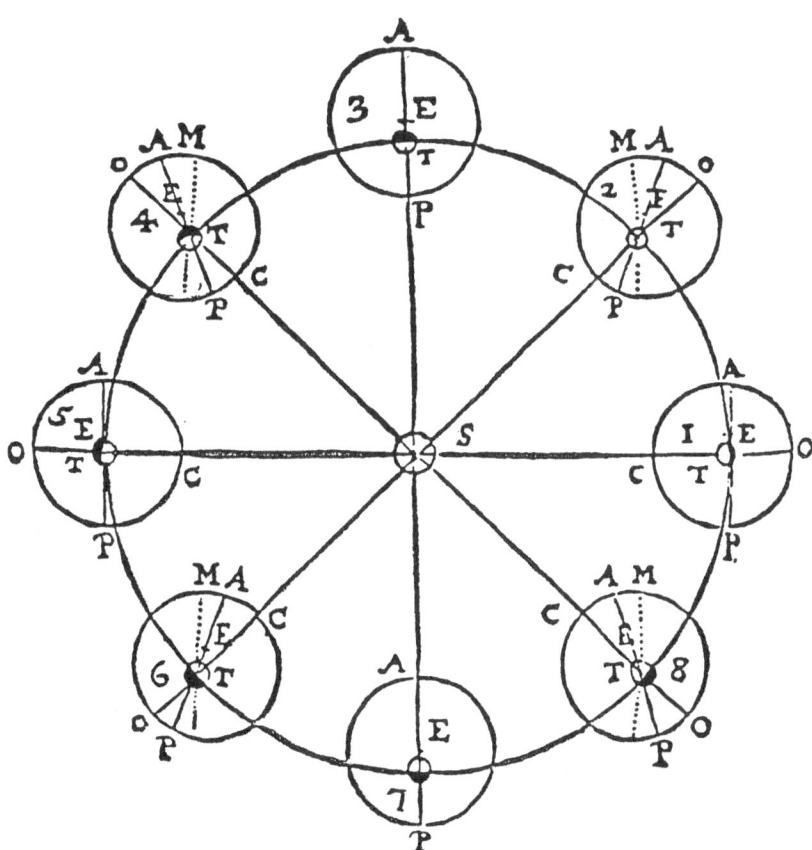

FIG. 1.5. *Horrocks's lunar theory. The diagram is taken from p. 471 of Jeremiah Horrocks, Opera Posthuma (London, 1673).*

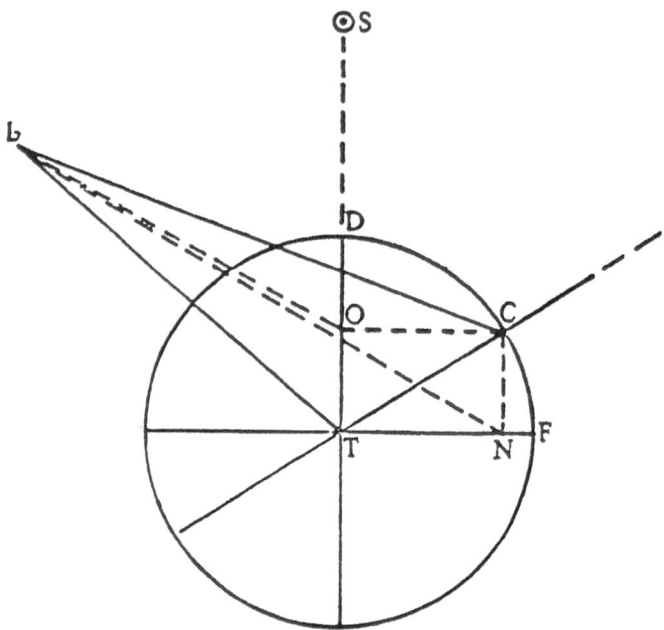

FIG. 1.6. *Kepler's final lunar theory (from Kepler, Gesammelte Werke, xi/1, 531).*

area of the triangle LOT, converted into time. By locating the base of the triangle in the line of syzygies, Kepler gets the second inequality to vanish when the Moon is in this line, twice a month. The second inequality does not affect the shape or orientation of the ellipse. In calculating its effect, Kepler first calculates the area of the triangle LCN, because it is easier to do, then subtracts or adds the area of the little triangle TCN to get the area of the triangle LOT. He calls the area of triangle TCN the *particula exsors*, the "little part specially chosen".

I come now to Jeremiah Horrocks, who after Kepler did more than anyone else in the 17th century to reduce the errors in Kepler's planetary and lunar tables. We don't know when he was born, but he matriculated at Emmanuel College, Cambridge in 1632, and the usual age for doing that was 14. He died on 3 January 1641 (O.S.) presumably at age 22. We can follow some of his steps in testing and refining Kepler's lunar theory in his letters to his friend William Crabtree. By September 1638, using his own observations plus observations reported by earlier astronomers, he had eliminated the *particula exsors*; it amounted at maximum to plus or minus 3 and a quarter arc-minutes; it was annoying to calculate, and the observations didn't confirm it. So he proceeded onward with a modified Keplerian theory, in which the second inequality was given by the area, not

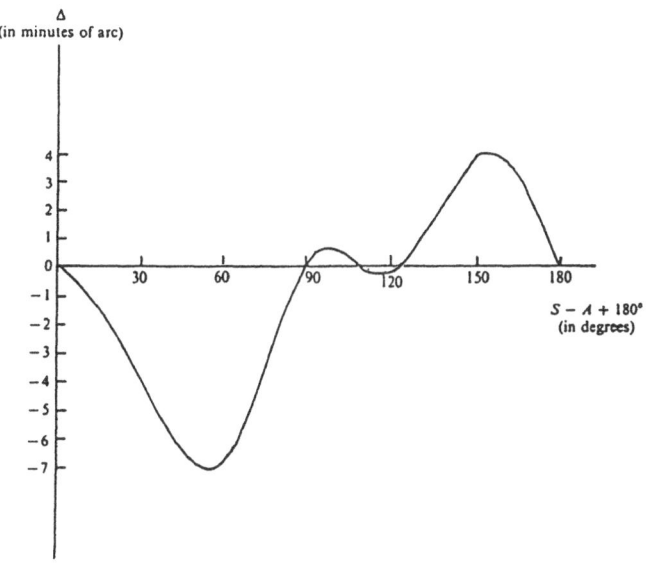

FIG. 1.7. *Horrock's - Kepler's Equations of Center*
(from my "On the Origin of Horrocks's Lunar Theory", Journal for the History of
Astronomy 18 (1987), p. 88; reprinted in my Astronomy from Kepler to Newton:
Historical Studies (Variorum Reprints, 1989)).

of triangle LOT, but of LCN, converted into time [6].

Horrocks made two further empirical discoveries, both from recorded observations of lunar eclipses. He reported the first discovery on 20 December 1638; it had to do with equations of center [7]. In Figure 1.5 consider how the equations of center in lunar eclipses change as the Earth moves from position 3 to position 7. In position 3 the Moon to be eclipsed has to be in apogee; in position 7 it has to be in perigee; in both positions the equation of center is zero. In between, the equation of center increases to a maximum at about position 5, then diminishes again. Horrocks calculated what the equation of center should be for intermediate positions, according to Kepler's theory, and found this theory in disagreement with observations, the discrepancies going as high as 7'. In Fig. 1.7 I have plotted the discrepancies, as differences between the final Horrocksian theory and the Keplerian theory [8]. First there is a negative difference: the equation of center is developing less rapidly than Kepler's theory requires. Then the observed equations of center catch up with Kepler's theory, and in the second quadrant of anomaly surpass the values implied by that theory. Horrocks obtained a theory in agreement with the observations by putting the line of apsides in oscillation (see Fig. 1.5): first the line of apsides advances beyond its mean position, then it moves back.

At this point, Horrocks had all the features of his final theory save one:

he was still treating the eccentricity as constant, and reckoning the second inequality as a physical equation only; that is, it affected the speed of the satellite on its ellipse, but not the shape of the ellipse.

Horrocks's final empirical discovery appears to have come from studying the durations of lunar eclipses. The time-interval from beginning to end of a lunar eclipse depends on the width of the Earth's shadow where the Moon crosses it, and on the Moon's speed. The narrowing of the Earth's shadow with distance from the Earth depends on solar parallax, for which Horrocks had a better value than anyone earlier. He was able to establish that the Moon at A in position 3 of the Earth is farther from the Earth than the Keplerian theory implied, and at P in position 7, closer to the Earth. The eccentricity of the lunar orbit was actually changing. And now all the effects of the first and second inequalities were deducible by way of Kepler's first two laws, elliptical orbit and area rule, from an ellipse changing rhythmically in shape and orientation, with a period of 7 months. It was a masterstroke of theoretical simplification, skillfully verified on the basis of somewhat rough data difficult to come by [9]. In the late 1660s and 1670s, Picard and Flamsteed would verify the changes in eccentricity using the micrometer [10].

It was the earliest theory of the Moon's first and second inequalities that was amenable to dynamical explanation in the Newtonian manner, and Newton did so explain it, in Corollaries 8 and 9 of Prop. 66 of Book I of the *Principia*. But the explanation was at best *qualitative*; Newton did not succeed in achieving a quantitative derivation [11]. The Horrocksian theory, with parameters fixed empirically, mirrored major features in the Moon's motions, as inferable from Newton's laws, and Newton made it the basis of his detailed lunar theory. From his law of gravitation, he was able to derive a few additional inequalities, for instance the annual variations in the motions of the node and the apse. Some small terms in his final theory may have had only an empirical basis: nowhere does he give a derivation of them [12]. The results were disappointing; the errors in Newton's lunar theory, Halley later found, can go as high as 10 arc-minutes, quite as high as in the unmodified Horrocksian theory.

Thus for Newton the Horrocksian theory proved a *cul-de-sac*. What was needed was a new start from Newton's laws. It would first be made in the 1740s.

Recently Victor Katz has pointed out that the calculus of the trigonometric functions was not codified, and made generally known, before Euler did so in his work on linear differential equations in 1739 [13]. This calculus was a prerequisite for an analytical treatment of the three-body problem. It was Euler, again, who in the early 1740s first undertook to derive the inequalities of the Moon by solving differential equations of motion. This earliest perturbational analysis of his has not survived; the resulting tables appeared in 1744 (according to Euler, in 1742) [14]. Euler's analytical procedure was first made public in an essay on the perturbations of Saturn by

Jupiter for the Paris Academy's prize competition of 1748. Here he solved the differential equations by the method of undetermined coefficients, and used multiple equations of condition, based on observations, to effect the differential correction of orbital elements and to fix coefficients of some perturbational terms.

The question for the Paris prize contest of 1748 had been chosen in a meeting of the prize commission in March 1746. Both Clairaut, at age 33, and d'Alembert, at 29, were members of this commission. From the moment the topic for the prize was agreed upon, each of them, unbeknownst to the other, set out to derive and solve differential equations representing the motion of a planet or satellite subject to perturbation. Being members of the Academy, they were ineligible to enter the contest. As the Marquise du Chatelet wrote to Francois Jacquier: "M. Clairaut and M. d'Alembert are after the system of the world, understandably they do not wish to be forestalled by the essays for the prize." Both men had developed a general procedure by early 1747, and both then turned to the lunar problem.

Thus by 1747 three mathematicians had undertaken to cope with the lunar problem analytically. All three assumed that it was necessary to proceed by approximation. This meant, as Euler later explained, [15] introducing an imaginary Moon whose motion approximated the motion of the true Moon, then determining perturbational terms expressing, to some order of approximation, the difference between the true and imaginary motions. The orbit and motion assumed for the imaginary Moon, Euler stressed, were arbitrary. In his earliest lunar theory, Euler used for his imaginary orbit an ellipse of fixed eccentricity and uniformly advancing apse.

The same is true for Clairaut, and I shall describe his theory because of the greater analytic development that Clairaut gives to it, and because of what its algebraic form led to.

Clairaut's differential equations were as follows [16]:

$$(1.1) \qquad rd^2v + 2drdv = \pi dx^2;$$

$$(1.2) \qquad rdv^2 - d^2r = (\frac{M}{r^2} + \phi)dx^2.$$

Here r is the radius vector, v the true longitude, x the time, M the sum of the masses of the central and revolving bodies, and ϕ and π the radial and transverse components of the perturbing force. By two adroit integrations, which won Euler's admiration, Clairaut obtained from the first of these equations an expression for dx, the differential of the time, in terms of dv, the differential of longitude; and with this he eliminated dx from the second equation, so as to obtain a differential equation in the variables r and v alone. These are familiar moves today, but Clairaut was the first to make them. By two integrations he then obtained r as a function of v:

$$\frac{f^2/M}{r} = 1 \quad - \quad g \sin v - c \cos v$$

(1.3) $$+ \ \sin v \int \Omega . \cos v . dv - \cos v \int \Omega . \sin v . dv,$$

where Ω has the rather complicated form

(1.4) $$\Omega = \frac{\frac{\phi r^2}{M} + \frac{\pi r}{M} \frac{dr}{dv} - \frac{2}{f^2} \int \pi r^3 dv}{1 + \frac{2}{f^2} \int \pi r^3 dv}.$$

Here f is a constant of integration.

Clairaut has segregated all the terms involving perturbation into Ω so that in the second line of equation (1.3) we have all the perturbational effects on the radius vector. Years later, in 1786, Laplace would derive a very similar expression for the perturbational variation of the radius vector:

$$\delta r = \frac{\left\{ \begin{array}{l} + \cos v \int ndt.r. \sin v \{2 \int DR + r. \frac{\partial R}{\partial r} \} \\ - \sin v \int ndt.r. \cos v \{2 \int DR + r. \frac{\partial R}{\partial r} \} \end{array} \right\}}{\frac{\mu}{a} \sqrt{1 - e^2}}.$$

Here Clairaut's Ω is replaced by a function of R, the perturbing function; the symbol "D" means differentiation solely with respect to the variables pertaining to the perturbed planet. Laplace was familiar with Clairaut's formula, and can hardly have been unaware of imitating it.

I want to go a little further in conjecture. In April, 1775, Lagrange wrote Laplace, to present the way he had been thinking about the secular variations in the eccentricity and aphelion of a perturbed planetary orbit. (Laplace had just stolen a march on him, using a Lagrangian ploy to derive these secular variations; Lagrange wanted to show his younger rival that he had come to the same result [17].) Lagrange started from Clairaut's equation for the radius vector, but rearranged the terms:

$$\frac{f^2/M}{r} = 1 \ - \ \sin v \{ g - \int \Omega . dv . \cos v \}$$

$$- \ \cos v \{ c + \int \Omega . dv . \sin v \}.$$

He then designated the bracketed expressions as the variables x and y:

$$x = e \sin I = g - \int \Omega . dv . \cos v,$$
$$y = e \cos I = c + \int \Omega . dv . \sin v,$$

where e is the variable eccentricity and I the variable longitude of the aphelion. The original equation thus becomes

$$\frac{f^2/M}{r} \ = \ 1 - x \sin v - y \cos v$$

$$= \ 1 - e \cos(v - I).$$

Taking differentials of all the variables on both sides of the equation, Lagrange shows that the two terms involving differentials of x and y cancel each other. In effect, x and y behave like constants with respect to the variables r and v, the variables that describe the motion of the planet or satellite on its momentary ellipse. Perturbed planetary motion is thus factored into motion *on* the ellipse, on the one hand, and motion and change of shape *of* the ellipse, on the other, and this factoring falls out of Clairaut's analysis. Lagrange's introduction of the variables x and y transforms Clairaut's formula into a direct expression of motion on an ellipse of variable eccentricity and apsidal line.

A year later, in 1776, Lagrange introduced the perturbing function: a function applying to any number of mutually attracting point-masses, and such that the force on any one of them in any direction was derivable by partial differentiation. All of Lagrange's and Laplace's further work in celestial mechanics would depend on it. Lagrange called it Ω. I suggest that this designation was a tacit tribute to Clairaut.

Over thirty years later, in 1808, Lagrange at 73 discovered the theorem that relates the time-rates of change of orbital elements to partial derivatives of the perturbing function with respect to those same orbital elements. It was a crucial step toward the canonical equations of Hamilton and Jacobi. He was still calling the perturbing function Ω.

Back to Clairaut and his equation.

$$\frac{f^2/M}{r} = 1 \quad - \quad g \sin v - c \cos v$$

(1.3)
$$+ \quad \sin v \int \Omega . \cos v . dv - \cos v \int \Omega . \sin v . dv.$$

From this equation one would like to obtain a value for r in terms of v, but Ω is itself a function of r, as well as of the position vector for the perturbing planet. This is, of course, a standard feature of perturbational analysis: to get what you want, you need to have it already, approximately. Now Clairaut's equation gives us, if we omit the terms involving Ω, the radius vector in a fixed ellipse. Should one use this value to substitute into Ω? Clairaut said No. The apsidal line of the Moon moves about $3°$ a month; the assumed expression for r would soon be wrong. Suppose instead, he proposed, that we substitute for r in Ω the expression for the radius vector in a moving ellipse:

(1.5)
$$\frac{k}{r} = 1 - e \cos mv.$$

Here k is a new latus rectum, e a new eccentricity, and m is a number to be determined, presumably a little less than 1, such that mv is the real anomaly, completing $360°$ when the planet or satellite returns to its apse. Use of this equation would be justified, Clairaut affirmed, if after the substitution into (1.3) the larger terms proved to have the form of (1.5), permitting k, e, and m to be identified in terms of other constants of the theory, while the other terms in the solution proved to be relatively small.

D'Alembert criticized this procedure, on the grounds that it took from observation a result that should be derived from the differential equations. In his own treatment of the problem, he assumed as his intermediate orbit a circle - an assumption which indeed borrows less from observation. But it was Clairaut, not d'Alembert, who made the next important discovery.

As is well known, all three of these mathematicians - Euler, Clairaut, and d'Alembert - found initially only about half the observed motion of the lunar apse. To Euler this was a sign that the inverse-square law was inexact, as he expected it to be, for he believed gravitational force to be due to hydrodynamic pressure in an aether, and in such an aether the inverse-square law could not be expected to hold exactly to all distances. D'Alembert supposed that an additional force, like magnetism, was responsible for the added motion of the Moon's apse. Clairaut at first proposed that the inverse-square law be modified by the addition of a small inverse-fourth-power term; and thereby involved himself in a verbal battle with Buffon, who claimed that such a two-term law was metaphysically repugnant.

From the beginning, Clairaut had recognized that a second-order approximation would be in order, to improve the accuracy of his theory. Substitution into Ω of the expression (1.5) led to a new, improved value for r, namely (1.6):

$$\frac{k}{r} = 1 \quad - \quad e \cos mv$$
$$+ \quad 0.0070988 \cos \frac{2v}{n}$$
$$- \quad 0.00949705 \cos(\frac{2}{n} - m)v$$
$$\text{(1.6)} \qquad\qquad + \quad 0.00018361 \cos(\frac{2}{n} + m)v.$$

Here the numerical values of k, e, and m have been determined, and as hoped, the last three terms on the right are much smaller than the first two. The second-order approximation would involve replacing the numerical coefficients here by letters and substituting the resulting formula back into Ω. Clairaut did not expect this second approximation to yield a greatly improved value for the apsidal motion, but he regarded it as necessary for the construction of lunar tables. He did not undertake this second approximation till late 1748, a year after he had first announced the discrepancy between the apsidal motion and Newton's law. And what he found surprised him. The first approximation had yielded, for the apsidal motion, 1°30′38" per sidereal revolution. The second approximation yielded 3°0′35", just 3′ shy of the observational value.

D'Alembert, in the lunar theory that he published in 1754, [18] carried
the iterations to the fourth approximation, obtaining four progressively
smaller terms, summing to 3°2′33″ about 1′ shy of the observational value.
In the *Encyclopedie* he would claim to be the first to have carried the
approximations beyond the second order; [19] but since he had started from
circular motion, his processes were not strictly comparable with Clairaut's.

Euler when he heard in mid-1749 of Clairaut's success in deriving the
apsidal motion was incredulous. He tried to locate the deficiency in his
own derivation, but in July wrote Clairaut to say he couldn't find it. [20]
Clairaut told him that the new result depended on the *square* of the perturbing force, but not till April 1752 did Euler succeed, by the more complete determination of a constant of integration, in deriving a motion of the
apsides in substantial agreement with observation. As he wrote Clairaut,

> ...since now two completely different methods lead to the
> same conclusion, no one will refuse to recognize the cor-
> rectness of your research .... I felicitate you on this happy
> discovery, and I even dare to say I regard this discovery as
> the most important and the most profound that has ever
> been made in mathematics. I ask your pardon a thousand
> times for having doubted the rightness of your retraction
> [of the inverse-fourth-power term], but I believe that my
> stubbornness will render your victory all the more bril-
> liant...[21]

Euler's wholehearted praise of Clairaut's achievement is in sharp contrast
with d'Alembert's repeated and often carping criticisms of Clairaut's pro-
cedures.

Successful lunar tables, accurate to 1 arc-minute, were first attained
by Tobias Mayer in 1753 [22]. Mayer developed the theoretical basis of his
tables from Euler's prize essay of 1748. In applying the theory, he made use
of his own very accurate observations of the Moon's distances from zodiacal
stars, whose positions he had carefully rectified. I suspect that his success
also depended crucially on Euler's method of equations of condition; that
is, on a statistical procedure bringing multiple observations to bear on the
determination of the constants of the theory.

As for Euler himself, his theoretical innovations were not at an end. In
an appendix to his *Theoria motus lunae* of 1753 he gave his first description
of the method of variation of orbital parameters; that is, a procedure which
assumes the Moon to be moving at each instant on an ellipse in accordance
with the area law - the aim being to reduce the difference between the
imaginary and actual motions [23]. In the early 1760s he made the first
assault ever on the restricted problem of three bodies, in which the Moon
is treated as massless, and exact solutions are sought [24]. In 1762 he made
another proposal: set aside all formal integration, and apply mechanical
quadrature directly to the differential equations [25] - a method now much
in vogue. In 1766 he showed how, from a series of equally spaced obser-

vations, by the calculus of finite differences, to obtain accurate values of
the components of position, velocity, and acceleration at a given instant,
to be used as a starting-point for numerical integration [26] - a method for
which Laplace is usually given credit. Finally, after becoming totally blind,
Euler commenced an entirely new assault on the lunar problem, using a
rectangular coordinate system rotating with the Moon's mean motion [27].
In the next century rotating coordinate systems would become standard
in assaults on the three-body problem. And not accidentally, the most
innovative of these, and the basis of the most accurate lunar theory ever
constructed, had its origin in the mind of G.W. Hill, a student of Euler's
writings.

Thus, in a long sequence of innovations, there are strands of continuity,
linking present-day lunar theory to the 18th-century mathematicians who
first formulated the Newtonian theory analytically, and linking their work,
in turn, to that of their 17th-century predecessors, who first transformed
Tycho's geometrical kinematics into a dynamical theory.

REFERENCES

[1] See O. Neugebauer, *A History of Ancient Mathematical Astronomy* (Springer-Verlag, 1975), I, pp. 68-80, 84-93.

[2] See N.M. Swerdlow and O. Neugebauer, *Mathematical Astronomy in Copernicus's De Revolutionibus* (Springer-Verlag, 1984), Part I, pp. 213-215, 233-241.

[3] In Chapter 58 of his *Astronomia Nova*, Kepler mistakenly thought he had disconfirmed the orbit he calls the *buccosa* as contrary to observations. See D.T. Whiteside, "Keplerian Eggs, Laid and Unlaid, 1600-1605", *Journal for the History of Astronomy 5* (1974), 1–21.

[4] For an accurate description of Kepler's celestial dynamics, see Bruce Stephenson, *Kepler's Physical Astronomy* (Springer-Verlag, 1987).

[5] For this stage of the development of Kepler's lunar theory see Johannes Kepler, *Gesammelte Werke*, Vol. XI:.1 (Munich: C.H. Beck, 1983), pp. 521 ff.

[6] On this stage of Horrocks's investigation, see his letter to Crabtree of 29 September 1638 is not given in the edition of 1673 of the *Opera Posthuma*(London, 1673), p. 315ff.

[7] The letter of 20 December 1638 is not given in the edition of 1673 of the *Opera Posthuma*, but is found in the edition of 1672 at p. 465.

[8] The diagram is taken from my "On the Origin of Horrocks's Lunar Theory", *Journal for the History of Astronomy 18* (1987), p. 88; reprinted in my *Astronomy from Kepler to Newton: Historical Studies* (Variorum Reprints, 1989).

[9] On this final stage in the development of Horrocks's lunar theory, see his *Philosophicall Exercises*, Royal Greenwich Observatory, Flamsteed MSS, 1xviii, fol. 83v.

[10] On Picard's verification of the increased eccentricity when the apsides are in the syzygies, see d'Alembert, *Encyclopedie*, T.IX, art. "Lune", p. 729; on Flamsteed's verification of the Horrocksian parallaxes of the Moon, see Horrocks, *Opera Posthuma*, 1673 ed., p. 489.

[11] For Laplace's comment on what Newton achieved here, see *Celestial Mechanics*, Vol. V (a reprint of Vol. V of the *Traité de Mécanique Céleste, Paris, 1825*; Chelsea Publishing Co., 1969), p. 391.

[12] Newton's final lunar theory may be found in I. Bernard Cohen, *Isaac Newton's Theory of the Moon's Motion (1702)* (London: Wm. Dawson & Sons, 1975).

For a study of this lunar theory, see D.T. Whiteside, "Newton's Lunar Theory: From High Hope to Disenchangment", *Vistas in Astronomy 19* (1977), 317–28.

[13] See Victor J. Katz, "The Calculus of the Trigonometric Functions", *Historia Mathematica 14* (1987), 311-24.

[14] Euler, "Reflexions sur les diverses manières dont on peut representer le mouvement de la lune", *Mémoires de Berlin 19* (1763), 180–193, para.7. The first publication of lunar tables by Euler that we know of appears in his *Tabulae Astronomicae Solis et Lunae*, published in 1744; see *Leonhardi Euleri Opera Omnia II*, 23, 1–10.

[15] *Mémoires de Berlin 19* (1763), p. 180 ff.

[16] My account is based on Clairaut, *Théorie de la Lune déduite du seul principe de l'attraction réciproquement proportionelle aux quarr'es de distances* (2d ed., Paris, 1765). It also owes much to Craig B. Waff, *Universal Gravitation and the Motion of the Moon's Apogee: The Establishment and Reception of Newton's Inverse-Square Law* (Ph.D. dissertation, The Johns Hopkins University, 1975).

[17] In September 1777 Lagrange would write Laplace: "If you are free of jealousy because of your success, I am not less so because of my character ..." (*Oeuvres de Lagrange*, XIV, 71).

[18] See d'Alembert, *Recherches sur differens points importans du systéme du monde*, Premiere Partie (Paris, 1754), pp. 113-18.

[19] *Encyclopédie*, T.IX, art. "Lune", p. 736.

[20] See Yu. Kh. Kopelevich, "The Petersburg Astronomy Contest in 1751", *Soviet Astronomy 9* (1966), pp. 655–56.

[21] Letter to Clairaut of 10 April 1751, in *Leonhardi Euleri Opera Omnia IV A, 5*.

[22] Tobias Mayer, "Novae Tabulae Motuum Solis et Lunae" *Commentarii societatis Regiae Scientiarum Gottingensis*, T. II (1752/53).

[23] *Opera Omnia*, II 23, 283ff.

[24] See *Opera Omnia II 25*, 246–57, 281–9; also, for the case of two fixed centers of force, *Opera Omnia II 6*, 209–46, 247–73, 274–94.

[25] *Mémoires de Berlin 19* (1763), 143–79.

[26] *Mémoires de Berlin 19* (1763), 221–34.

[27] See *Opera Omnia II, 22*.

A LIMITING ABSORPTION PRINCIPLE FOR SEPARATED DIRAC OPERATORS WITH WIGNER VON NEUMANN TYPE POTENTIALS

HORST BEHNCKE* AND PETER REJTO†

1. Introduction. Several authors formulated a limiting absorption principle for several classes of operators. For the physical significance of this principle we refer to the paper of Eidus [Ei] and for brevity, for additional references we only refer to the AMS Memoir of Ben-Artzi and Devinatz [BD].

In this paper we establish the principle of limiting absorption for a class of separated Dirac operators corresponding to a class of potentials which contains the one of von Neumann - Wigner [VW], [RS].

In Section 2 we define this class of potentials and formulate our conditions on the intervals. Then in Theorem 2.1, which is our main theorem, we state the principle of limiting absorption. We prove Theorem 2.1 in the four sections that follow.

In Section 3 we construct approximate solutions to our basic system near infinity. The result of this construction is described in Theorem 3.1. To prove Theorem 3.1, first we follow Levinson [Ea] and in Lemma 3.2 we diagonalize the long range part of the potential. Then in Lemma 3.3 we follow Harris-Lutz [HL] and with the help of an approximate solution to a Riccati equation we achieve that the 21-element of the resulting system is short range. To prove Lemma 3.3 we adapt the notion of a slowly varying function to our class of potentials. Then we seek an approximate solution to the basic Riccati equation with the help of slowly varying functions. This construction is similar to the ones of [R], [DR] and [DMR].

In Section 4 we formulate estimates for the solutions of the basic system near zero. First, we construct approximate solutions and then use a result of Love-Erdelyi-Olver [L],[Er],[O],[RT] to show that the same estimates hold for the exact solutions.

In Section 5 we formulate estimates for the basic system near infinity. In Theorem 5.1 we show that the exact solutions satisfy the same estimates as the approximate solutions of Theorem 3.1.

In Section 6, in Theorem 6.1 we formulate the weighted resolvent estimates which imply the principle of limiting absorption. The proof of Theorem 6.1 is based on the Weyl-Weidmann formula [We] for the resolvent kernel and on the Schur-Holmgren-Carleman bound of an integral operator [F], [Ok]. Since this construction is simpler under the additional assumption (6.3), in this section we assume it. In Theorem 6.2 we choose a weight and then show that the Schur-Holmgren-Carleman bound of the weighted

---

* Fachbereich Mathematik, 4500 Osnabrück, Germany.
† School of Mathematics, University of Minnesota, Minneapolis, MN 55455.

resolvent operator with respect to this weight, is uniformly bounded. Here
we use the same Schur-Holmgren-Carleman weight as in [LR]. To prove
Theorem 6.2, we need global upper estimates for each of these two normalized solutions and a lower estimate for their Wronskian. These estimates
are formulated in Lemma 6.3. We prove the upper estimates by giving upper estimates for the connection constants. We prove the lower estimates
using an adaptation of Lemma 3.3 of [DMR].

In Section 7 we complete the proof of Theorem 6.1 and hence of the
main Theorem 2.1 by removing the additional assumption (6.3).

For recent work on absolute continuity we refer to the papers of Behncke
[B], [B], Gilbert-Pearson [GP] and [St]. For the theory of embedded half-
bound states we refer Hinton-Klaus-Shaw [HKS] and for the question of
asymptotics near resonance points we refer to the paper of Klaus [K].

It is a pleasure to thank Professors Devinatz, Harris, McCarthy and
the referee for their valuable suggestions.

2. Formulation of the result.
We consider Dirac operators with
spherically symmetric potentials and assume that they are real valued and
locally integrable on $(0, \infty)$. The separated Dirac operator with electric
potential p and mass m can be written, at least formally, as

$$K = \begin{bmatrix} p - m & \ell/t \\ \ell/t & p + m \end{bmatrix} + DJ$$

where, $\ell = \pm 1 \pm 2, \dots$, $J = \begin{bmatrix} 0 & -1 \\ 1 & 0 \end{bmatrix}$ and $D = \frac{d}{dt}$. Note that the off
diagonal elements of this matrix are equal. Hence this operator K is a
particular case of an operator of the form,

$$(2.1) \qquad\qquad K = V + DJ,$$

where the matrix V is of the form

$$(2.2) \qquad\qquad V = \begin{bmatrix} V_1 & V_2 \\ V_2 & V_3 \end{bmatrix}.$$

From such a formal operator K selfadjoint operators can be derived in
a standard fashion, see e.g. [We] and we call these selfadjoint operators
Hamiltonians.

We start the formulation of the principle of limiting absorption for our
Hamiltonians by formulating assumptions on the potentials. Specifically,
we assume that in some neighborhood of infinity, say (c, ∞), each V_j admits
a decomposition of the form,

$$(2.3) \qquad V_j = S_j + P_j + W_j, \quad j = 1, 2, 3, \text{ or } V = S + P + W,$$

where the S_j are short range in the sense that,

$$(2.4) \qquad\qquad S_j \in \mathcal{L}^1(c, \infty),$$

and P_j are long range in the sense that,

(2.5) $$P'_j \in \mathcal{L}^1(c, \infty), \quad j = 1, 2, 3.$$

Clearly, assumption (2.5) implies that the matrix potential P converges to a limiting matrix $P(\infty)$ for $t \to \infty$, and we assume that $P(\infty)$ is such that,

(2.6) $$P_1(\infty) = -m, \; P_3(\infty) = m > 0 \text{ and } P_2(\infty) = 0.$$

Then, we assume that each W_j can be written in the form,

(2.7) $$W_j = f_j \sin g_j, \quad j = 1, 2, 3;$$

where, each f_j is such that

(2.8) $f_j \in \mathcal{L}^2(c, \infty)$ and it is differentiable and $f'_j \in \mathcal{L}^1(c, \infty)$

and g_j is such that,

(2.9) $\lim\limits_{t \to \infty} g'_j(t)$ exists and $g'_j(\infty) > 0$, or $g'_j(t)$ tends to infinity.

Furthermore, each g_j is 2-times differentiable and

(2.10) $$\frac{g''_j}{(g'_j)^2} f_j \in \mathcal{L}^1(c, \infty).$$

Note that these assumptions define a class of potentials which contains the oscillating part of the von Neumann Wigner potential,

$$W_j(t) = (1 + t)^{-1} \sin t.$$

Finally, we assume that on the interval $(0, c)$ the matrix V admits a decomposition of the form

(2.11) $$V(t) = t^{-1} \cdot JC_0(e, \ell) + B(t),$$

where

(2.12) $$C_0(e, \ell) = \begin{bmatrix} -e & \ell \\ \ell & -e \end{bmatrix}, \quad e \in \mathbf{R}^1,$$

and

(2.13) $$\|B\|_{\mathcal{L}^\infty(0,c)} < \infty.$$

We continue the formulation of the principle of limiting absorption for our Hamiltonians by formulating assumptions on the interval $\mathcal{I}$. First, we we assume that for each $\lambda \in \mathcal{I}$ the limit matrix, $P(\infty) - \lambda$ has two distinct imaginary eigenvalues;

(2.14) $Spec\ (P(\infty) - \lambda) = \{i\mu(\infty), \; -i\mu(\infty)\}, \; \mu(\infty) \neq 0, \; \mu(\infty) \in \mathcal{R}.$

Our second assumption is that for each $j = 1, 2, 3$,

(2.15) $\inf\limits_{t > c} \inf\limits_{\lambda \in \mathcal{I}} |\, 2\mu(\infty) - g_j'(t)| \, > \, 0.$

We complete the formulation of a principle of limiting absorption for our Hamiltonians by introducing the family of operators,

(2.16) $F_s^{\pm}(\lambda) = m_s^{1/2} R(\lambda) m_s^{1/2}, \, Im\lambda \neq 0.$

Here $R(\lambda)$ is the resolvent of a given selfadjoint extension of the formal operator of the definition (2.1),

(2.17) $R(\lambda) = (\lambda I - K)^{-1}, \, Im\lambda \neq 0,$

and m_s is the operator of multiplication by the function,

(2.18) $m_s(\xi) = (1 + \xi^2)^{-s/2}, \quad \xi \in \mathbf{R}^+, \quad s \in \mathbf{R}.$

THEOREM 2.1. *Let the given potential V satisfy assumptions (2.3) ... (2.14) and let the given interval $\mathcal{I}$ satisfy assumption (2.15) with respect to it. Then, for each selfadjoint extension of the operator of the definition (2.1) the principle of limiting absorption holds over the interval $\mathcal{I}$. More specifically, there are two continuous operator valued functions $F_s^{\pm}(\lambda)$ of the variable $\lambda \in \mathcal{I}$ such that,*

(2.19) $\lim\limits_{\epsilon = 0} \sup\limits_{\lambda \in \mathcal{R}_{\pm}(\mathcal{I})} \|F_s^{\pm}(\lambda) - F_s^{\pm}(\lambda \pm \epsilon)\| = 0 \quad s > 1/2.$

To motivate assumption (2.14) we observe that assumption (2.6) and formula (3.5), to be stated, show that for $Re\lambda = \lambda$ assumption (2.14) is equivalent to

(2.20) $|Re\lambda| > m.$

This assumption, in turn, is equivalent to the assumption that for such a λ the basic system corresponding to the operator (2.1),

(2.21) $u' = \begin{bmatrix} -V_2 & -V_3 + \lambda \\ V_1 - \lambda & V_2 \end{bmatrix} u = (V - \lambda J)u,$

has oscillatory solutions. In other words, λ is in the continuous spectrum of the Hamiltonian.

We shall prove the main Theorem 2.1 in the four sections that follow. Our proof will make essential use of the Weyl-Weidmann formula [We], which for $Im\lambda \neq 0$ allows us to study the resolvent of the definition (2.17) with the help of two solutions of the basic system (2.21) and their Wronskian. More specifically, this formula allows us to study the asymptotic properties of this resolvent kernel with the help of the asymptotic properties of two solutions of the basic system (2.21) and their Wronskian. The technical part of the study of the asymptotic properties of such solutions is isolated in the following Section 3. In it, we construct approximate solutions to the basic system (2.21).

3. Construction of approximate solutions to the basic system (2.21) near infinity

THEOREM 3.1. *Let the potential V and interval $\mathcal{I}$ satisfy the assumptions of Theorem 2.1. Then for each λ with $\operatorname{Re}\lambda$ in $\mathcal{I}$ there are approximate solutions z_r and y_r to the system (2.21) in the sense that there is a matrix E_r such that*

$$(3.1) \qquad \sup_{\operatorname{Re}\lambda \in \mathcal{I}} \|E_r(.,\lambda)\|_{\mathcal{L}^1(c,\infty)} =< \infty.$$

and z_r and y_r satisfy,

$$(3.2) \qquad y_r' = [(V - \lambda J + E_r)]y_r.$$

Furthermore,

$$(3.3) \qquad y_r \in L^2((c,\infty), \mathcal{C}_2) \quad for\ \operatorname{Im}\lambda \neq 0.$$

We prove Theorem 3.1 by constructing an approximate fundamental matrix to the system (2.21) by a repeated change of the dependent variable. As is well known if u is a solution of the system (2.21) and if A is a given smooth and invertible matrix, then the function $v = A^{-1}u$ satisfies,

$$(3.4) \qquad v' = [A^{-1}(V - \lambda J)A - A^{-1}A']v.$$

As a first step of the proof of Theorem 3.1 we follow Levinson [Ea] and choose A to be a diagonalizing transformation for the the matrix $P - \lambda J$, the long range part of $V - \lambda J$. To do this, note that assumptions (2.6) and (2.14) allow us to choose a far out region $[c, \infty)$ so that for t in $[c, \infty)$ this matrix has two distinct eigenvalues, say $\pm i\mu$. Then elementary algebra yields the formula,

$$(3.5) \qquad \mu = \mu(t, \lambda) = ((P_1(t) - \lambda)(P_3(t) - \lambda) - P_2^2(t))^{1/2}.$$

For future reference we choose the branch of the square root function so that,

$$(3.6) \qquad \operatorname{Re} i\mu \leq 0,\ for\ \operatorname{Im}\lambda \geq 0.$$

The lemma that follows formulates an elementary property of our class of potentials. It was also verified in [BR].

LEMMA 3.1. *Let the potential P and interval $\mathcal{I}$ satisfy the assumptions of Theorem 2.1. Then for each λ with $\operatorname{Re}\lambda$ in $\mathcal{I}$ there is a transformation $A = A(t, \lambda)$ such that,*

$$(3.7) \qquad A^{-1}(P - \lambda J)A = \operatorname{diag}(i\mu, -i\mu).$$

Furthermore,

$$(3.8) \qquad \sup_{\operatorname{Re}\lambda \in \mathcal{I}} \|A(.,\lambda)\|_{\mathcal{L}^\infty(c,\infty)} + \sup_{\operatorname{Re}\lambda \in \mathcal{I}} \|A^{-1}(.,\lambda)\|_{\mathcal{L}^\infty(c,\infty)} < \infty,$$

and

$$(3.9) \qquad \sup_{Re\lambda \in \mathcal{I}} \|A'(.,\lambda)\|_{\mathcal{L}^1(c,\infty)} < \infty.$$

Conclusions (3.9) and (3.8) of Lemma 3.1 allow us to choose A in formula (3.4) by this lemma. For future reference we define:

$$(3.10) \qquad R(1) = A^{-1}SA - A^{-1}A',$$

and

$$(3.11) \qquad U(1) = [\text{diag}(i\mu, -i\mu) + A^{-1}WA].$$

Then combining the definitions (3.11) and (3.10) with conclusion (3.7) and with assumption (2.3) we find,

$$(3.12) \qquad A^{-1}(V - \lambda J)A - A^{-1}A' = U(1) + R(1).$$

Hence replacing v by u_1 the system (3.4) can be written in the form,

$$(3.13) \qquad u_1' = [U(1) + R(1)]u_1.$$

To motivate the definition (3.10) note that conclusions (3.8) and (3.9) together with assumption (2.4) show that $R(1)$ is short range. More specifically, they show that

$$(3.14) \qquad \sup_{Re\lambda \in \mathcal{I}} \|R(1)(.,\lambda)\|_{\mathcal{L}^1(c,\infty)} < \infty.$$

As a second step of the proof of Theorem 3.1 we choose A in this formula so that the (2,1) element of the system resulting from (3.13) is short range. We do this with the help of a transformation suggested by Harris-Lutz [HL]. It is of the form $I + Q_\ell$ where

$$(3.15) \qquad Q_\ell = q_\ell \cdot \begin{bmatrix} 0 & 0 \\ 1 & 0 \end{bmatrix}$$

and the properties of q_ℓ are described in the lemma that follows.

LEMMA 3.2. *Let the potentials P, W and interval $\mathcal{I}$ satisfy the assumptions of Theorem 2.1 and let the matrix $U(1)$ be given by the definition (3.11). Then, for each λ with $Re\lambda \in \mathcal{I}$ there is a function $q_\ell = q_\ell(t, \lambda)$ such that, for the corresponding matrix function of the definition (3.15),*

$$(3.16)\sup_{Re\lambda \in \mathcal{I}} \|[(I + Q_\ell)^{-1}U(1)(I + Q_\ell) - (I + Q_\ell)^{-1}Q_\ell']_{21}\|_{\mathcal{L}^1(c,\infty)} < \infty.$$

Furthermore,

$$(3.17) \qquad \sup_{Re\lambda \in \mathcal{I}} \|q_\ell(.,\lambda)\|_{\mathcal{L}^\infty(c,\infty)} < \infty,$$

and

(3.18)
$$\sup_{Re\lambda\in\mathcal{I}} \|q_\ell(.,\lambda)\|_{\mathcal{L}^2(c,\infty)} < \infty.$$

We start the proof of conclusion (3.16) with the formula,

(3.19)
$$[(I+Q_\ell)^{-1}U(1)(I+Q_\ell) - (I+Q_\ell)^{-1}Q'_\ell]_{21}$$
$$= -q'_\ell + U(1)_{21} - 2U(1)_{11}q_\ell - U(1)_{12}q_\ell^2.$$

To prove this formula, we show that the definition (3.15) implies that for each matrix B,

$$(I+Q_\ell)^{-1}B(I+Q_\ell) - (I+Q_\ell)^{-1}Q'_\ell =$$

(3.20)
$$\begin{bmatrix} B_{11} + q_\ell B_{12} & B_{12} \\ -q'_\ell + B_{21} - q_\ell(B_{11} - B_{22}) - q_\ell^2 B_{12} & B_{22} - q_\ell B_{12} \end{bmatrix}$$

To verify formula (3.20), note that the definition (3.15) yields,

(3.21)
$$Q_\ell^2 = 0 \text{ and so } (I+Q_\ell)^{-1} = I - Q_\ell.$$

Formula (3.21), in turn, yields,

$$(I+Q_\ell)^{-1}B(I+Q_\ell) = B + BQ_\ell - Q_\ell B - Q_\ell BQ_\ell.$$

Similarly, it follows from the definition (3.15) that

$$-(I+Q_\ell)^{-1}Q'_\ell = -Q'_\ell.$$

Combining these two formula with the definition (3.15) we find formula (3.20). Then, applying formula (3.20) to the matrix $B = U(1)$ and using that according to the definition (3.11) trace $U(1) = 0$, we find formula (3.19).

We continue the proof of conclusion (3.16) by constructing an approximate solution to the Riccati equation,

(3.22)
$$- q'_\ell + U(1)_{21} - 2U(1)_{11}q_\ell - U(1)_{12}q_\ell^2 = 0,$$

which we obtained by setting the right member of formula (3.19) equal to zero. Now we observe that there are functions $c_{\pm j}^{rs}$, $r = 1, 2$, $s = 1, 2$, and $j = 1, 2, 3$ with the property that for $r \neq s$

(3.23)
$$U(1)_{rs} = \sum_{j=1}^{j=3}[c_{rs}^{+j} f_j \cdot \exp(ig_j) + c_{rs}^{-j} f_j \cdot \exp(-ig_j)],$$

and each of these reduced coefficient functions is slowly varying in the sense that,

(3.24)
$$\sup_{Re\lambda\in\mathcal{I}} \|(c_{rs}^{\pm j})'\|_{\mathcal{L}^1(c,\infty)} < \infty.$$

To verify formula (3.23) we note that combining the definitions (3.11), (2.3) and (2.21) with assumption (2.7) we find constants, δ_{rsjk}, such that,

$$U(1)_{rs} = \frac{1}{2i} \cdot \sum_{j=1}^{j=3} \sum_{k=1}^{k=2} \delta_{rsjk} \cdot (A^{-1})_{rj} \left[\exp(ig_j) - \exp(-ig_j) \right] A_{ks},$$

$$\text{where } \delta_{rsjk} = \pm 1 \text{ or } 0.$$

Hence formula (3.23) holds with

$$c_{rs}^{\pm j} = \pm \frac{1}{2i} \cdot \sum_{k=1}^{k=2} \delta_{rsjk} \cdot (A^{-1})_{rj} A_{ks}.$$

To establish estimate (3.24), we combine this formula for the reduced coefficient functions with conclusions (3.9) and (3.8) of Lemma 3.2 and use that the product of two slowly varying functions is slowly varying. To see that the product of two slowly varying functions is slowly varying note that estimate (3.24) implies that for each j

$$\sup_{Re\lambda \in I} \| c_{rs}^{\pm j}(., \lambda) \|_{L^\infty(c,\infty)} < \infty.$$

Similarly, it follows that

$$(3.25) \quad U(1)_{11} = i\mu + \sum_{j=1}^{j=3} [c_{11}^{+j} f_j \cdot \exp(ig_j) + c_{11}^{-j} f_j \cdot \exp(-ig_j)],$$

where, as before, the reduced coefficient functions $c_{11}^{\pm j}$, are slowly varying. Inserting formula (3.23) into the Riccati equation (3.22) and using formula (3.25) to split the term $2U(1)_{11}q_\ell$ into two we find,

$$(3.26) \quad \begin{aligned} &-q'_\ell + \sum_{j=1}^{j=3} [c_{21}^{+j} f_j \cdot \exp(ig_j) + c_{21}^{-j} \cdot f_j \exp(-ig_j)] - 2i\mu q_\ell \\ &-2 \sum_{j=1}^{j=3} [c_{11}^{+j} f_j \cdot \exp(ig_j) + c_{11}^{-j} f_j \cdot \exp(-ig_j)] q_\ell \\ &- \sum_{j=1}^{j=3} [c_{12}^{+j} f_j \cdot \exp(ig_j) + c_{12}^{-j} f_j \cdot \exp(-ig_j)] q_\ell^2 = 0. \end{aligned}$$

We seek an approximate solution to the Riccati equation (3.26) in the class of functions of the form,

$$(3.27) \quad q_\ell = \sum_{j=1}^{j=3} [a^{+j} f_j \cdot \exp(ig_j) + a^{-j} f_j \cdot \exp(-ig_j)],$$

where the reduced coefficient functions, $a^{\pm j}$ are slowly varying;

$$(3.28) \qquad \sup_{\mathrm{Re}\lambda \in I} \|(a^{\pm j})'(.,\lambda)\|_{\mathcal{L}^1(c,\infty)} < \infty.$$

Next we observe that assumption (2.8) implies that

$$(3.29) \qquad \begin{aligned} & f_i \cdot f_j \in \mathcal{L}^1(c,\infty) \ \text{ and } \ f_i \cdot f_j \cdot f_k \in \mathcal{L}^1(c,\infty) \\ & \text{for, } i = 1,2,3; \ j = 1,2,3; \ k = 1,2,3. \end{aligned}$$

Combining estimates (3.29), (3.28) and (3.24) with assumption (3.27) we see that the second and third order terms in the functions f_j are short range. Hence, inserting assumption (3.27) into the Riccati equation (3.26) and dropping these short range terms in the resulting equation we find,

$$
\begin{aligned}
(3.30) \qquad & -\sum_{j=1}^{j=3}[a^{+j}ig'_j f_j \cdot \exp(ig_j) - a^{-j}ig'_j f_j \cdot \exp(-ig_j)] \\
& + \sum_{j=1}^{j=3}[c_{21}^{+j} f_j \cdot \exp(ig_j) + c_{21}^{-j} f_j \cdot \exp(-ig_j)] \\
& -2\sum_{j=1}^{j=3}[i\mu a^{+j} f_j \cdot \exp(ig_j) + i\mu a^{-j} f_j \cdot \exp(-ig_j)] = 0.
\end{aligned}
$$

Rearranging the terms in equation (3.30) and setting the coefficients of each of the functions $\exp(ig_j)$ and $\exp(-ig_j)$ equal to zero leads to the formula,
(3.31$\pm$)
$$a^{\pm j} f_j = (2i\mu \pm ig'_j)^{-1} f_j c_{21}^{\pm j}, \ \text{ or } \ a^{\pm j} = (2i\mu \pm ig'_j)^{-1} \cdot c_{21}^{\pm j}, \qquad \text{for } j = 1,2,3.$$

From now on we shall define these reduced coefficient functions by the second set of formula of (3.31$\pm$).

We complete the proof of conclusion (3.16) by combining these assumptions with the ones of Theorem 2.1. More specifically, define

$$(3.32) \qquad e_{21} = [(I+Q_\ell)^{-1}U(1)(I+Q_\ell) - (I+Q_\ell)^{-1}Q'_\ell]_{21}.$$

Then, since the left members of the equations (3.26) and (3.22) are equal, we see from the equation (3.30), from assumption (3.27) and from formula (3.19) that

$$
\begin{aligned}
(3.33) \qquad e_{21} = & -\sum_{j=1}^{j=3}[(a^{+j} f_j)' \cdot \exp(ig_j) + (a^{-j} f_j)' \cdot \exp(-ig_j)] \\
& -2\sum_{j=1}^{j=3}[c_{11}^{+j} f_j \cdot \exp(ig_j) + c_{11}^{-j} f_j \cdot \exp(-ig_j)]q_\ell \\
& -\sum_{j=1}^{j=3}[c_{12}^{+j} f_j \cdot \exp(ig_j) + c_{12}^{-j} f_j \cdot \exp(-ig_j)]q_\ell^2.
\end{aligned}
$$

Estimates (3.29) and (3.24) together with assumption (2.8) show that the short range character of each of these three terms is implied by estimate (3.28). Combining assumptions (2.15), (2.10), (2.9), and (2.5) with formula (3.5) we see that the first factors of the definitions $(3.31_\pm)$ are slowly varying. Since according to estimate (3.24) so are the second factors, estimate (3.28) follows. Hence, each of the three terms in formula (3.33) is short range and this completes the proof of conclusion (3.16). Similarly, we see that assumptions (2.15), (2.9) and (2.8) imply conclusions (3.17) and (3.18). Thus, the proof of Lemma 3.2 is complete.

Applying formula (3.20) to the matrix $B = U(1) + R(1)$ and using the definition (3.32) we find,

$$(I + Q_\ell)^{-1}(U(1) + R(1))(I + Q_\ell) - (I + Q_\ell)^{-1}Q'_\ell =$$

(3.34)
$$\begin{bmatrix} (U(1) + R(1))_{11} & (U(1) + R(1))_{12} \\ e_{21} + R(1)_{21} & (U(1) + R(1))_{22} \end{bmatrix} +$$

$$q_\ell \begin{bmatrix} (U(1) + R(1))_{12} & 0 \\ -R(1)_{11} + R(1)_{22} - q_\ell R(1)_{12} & -(U(1) + R(1))_{12} \end{bmatrix}$$

Next define,

$$R(2) = \begin{bmatrix} R(1)_{11} + q_\ell R(1)_{12} & R(1)_{12} \\ e_{21} + R(1)_{21} - q_\ell(R(1)_{11} - R(1)_{22}) - q_\ell^2 R(1)_{12} & R(1)_{22} - q_\ell R(1)_{12} \end{bmatrix}$$
(3.35)

and

(3.36) $$U(2) = \begin{bmatrix} U(1)_{11} + q_\ell U(1)_{12} & U(1)_{12} \\ 0 & -(U(1)_{11} + q_\ell U(1)_{12}) \end{bmatrix}$$

Then using that according to the definition (3.11) trace $U(1) = 0$, we see from formula (3.34) and from the definitions (3.36) and (3.35) that

(3.37) $(I + Q_\ell)^{-1}(U(1) + R(1))(I + Q_\ell) - (I + Q_\ell)^{-1}Q'_\ell = U(2) + R(2).$

We see from formula (3.37), in turn, that the transformation

(3.38) $u_2 = (I + Q_\ell)^{-1}u_1$

carries the system (3.13) into

(3.39) $u'_2 = [U(2) + (R(2)]u_2.$

As a third step of the proof of Theorem 3.1 we choose A in this formula so that the (12) element of the system resulting from (3.39) is short range. We do this with the help of a transformation suggested by Harris-Lutz [HL]. It is of the form $I + Q_u$ where,

(3.40) $$Q_u = q_u \cdot \begin{bmatrix} 0 & 1 \\ 0 & 0 \end{bmatrix}$$

and q_u will be an approximate solution of the linear differential equation (3.46) to be stated in the proof of the following lemma.

LEMMA 3.3. *Let the potentials P, W and interval $\mathcal{I}$ satisfy the assumptions of Theorem 2.1 and let the matrix $U(2)$ be given by the definition (3.36). Then, to each λ with $\operatorname{Re}\lambda \in \mathcal{I}$ there is a function $q_u = q_u(t, \lambda)$ such that, for the corresponding matrix function of the definition (3.40),*

$$(3.41) \quad \sup_{\operatorname{Re}\lambda \in \mathcal{I}} \|[(I+Q_u)^{-1}U(2)(I+Q_u)-(I+Q_u)^{-1}Q'u]_{12}\|_{\mathcal{L}^1(c,\infty)} < \infty.$$

Furthermore,

$$(3.42) \qquad \sup_{\operatorname{Re}\lambda \in \mathcal{I}} \|q_u(., \lambda)\|_{\mathcal{L}^\infty(c,\infty)} < \infty.$$

We start the proof of conclusion (3.41) with the formula,

$$[(I+Q_u)^{-1}U(2)(I+Q_u)-(I+Q_u)^{-1}Q'_u]_{12} = -q'_u+U(2)_{12}+2U(2)_{11}q_u.$$
$$(3.43)$$

To prove this formula, note that the definition (3.40) shows that formula (3.21) also holds for Q_u,

$$(3.44) \qquad Q_u^2 = 0 \text{ and so } (I+Q_u)^{-1} = I - Q_u.$$

This formula allows us to repeat the proof of formula (3.20) and conclude that for each matrix B,

$$(I+Q_u)^{-1}B(I+Q_u) - (I+Q_u)^{-1}Q'_u =$$
$$(3.45) \quad \begin{bmatrix} B_{11} - q_u B_{21} & -q'_u + B_{12} + q_u(B_{11} - B_{22}) - q_u^2 B_{21} \\ B_{21} & B_{22} + q_u B_{21} \end{bmatrix}$$

Applying formula (3.45) to the matrix $B = U(2)$ and using that according to the definition (3.36) trace $U(2) = 0$ and $U(2)_{21} = 0$, we find formula (3.43).

We continue the proof of conclusion (3.41) by constructing an approximate solution to the linear differential equation,

$$(3.46) \qquad\qquad - q'_u + U(2)_{12} + 2U(2)_{11}q_u = 0,$$

which we obtained by setting the right member of formula (3.43) equal to zero. Inserting the definition (3.36) and formula (3.23) into the linear differential equation (3.46) and using formula (3.25) to split the term $2U(1)_{11}q_u$ into two, we find,

$$(3.47) \begin{aligned} -q'_u + &\sum_{j=1}^{j=3} [c_{12}^{+j} f_j \cdot \exp(ig_j) + c_{12}^{-j} f_j \cdot \exp(-ig_j)] + 2i\mu q_u \\ +&2\sum_{j=1}^{j=3} \{[c_{11}^{+j} f_j \cdot \exp(ig_j) + c_{11}^{-j} f_j \cdot \exp(-ig_j))] + \\ &+ [c_{12}^{+j} f_j \cdot \exp(ig_j) + c_{12}^{-j} f_j \cdot \exp(-ig_j)] q_\ell \} q_u = 0. \end{aligned}$$

We seek an approximate solution to the equation (3.47) in the class of functions of the form,

$$(3.48) \qquad q_u = \sum_{j=1}^{j=3} [b^{+j} f_j \cdot \exp(ig_j) + b^{-j} f_j \cdot \exp(-ig_j)]$$

where, the functions $b^{\pm j}$ are slowly varying. Inserting the definition (3.48) into the linear differential equation (3.47) and dropping the derivatives of these slowly varying functions and the second and third order terms in the functions f_j we find,

$$
\begin{aligned}
(3.49) \qquad & \sum_{j=1}^{j=3} \{[-ib^{+j} g'_j f_j \cdot \exp(ig_j) + ib^{-j} g'_j f_j \cdot \exp(-ig_j)] + \\
& \qquad + [c_{12}^{+j} f_j \cdot \exp(ig_j) + c_{12}^{-j} f_j \cdot \exp(-ig_j)]\} + \\
& +2 \sum_{j=1}^{j=3} [i\mu b^{+j} f_j \cdot \exp(ig_j) + i\mu b^{-j} f_j \cdot \exp(-ig_j)] = 0.
\end{aligned}
$$

Rearranging the terms in equation (3.49) and setting the coefficients of each of the functions $\exp(ig_j)$ and $\exp(-ig_j)$ equal to zero leads to the formula,

$$(3.50_\pm) \qquad b^{\pm j} f_j = -(2i\mu \pm ig'_j)^{-1} f_j \cdot c_{12}^{\pm j} \cdot \qquad \text{for } j = 1, 2, 3.$$

From now on we shall define these coefficient functions by formula $(3.50_\pm)$.

We complete the proof of conclusion (3.41) by inserting the definitions (3.48), (3.36) and formula (3.25), (3.23) into the formula (3.43). Then defining,

$$(3.51) \qquad e_{12} = [(I + Q_u)^{-1} U(2)(I + Q_u) - (I + Q_u)^{-1} Q'_u]_{12},$$

and using the equation (3.49) we find,

$$
\begin{aligned}
(3.52) \qquad e_{12} = & -\sum_{j=1}^{j=3} [(b^{+j} f_j)' \cdot \exp(ig_j) + (b^{-j} f_j)' \cdot \exp(-ig_j)] \\
& +2 \sum_{j=1}^{j=3} \{[c_{11}^{+j} f_j \cdot \exp(ig_j) + c_{11}^{-j} f_j \cdot \exp(-ig_j))] + \\
& \qquad [c_{12}^{+j} f_j \cdot \exp(ig_j) + c_{12}^{-j} f_j \cdot \exp(-ig_j)] q_\ell \} q_u
\end{aligned}
$$

Similarly to the way that formula (3.33), assumptions (2.15), (2.10), (2.9), (2.8) and (2.5) and the definitions $(3.31_\pm)$, (3.27) implied conclusion (3.16), we see that formula (3.52), the definitions $(3.50_\pm)$, (3.48) and these assumptions imply conclusion (3.41).

At the same time, we see that assumptions (2.15), (2.9) and (2.8) imply conclusion (3.42). Thus the proof of Lemma 3.4 is complete.

Applying formula (3.45) to the matrix $B = U(2) + R(2)$ and using the definition (3.51) we find,

$$(I + Q_u)^{-1}(U(2) + R(2))(I + Q_u) - (I + Q_u)^{-1}Q_u' =$$

(3.53)
$$\begin{bmatrix} (U(2) + R(2))_{11} & e_{12} + R(2)_{12} \\ (U(2) + R(2))_{21} & (U(2) + R(2))_{22} \end{bmatrix} +$$

$$+ q_u \begin{bmatrix} (U(2) + R(2))_{21} & R(2)_{11} - R(2)_{22} - q_u R(2)_{21} \\ 0 & -(U(2) + R(2))_{21} \end{bmatrix}.$$

Next define,

$$R(3) = \begin{bmatrix} R(2)_{11} & e_{12} + R(2)_{12} \\ R(2)_{21} & R(2)_{22} \end{bmatrix} +$$

(3.54)
$$+ q_u \begin{bmatrix} R(2)_{21} & R(2)_{11} - R(2)_{22} - q_u R(2)_{21} \\ R(2)_{21} & R(2)_{22} - R(2)_{21} \end{bmatrix} + q_\ell \begin{bmatrix} U(1)_{12} & 0 \\ 0 & U(1)_{12} \end{bmatrix}.$$

and

(3.55)
$$U(3) = \begin{bmatrix} U(1)_{11} & 0 \\ 0 & -U(1)_{11} \end{bmatrix}.$$

Combining the definitions (3.55), (3.54) and (3.36) with formula (3.53) we have,

(3.56) $(I + Q_u)^{-1}(U(2) + R(2))(I + Q_u) - (I + Q_u)^{-1}Q_u' = U(3) + R(3).$

We see from formula (3.56) that the transformation

(3.57) $u_3 = (I + Q_u)^{-1}u_2$

carries the system (3.39) into

(3.58) $u_3' = [U(3) + (R(3)]u_3.$

As a fourth and final step of the proof of Theorem 3.1 we define an approximate fundamental matrix to the basic system (2.21) by,

$$Y_r = Y_r(t, \lambda) = A(t)(I + Q_\ell(t))(I + Q_u(t)) \cdot \exp[\int_c^t U(3)(s)ds], \ c \le t.$$
(3.59)

Then we see from the product rule of differentiation that the definition

(3.60) $\begin{aligned} C_r &= A'A^{-1} + AQ_\ell'A^{-1} + A(I + Q_\ell)Q_u'(I + Q_\ell)^{-1}A^{-1} + \\ &\quad + A(I + Q_\ell)(I + Q_u)U(3)(I + Q_u)^{-1}(I + Q_\ell)^{-1}A^{-1} \end{aligned}$

implies,

$$(3.61) \qquad\qquad Y'_r = C_r Y_r.$$

Next define,

$$(3.62) \qquad E_r = -A(I + Q_\ell)(I + Q_u)R(3)(I + Q_u)^{-1}(I + Q_\ell)^{-1}A^{-1}.$$

Then combining the definitions (3.62), (3.60) with formula (3.56), (3.37) and (3.12) we find,

$$(3.63) \qquad\qquad C_r = V - \lambda J + E_r$$

and so, we see from the system (3.61) and from assumption (2.3) that,

$$(3.64) \qquad\qquad Y'_r = [(V - \lambda J) + E_r)]Y_r.$$

Finally, using the definition (3.59) we define

$$(3.65) \qquad\qquad z_r = Y_r \cdot \begin{bmatrix} 1 \\ 0 \end{bmatrix}$$

and

$$(3.66) \qquad\qquad y_r = Y_r \cdot \begin{bmatrix} 0 \\ 1 \end{bmatrix}$$

Then, clearly conclusion (3.2) of Theorem 3.1 holds for each of these two vectors.

We start the proof of conclusion (3.1) by showing that,

$$(3.67) \qquad\qquad \sup_{Re\lambda \in \mathcal{I}} \|R(3)(.,\lambda)\|_{\mathcal{L}^1(c,\infty)} < \infty.$$

To verify estimate (3.67) first, we combine the definition (3.35) with conclusions (3.17) and (3.16) of Lemma 3.2 and with estimate (3.14). This yields,

$$\sup_{Re\lambda \in \mathcal{I}} \|R(2)(.,\lambda)\|_{\mathcal{L}^1(c,\infty)} < \infty.$$

Combining this estimate with conclusions (3.42) and (3.41) of Lemma 3.3 we find,

$$\sup_{Re\lambda \in \mathcal{I}} \|(q_u R(2))(.,\lambda)\|_{\mathcal{L}^1(c,\infty)} < \infty.$$

Second, we claim that

$$(3.68) \qquad\qquad \sup_{Re\lambda \in \mathcal{I}} \|q_\ell U(1)_{12}(.,\lambda)\|_{\mathcal{L}^1(c,\infty)} < \infty.$$

To see estimate (3.68) we need that formulae (3.27) and (3.23) together yield,

$$q_\ell U(1)_{12} = \sum_{j=1}^{j=3} \sum_{k=1}^{k=3} \{ [a^{+j} f_j \cdot \exp(ig_j) + a^{-j} f_j \exp(-ig_j)] \cdot$$
$$\cdot [c_{12}^{+k} f_k \cdot \exp(ig_k) + c_{12}^{-k} f_k \cdot \exp(-ig_k)] \}.$$

Note that this formula contains only second order terms in the functions f_j, and so, combining it with estimates (3.29), (3.28) and (3.24) we obtain estimate (3.68). Then, inserting these estimates into the definition (3.54) we arrive at estimate (3.67).

We continue the proof of conclusion (3.1) by combining conclusion (3.42) of Lemma 3.3 with the definition (3.40) and by combining conclusion (3.17) of Lemma 3.2 with the definition (3.15). This yields,

$$\sup_{Re\lambda \in \mathcal{I}} \|(I + Q_\ell(., \lambda))(I + Q_u(., \lambda))\|_{\mathcal{L}^\infty(c,\infty)} < \infty.$$

and so, we see from the first half of conclusion (3.8) of Lemma 3.1 that

$$(3.69) \quad \sup_{Re\lambda \in \mathcal{I}} \|A(., \lambda)(I + Q_\ell(., \lambda)(I + Q_u(., \lambda))\|_{\mathcal{L}^\infty(c,\infty)} < \infty.$$

Combining estimate (3.69), in turn, with formula (3.44), (3.21) and with the second half of conclusion (3.8) of Lemma 3.1 we find,

$$(3.70) \sup_{Re\lambda \in \mathcal{I}} \|(I + Q_u(., \lambda))^{-1}(I + Q_\ell(., \lambda))^{-1}A(., \lambda)^{-1}\|_{\mathcal{L}^\infty(c,\infty)} < \infty.$$

Finally, combining estimates (3.70), (3.69) and (3.67) with the definition (3.62) we obtain conclusion (3.1) of Theorem 3.1

To verify conclusion (3.3) we note that it is an immediate consequence of the definitions (3.66) and (3.6). In fact, this property motivated the choice of the branch in the definition (3.6).

4. Estimates for the solutions of the basic system (2.21) near zero. We start this section by describing approximate solutions to the basic system (2.21) on the interval $(0, c)$. For this purpose, first with the help of the definition (2.12) define an approximate system by,

$$(4.1) \qquad\qquad Y_\ell'(t) = \frac{1}{t} J C_0(e, \ell) Y_\ell(t).$$

To motivate this choice, note that by assumptions (2.13) and (2.12) the coefficient matrices of the systems (4.1) and (2.21) differ by a bounded matrix. Since the function values of the coefficient matrix of the system (4.1) commute, it has a fundamental matix of the form,

$$(4.2) \qquad\qquad Y_\ell(t) = \exp[\int_c^t \frac{1}{\sigma} J C_0(e, \ell) d\sigma].$$

Second, we need the elementary fact that the spectrum of this coefficient matrix is given by,

$$(4.3) \qquad Spec\,(JC_0(e,\ell)) = \{-(\ell^2 - e^2)^{1/2}, (\ell^2 - e^2)^{1/2}\}.$$

Hence there are vectors $a_\pm$, such that

$$(4.4) \qquad JC_0(e,\ell)a_\pm = \pm(\ell^2 - e^2)^{1/2}a_\pm.$$

Then, we define the approximate solution y_ℓ by,

$$(4.5) \qquad y_\ell = Y_\ell(t)a_+,$$

and the corresponding weight function by

$$(4.6) \qquad w_\ell(t) = |y_\ell(t)|.$$

With the help of this weight function for each f in $C(0,c)$ we define the norm,

$$(4.7) \qquad \|f\|_{w_\ell} = \sup_{0<t<c} |f(t)|w_\ell(t)^{-1},$$

and denote by B_{w_ℓ} the space of those functions for which this norm is finite.

In the following theorem we show that the basic system (2.21) also admits a solution which is asymptotic to $y_\ell(t,\lambda)$ at $t = 0$ and for which this norm is finite.

THEOREM 4.1. *Let the assumptions of Theorem 2.1 hold, let the approximate solution y_ℓ be given by the definition (4.5) and let the space B_{w_ℓ} be given with the help of the definition (4.7). Then, on the interval $(0,c)$ the basic system (2.21) admits a solution f_ℓ in B_{w_ℓ} such that,*

$$(4.8) \qquad \sup_{Re\lambda\in\mathcal{I}} \|f_\ell(\cdot,\lambda)\|_{w_\ell} < \infty$$

and

$$(4.9) \qquad \sup_{Re\lambda\in\mathcal{I}} \lim_{t\to 0} |f_\ell(t,\lambda) - y_\ell(t,\lambda)| \cdot |y_\ell(t,\lambda)|^{-1} = 1.$$

We start the proof of Theorem 4.1 by deriving a Volterra equation for such a solution of the basic system (2.21). To do this, first define the error potential to be the difference of the coefficient matrices of the original system (2.21) and of the approximate system (4.1),

$$(4.10) \qquad E_\ell(t) = (V(t) - \lambda J) - \frac{1}{t}JC_0(e,\ell).$$

Then adding $E_\ell u$ to both sides of the basic system (2.21) we find,

$$(4.11) \qquad u' - \frac{1}{t}JC_0(e,\ell)u = E_\ell u.$$

Second, with the help of the definitions (4.4) and (4.2) define,

$$(4.12) \qquad z_\ell = Y_\ell(t)a_-.$$

Then it is clear from the definitions (4.5) and (4.12) that each of these two approximate solutions satisfies the homogeneous equation corresponding to the system (4.11). Third, with the notation of [LR], define the kernel

$$(4.13) \; Q_\ell(t,s) = \big(y_\ell(t) >< Jz_\ell(s) - z_\ell(t) >< Jy_\ell(s)\big) < y_\ell, Jz_\ell >^{-1} E_\ell(s)$$

and the corresponding operator

$$(4.14) \qquad Q_\ell f(t) = \int_0^t Q_\ell(t,s)f(s)ds, \quad f \in C_0^\infty((0,c),\mathcal{C}_2).$$

Then, we know that (e.g. [LR]

$$(4.15) \qquad [(D - (V - \lambda J) + E_\ell]Q_\ell f = E_\ell f, \quad f \in C_0^\infty((0,c),\mathcal{C}_2).$$

We see from the definitions (4.10) and (4.5) that

$$(4.16) \qquad [(D - (V - \lambda J) + E_\ell]y_\ell = 0,$$

and so, relation (4.15) shows that each solution f_ℓ of the Volterra equation

$$(4.17) \qquad f_\ell = y_\ell + Q_\ell f_\ell,$$

also satisfies the basic system (4.11).

We continue the proof of Theorem 4.1 by making essential use of a result of Love [L], Erdelyi [Er] and Olver [O], [RT]. To formulate it, following them we define

$$(4.18) \qquad \|Q_\ell\|(LEO, w_\ell) = \int_{\mathcal{I}} \sup_{\xi > \eta} w_\ell(\xi)^{-1} w_\ell(\eta)|Q_\ell(\xi,\eta)|d\eta.$$

LEMMA 4.1. *Let the Volterra operator Q_ℓ be given by the definition (4.14) and let the space $\mathbf{B}_{w_\ell}$ be given with the help of the definition (4.7). Assume that*

$$(4.19) \qquad \|Q_\ell\|(LEO, w_\ell) < \infty.$$

Then,

$$(4.20) \qquad Q_\ell \in \mathbf{B}(\mathbf{B}_{w_\ell}) \quad and \quad \|Q_\ell\|(w_\ell) \le \|Q_\ell\|(LEO, w_\ell).$$

Furthermore, for each $\lambda \ne 0$ in $\mathbf{C}$,

$$(4.21) \qquad (\lambda I - Q_\ell)^{-1} \in \mathbf{B}(\mathbf{B}_{w_\ell})$$

and

$$(4.22) \qquad \|(\lambda I - Q_\ell)^{-1}\|(w_\ell) \le |\lambda|^{-1} \exp(|\lambda|^{-1}\|Q_\ell\|(LEO, w_\ell)).$$

We complete the proof of Theorem 4.1 by noting that application of the LEO Lemma 4.1 to the Volterra operator of the definition (4.14) yields Theorem 4.1.

5. Estimates for the solutions of the basic system (2.21) near infinity. We start this section by defining a weight function with the help of the approximate solution of the definition (3.66). More specifically, we define,

$$(5.1) \qquad\qquad w_r(t) = |y_r(t)|$$

Then, with the help of this weight function for each f in $C(c, \infty)$ define the norm,

$$(5.2) \qquad\qquad \|f\|_{w_r} = \sup_{t>c} |f(t)| w_r(t)^{-1},$$

and denote by $B(w_r)$ the space of those functions for which this norm is finite. In the following theorem we show that the basic system (2.21) admits a solution which is asymptotic to $y_r(t, \lambda)$ at $t = \infty$ and for which this norm is finite.

THEOREM 5.1. *Let the assumptions of Theorem 2.1 hold, let the approximate solution y_r be given by the definition (3.66) and let the space $\mathbf{B}_r$ be given with the help of the definition (5.2). Then, the basic system (2.21) admits a solution f_r in $\mathbf{B}_r$ such that,*

$$(5.3) \qquad\qquad \sup_{Re\lambda \in \mathcal{I}} \|f(\cdot, \lambda)\|_{w_r} < \infty$$

and

$$(5.4) \qquad\qquad \sup_{Re\lambda \in \mathcal{I}} \lim_{t\to\infty} |f_r(t, \lambda) - y_r(t, \lambda)| \cdot |y_r(t, \lambda)|^{-1} = 1.$$

We start the proof of conclusion (5.3) by deriving Volterra equations for the solutions of the basic system (2.21). To do this, first note that adding an arbitrary function $E_r u$ to both sides of the basic system (2.21) we find,

$$(5.5) \qquad\qquad u' - (V - \lambda J)u + E_r u = E_r u.$$

Second, we choose this arbitrary function by the definition (3.62) and define the kernel

$$Q_r(t, s) = \big(y_r(t) >< J z_r(s) - z_r(t) >< J y_r(s) \big) < y_r(s), J z_r(s) >^{-1} E_r(s)$$
$$(5.6)$$

and the corresponding operator

$$(5.7) \qquad Q_r f(t) = \int_t^\infty Q_r(t, s) f(s) ds, \quad f \in C_0^\infty((c, \infty), C_2).$$

Then, it is not difficult to show that

$$(5.8) \qquad [(D - (V - \lambda J) + E_r] Q_r f = E_r f, \quad f \in C_0^\infty((c, \infty), C_2).$$

We see from conclusion (3.2) of Theorem 3.1 that

(5.9) $$[(D - (V - \lambda J) + E_r]y_r = 0,$$

and so, relation (5.8) shows that each solution f_r of the Volterra equation

(5.10) $$f_r = y_r + Q_r f_r,$$

also satisfies the system (5.5).

We continue the proof of conclusion (5.3) by showing that the Volterra operator of the definition (5.7) satisfies the assumptions of the LEO Lemma 4.1, which is the statement of the lemma that follows. Here of course, we use Q_r in place of Q_ℓ and w_r in place of w_ℓ. For completeness, we display the definition (4.18) with these two replacements,

(5.11) $$\|Q_r\|(LEO, w_r) = \int_I \sup_{\xi > \eta} w_r(\xi)^{-1} w_r(\eta) |Q_r(\xi, \eta)| d\eta.$$

LEMMA 5.1. *Let the Volterra operator Q_r be given by the definition (5.7) and let the space $\mathbf{B}(w_r)$ be given with the help of the definitions (5.2) and (3.66). Then,*

(5.12) $$\sup_{Re\lambda \in I} \|Q_r\|(LEO, w_r) < \infty.$$

We start the proof of Lemma 5.1 by showing that it holds for the first term of formula (5.6);

(5.13) $$\begin{aligned} \int_c^\infty \sup_{t > s} w_r(t)^{-1} w_r(s) |y_r(t) > \\ < Jz_r(s)| \cdot | < y_r(s), Jz_r(s) >^{-1} | \cdot |E_r(s)| ds < \infty. \end{aligned}$$

To prove estimate (5.13), first we show that,

(5.14) $$\sup_{s > c} w_r(s) |z_r(s)| < \infty.$$

To see estimate (5.14) note that according to the definition (3.55) the matrix $U(3)$ is diagonal and trace $U(3) = 0$. Hence the definition (3.66) yields,

(5.15) $$y_r(t) = A(t)(I + Q_\ell(t))(I + Q_u(t)) \cdot \exp[\int_c^t U(3)_{22}(\sigma) d\sigma] \cdot \begin{bmatrix} 0 \\ 1 \end{bmatrix},$$

and the definition (3.65) yields,

(5.16) $$z_r(s) = A(s)(I + Q_\ell(s))(I + Q_u(s)) \cdot \exp[-\int_c^s U(3)_{22}(\sigma) d\sigma] \cdot \begin{bmatrix} 1 \\ 0 \end{bmatrix}.$$

Taking the absolute values of formula (5.16) and (5.15) and multiplying the resulting formula together we find,

(5.17) $$|y_r(s)| \cdot |z_r(s)| \leq \|A(s)(I + Q_\ell(s))(I + Q_u(s))\|^2.$$

Combining estimate (5.17) with the definition (5.1) and with estimate (3.69) we find estimate (5.14). To prove estimate (5.13), second we show that,

$$(5.18) \qquad \inf_{Re\lambda \in \mathcal{I}} \inf_{s>c} | < y_r(s), Jz_r(s) > | \neq 0.$$

To see estimate (5.18) note that formula (5.16) and (5.15) together yield,

$$< y_r(s), Jz_r(s) >= \left\{ < A(s)(I + Q_\ell(s))(I + Q_u(s)) \begin{bmatrix} 1 \\ 0 \end{bmatrix}, \right.$$
$$\left. JA(s)(I + Q_\ell(s))(I + Q_u(s)) \begin{bmatrix} 0 \\ 1 \end{bmatrix} > \right\}.$$

Since an invertible matrix maps linearly independent vectors into linearly independent vectors, we see from this formula and from estimate (3.70) that estimate (5.18) holds. Finally, we note that combining estimates (5.18) and (5.14) with conclusion (3.1) of Theorem 3.1 and with the definition (5.1) we find estimate (5.13).

We continue the proof of Lemma 5.1 by showing that it also holds for the second term of formula (5.6);

$$(5.19) \qquad \int_c^\infty \sup_{t>s} w_r(t)^{-1} w_r(s) |z_r(t) >< Jy_r(s)| \cdot$$
$$\cdot | < y_r(s), Jz_r(s) >^{-1} | \cdot |E_r(s)| ds < \infty.$$

As a first step of the proof of estimate (5.19) we show the existence of a constant γ, such that,

$$(5.20) \qquad w_r(s) \cdot |y_r(s)| \leq \gamma \cdot \exp[-Re2 \int_c^s i\mu(\sigma)d\sigma]$$

To see this, we note that similarly to estimate (5.17),

$$w_r(s) \cdot |y_r(s)| \leq \|A(s)(I+Q_\ell(s))(I+Q_u(s))\|^2 \cdot \exp[Re2 \int_c^s U(3)_{22}(\sigma)d\sigma].$$
$$(5.21)$$

Inserting estimate (3.69) into this one we find a constant γ such that

$$(5.22) \qquad w_r(s) \cdot |y_r(s)| \leq \gamma \cdot \exp[Re2 \int_c^s U(3)_{22}(\sigma)d\sigma].$$

Next we claim that there is a possibly different constant γ such that,

$$(5.23) \qquad \exp[Re2 \int_c^s U(3)_{22}(\sigma)d\sigma] \leq \gamma \cdot \exp[-Re2 \int_c^s i\mu(\sigma)d\sigma].$$

To prove estimate (5.23) note that the definition (3.55) and formula (3.25) together yield,

$$\exp[2 \int_c^s U(3)_{22}(\sigma)d\sigma] = \exp[-2 \int_c^s i\mu(\sigma)d\sigma] \cdot$$

$$\cdot \exp[2 \int_c^s \sum_{j=1}^{j=3} [c_{11}^{+j} f_j \cdot \exp(ig_j) + c_{11}^{-j} \cdot f_j \exp(-ig_j)](\sigma)d\sigma].$$

An integration by parts together with estimate (3.24) and assumptions (2.9) and (2.8) shows that the supremum of the absolute value of the second factor is finite. Hence, denoting this supremum by γ and taking the absolute value of this formula we find estimate (5.23).

As a second step of the proof of estimate (5.19) we show that there is a constant γ such that

(5.24) $$w_r(t)^{-1} \cdot |z_r(t)| \leq \gamma \exp[Re2 \int_c^t i\mu(\sigma)d\sigma].$$

In fact, combining the definition (5.1) with formula (5.15), (5.16) we find that the constant γ of estimate (3.69) is such that

$$w_r(t)^{-1} \cdot |z_r(t)| \leq \gamma \exp[-Re2 \int_c^t U(3)_{22}(\sigma)d\sigma].$$

The proof of estimate (5.23) also shows that

(5.25) $$\exp[-Re2 \int_c^s U(3)_{22}(\sigma)d\sigma] \leq \gamma \cdot \exp[Re2 \int_c^s i\mu(\sigma)d\sigma].$$

Now combining estimate (5.25) with the previous one, we find estimate (5.24).

As a third and final step of the proof of estimate (5.19) we multiply estimates (5.24) and (5.20) together. This yields,

(5.26) $$w_r(t)^{-1} \cdot |z_r(t)|w_r(s) \cdot |y_r(s)| \leq \gamma^2 \exp[Re2 \int_s^t i\mu(\sigma)d\sigma]$$

Now we need that according to the definition (3.6) the integrand on the right of estimate (5.26) is negative. In fact, this property motivated the choice of the branch of the square root function in the definition (3.5). Hence the exponential in estimate (5.26) is majorized by 1, and so,

(5.27) $$w_r(t)^{-1} \cdot |z_r(t)|w_r(s) \cdot |y_r(s)| \leq \gamma^2.$$

Combining estimates (5.27) and (5.18) with conclusion (3.1) of Theorem 3.1 we obtain estimate (5.19).

We complete the proof of Lemma 5.1 by noting that inserting estimates (5.19) and (5.13) and the definition (5.6) into the definition (5.11) we find conclusion (5.12).

We complete the proof of conclusion (5.3) by noting that Lemma 5.1 allows us to apply the LEO Lemma 4.2 to the Volterra operator of the definition (5.7) and to the weight function of the definition (5.1). Then conclusion (4.21) with $\lambda = 1$, yields the invertibility of the Volterra equation (5.10) and conclusion (4.21) yields,

$$(5.28) \qquad \|f(\cdot, \lambda)\|_{w_r} \leq \exp(\|Q_r\|(LEO, w_r)) \cdot \|y_r\|_{w_r}.$$

We see from the definitions (5.2) and (5.1) that $\|y_r\|_{w_r} \leq 1$ and so, combining estimate (5.28) with conclusion (5.12) of Lemma 5.1, we obtain conclusion (5.3).

To prove conclusion (5.3) note that the already established conclusion (5.3) allows us to conclude that

$$\lim_{t \to \infty} |Q_r f_r(t)| \cdot |y_r(t)|^{-1} = 0.$$

Then combining this estimate with the Volterra equation (5.10) we obtain conclusion (5.4). This completes the proof of Theorem 5.1.

6. Proof of the main theorem 2.1 under the additional assumption (6.3). In this section we prove the main Theorem 2.1 under the additional assumption (6.3), to be stated. The considerations of [DR] show that it is implied by the following Theorem 6.1. In it, for a given angle $0 < \theta \leq \pi$ we define

$$(6.1) \qquad \mathcal{R}_{\pm}(\mathcal{I}) = \{\lambda \in C : |Re\lambda \in \mathcal{I}, 0 < \pm arg\lambda < \theta\}.$$

THEOREM 6.1. *Let the assumptions and notations of the main Theorem 2.1 hold. Then, there are regions of the form (6.1) such that,*

$$(6.2) \qquad \sup_{\lambda \in \mathcal{R}_{\pm}(\mathcal{I})} \|m_s^{1/2} R(\lambda) m_s^{1/2}\| < \infty, \; s > 1/2.$$

To prove Theorem 6.1 first we need the Weyl-Weidmann construction [We], [DS] for the resolvent kernel of the operator of the basic system (2.21). To describe this construction we make the additional assumption that the constant e of the definition (2.12) is such that,

$$(6.3) \qquad |e| < (\ell^2 - 1/4)^{1/2}$$

and denote by f_r and f_ℓ the solutions of Theorem 5.1 and of Theorem 4.1, respectively, extended to all of $\mathbf{R}^+$. Then according to this construction, in the notation of [LR] this resolvent kernel is given by;

$$(6.4) \; R(\lambda)(\xi, \eta) = <f_\ell, Jf_r>^{-1} \cdot \begin{cases} f_r(\xi, \lambda) >< f_\ell(\eta, \lambda), & \text{for } \eta < \xi \\ f_r(\eta, \lambda) >< f_\ell(\xi, \lambda), & \text{for } \eta > \xi. \end{cases}$$

To prove Theorem 6.1 second we need the Schur-Holmgren-Carleman bound of a given integral operator R with reference to a given positive measurable function t, [F] [Ok]. This is defined by,

$$\|R\|(t) = \left(\sup t(\xi)^{-1} \int_{\mathcal{I}} \|R(\xi, \eta)\| t(\eta) d\eta \cdot \sup t(\eta)^{-1} \int_{\mathcal{I}} \|R(\xi, \eta)\| t(\xi) d\xi\right)^{-1/2},$$
(6.5)

where the supremum is taken over the support of t. According to their result [F],[Ok], if the support of $t(\xi)t(\eta)$ contains the support of $R(\xi, \eta)$, then

(6.6) $$\|R\| \leq \|R\|(t)),$$

where the left member is the operator norm. In view of the bound (6.6) Theorem 6.1 is implied by the following theorem.

THEOREM 6.2. *Let the assumptions of Theorem 6.1 hold, let the real number e satisfy assumption (6.3) and let the function t be given by,*

(6.7) $$t(\eta) = \eta^{-1/2}.$$

Then, there are regions of the form (6.1) such that,

(6.8) $$\sup_{\lambda \in \mathcal{R}_\pm(\mathcal{I})} \|m_s^{1/2} R(\lambda) m_s^{1/2}\|(t) < \infty.$$

In the following lemma we isolate the key estimates that we need to prove Theorem 6.2. In it, we extend the weight function of the definition (4.6) to all of $\mathbf{R}^+$. More specifically, with the help of the approximate solutions of the definitions (4.5) and (3.65) we define,

(6.9) $$w_{\ell,e}(t) = \begin{cases} |y_\ell(t)||z_r(c)|, & \text{for } t \in (0, c) \\ |z_r(t)||y_\ell(c)|, & \text{for } t \in (c, \infty) \end{cases}.$$

Similarly, we extend the weight function of the definition (5.1) to all of $\mathbf{R}^+$,

(6.10) $$w_{r,e}(t) = \begin{cases} |z_\ell(t)||y_r(c)|, & \text{for } t \in (0, c) \\ |y_r(t)||z_\ell(c)|, & \text{for } t \in (c, \infty) \end{cases}.$$

LEMMA 6.1. *Let the assumptions of the main Theorem 2.1 hold and let the f_ℓ and f_r be the solutions of Theorems 4.1 and 5.1 respectively. Then, the weight functions of the definitions (6.9) and (6.10) are such that,*

(6.11) $$\sup_{Re\lambda \in \mathcal{I}} \|f_\ell(\cdot, \lambda)\|_{w_{\ell,e}} < \infty$$

and

(6.12) $$\sup_{Re\lambda \in \mathcal{I}} \|f_r(\cdot, \lambda)\|_{w_{r,e}} < \infty.$$

*Furthermore, the boundary value of their Wronskian, as $\lambda + i\epsilon \to \lambda \in \mathcal{I}$
from above, is such that,*

$$(6.13) \qquad \inf_{\lambda \in \mathcal{I}} | < f_\ell(\lambda), Jf_r(\lambda) > | \neq 0.$$

As a first step of the proof of conclusion (6.11) define the weight function

$$(6.14) \qquad w_{r,1}(t) = |z_r(t)|, \quad \text{for } t \in (c, \infty).$$

Then, similarly to the proof of Theorem 5.1 we see that the basic system
(2.21) admits a solution g_r such that,

$$(6.15) \qquad \sup_{Re\lambda \in \mathcal{I}} \|g_r(\cdot, \lambda)\|_{w_{r,1}} < \infty$$

and

$$(6.16) \qquad \sup_{Re\lambda \in \mathcal{I}} \lim_{t \to \infty} (|g_r(t, \lambda) - z_r(t, \lambda)| \cdot |z_r(t, \lambda)|^{-1}) = 1.$$

The asymptotic formula (6.16), conclusion (5.4) of Theorem 5.1 and the
definitions (3.66), (3.65) together show that the solutions f_r and g_r are
linearly independent. Hence there are constants $\alpha_r(\lambda)$ and $\beta_r(\lambda)$, such
that

$$(6.17) \qquad f_\ell(t, \lambda) = \alpha_r(\lambda)f_r(t, \lambda) + \beta_r(\lambda)g_r(t, \lambda) \quad \text{for } t \in (c, \infty).$$

As a second step of the proof of conclusion (6.11) we show that

$$(6.18) \qquad \sup_{Re\lambda \in \mathcal{I}} |\alpha_r(\lambda)| < \infty.$$

To prove estimate (6.18) first note that formula (6.17) yields,

$$(6.19) \quad \alpha_r(\lambda) = < f_\ell(c, \lambda), Jg_r(c, \lambda) > \cdot (< f_r(c, \lambda), Jg_r(c, \lambda) >)^{-1}.$$

Combining conclusion (4.8) of Theorem 4.1 with the definition (4.6) and
combining estimate (6.15) with the definition (6.14), we see that there is a
constant γ such that,

$$(6.20) \qquad | < f_\ell(c, \lambda), Jg_r(c, \lambda) > | \leq \gamma |y_\ell(c, t)| |z_r(c, t)|.$$

Combining this estimate, in turn, with the definitions (4.5), (3.65) and with
estimate (3.69) we find,

$$(6.21) \qquad \sup_{Re\lambda \in \mathcal{I}} | < f_\ell(c, \lambda), Jg_r(c, \lambda) > | < \infty.$$

In other words, estimate (6.18) holds for the first factor of formula (6.19). To see that it also holds for the second factor we need that the trace of the coefficient matrix of the basic system (2.21) is 0, and so,

$$< f_r(c, \lambda), Jg_r(c, \lambda) >= \lim_{t \to \infty} < f_r(t, \lambda), Jg_r(t, \lambda) >$$

Conclusion (5.4) of Theorem 5.1 and the asymptotic formula (6.16) together show that

$$\lim_{t \to \infty} < f_r(t, \lambda), Jg_r(t, \lambda) >= \lim_{t \to \infty} < y_r(t, \lambda), Jz_r(t, \lambda) > .$$

Combining this formula with estimate (5.18) we find,

$$(6.22) \qquad \inf_{Re\lambda \in \mathcal{I}} |(< f_r(c, \lambda), Jg_r(c, \lambda) >) \neq 0.$$

Hence estimate (6.18) holds for the second factor of formula (6.19), and this completes the proof of estimate (6.18).

As a third step of the proof of conclusion (6.11) we show that

$$(6.23) \qquad \sup_{Re\lambda \in \mathcal{I}} |\beta_r(\lambda)| < \infty.$$

To prove estimate (6.23) first note that formula (6.17) yields,

$$(6.24) \quad \beta_r(\lambda) =< f_\ell(c, \lambda), Jf_r(c, \lambda) > \cdot(< g_r(c, \lambda), Jf_r(c, \lambda), >)^{-1}.$$

Similarly to the way that conclusion (4.8) of Theorem 4.1 did imply estimate (6.21), we see that conclusion (5.3) of Theorem 5.1 implies,

$$(6.25) \qquad \sup_{Re\lambda \in \mathcal{I}} | < f_\ell(c, \lambda), Jf_r(c, \lambda) > | < \infty.$$

Inserting estimates (6.25) and (6.22) into formula (6.24) we find estimate (6.23).

As a fourth step of the proof of conclusion (6.11) we note that combination of estimates (5.25) and (5.23) with formula (5.16) and (5.15) and with the definition (3.6) yields,

$$(6.26) \qquad |y_r(t)| \leq \gamma \cdot |z_r(t)|, \quad \text{for } t \in (c, \infty).$$

Combining estimate (6.26), in turn, with conclusion (5.3) of Theorem 5.1 and with the definition (6.14) we find,

$$(6.27) \qquad \sup_{Re\lambda \in \mathcal{I}} \|f_r(\cdot, \lambda)\|_{wr,1} < \infty.$$

Inserting estimates (6.27), (6.23), (6.18) and (6.15) into formula (6.17) we obtain,

$$(6.28) \qquad \sup_{Re\lambda \in \mathcal{I}} \|f_\ell(\cdot, \lambda)\|_{wr,1} < \infty.$$

Finally, combining estimate (6.28) with conclusion (4.8) of Theorem 4.1 and with the definition (6.9), we arrive at conclusion (6.11).

The proof of conclusion (6.12) is similar and for brevity we skip the details.

We start the proof of conclusion (6.13) by showing that for large enough t,

$$(6.29) \qquad f_{r1}(t) \neq 0 \quad and \quad f_{r2}(t) \neq 0.$$

Here, of course the subscripts $1, 2$ denote the components of this vector. We see from assumption (2.14) that to show relation (6.29) it suffices to show that

$$(6.30) \qquad \lim_{t \to \infty} f_{r2}(t) f_{r1}(t)^{-1} = (i\mu(\infty) + m) \cdot \lambda^{-1}.$$

To see relation (6.30), we need that the definitions (3.66), (3.59) and (3.55) together with conclusions (3.42) and (3.17) of Lemmata 3.3 and 3.2 yield,

$$(6.31) \quad y_r(t) \sim (-\exp \int_c^t U(1)_{11}(\sigma) d\sigma) \cdot A(\infty) \begin{bmatrix} 0 \\ 1 \end{bmatrix}, \quad for \quad t \to \infty.$$

We see from conclusion (3.7) of Lemma 3.1 that the vector on the right is an eigenvector of the matrix $P(\infty) - \lambda J$ with eigenvalue $-i\mu(\infty)$. This fact, assumption (2.6) and the asymptotic formula (6.31) together allow us to verify the asymptotic formula (6.30) for the approximate solution y_r. Combining this asymptotic formula with conclusion (5.4) of Theorem 5.1 we find the asymptotic formula (6.30) for the solution f_r itself.

We continue the proof of conclusion (6.13) by showing that if t is so large that relation (6.29) holds, then

$$(6.32) \quad \begin{aligned} &Im \{< f_\ell(\lambda), J f_r(\lambda) > f_{\ell 1}(t)^{-1} f_{r1}(t)^{-1}\} = Im \{f_{r2}(t) f_{r1}(t)^{-1}\}, \\ &for \ f_{\ell 1}(t)^{-1} \neq 0, \end{aligned}$$

and

$$(6.33) \quad \begin{aligned} &Im \{< f_\ell(\lambda), J f_r(\lambda) > f_{\ell 2}(t)^{-1} f_{r21}(t)^{-1}\} = Im \{f_{r1}(t) f_{r2}(t)^{-1}\} \\ &for \ f_{\ell 2}(t)^{-1} \neq 0, \end{aligned}$$

For brevity we verify formula (6.32) only. The definition of J yields,

$$\begin{aligned} &< f_\ell(\lambda), J f_r(\lambda) > f_{\ell 1}(t)^{-1} f_r(t)^{-1} = f_{r2}(t) f_{r1}(t)^{-1} - f_{\ell 2}(t) f_{\ell 1}(t)^{-1}, \\ &for \ f_{\ell 1}(t) \neq 0 \end{aligned}$$
(6.34)

We see from the additional assumption (6.3) that,

$$(6.35) \qquad Im \, (\ell^2 - e^2)^{1/2} = 0$$

and so, by formula (4.3) the eigenvalues of the matrix $JC(e, \ell)$ are real. Since according to the definition (2.12) this matrix is real, it follows that

the eigenvectors are also real. Combining these two facts with the definition
(4.5) we find that the approximate solution y_ℓ is real;

$$(6.36) \qquad\qquad Im\, y_\ell = 0.$$

Combining this relation, in turn, with conclusion (4.9) of Theorem 4.1 and
with the fact that the coefficients of the basic system (2.21) are real, we
obtain that the solution f_ℓ is also real;

$$(6.37) \qquad\qquad Im\, f_\ell = 0.$$

Inserting formula (6.37) into formula (6.34) we arrive at formula (6.32).
We see from the uniqueness of the Cauchy problem of the basic system
(2.21) and from conclusion (4.9) of Theorem 4.1 and the definition (4.5)
that for each t, $f_r(t) \neq 0$. Hence one of the components of this vector is
not 0, and so either formula (6.32) or formula (6.33) holds. Combining this
fact with the the asymptotic formula (6.30) and with assumption (2.14) we
obtain,

$$(6.38) \qquad\qquad < f_\ell(\lambda), J f_r(\lambda) > \neq 0.$$

We see from Theorems 4.1 and 5.1 that the left member of relation (6.38)
depends continuously on $\lambda \in \mathcal{I}$ and so, conclusion (6.13) follows. This
completes the proof of Lemma 6.2.

Now it is straightforward to show that Lemma 6.2 implies Theorem
6.2. Since in Lemma 5.1 of [LR] we proved a similar implication, we omit
the details and consider the proof of Theorem 6.2 complete. This also
completes the proof of the main Theorem 2.1

7. Removal of the additional assumption 6.3. In this section we
remove the additional assumption (6.3). Accordingly, let

$$(7.1) \qquad\qquad |e| \geq (\ell^2 - 1/4)^{1/2}.$$

Then the closure of the formal operator of the definition (2.1) is no longer
self-adjoint and the Weyl-Weidmann construction leading to formula (6.4)
breaks down. However, we can use this construction for the resolvent of a
given self-adjoint extension of this formal operator.

To describe this construction we need that assumption (7.1) yields,
$|Re\,(\ell^2 - e^2)^{1/2}| < 1/2$ and so, the definitions (4.12), (4.4) and (4.2) together
show that,

$$(7.2) \qquad\qquad y_\ell \in \mathcal{L}^2(0, c) \text{ and } z_\ell \in \mathcal{L}^2(0, c).$$

For brevity we only consider the case of,

$$(7.3) \qquad\qquad |e| < \ell,$$

since the other case is quite similar. In this case relation (6.35) is still valid and hence so is formula (6.36). Replacing the definition (4.5) by the definition (4.12) in the proof of formula (6.36) we find,

$$(7.4) \qquad Im\, z_\ell = 0.$$

Hence, each linear combination with real coefficients of these two approximate solutions is also real. To be specific let $\alpha \in \mathcal{R}$ be given and define,

$$(7.5) \qquad y_{\ell,\alpha} = \sin \alpha \cdot y_\ell + \cos \alpha \cdot z_\ell.$$

Replacing the approximate solution y_ℓ by $y_{\ell,\alpha}$ in the proof of Theorem 4.1 we see that the basic system (2.21) admits a solution $f_{\ell,\alpha}$ for which conclusions (4.8) and (4.9) hold. Since $y_{\ell,\alpha}$ satisfies a real boundary condition at zero, it follows from conclusion (4.9) that so does $f_{\ell,\alpha}$. Similarly, it follows that we can also replace f_ℓ by $f_{\ell,\alpha}$ in conclusion (6.13) of Lemma 6.3. Combining these two facts we see that we can make the same replacement in the Weyl-Weidmann formula (6.4). This yields;

$$R_\alpha(\lambda)(\xi, \eta) =< f_{\ell,\alpha}(\lambda), Jf_r(\lambda) >^{-1} \cdot \begin{cases} f_r(\xi, \lambda) >< f_{\ell,\alpha}(\eta, \lambda), & \text{for } \eta < \xi \\ f_r(\eta, \lambda) >< f_{\ell,\alpha}(\xi, \lambda), & \text{for } \eta > \xi \end{cases}.$$
(7.6)

Then we know [We] that the corresponding operator is the resolvent of the self-adjoint extension of the formal operator (2.2) given by the boundary condition of $f_{\ell,\alpha}(\lambda)$. We denote this resolvent operator by $R_\alpha(\lambda)$. Similarly, it follows that we can also replace f_ℓ by $f_{\ell,\alpha}$ in conclusion (6.11) of Lemma 6.2. Thus, we can replace f_ℓ by $f_{\ell,\alpha}$ in each of the two conclusions of Lemma 6.2. This fact allows us to repeat the proof of Theorem 6.1 under assumptions (7.1) and (7.3) and conclude that

$$(7.7) \qquad \sup_{\lambda \in \mathcal{R}_\pm(\mathcal{I})} \|m_s^{1/2} R_\alpha(\lambda) m_s^{1/2}\| < \infty, \quad s > 1/2.$$

Since we have seen in Section 6 that Theorem 6.1 implies the main Theorem 2.1, estimate (7.7) yields the Main Theorem 2.1 under assumptions (7.1) and (7.3).

REFERENCES

[A] ATKINSON, F. V. , *The asymptotic solution of second-order differential equations*, Ann. Mat. Pura-Appl. **37** (1954), pp. 347–378; Oberwolfach Tagungsbericht 20.1-26.1, 1991.

[B] BEHNCKE, HORST, *Absolute continuity of Hamiltonians with von Neumann Wigner potentials*, Proc. Amer. Math. Soc. **111** (1991), pp. 373–384; *Absolute continuity of Hamiltonians with von Neumann Wigner potentials II*, Manuscripta Math. **71** (1991), pp. 163–181.

[BR] BEHNCKE, H. AND REJTO, P., *Schrödinger and Dirac operators with oscillating potentials*, Univ. of Minnesota Math. Report #88-111, 1988; *Schrödinger and Dirac operators with von Neumann Wigner type potentials*, Preprint, Universität Osnabrück, 1990.

[BD] BEN-ARTZI, MATANIA AND DEVINATZ, ALLEN, " The limiting absorption principle for partial differential operators," American Mathematical Society,
Memoir; Volume 66–Number 364 (1987).

[DMR] DEVINATZ, ALLEN, MOECKEL, RICHARD, AND REJTO, PETER, *A limiting absorption principle for Schrödinger operators with von Neumann-Wigner type
potentials*, Integral Equations Operator Theory **14** (1991), pp. 13–68.

[DR] DEVINATZ, ALLEN AND REJTO, PETER, *A limiting absorption principle for
Schrödinger operators with oscillating potentials I*, J. Diff. Equations **49**
(1983), pp. 29–84; II, J. Diff. Equations **49** (1983), pp. 85–104; *Some absolutely continuous ordinary differential operators with oscillating potentials*,
in "Qualitative properties of differential equations," University of Alberta
Press (1987), pp. 95–106. Proceedings of the 1984 Edmonton Conference.
(Ed) Allegretto, W. and Butler, G.J.

[DS] DUNFORD, NELSON AND SCHWARTZ, JACOB T., "Linear Operators," Wiley-
Interscience, New York, London (1962) Part II.

[Ea] EASTHAM, M. S. P., "Asymptotic Solutions of Linear Differential Systems: Application of Levinson theorem," Oxford, Clarendon Press, (1989).

[Ei] EIDUS, D. M., *The principle of limiting absorption*, (in Russian); Amer.
Math.Soc.Transl. **47** (2) (1965), pp. 157-191. See the proof of Theorem
4, Mat. Sb. **57** (1962), pp. 1–99.

[En] ENZ, C.P., (Ed), "Pauli Lectures on Physics, Vol.5 Wave Mechanics," MIT
Press, Cambridge, MA, 1977. See Section 27, The WKB-method.

[Er] ERDELYI, A., *Asymptotic solutions of differential equations with transition
points or singularities*, See Section 4, J. Math. Phys. **1** (1960), pp. 16–
26.

[F] FRIEDRICHS, K. O., "Spectral theory of operators in Hilbert space," Springer-
Verlag, Lecture Notes in Applied Mathematical Sciences, Vol 9, 1973. See
Section 20.

[GP] GILBERT, D. J. AND PEARSON, D. B., *On subordinacy and analysis of the spectrum of one-dimensional Schrödinger operators*, J. Math. Anal. Appl. **128**,
(1987), pp. 30–56.

[H] HEINZ, E., *Über das absolut stetige Spektrum singulärer Differentialgle-
ichungssysteme*, Nr 1, Nachr. Akad. Wiss. Göttingen II. Math-Phys. Kl
(1982), pp. 1–9.

[HKS] HINTON, D. B., KLAUS, M., SHAW, J. K., *Tichmarsh-Weyl-Levinson theory
for a four-dimensional Hamiltonian system*, Proc. London Math. Soc. **59**,
(1989), pp. 339–372; *Embedded half-bound states for potentials of Wigner-
von Neumann type*, Proc. London Math. Soc. **62** (1991), pp. 607–646.

[HL] HARRIS, W.A. AND LUTZ, D.A., *Asymptotic integration of adiabatic oscillators*,
J. Math. Anal. Appl. **51** (1975), pp. 76–93; *A unified theory of asymptotic
integration*, J. Math. Anal. Appl. **57** (1977), pp. 571–586.

[HS] HINTON, D. B. AND SHAW, J. K., *Absolutely continuous spectra of Dirac systems
with long range, short range and oscillating potentials*, Quart. J. Math.
Oxford-Ser. (2) **36** (1985), pp. 183–213; *Absolutely continuous spectra of
second order differential operators with short range and long range potentials*, SIAM J. Math. Anal **17** (1986), pp. 182–196.

[K] KLAUS, MARTIN, *Asymptotic behavior of Jost functions near resonance points
for Wigner-von Neumann type potentials*, J. Math. Phys. **32** (1991), pp.
163–174.

[L] LOVE, C. E., *Singular integral equations of the Volterra type*, Trans. Amer.
Math. Soc. **15** (1914), pp. 467–476.

[LR] LANDGREN, J. J. AND REJTO, P. A., *An application of the maximum principle
to the study of essential selfadjointness of Dirac operators. I*, J. Math.
Phys. **20** (1979), pp. 2204–2211.

[LRK] LANDGREN, J. J. AND REJTO, P. A.; APPENDIX BY KLAUS, MARTIN, *An application of the maximum principle to the study of essential selfadjointness*

of Dirac operators II, J. Math. Phys. **21** (1980), pp. 1210–1217.

[Ok] OKIKIOLU, G. O., "Aspects of the theory of bounded integral operators," Academic Press, 1975.

[O] OLVER, F. J. W., "Asymptotics and Special Functions," Academic Press, 1974.

[P] PEARSON, D. B., *Scattering theory for a class of oscillating potentials*, Helv. Phys. Acta **52** (1979), pp. 541–554.

[R] REJTO, P. A., *An application of the third order JWKB-approximation method to prove absolute continuity I. The construction*, Helv. Phys. Acta **50** (1977), pp. 479–494; II. The estimates, Helv. Phys. Acta **50** (1977), pp. 495–508; *On a theorem of Tichmarsh-Kodaira-Weidmann concerning absolutely continuous operators, I*, J. Approx. Theory **21** (1977), pp. 333–351; *On a theorem of Tichmarsh-Kodaira-Weidmann concerning absolutely continuous operators, II*, Indiana Univ. Math. J. **25** (1976), pp. 629–658.

[RS] REED, M. AND SIMON, B., "Analysis of operators, Methods of Modern Mathematical Physics, Vol. IV," Academic Press, 1978. See Section XIII.13, Example 1.

[RT] REJTO, P. AND TABOADA, M., *Weighted resolvent estimates for Volterra operators on unbounded intervals*, J. Math. Anal. Appl. **160**, 1 (1991), pp. 223–235.

[Si] SIBUYA, YASUTAKA, "Global Theory of a second order linear ordinary differential equation with a polynomial coefficient," North Holland Mathematical Studies, **18**, 1975.

[St] STOLZ, GÜNTER, *On the absolutely continuous spectrum of perturbed periodic Sturm-Liouville operators*, J. Reine Angew Math. **416** (1991), pp. 1–23; *Bounded solutions and absolute continuity of Sturm-Liouville operators*, Johann Wolfgang Goethe-Universität Frankfurt, February 1991.

[VW] VON NEUMANN, J. AND WIGNER, E., *Über merkwürdige diskrete Eigenwerte*, Phys. Z. **30** (1929), pp. 465–467.

[We] WEIDMANN, J., "Spectral Theory of Ordinary Differential Operators," Springer-Verlag Lecture Notes in Mathematics # 1258, 1987.

[Wh] WHITE, D. A. W., *Schroedinger operators with rapidly oscillating central potentials*, Trans. Amer. Math. Soc. **275** (1983), pp. 641–677.

LAX PAIRS IN THE HENON-HEILES AND RELATED FAMILIES

R.C. CHURCHILL* AND G.T. FALK[†]

A decade ago the impressive success of the Painleve test in locating integrable systems within parameterized families of ordinary differential equations led to speculation, by some, that passing the test was a necessary condition for integrability. A subsequent infinite list of examples produced by A. Ramani et. al. [R-D-G] forced a qualification of this conjecture, and counterexamples to that qualification have now appeared (e.g. see [K-C-R]). With the benefit of hindsight it seems fair to say that the test is more appropriately associated with 'algebraic' integrability (see [A-vM2, vMo]) and, more generally, with the possibility of augmenting isoenergy surfaces or the phase space so as to realize the flow within a complete $\mathbb{C}$-action. In the algebraically integrable case this augmentation is most elegantly accomplished by means of the spectral curve of a Lax pair. (We do not assume Lax pairs and associated spectral curves are familiar to the reader.) Unfortunately, Lax pairs admitting useful spectral curves can be difficult to produce.

The Ramani examples include the case $(A, B, \lambda) = (0, 0, 1/6)$ of the Henon-Heiles family

$$(0.1) \quad H(q_1, q_2, p_1, p_2) = (1/2)(p_1^2 + p_2^2) + (1/2)(Aq_1^2 + Bq_2^2) + (1/3)q_1^3 + \lambda q_1 q_2^2$$

on $\mathbb{C}^4 = \{(q_1, q_2, p_1, p_2)\}$. For this choice the corresponding vector field X_H passes the (original) Painleve test, and an associated Lax pair of 2×2 matrices is constructed in [N-T-Z]. (In fact this case of (0.1) is closely related to the Kowalevski top and geodesic flows on $SO(4)$ [A-vM1].) In this note we construct 2×2 Lax pairs for all the Ramani examples, for the more general Painleve case $(A, B, 1/6)$ of (0.1), and for the only other $(\lambda \neq 0)$ parameter choice in (0.1) for which X_H passes the test, i.e. $(A, B, \lambda) = (A, A, 1)$.

Our treatment of 2×2 Lax pairs is formulated in §1 in terms of differential algebras, for this context seems to best reveal the completely elementary nature of the ideas. The search for a Lax pair for a particular vector field then becomes an exercise involving undetermined coefficients, and when the corresponding equations admit a non-trivial solution the procedure becomes equivalent to a construction of D. Mumford in ([Mum, pp.

---

* Department of Mathematics, Hunter College, 695 Park Ave., New York, NY 10021. Author's research supported in part by NSF Grant # DMS8802911.

[†] Department of Mathematics, University of Calgary, Calgary, Alberta T2N 1N4 Canada. Current address: Department of Mathematics, University of Rochester, Rochester, NY 14627. Author's research supported in part by the Natural Sciences and Engineering Research Council of Canada.

3.43]), recently generalized by A. Beauville [Bea]. The examples are presented in §2, and the use of Lax pairs for linearizing isoenergetic flows is
reviewed within our context in §3. (This construction is standard, but is
so straightforward in our framework that it seems worth repeating. For a
general treatment of linearization using Lax pairs see [Gri].)

The work of this paper was carried out at our home institutions and,
with the support of the National Science Foundation, at the Institute for
Mathematics and its Applications in Minneapolis. Conversations with A.
Baider and M. Tabor on this material are gratefully acknowledged. In
particular, the viewpoint in [Bai] strongly influenced our approach to linearization. We also wish to thank an anonymous referee for constructive
criticism and for pointing out references [Bea], [Mum], and [R-D-G].

1. Algebraic preliminaries. $\mathcal{R}$ is an integral domain, extending a
field K of characteristic zero, and $\delta : \mathcal{R} \to \mathcal{R}$ is a derivation over K (i.e.
$K \subset \ker \delta$). For any $r \in \mathcal{R}$ we write $\delta(r)$ as r', so as to express the
derivation conditions in the familiar forms $(r + s)' = r' + s'$ and $(rs)' =
rs' + r's$. Whenever convenient we (uniquely) extend δ to the quotient
field F of $\mathcal{R}$ (when $F \neq \mathcal{R}$) by invoking the quotient rule. δ is assumed
nontrivial in the sense that the second inclusion in $K \subset \ker \delta \subset \mathcal{R}$ is proper.
Elements of $\ker \delta$ (i.e. satisfying $r' = 0$) are the *constants* of δ: those in K
are *trivial*, or *scalars*; those in $(\ker \delta)\backslash K$ are *nontrivial*, or *integrals* (of δ).
The latter terminology is motivated by the following example.

EXAMPLE 1.1. *Let X be a distinguished vector field on a complex
analytic manifold M and $\mathcal{A}$ a ring of meromorphic functions on M closed
under the derivation $\delta : a \mapsto a' := X(a)$. Extend δ to $\mathcal{R} := \mathcal{A}[x]$, where x
is any indeterminant over $\mathcal{A}$, by setting $x' = 0$. In this context integrals of
δ within $\mathcal{A}$ are simply nontrivial integrals, in the usual sense, of X.*

When $S = (s_{ij})$ is a matrix with entries in $\mathcal{R}$ we let $S' := (s'_{ij})$. A
Lax pair (for δ) is then an ordered pair (Q, P) of $m \times m$ matrices over $\mathcal{R}$
satisfying

$$(1.1) \qquad\qquad Q' = [P, Q] \neq 0,$$

where $[P, Q] := PQ - QP$ is the usual commutator. To see an example
choose any nonconstant r and constant k in $\mathcal{R}$ and set

$$(1.2) \qquad Q = \begin{pmatrix} r & 1 \\ -(r^2 + k) & -r \end{pmatrix}, \qquad P = \begin{pmatrix} 0 & 0 \\ -r' & 0 \end{pmatrix}.$$

Notice that $\det(Q) = k$. When X in Example 1.1 admits nontrivial integrals $f_1, ..., f_n$ one can take $k = \Sigma_{j=0}^n f_j x^j$ in (1.2) and thereby reconstruct
all the integrals from Q.

The following result is standard; a proof is included for completeness.

PROPOSITION 1.2. *For any Lax pair (Q, P) for δ all symmetric functions of the eigenvalues of Q over the rational field $\mathbb{Q}$ are constants.*

In particular the coefficients of powers of y in the characteristic polynomial $\det(Q - yI)$ of Q, which include $\det(Q)$ and $\operatorname{tr}(Q)$, must be constants of δ.

Proof. From (1.1) we have $(Q^n)' = \Sigma_{j=0}^n Q^{n-j} Q' Q^{j-1} = \Sigma_{j=0}^n (Q^{n-j} P Q^j - Q^{n-j+1} P Q^{j-1})$, whence $\operatorname{tr}(AB) = \operatorname{tr}(BA)$ and $\operatorname{tr}(A + B) = \operatorname{tr}(A) + \operatorname{tr}(B)$ imply $(\operatorname{tr}(Q^n))' = \operatorname{tr}((Q^n)') = 0$. But in terms of the spectrum $\{\lambda_j\}_{j=0}^m$ of Q we have $\operatorname{tr}(Q^n) = \Sigma_{j=0}^n \lambda_j^n$, and (by the characteristic zero assumption) these power sums generate all symmetric functions of the λ_j over $\mathbb{Q}$ (e.g., see [B-P, pp. 170-1]). $\square$

COROLLARY 1.3. *If the characteristic polynomial* $\det(Q - yI)$ *can be written as a polynomial in* y *and a constant* $x \in \mathcal{R}$, *then all coefficients of this polynomial are constants.*

In fact one can replace x in this statement by constants $x_1, ..., x_r$.

EXAMPLE 1.4. (adapted from [N-T-Z]): *In Example 1.1 let* $M = \mathbb{C}^4$, *let* A *denote the polynomial algebra over* $\mathbb{C}$ *in the canonical variables* (q_1, q_2, p_1, p_2), *and let* $X = X_H$ *correspond to the case* $(A, B, \lambda) = (0, 0, 1/6)$ *in (0.1). Then* (Q, P), *where*

$$Q = \begin{pmatrix} -\frac{1}{6}p_1 x + \frac{1}{72}q_2 p_2 & 2\left(x^2 + \frac{1}{6}q_1 x - \frac{1}{144}q_2^2\right) \\ 2x^3 - \frac{1}{3}q_1 x^2 + \frac{1}{72}(4q_1^2 + q_2^2)x + \frac{1}{72}p_2^2 & \frac{1}{6}p_1 x - \frac{1}{72}q_2 p_2 \end{pmatrix},$$

$$P = \begin{pmatrix} 0 & 1 \\ x - \frac{1}{3}q_1 & 0 \end{pmatrix}$$

is a Lax pair for (the derivation associated with) X_H. *Here*

$$\det(Q - yI) = y^2 - 4x^5 - \frac{1}{18}Hx^2 + \frac{1}{216}Gx,$$

where

$$G = G(q, p) = p_2(q_2 p_1 - q_1 p_2) + (1/6)q_2^2(q_1^2 + (1/4)q_2^2),$$

and so G *must be an integral of* X_H. *One checks easily that* G *and* H *are (functionally) independent.*

The Lax pair of the preceding example, and all those we will encounter, have the structure considered in the next result, which is our formulation of the equations beginning page 3.43 of [Mum].

PROPOSITION 1.5. *Let*

$$Q = \begin{bmatrix} * & 2q \\ * & * \end{bmatrix}, \qquad P = \begin{bmatrix} 0 & 1 \\ p & 0 \end{bmatrix}$$

be matrices over $\mathcal{R}$ *with* $\operatorname{tr}(Q) = 0$. *Then* $Q' = [P, Q]$ *iff*

(i)
$$Q = \begin{bmatrix} -q' & 2q \\ -q'' + 2pq & q' \end{bmatrix}$$

and

(ii) $$q''' = 4pq' + 2p'q.$$

Moreover, when this is the case (Q, P) is a Lax pair for δ, i.e. $Q' \neq 0$, iff $q' \neq 0$, i.e., iff q is not a constant.

All assertions are verified with straightforward calculations.

Thus the existence of Lax pairs of the indicated form reduces to the existence of nontrivial solutions in $\mathcal{R}$ of

(1.3) $$q''' = 4pq' + 2p'q,$$

but such solutions are notoriously difficult to find. It is interesting to note, however, that nontrivial solutions can always be constructed in the quotient field F of $\mathcal{R}$. This is a simple consequence of the following observation.

PROPOSITION 1.6. *Given $0 \neq q \in F$ write $q = s^2$ and set $\eta := s'/s$. Then for any $p \in F$ the following statements are equivalent.*
(i). $q''' = 4pq' + 2p'q$.
(ii) $s^3(s'' - ps)$ is a constant.
(iii) $e^{4\int \eta}(\eta' + \eta^2 - p)$ is a constant.
In particular, if η, p satisfy the Riccati equation $\eta' + \eta^2 = p$ then q, p, where $q := e^{4\int \eta}$, satisfy (1.3).

When s and/or $e^{4\int \eta}$ are not in F the elements in (ii) and (iii) must be interpreted as belonging to an appropriate extension field. Note, however, that $\eta \in F$.

Proof. When $q = s^2$ equation (1.3) is equivalent to $ss''' + 3s's'' = 4pss' + p's^2$, which is more conveniently written $s(s'' - ps)' = -3s'(s'' - ps)$. The equivalence of (i) and (ii) follows easily, and that of (ii) and (iii) is trivially verified. $\square$

COROLLARY 1.7. *Choose any nonconstant $s \in F$ and let $q := s^2, p := s''/s$. Then q, p is a non-trivial solution of (1.3).*

To generate additional non-trivial solutions in F simply note that $q, p \in F$ with $q \neq 0$ satisfy (1.3) iff this is the case for $q, p + cq^{-2}$ for any constant $c \in F$.

2. The examples. The first three examples are straightforward applications of Proposition 1.5 to the vector field $X_H = X_{H(A,B,\lambda)}$ on $\mathbb{C}^4$ defined by the Henon-Heiles Hamiltonian

(2.1) $$H(q_1, q_2, p_1, p_2) = \frac{1}{2}(p_1^2 + p_2^2) + \frac{1}{2}(Aq_1^2 + Bq_2^2) + \frac{1}{3}q_1^3 + \lambda q_1 q_2^2.$$

In each case the indicated solution of

(2.2) $$q''' = 4pq' + 2p'q$$

was uncovered by assuming

(2.3) $$q = x^2 + a_1 x + a_0, \quad p = x + b_1;$$

then using undetermined coefficients together with the MAPLE symbolic computation package to solve for the polynomials $a_j = a_j(q_1, q_2, p_1, p_2)$ and $b_1 = b_1(q_1, q_2, p_1, p_2)$.

Example 1 $(A, B, \lambda) = (A, B, 1/6)$: Take

$$q = x^2 + \tfrac{1}{6}q_1 x - \tfrac{1}{64}(4B - A)^2 + \tfrac{1}{48}(4B - A)q_1 - \tfrac{1}{144}q_2^2$$
$$p = x - \tfrac{1}{3}q_1 - \tfrac{1}{8}(4B + A).$$

Here

$$\det(Q - yI) = y^2 - 4x^5 + \tfrac{1}{2}(4B + A)x^4 + \tfrac{1}{8}(4B - A)^2 x^3$$
$$- \left(\tfrac{1}{18}H + \tfrac{1}{64}(4B + A)(4B - A)^2\right)x^2 + f_1(G, H)x + f_0(G, H),$$

where G denotes the independent integral

$$G := \frac{3}{2}(4B - A)(Bq_2^2 + p_2^2) + Bq_1 q_2^2 + p_2(q_2 p_1 - q_1 p_2) + \frac{1}{6}q_2^2\left(q_1^2 + \frac{1}{4}q_2^2\right),$$

and where the coefficients of the functionally independent polynomials f_1 and f_0 involve A and B.

Example 1.4 is the special case $A = B = 0$.

Example 2 $(A, B, \lambda) = (A, A, 1)$: Take

$$q = x^2 + \tfrac{1}{6}(q_1 + q_2 + cH)x - \tfrac{1}{48}(A - 2cH)(2q_1 + 2q_2 - 4cH + 3A),$$
$$p = x - \tfrac{1}{3}(q_1 + q_2 + cH),$$

where c is an arbitrary complex scalar. For $c \neq 0$ we have

$$\det(Q - yI) = y^2 - 4x^5 + f_3(G, H)x^3 + f_2(G, H)x^2 + f_1(G, H)x + f_0(G, H),$$

where

$$G = p_1 p_2 + q_1 q_2(A + q_1) + \frac{1}{3}q_2^2$$

is an independent integral and the coefficients of the pairwise independent polynomials f_j involve c and A. When $c = 0$ the coefficients in $\det(Q - yI)$ are functionally dependent, i.e., the corresponding Lax pair does not produce two independent integrals.

Example 3 $(A, B, \lambda) = (A, 0, 0)$: Take

$$q = x^2 + \left(\tfrac{1}{6}q_1 - cp_2 + \tfrac{A}{12}\right)x - \tfrac{1}{4}cp_2(2q_1 + 24cp_2 + A),$$
$$p = x - \left(\tfrac{1}{3}q_1 - 2cp_2 + \tfrac{A}{6}\right)$$

where c is an arbitrary complex scalar. For $c \neq 0$ we have

$$\det(Q - yI) = y^2 - 4x^5 + f_3(G)x^3 + f_2(G, H)x^2 + f_1(G, H)x + f_0(G, H),$$

where $G = p_2$ is an independent integral and the coefficients of the pairwise independent polynomials f_j involve c and A. In this instance the choice $c = 0$ does produce independent integrals within the coefficients of $\det(Q - yI)$.

Example 4: To produce the Ramani examples we generalize from (2.3) to the form

$$(2.4) \qquad q = x^2 + a_1 x + a_0, \quad p = \sum_{j=0}^{n} b_j x^{n-j}, \quad n \geq 1.$$

Moreover, to begin we may assume we are working with a derivation δ on an integral domain $\mathcal{R}$ of the form $K[x]$, where x is an indeterminant satisfying $x' = 0$.

For any $(n+2)$-tuple $c = (x_0, \ldots, c_{n+1})$ of constants in K let

$$\phi_c(\ell) = \sum_{0 \leq i+2j \leq \ell} (-1)^{i+j}(i+j+1) \binom{i+j}{i} c_{\ell-(i+2j)} a_1^i a_0^j, \quad \ell = 0, \ldots, n+1.$$
(2.5)

PROPOSITION 2.1. q, p as in (2.4) satisfy (2.2) iff there is an $(n+2)$-tuple $c = (c_0, \ldots, c_{n+1})$ of constants in K satisfying
(i) $a_1''' = -2\phi_c(n+1)$
(ii) $a_0''' = 4b_n a_0' + 2b_n' a_0$
(iii) $b_\ell = \phi_c(\ell), 0 \leq \ell \leq n$.

Proof. Substituting (2.4) into (2.2) and equating x-coefficients yields (ii),

(iv) $b_0' = 0$,
(v) $b_1' = -2b_0 a_1'$,
(vi) $a_1''' = 4(b_{n-1}a_0' + b_n a_1') + 2(b_{n-1}'a_0 + b_n'a_1)$,

and, when $n \geq 2$,

(vii) $b_\ell' = -2(a_1' b_{\ell-1} + a_0' b_{\ell-2}) - (a_1 b_{\ell-1}' + a_0 b_{\ell-2}'), \quad \ell = 2, \ldots, n$.

By (iv) and (v) the quantities $c_0 := b_0$ and $c_1 := b_1 + 2c_0 a_1$ are constants, thus establishing (iii) for $\ell = 0, 1$.

Suppose there is an ℓ-tuple $c = (c_0, \ldots, c_{\ell-1})$ such that $b_j = \phi_c(j)$ for $j < \ell$. Using (vii) a straightforward calculation shows that $b_\ell' = \phi_c'(\ell)$, and so there is a constant c_ℓ such that for $d := (c_0, \ldots, c_{\ell-1}, c_\ell)$ we have $b_\ell = \phi_c(\ell) + c_\ell = \phi_d(\ell)$. To verify (i) use these expressions for b_ℓ' and b_ℓ in (vi).

The converse is clear from the above argument, or by direct calculation.
□

Now, specialize to the case of a Hamiltonian vector field on $\mathbb{C}^4 = \{(q_1, q_2, p_1, p_2)\}$. For any $n \geq 1$ and any $(n+2)$-tuple $c = (c_0, \ldots, c_{n+1})$

of complex scalars set

$$c_{n+2} := 0,$$
$$a_0 := -q_2^2,$$
$$a_1 := 2q_1,$$
$$\psi_c(k) = \sum_{0 \le i+2j \le \ell}(-1)^{i+j} \binom{i+j}{i} c_{k-(i+2j)}a_1^i a_0^j, \qquad k = 0, ...n+2,$$
$$q := x^2 + a_1 x + a_0,$$
$$p := \sum_{k=0}^{n} \phi_c(k)x^{n-k} \qquad \text{see}((2.5)).$$

Then (Q, P) as in Proposition 1.4 is a Lax pair for the vector field X_H on $\mathbb{C}^4$ with Hamiltonian

(2.6) $$H(q_1, q_2, p_1, p_2) = \frac{1}{2}(p_1^2 + p_2^2) - \frac{1}{2}\psi_c(n+1).$$

Moreover,

$$\det(Q - yI) = y^2 - 4\sum_{j=0}^{3} c_j x^{n+4-j} - 8Hx^2 + 8Gx,$$

where

(2.7) $$G = p_2(q_2 p_1 - q_1 p_2) + \frac{1}{2}q_2^2 \psi_c(n+1)$$

Indeed, straightforward calculation gives $q_1'' = -\psi_c(n+1)$, $q_2'' = q_2\psi_c(n)$. But then $a_1'' = -2\psi_c(n+1)$, and Proposition 1.5 with $s = q_2$ then gives $a_0''' = 4a_0'\psi_c(n) + 2a_0\psi_c'(n)$. It follows from Proposition 2.1 that (Q, P) is a Lax pair for X_H. To check the asserted form of $\det(Q - yI)$ calculate coefficients of the various powers of x using the identities $\phi_c(k+1) + a_1\phi_c(k) + a_0\phi_c(k-1) = \psi_c(k+1)$ and $a_0\psi_c(k) + a_1\psi_c(k+1) = -\psi_c(k+2)$. To verify independence note that the coefficient of $dp_1 \wedge dp_2$ in $dG \wedge dH$ is $q_2(p_1^2 - p_2^2) - 2q_1 p_1 p_2$ for any $(k+1)$-tuple c.

The Hamiltonian (2.6) and the corresponding integrals (2.7) are those introduced by Ramani et al [R-D-G].

Example 5 [K-C-R]: The Hamiltonian

$$H(q, p) = p_1(q_1^2 + p_1^2) + \epsilon q_1(q_2^2 + p_2^2)$$

admits $G = q_2^2 + p_2^2$ as an independent integral. Letting r be any nonconstant and $k = x^2 + Gx + H$ we can use (1.2) to produce a Lax pair (Q, P) satisfying $\det(Q) = x^2 + Gx + H$. Of course this method generalizes to any integrable n-degree of freedom system. This particular example is of interest because for $\epsilon \in \mathbb{R} \setminus \mathbb{Q}$ it fails the Painleve test by virtue of the presence of irrational 'Kovalevski exponents', i.e. it is a counterexample to the generalized Painleve conjecture of [R-D-G].

3. Linearization. Let H be a polynomial two degree of freedom Hamiltonian on $\mathbb{C}^4$ of the classical form $H(q_1, q_2, p_1, p_2) = \frac{1}{2}(p_1^2 + p_2^2) + V(q_1, q_2)$, where the degree of the potential function V is the least three. It proves convenient to express this degree as

$$(3.1) \qquad \deg(V(q_1, q_2)) = 2(g-1) + \epsilon, \quad 1 < g \in \mathbb{Z}, \epsilon = 1 \text{ or } 2.$$

Suppose (Q, P) is a 2×2 Lax pair for X_H having the form

$$(3.2) \qquad Q = \begin{pmatrix} -q' & 2q \\ -q'' + 2pq & q' \end{pmatrix}, \qquad P = \begin{pmatrix} 0 & 1 \\ p & 0 \end{pmatrix}$$

considered in Proposition 1.5, together with the following additional properties: p and q are polynomials in x with $n := \deg(g) \geq 2$; and

$$(3.3) \qquad \det(Q(m, x)) = \sum_{j=0}^{2g+\epsilon} c_j(m) x^j, \quad m \in \mathbb{C}^4, \quad c_{2g+\epsilon} \neq 0,$$

where $g > 1$ is as in (3.1) and the constants $c_1(m)$ are polynomials in two independent integrals G_1, G_2 of X_H. These hypotheses are satisfied by all the Lax pairs encountered in §2.

Assume values g_i for the respective G_i are chosen in such a way that the intersection of the two level surfaces $G_i = g_i = c_i$ is a smooth submanifold $\Sigma \subset \mathbb{C}^4$ and the polynomial

$$(3.4) \qquad \det(Q(x)) := \sum_{j=0}^{2g+\epsilon} c_j x^j$$

is separable (i.e. all zeros are simple). Now set $\Sigma' := \Sigma \backslash \Delta$, where $\Delta \subset \mathbb{C}^4$ is the discriminant locus of $q = q(m, x)$. This last adjustment is to insure that $q(m, x)$, for any $m \in \Sigma'$, has the form

$$(3.5) \qquad q(m, x) = \prod_{j=1}^{n} (x - \mu_j),$$

where all μ_j are distinct.

Under the assumptions detailed in the preceding two paragraphs there exists a complex g-dimensional torus T and an analytic equivariant mapping $\eta : \Sigma' \to T$ between $X_H | \Sigma'$ and a linear flow on T.

To see this first note that for any $m \in \mathbb{C}^4$ we have from (3.2) that

$$(3.6) \quad (q'(m, x))^2 + \det(Q(m, x)) = 2q(m, x)[q''(m, x) - 2p(m, x)q(m, x)],$$

and so for $m \in \Sigma'$ any zero μ of $q(m, x)$ satisfies $(q'(m, \mu))^2 + \det(Q(\mu)) = 0$. But then $(\mu, q'(m, \mu)) \in C$, where $C \subset \mathbb{C}P^2$ is the nonsingular hyperelliptic curve defined by $y^2 + \det(Q(x)) = 0$. (C is the *spectral curve* which (Q, P)

associates with $X_H|\Sigma$.) An ordering $(\mu_1, ..., \mu_n)$ of the zeros of $q(m, x)$ would define a mapping of Σ' into $C^n := C \times \cdots \times C$, but of course any imposed ordering would be artificial. To remedy the situation pass to the n-fold symmetric product $S^n C := C^n/S_n$, where the quotient is by the obvious action of the symmetric group S_n on C^n. The assignment $m \mapsto ((\mu_1, q'(m, \mu_1)), ..., (\mu_n, q'(m, \mu_n)))/S_n$ then defines a smooth analytic mapping $\beta : \Sigma' \rightarrow S^n C$, and by differentiating (3.5) and using (3.6) one sees that β is equivariant between $X_H'|\Sigma'$ and the vector field

$$(3.7) \qquad \mu_i'(-\det(Q(\mu_i)))^{-\frac{1}{2}} = \left(\prod_{j \neq i}(\mu_i - \mu_j)\right)^{-1}, \quad i = 1, ..., n.$$

on $S^n C$.

Now consider the Jacobi variety $\mathrm{Jac}(C)$ of C, i.e. $\mathbb{C}^g$ modulo the periods of the standard basis

$$\omega_j = z^{j-1}(-\det(Q(z)))^{-1/2}dz, \quad j = 1, ..., g,$$

of abelian differentials on C. The Jacobi mapping $\alpha : S^n C \rightarrow \mathrm{Jac}(C)$, defined modulo periods by

$$\alpha_k = \sum_{j=1}^{n} \int^{\mu_j} \omega_k, k = 1, ..., g,$$

is then equivariant between (3.7) and the linear field

$$\alpha_1' = \cdots = \alpha_{g-1}' = 0, \alpha_g' = 1$$

on $\mathrm{Jac}(C)$. The mapping $\eta := \alpha*\beta : \Sigma' \rightarrow T := \mathrm{Jac}(C)$ is therefore equivariant between $X_H|\Sigma'$ and a linear flow on a torus.

Remark: When the degree of $V(Q_1, q_2)$ is five or greater we have $\dim(T) > 2 = \dim(\Sigma')$. For our examples this difference in dimension is measured by the fractional exponent encountered (as in [R-D-G]) in the Painleve test.

For additional discussions of the Henon-Heiles family and the Painleve test see [Bai], [C-T-W], and [Tab]. For a discussion of the non-integrable parameter values within the family see [Ito] and [Rod].

REFERENCES

[A-vM1] ADLER, M., AND VAN MOERBEKE, P., *The Kowalewski and Henon-Heiles Motions as Manakov Geodesic Flows on SO(4) - a Two Dimensional Family of Lax Pairs*, Comm. Math. Phys. 113 (1988), pp. 659–700.

[A-vM2] ADLER, M., AND VAN MOERBEKE, P., *The Complex Geometry of the Kowalewski-Painleve Analysis*, Invent. Math. 97 (1989), pp. 3–51.

[Bai] BAIDER, A., *Completion of Vector Fields and the Painleve Property*, J. Diff. Equations, 97 (1992), pp. 27–53.

[Bea] BEAUVILLE, A., *Jacobiennes des Courbes Spectrales et Systemes Hamiltoniens Completement Integrables*, Acta Math. 164 (1990), pp. 211–235.

[B-P] BURNSIDE, W.S., AND PANTON, A.W., *The Theory of Equations with an Introduction to the Theory of Binary Algebraic Forms*, Vol. I, Dover, New York, 1960.

[C-T-W] CHANG, Y.F. TABOR, M., AND WEISS, J., *Analytic Structure of the Henon-Heiles Hamiltonian in Integrable and Non-integrable Regimes*, J. Math. Phys., 23 (4) (1982), pp. 531–8.

[Gri] GRIFFITHS, P., *Linearizing Flows and a Cohomological Interpretation of Lax Equations*, Am. J. Math. 107 (1985), pp. 1445–83.

[Ito] ITO, H., *A Criterion for Non-integrability of Hamiltonian Systems with Non-homogeneous Potentials*, Z. Angew. Math. Phys., 38 (1987), pp. 459–476.

[K-C-R] KUMMER, M., CHURCHILL, R.C., AND ROD, D.L., *On Kovalevski Exponents*, in *Essays on Classical and Quantum Dynamics*, (A Festschrift in Honor of Albert W. Saenz) (J.A. Ellison and H. Uberall, Eds.), Gordon and Breach, New York, 1992.

[Mum] MUMFORD, D., *Tata Lectures on Theta II*, Progress in Math. 43, Birkhauser, Boston, 1984.

[N-T-Z] NEWELL, A.C., TABOR, M., AND ZENG, Y.B., *A Unified Approach to Painleve Expansions*, Physica 29D (1989), pp. 1–68.

[R-D-G] RAMANI, A., DORIZZI, B., AND B. GRAMMATICOS, *Painleve Conjecture Revisited*, Phys. Rev. Letters 49 (1982), pp. 1539–1541.

[Rod] ROD, D.L., *On a Theorem of Ziglin in Hamiltonian Dynamical Systems*, (K.R. Meyer and D.E. Saari, Eds.), Contemp. Math., 81, American Mathematical Society, Providence, RI, 1988, pp. 259–270.

[Tab] TABOR, M., *Chaos and Integrability in Nonlinear Dynamics*, John Wiley, New York, 1989.

[vMo] VAN MOERBEKE, P., *The Geometry of Painleve Analysis*, in *Proceedings of the Workshop on Finite Dimensional Integrable Nonlinear Dynamical Systems*, (P.G.L. Leach and W.H. Steeb, Eds.), 1-33, World Scientific, Singapore, 1988.

POINCARÉ COMPACTIFICATION OF HAMILTONIAN POLYNOMIAL VECTOR FIELDS[*]

J. DELGADO[†], E.A. LACOMBA[†], J. LLIBRE[‡], AND E. PÉREZ[†]

Introduction. There exists an extensive literature on changes of variables which transform the equations of motion of interesting problems in Celestial Mechanics into polynomial form (see [Heg]). In most cases this is achieved by regularizing double collisions or introducing redundant variables, or both. In previous works [DLLP1,2] we have exploited this idea to extend the equations of motion of the collinear three body problem to a compact manifold by means of the Poincaré compactification. Such compactification was introduced to study the behaviour at infinity of polynomial vector fields (see for instance [CL]). In the Poincaré compactification of some n-body problems the critical points which appear there are extremely degenerate. In this paper we focus our attention on generic properties of arbitrary Hamiltonian polynomial vector fields, especially at infinity.

In the first Section, we review the Poincaré compactification for a general polynomial vector field. Since this is fundamental for our computations, we present all the details with a slight modification in the notation from the construction outlined in [CL].

In Section 2, we give the global expressions for the Poincaré compactification of a Hamiltonian polynomial vector field. The explicit computations of the expressions of the vector field in local charts are also given. All results from Section 3 to 5 refer to the Poincaré compactification of Hamiltonian polynomial vector fields.

Section 3 summarizes the behavior of the vector field at infinity. We see that all the energy levels extend to the same invariant set at infinity. Section 4 is devoted to generic properties. We prove that generically, in the coefficient topology, the above invariant set at infinity is a smooth manifold and all the critical points in the Poincaré sphere are hyperbolic. In the final section, we study two simple cases of Hamiltonian polynomial vector fields. A fairly complete description of the flow can be given in these two cases.

We remark that the majority of Hamiltonian polynomial vector fields appearing in Classical Mechanics, like for example the Henon-Heiles and Contopoulos, are non generic in the above sense.

1. Poincaré compactification for polynomial vector fields. As far as we know the Poincaré compactification for a polynomial vector field

---

[*] The authors were supported by a DGCICYT grant number PB90-0695 and the second and third authors were also supported by a CONACYT grant.

[†] Departamento de Matemáticas, Universidad Autónoma Metropolitana–Iztapalapa, Apdo. 55534, 09340 México, D.F., México.

[‡] Departament de Matemàtiques, Universitat Autònoma de Barcelona, 08193 Bellaterra, Spain.

in $\mathbb{R}^n$ was introduced for the first time in [CL], following the essential ideas given by Poincaré in $\mathbb{R}^2$. Since this compactification plays a main role in this paper we reproduce with more detail the description given in [CL].

Let $X = (P^1, \ldots, P^n)$ be a polynomial vector field in $\mathbb{R}^n$. We identify $\mathbb{R}^n$ with the hyperplane $\Pi = \{y \in \mathbb{R}^{n+1} \mid y_{n+1} = 1\}$ tangent to the *Poincaré sphere* $S^n = \{y \in \mathbb{R}^{n+1} \mid \sum_{i=1}^{n+1} y_i^2 = 1\}$ at the north pole. The Poincaré compactification of X consists in making two copies of the flow of X, one on the open northern hemisphere H^+ of S^n and the other one on the open southern hemisphere H^- of S^n, through the central projection. That is, for each point in Π we draw the straight line through this point and the origin of $\mathbb{R}^{n+1}$ obtaining in this way two antipodal points in S^n, one in H^+ and the other in H^-. More concretely, this construction defines the following two diffeomorphisms

$$\phi^+ : \mathbb{R}^n \to H^+ \qquad and \qquad \psi^- : \mathbb{R}^n \to H^-$$

given by

$$\phi^+(x) = \frac{1}{\Delta(x)}(x_1, \ldots, x_n, 1),$$

$$\psi^-(x) = -\frac{1}{\Delta(x)}(x_1, \ldots, x_n, 1),$$

where

$$\Delta(x) = (1 + \sum_{i=1}^{n} x_i^2)^{1/2}.$$

In this form X induces a vector field $\widehat{X}$ in $H^+ \bigcup H^-$ defined by

$$\widehat{X}(y) = \begin{cases} (D\phi^+)_x X(x) & \text{if } y = \phi_{n+1}(x), \\ (D\psi^-)_x X(x) & \text{if } y = \psi_{n+1}(x). \end{cases}$$

The expression for $\widehat{X}(y)$ on $H^+ \bigcup H^-$ is

$$\widehat{X}(y) = y_{n+1} \begin{pmatrix} 1 - y_1^2 & -y_1 y_2 & \cdots & -y_1 y_n \\ -y_2 y_1 & 1 - y_2^2 & \cdots & -y_2 y_n \\ \vdots & & & \\ -y_n y_1 & -y_n y_2 & \cdots & 1 - y_n^2 \\ -y_{n+1} y_1 & -y_{n+1} y_2 & \cdots & -y_{n+1} y_n \end{pmatrix} \begin{pmatrix} \widehat{P}^1 \\ \widehat{P}^2 \\ \vdots \\ \widehat{P}^{n-1} \\ \widehat{P}^n \end{pmatrix}$$

where $\widehat{P}^i(y_1, \ldots, y_{n+1}) = P^i(y_1/y_{n+1}, \ldots, y_n/y_{n+1})$.

The equator $S^{n-1} = \{y \in S^n \mid y_{n+1} = 0\}$ of the Poincaré sphere corresponds to the infinity of $\mathbb{R}^n$, and the key point of the Poincaré compactification is the possibility of extending the flow given by $\widehat{X}$ on $S^n \setminus S^{n-1}$ to S^n. In this way we will be able to study the orbits of X going to or coming from the infinity of $\mathbb{R}^n$. This extension is possible due to the polynomial character of X. Thus, the vector field

$$\widetilde{X}(y) = y_{n+1}^{m-1}\,\widehat{X}(y) = \begin{pmatrix} 1-y_1^2 & -y_1 y_2 & \cdots & -y_1 y_n \\ -y_2 y_1 & 1-y_2^2 & \cdots & -y_2 y_n \\ \vdots & & & \\ -y_n y_1 & -y_n y_2 & \cdots & 1-y_n^2 \\ -y_{n+1} y_1 & -y_{n+1} y_2 & \cdots & -y_{n+1} y_n \end{pmatrix} \begin{pmatrix} \widetilde{P^1} \\ \widetilde{P^2} \\ \vdots \\ \widetilde{P^{n-1}} \\ \widetilde{P^n} \end{pmatrix}.$$

(1.1)
is analytical on the whole S^n. Here $m = \max\{\,\text{degree}(P^1), \ldots, \text{degree}(P^n)\}$ is the degree of X and

(1.2) $$\widetilde{P^k}(y_1, \ldots, y_{n+1}) = y_{n+1}^m P^k(y_1/y_{n+1}, \ldots, y_n/y_{n+1}).$$

Notice that $\widetilde{P^k}$ are homogeneous polynomials of degree m. The vector field $\widetilde{X}$ is called *The Poincaré compactification* of X.

To obtain the analytical expression for $\widetilde{X}$ we will consider S^n as a differentiable manifold. Thus we cover it by $2(n+1)$ local charts

$$(U_i, \phi_i), \qquad (V_i, \psi_i) \qquad \text{for } i = 1, \ldots, n+1, \text{ where}$$

$$U_i = \{y \in S^n \mid y_i > 0\}, \qquad V_i = \{y \in S^n \mid y_i < 0\},$$

and $\phi_i : U_i \to \mathbb{R}^n$, $\psi_i : V_i \to \mathbb{R}^n$ are given by the same expression

$$(1.3)\,(y_1, \ldots, y_{n+1}) \to \left(\frac{y_1}{y_i}, \ldots, \frac{y_{i-1}}{y_i}, \frac{y_{i+1}}{y_i}, \ldots, \frac{y_{n+1}}{y_i} \right) = (z_1, \ldots, z_n).$$

The local coordinates are denoted by $(z_1, \ldots, z_n)$, although these variables have a different meaning in each chart. In any local chart the points of infinity have the coordinate z_n equal to zero. Notice that all the local charts contain points of infinity except (U_{n+1}, ϕ_{n+1}) and (V_{n+1}, ψ_{n+1}). Furthermore, $U_{n+1} = H^+, V_{n+1} = H^-$ and $\phi_{n+1} = (\phi^+)^{-1}$, $\psi_{n+1} = (\psi^-)^{-1}$.

We will now get the analytical expression of $\widetilde{X}$ in each local chart. We start doing the computation for U_1. Let $y \in U_1 \cap H^+$. Thus, since the differential $(D\phi_1)_y$ goes from $T_y U_1$ to $T_{\phi_1(y)}\mathbb{R}^n$ we get

$$\begin{aligned} (D\phi_1)_y(\widetilde{X}(y)) &= (D\phi_1)_y(y_{n+1}^{m-1}\,\widetilde{X}(y)) \\ &= y_{n+1}^{m-1}(D\phi_1)_y(D\phi^+)_x X(x) \\ &= y_{n+1}^{m-1} D(\phi_1 \circ \phi^+)_x X(x), \end{aligned}$$

where $y = \phi^+(x)$. Then

$$D(\phi_1 \circ \phi^+)_x X(x) = \frac{1}{x_1^2}(-x_2 P^1 + x_1 P^2, -x_3 P^1 + x_1 P^3, \ldots, -x_n P^1 + x_1 P^n, -P^1),$$

where $P^i \equiv P^i(x_1, \ldots, x_n)$.

In the local chart U_1 we have

$$(z_1, \ldots, z_n) = \phi_1(y_1, \ldots, y_{n+1}) = \left(\frac{y_2}{y_1}, \ldots, \frac{y_{n+1}}{y_1}\right),$$

and we get

$$D(\phi_1 \circ \phi^+)_x X(x) = z_n(-z_1 P^1 + P^2, -z_2 P^1 + P^3, \ldots, -z_{n-1} P^1 + P^n, -z_n P^1),$$

where $P^i \equiv P^i(1/z_n, z_1/z_n, \ldots, z_{n-1}/z_n)$. Since $y_{n+1}^{m-1} = [z_n/\Delta(z)]^{m-1}$ the vector field $\tilde{X}$ becomes

$$\frac{z_n^m}{\Delta(z)^{m-1}}(-z_1 P^1 + P^2, -z_2 P^1 + P^3, \ldots, -z_{n-1} P^1 + P^n, -z_n P^1).$$

If $y \in U_1 \cap H^-$ we get the same expression as above. Scaling the expression of $\tilde{X}$ by $\Delta(z)^{m-1} > 0$ in each local chart, and doing similar computations for the other local charts we obtain the following expressions for $\tilde{X}$.

In U_1: $(-z_1 \tilde{P}^1 + \tilde{P}^2, -z_2 \tilde{P}^1 + \tilde{P}^3, \ldots, -z_{n-1} \tilde{P}^1 + \tilde{P}^n, -z_n \tilde{P}^1)$

where $\tilde{P}^i \equiv \tilde{P}^i(1, z_1, \ldots, z_n)$;

in U_2: $(-z_1 \tilde{P}^2 + \tilde{P}^1, -z_2 \tilde{P}^2 + \tilde{P}^3, \ldots, -z_{n-1} \tilde{P}^2 + \tilde{P}^n, -z_n \tilde{P}^2)$

where $\tilde{P}^i \equiv \tilde{P}^i(z_1, 1, \ldots, z_n)$;

$$\vdots$$

in U_n: $(-z_1 \tilde{P}^n + \tilde{P}^1, -z_2 \tilde{P}^n + \tilde{P}^2, \ldots, -z_{n-1} \tilde{P}^n + \tilde{P}^{n-1}, -z_n \tilde{P}^n)$

where $\tilde{P}^i \equiv \tilde{P}^i(z_1, \ldots, z_n, 1)$;

in U_{n+1}: $(P^1, P^2, \ldots, P^n)$

where $P^i \equiv \tilde{P}^i(z_1, z_2, \ldots, z_n)$.

On the other hand, the expression for $\tilde{X}$ in the local chart V_i is the same as in U_i multiplied by $(-1)^{m-1}$. This is due to the fact that in V_i, $y_{n+1}^{m-1} = [-z_n/\Delta(z)]^{m-1} = (-1)^{m-1}[z_n/\Delta(z)]^{m-1}$.

From the expressions of $\tilde{X}$ in the local charts, we deduce that the infinity is invariant by the flow of $\tilde{X}$.

We will call *finite* (respectively *infinite*) *critical points* of X or $\tilde{X}$, the critical points of $\tilde{X}$ which lie on $S^n \setminus S^{n-1}$ (respectively S^{n-1}).

Notice that the orbits in S^n are always symmetric with respect to the origin of $\mathbb{R}^{n+1}$, but the vector field $\tilde{X}$ is only symmetric when m is odd. Furthermore, due to this symmetry, if $y \in S^{n-1}$ is an infinite critical point then $-y$ is another one.

2. Poincaré compactification for Hamiltonian polynomial vector fields.

A polynomial function $H : \mathbb{R}^{2d} \to \mathbb{R}$ will be called a *polynomial Hamiltonian*, and

$$X_H = \left(\frac{\partial H}{\partial y_{d+1}}, \ldots, \frac{\partial H}{\partial y_{2d}}, -\frac{\partial H}{\partial y_1}, \ldots, -\frac{\partial H}{\partial y_d} \right)$$

will be called the *Hamiltonian polynomial vector field associated to H*.

As we would expect in the case of a Hamiltonian polynomial vector field the Poincaré compactification can also be described in terms of the derivatives of H, as it is shown in the following result.

PROPOSITION 2.1. *Let $H(y_1, \ldots, y_n)$ be a polynomial Hamiltonian with $n = 2d$ variables and degree $m + 1$. Then the following statements hold.*

(a) The Poincaré compactification of X_H is

$$\tilde{X}_H = \left(\frac{\partial H^*}{\partial y_{d+1}} + \lambda y_1, \ldots, \frac{\partial H^*}{\partial y_{2d}} + \lambda y_d, -\frac{\partial H^*}{\partial y_1} + \lambda y_{d+1}, \ldots, \right.$$
$$\left. -\frac{\partial H^*}{\partial y_d} + \lambda y_{2d}, \lambda y_{n+1} \right)$$

where

$$H^*(y_1, \ldots, y_{n+1}) = y_{n+1}^{m+1} H\left(\frac{y_1}{y_{n+1}}, \ldots, \frac{y_n}{y_{n+1}} \right),$$

and

$$\lambda = \sum_{i=1}^{d} \left(y_{i+d} \frac{\partial H^*}{\partial y_i} - y_i \frac{\partial H^*}{\partial y_{i+d}} \right).$$

(b) The energy level $H^* = 0$ is invariant under the flow of $\tilde{X}_H$.

Proof: From (1.2) and the definition of H^* we get

$$(2.1) \qquad \widetilde{\frac{\partial H}{\partial y_i}} = y_{n+1}^{m} \frac{\partial H}{\partial y_i}\left(\frac{y_1}{y_{n+1}}, \ldots, \frac{y_n}{y_{n+1}} \right) = \frac{\partial H^*}{\partial y_i}.$$

From (1.1) and the above equality, (a) follows easily.

Computing the derivative of H^* along solutions curves of $\tilde{X}_H$ one gets

$$\frac{dH^*}{dt} = \sum_{i=1}^{d} \left[\frac{\partial H^*}{\partial y_i} y_i' + \frac{\partial H^*}{\partial y_{i+d}} y_{i+d}' \right] + \frac{\partial H^*}{\partial y_{n+1}} y_{n+1}'$$

$$= \sum_{i=1}^{d} \left[\frac{\partial H^*}{\partial y_i} \left(\frac{\partial H^*}{\partial y_{i+d}} + \lambda y_i \right) + \frac{\partial H^*}{\partial y_{i+d}} \left(-\frac{\partial H^*}{\partial y_i} + \lambda y_{i+d} \right) \right] + \frac{\partial H^*}{\partial y_{n+1}} \lambda y_{n+1}$$

$$= \lambda \left(\sum_{i=1}^{d} \left[\frac{\partial H^*}{\partial y_i} y_i + \frac{\partial H^*}{\partial y_{i+d}} y_{i+d} \right] + \frac{\partial H^*}{\partial y_{n+1}} y_{n+1} \right)$$

$$= \lambda(m+1) H^*,$$

where we have applied Euler's theorem for homogeneous functions to H^*. From these equalities it follows that the level $H^* = 0$ is invariant under the flow of $\widetilde{X}_H$. Hence (b) is proved.

         ∎

We remark that $H^*(y_1, \ldots, y_{n+1})$ is a homogeneous polynomial of degree $m+1$. We will now get the analytical expressions of $\widetilde{X}_H$, in each local chart, and again we can describe them in terms of the derivatives of H^*. We start doing the computations for the chart U_1. Thus, if $(z_1, \ldots, z_n)$ are the coordinates of U_1, from (1.3) we have

$$H^*(y_1, \ldots, y_{n+1}) = y_1^{m+1} H^*\left(1, \frac{y_2}{y_1}, \ldots, \frac{y_{n+1}}{y_1}\right) \equiv y_1^{m+1} \Gamma(z_1, \ldots, z_n).$$

From the general expression of a polynomial vector field in the local chart U_1 and (2.1) we get

$$\left(-z_1 \frac{\partial H^*}{\partial y_{d+1}} + \frac{\partial H^*}{\partial y_{d+2}}, \ldots, -z_{d-1} \frac{\partial H^*}{\partial y_{d+1}} + \frac{\partial H^*}{\partial y_{2d}}, -z_d \frac{\partial H^*}{\partial y_{d+1}} - \frac{\partial H^*}{\partial y_1}, \right.$$
$$\left. -z_{d+1} \frac{\partial H^*}{\partial y_{d+1}} - \frac{\partial H^*}{\partial y_2}, \ldots, -z_{2d-1} \frac{\partial H^*}{\partial y_{d+1}} - \frac{\partial H^*}{\partial y_d}, -z_{2d} \frac{\partial H^*}{\partial y_{d+1}}\right)\Bigg|_{(1, z_1, \ldots, z_n)}.$$

Since

$$\frac{\partial H^*}{\partial y_1} = y_1^m \left((m+1)\Gamma - \sum_{i=1}^n z_i \frac{\partial \Gamma}{\partial z_i}\right),$$
$$\frac{\partial H^*}{\partial y_i} = y_1^m \frac{\partial \Gamma}{\partial z_{i-1}},$$

for $i = 2, \ldots, n$, we obtain

$$\frac{\partial H^*}{\partial y_1}(1, z_1, \ldots, z_n) = (m+1)\Gamma - \sum_{i=1}^n z_i \frac{\partial \Gamma}{\partial z_i},$$
$$\frac{\partial H^*}{\partial y_i}(1, z_1, \ldots, z_n) = \frac{\partial \Gamma}{\partial z_{i-1}},$$

for $1 = 2, \ldots, n$. Therefore, substituting these equalities in the expression of $\widetilde{X}$ for U_1:

$$\left(-z_1 \frac{\partial \Gamma}{\partial z_d} + \frac{\partial \Gamma}{\partial z_{d+1}}, \ldots, -z_{d-1} \frac{\partial \Gamma}{\partial z_d} + \frac{\partial \Gamma}{\partial z_{2d-1}}, -(m+1)\Gamma + \sum_{d \neq i=1}^n z_i \frac{\partial \Gamma}{\partial z_i}, \right.$$
$$\left. -z_{d+1} \frac{\partial \Gamma}{\partial z_d} - \frac{\partial \Gamma}{\partial z_1}, \ldots, -z_{2d-1} \frac{\partial \Gamma}{\partial z_d} - \frac{\partial \Gamma}{\partial z_{d-1}}, -z_{2d} \frac{\partial \Gamma}{\partial z_d}\right).$$

Doing similar computations for the other local charts we get for U_2:

$$\left(-z_1 \frac{\partial \Gamma}{\partial z_{d+1}} + \frac{\partial \Gamma}{\partial z_d}, -z_2 \frac{\partial \Gamma}{\partial z_{d+1}} + \frac{\partial \Gamma}{\partial z_{d+2}}, \ldots, -z_{d-1} \frac{\partial \Gamma}{\partial z_{d+1}} + \frac{\partial \Gamma}{\partial z_{2d-1}}, \right.$$

$$-z_d\frac{\partial\Gamma}{\partial z_{d+1}} - \frac{\partial\Gamma}{\partial z_1}, -(m+1)\Gamma + \sum_{d+1\neq i=1}^{n} z_i\frac{\partial\Gamma}{\partial z_i},$$

$$-z_{d+2}\frac{\partial\Gamma}{\partial z_{d+1}} - \frac{\partial\Gamma}{\partial z_2}, \ldots, -z_{2d-1}\frac{\partial\Gamma}{\partial z_{d+1}} - \frac{\partial\Gamma}{\partial z_{d-1}}, -z_{2d}\frac{\partial\Gamma}{\partial z_{d+1}}\Bigg);$$

$$\vdots$$

for U_d :

$$\left(-z_1\frac{\partial\Gamma}{\partial z_{2d-1}} + \frac{\partial\Gamma}{\partial z_d}, -z_2\frac{\partial\Gamma}{\partial z_{2d-1}} + \frac{\partial\Gamma}{\partial z_{d+1}}, \ldots, -z_{d-1}\frac{\partial\Gamma}{\partial z_{2d-1}} + \frac{\partial\Gamma}{\partial z_{2d-2}},\right.$$

$$-z_d\frac{\partial\Gamma}{\partial z_{2d-1}} - \frac{\partial\Gamma}{\partial z_1}, \ldots, -z_{2d-2}\frac{\partial\Gamma}{\partial z_{2d-1}} - \frac{\partial\Gamma}{\partial z_{d-2}},$$

$$-(m+1)\Gamma + \sum_{2d-1\neq i=1}^{n} z_i\frac{\partial\Gamma}{\partial z_i}, -z_{2d}\frac{\partial\Gamma}{\partial z_{2d-1}}\Bigg);$$

for U_{d+1} :

$$\left((m+1)\Gamma - \sum_{1\neq i=1}^{n} z_i\frac{\partial\Gamma}{\partial z_i}, z_2\frac{\partial\Gamma}{\partial z_1} + \frac{\partial\Gamma}{\partial z_{d+1}}, \ldots, z_d\frac{\partial\Gamma}{\partial z_1} + \frac{\partial\Gamma}{z_{2d-1}},\right.$$

$$z_{d+1}\frac{\partial\Gamma}{\partial z_1} - \frac{\partial\Gamma}{\partial z_2}, z_{d+2}\frac{\partial\Gamma}{\partial z_1} - \frac{\partial\Gamma}{\partial z_3}, \ldots, z_{2d-1}\frac{\partial\Gamma}{\partial z_1} - \frac{\partial\Gamma}{\partial z_d}, z_{2d}\frac{\partial\Gamma}{\partial z_1}\Bigg);$$

$$\vdots$$

for U_{2d}

$$\left(z_1\frac{\partial\Gamma}{\partial z_d} + \frac{\partial\Gamma}{\partial z_{d+1}}, z_2\frac{\partial\Gamma}{\partial z_d} + \frac{\partial\Gamma}{\partial z_{d+2}}, \ldots, z_{d-1}\frac{\partial\Gamma}{\partial z_d} + \frac{\partial\Gamma}{\partial z_{2d-1}},\right.$$

$$(m+1)\Gamma - \sum_{d\neq i=1}^{n} z_i\frac{\partial\Gamma}{\partial z_i}, z_{d+1}\frac{\partial\Gamma}{\partial z_d} - \frac{\partial\Gamma}{\partial z_1}, z_{d+2}\frac{\partial\Gamma}{\partial z_d} - \frac{\partial\Gamma}{\partial z_2}, \ldots,$$

$$z_{2d-1}\frac{\partial\Gamma}{\partial z_d} - \frac{\partial\Gamma}{\partial z_{d-1}}, z_{2d}\frac{\partial\Gamma}{\partial z_d}\Bigg);$$

for U_{2d+1} :

$$\left(\frac{\partial\Gamma}{\partial z_{d+1}}, \ldots, \frac{\partial\Gamma}{\partial z_{2d}}, -\frac{\partial\Gamma}{\partial z_1}, \ldots, -\frac{\partial\Gamma}{\partial z_d}\right).$$

3. The flow at infinity. In the following theorem we study how the energy levels $E_h = H^{-1}(h)$ are transformed into the Poincaré sphere S^n. To do this we start by expressing the compactified vector field restricted to infinity in terms of the highest degree homogeneous part of H. Then we show that the boundary of the image of any energy level under the central projection is the same subset of the equator S^{n-1} of the Poincaré sphere, denoted by E^∞. Moreover, we give sufficient conditions for E^∞ to be a smooth $(n-2)$-dimensional manifold, and we show that all infinite critical points are in E^∞.

In what follows for any $h \in \mathbb{R}$ we will consider our Hamiltonian H of degree $m+1$ in the form

$$H = h + H_1 + \ldots + H_{m+1}$$

where H_i is the homogeneous part of H of degree i for $i = 1, 2, \ldots, m+1$. Notice that with this formulation it is sufficient to study the energy level $H = 0$ for studying any arbitrary level of the original Hamiltonian system. Recall that we have identified $\mathbb{R}^{2d}$ with the hyperplane $\Pi = \{y \in \mathbb{R}^{n+1} \mid y_{m+1} = 1\}$ tangent to the Poincaré sphere S^n at the north pole.

THEOREM 3.1. *Let $H : \mathbb{R}^{2d} \to \mathbb{R}$ be a polynomial Hamiltonian of degree $m+1$ depending on $n = 2d$ variables $(y_1, \ldots, y_n)$, X_H the Hamiltonian vector field associated to H on Π. Let $\widetilde{X}_H$ be the Poincaré compactification of X_H. Then the following statements hold:*
(a) The energy level $E_h = H^{-1}(0)$ is mapped diffeomorphically by the central projection ϕ^+ (respectively ψ^-) onto a subset E_h^+ (respectively E_h^-) of the northern (respectively southern) hemisphere. The sets E_h^+ and E_h^- are invariant under the flow of $\widetilde{X}_H$.
(b) Let $S^{n-1} = S^n \cap \{y_{n+1} = 0\}$. Then the restriction $\widetilde{X}_H|S^{n-1}$ is given by

$$\left(\frac{\partial H_{m+1}}{\partial y_{d+1}} + \lambda y_1, \ldots \frac{\partial H_{m+1}}{\partial y_{2d}} + \lambda y_d, \frac{-\partial H_{m+1}}{\partial y_1} + \lambda y_{d+1}, \ldots \frac{-\partial H_{m+1}}{\partial y_d} + \lambda y_{2d} \right)$$

where

$$\lambda = \sum_{i=1}^{d} \left(y_{i+d} \frac{\partial H_{m+1}}{\partial y_i} - y_i \frac{\partial H_{m+1}}{\partial y_{i+d}} \right).$$

(c) The flow of $\widetilde{X}_H$ leaves the subset

$$E^\infty = \{y \in S^{n-1} \mid H_{m+1}(y_1, \ldots, y_n) = 0\}$$

invariant.
(d) The boundary of E_h^+ (respectively E_h^-) is contained in E^∞.
(e) If 0 is a regular value of $H_{m+1}|S^{n-1}$, then E^∞ is an $(n-2)$-dimensional smooth manifold.
(f) The critical points of $\widetilde{X}_H|S^{n-1}$ are contained in E^∞.

Proof: Statement (a) follows easily from the definition of the Poincaré compactification (see Sections 1 and 2). We know from the Poincaré compactification that the infinity $S^{n-1} \subset \{y_{n+1} = 0\}$ is invariant under the

flow of $\widetilde{X}_H$. Therefore, (b) follows easily from Proposition 2.1. To prove
(c) it is sufficient to show that the energy level $H_{m+1} = 0$ is invariant under
the flow of $\widetilde{X}_H|S^{n-1}$ and this follows in a similar way as in the proof of
Proposition 2.1(b). Since $\{H^* = 0\} \cap S^{n-1} = E^\infty$, from Proposition 2.1(b)
and the definition of E^∞ it follows (d). Since $E^\infty = \left(H_{m+1}|S^{n-1}\right)^{-1}(0)$
from the Preimage Theorem (see [GP]), (e) follows. The critical points of
$\widetilde{X}_H|S^{n-1}$ must satisfy the system

$$\frac{\partial H_{m+1}}{\partial y_{d+i}} + \lambda y_i \quad = \quad 0,$$

$$\frac{\partial H_{m+1}}{\partial y_i} - \lambda y_{d+i} \quad = \quad 0,$$

for $i = 1, 2, \ldots, d$. Multiplying the first equation by y_{d+i}, the second by y_i,
adding with respect to i and by using Euler's Theorem for homogeneous
functions it follows $(m+1)H_{m+1} = 0$. Therefore, any critical point of
$\widetilde{X}_H|S^{n-1}$ is in E^∞. Hence (f) is proved.

$\blacksquare$

4. Generic properties. Let $H = h + H_1 + \ldots + H_{m+1}$ be a polynomial
Hamiltonian. We have shown in the previous sections that the zero energy
level of H^* intersects the equator of the Poincaré sphere in the same set
E^∞ for any value of h.

In this section we want to show first that generically E^∞ is a smooth
manifold. Afterwards we will prove also generically that all finite and
infinite critical points of $\widetilde{X}_H$ are hyperbolic.

Let $\mathcal{H}^{m+1}$ be the set of all polynomial Hamiltonians of degree exactly
$m+1$ in the variables $(y_1, \ldots, y_n)$ with $n = 2d$. We identify $\mathcal{H}^{m+1}$ with an
open subset of $\mathbb{R}^N$ where N is the maximal number of coefficients of the
polynomials of $\mathcal{H}^{m+1}$. Explicitly, let $\phi : \mathcal{H}^{m+1} \to \mathbb{R}^N$ be a map associating
to each polynomial Hamiltonian of degree $m+1$, its vector of coefficients in
a given fixed order; then the polynomial H is identified with $\phi(H) \in \mathbb{R}^N$.
We cannot identify $\mathcal{H}^{m+1}$ with the whole $\mathbb{R}^N$ because the polynomials of
$\mathcal{H}^{m+1}$ have degree at most $m+1$, and at least one of the coefficients of
degree $m+1$ is non zero. We endow $\mathcal{H}^{m+1}$ with the Euclidean topology of
$\mathbb{R}^N$ usually called the *coefficient topology*.

Let $\mathcal{G}^{m+1} = \{H \in \mathcal{H}^{m+1} \mid E^\infty \text{ is a smooth manifold}\}$. We want to
prove that $\mathcal{G}^{m+1}$ is open an dense in $\mathcal{H}^{m+1}$ with the coefficient topology.
In fact we will prove that its complement is contained in an algebraic
hypersurface of $\mathbb{R}^N$. For doing this we need the notion of the resultant of
two polynomials.

Let P and Q be two polynomials in the variable x and coefficients in
$\mathbb{R}$. Then the resultant of P and Q, denoted by $Res_x(P, Q)$, is a polynomial
in the coefficients of P and Q such that if P and Q have a common root
then $Res_x(P, Q)$ is zero.

Let us consider the two polynomials of degree n and m respectively, i.e.

$$P(x) = a_0 x^n + a_1 x^{n-1} + \cdots + a_{n-1} x + a_n,$$
$$Q(x) = b_0 x^m + b_1 x^{m-1} + \cdots + b_{m-1} x + b_m,$$

with a common real root x^*.

Then by multiplying the first equation succesively by $x^{m-1}, x^{m-2}, \ldots,$ x^2, x and the second by $x, x^2, \ldots, x^{n-1}$ we get the linear system of equations

$$a_0 x^{n+m-1} + a_1 x^{n+m} + \cdots + a_{n-1} x^m + a_n x^{m-1} = 0,$$
$$\vdots$$
$$a_0 x^{n+1} + a_1 x^n + \cdots + a_{n-1} x^2 + a_n x = 0,$$
$$a_0 x^n + a_1 x^{n-1} + \cdots + a_{n-1} x + a_n = 0,$$
$$b_0 x^m + b_1 x^{m-1} + \cdots + b_{m-1} x + b_m = 0,$$
$$b_0 x^{m+1} + b_1 x^m + \cdots + b_{m-1} x^2 + b_m x = 0,$$
$$\vdots$$
$$b_0 x^{m+n-1} + b_1 x^{m+n} + \cdots + b_{m-1} x^n + b_m x^{n-1} = 0.$$

The determinant of the system is called the resultant of P and Q. Since there exists the nontrivial solution $(1, x^*, \ldots, (x^*)^{n+m-1})$ the resultant is zero, see [B].

THEOREM 4.1. *There exists a polynomial function* $\mathcal{R}: \mathbb{R}^N \to \mathbb{R}$ *such that if* $H \in \mathcal{H}^{m+1} \setminus \mathcal{G}^{m+1}$, *then* $\mathcal{R}(\phi(H)) = 0$.

Proof: If $H \notin \mathcal{G}^{m+1}$ then $\left(H_{m+1} | S^{n-1} \right)^{-1} (0)$ has a critical point $(y_1^*, y_2^*, \ldots, y_n^*)$ (otherwise by the Implicit Function Theorem E^∞ would be a smooth manifold). It follows that the system of equations with a Lagrange multiplier μ

$$\frac{\partial H_{m+1}}{\partial y_i}(y_1^*, y_2^*, \ldots, y_n^*) + \mu y_i^* = 0,$$

for $i = 1, \ldots, n$ and

$$\sum_{i=1}^n (y_i^*)^2 - 1 = 0$$

has a non trivial solution. By multiplying the ith equation by y_i^* and adding together we get

$$\sum_{i=1}^n y_i^* \frac{\partial H_{m+1}}{\partial y_i}(y_1^*, y_2^*, \ldots, y_n^*) + \mu \sum_{i=1}^n (y_i^*)^2 = 0.$$

Then, from the homogeneity of H_{m+1} and using that the critical point is in $S^{n-1} \cap H_{m+1}^{-1}(0)$, we get that $\mu = 0$. Therefore we get a set of homogeneous polynomial equations

$$(4.1) \qquad \frac{\partial H_{m+1}}{\partial y_i}(y_1^*, y_2^*, \ldots, y_n^*) = 0,$$

which for $i = 1, \ldots, n$, has a non trivial solution $(y_1^*, y_2^*, \ldots, y_n^*) \in S^{n-1}$.

Let $L_k = \partial H_{m+1}/\partial y_k$ and define inductively

$$
\begin{aligned}
R_k(y_3, \cdots, y_n) &= \operatorname{Res}_{y_2}\left(L_1(1, y_2, y_3, \cdots, y_n), L_k(1, y_2, y_3, \cdots, y_n)\right), \\
&\qquad k = 2, \cdots, n \\
R_{2k}(y_4, \cdots, y_n) &= \operatorname{Res}_{y_3}\left(R_2(y_3, y_4, \cdots, y_n), R_k(y_3, y_4, \cdots, y_n)\right), \\
&\qquad k = 3, \cdots, n \\
R_{23k}(y_5, \cdots, y_n) &= \operatorname{Res}_{y_4}\left(R_{23}(y_4, y_5, \cdots, y_n), R_{2k}(y_4, y_5, \cdots, y_n)\right), \\
&\qquad k = 4, \cdots, n \\
&\vdots \\
R_{23 \cdots n} &= \operatorname{Res}_{y_n}\left(R_{23\cdots(n-1)}(y_n), R_{23\cdots(n-2)n}(y_n)\right),
\end{aligned}
$$

where Res_{y_j} is the resultant of polynomials with respect to the variable y_j. We want to show that there exists a polynomial function $\mathcal{R} : \mathbb{R}^N \to \mathbb{R}$ such that if the polynomials L_k have a common real root, then $\mathcal{R} = 0$.

At the final step $R_{23\cdots n}$ is just a number which is a polynomial function of the coefficients of the polynomial H. We claim that if the solution $(y_1^*, \ldots, y_n^*)$ has $y_1^* \neq 0$ then $R_{23\cdots n}$ must be zero. For if $y_1^* \neq 0$ then by homogenity of L_k it follows that if $\bar{y}_k = y_k^*/y_1^*$, for $k = 2, \ldots, n$, then the systems $L_1(1, \xi, \bar{y}_3, \cdots, \bar{y}_n) = 0$ and $L_k(1, \xi, \bar{y}_3, \cdots, \bar{y}_n) = 0$ for $k = 2, \cdots n$ have the common solution $\xi = y_2^*/y_1^*$. Therefore $R_k(\bar{y}_3, \cdots, \bar{y}_n) = 0$, for $k = 2, \cdots n$. Again, considering the polynomials $R_k(\xi, \bar{y}_4, \cdots, \bar{y}_n)$, they have the common root $\xi = \bar{y}_3$, therefore $R_{2k}(\bar{y}_4, \cdots, \bar{y}_n) = 0$. In this way we obtain that $R_{23\cdots n} = 0$. In general we rename $R_{23\cdots n}$ as $\mathcal{R}_1$, which is not necessarily zero if $y_1^* = 0$.

Now if the solution has $y_j \neq 0$ we just replace $L_k(1, y_2, \cdots, y_n)$ by $L_k(y_1, \cdots, 1, \cdots, y_n)$ where the 1 appears in the jth position in the definition of R_k; then we proceed as above to get $\mathcal{R}_j$. Finally defining $\mathcal{R} = \Pi_{j=1}^n \mathcal{R}_j$ we have that for a non trivial solution of system (4.1) one must have $\mathcal{R} = 0$. Now this polynomial depends only on the coefficients of the highest degree terms of H, but trivially it can be considered as a polynomial in all the coefficients. This completes the proof. ∎

From Theorem 4.1 it follows that the complement of $\mathcal{G}^{m+1}$ in $\mathcal{H}^{m+1}$ is contained in an algebraic hypersurface. In particular, it follows that $\mathcal{G}^{m+1}$ contains an open and dense set in $\mathcal{H}^{m+1}$.

Now we will show that for generic polynomial Hamiltonians the compactification of their corresponding Hamiltonian systems have all their finite and infinite critical points hyperbolic. This result follows inmediately from the next theorem.

LEMMA 4.1. *Given the real polynomial $q(x)$, define the real polynomials $q_1(x)$ and $q_2(x)$ by $q(ix) = q_1(x) + iq_2(x)$, where $i = \sqrt{-1}$. If $q(x)$ has a pure imaginary root ib then the resultant $R(q_1, q_2) = 0$.*

Proof: Since $q(ib) = 0$, then $q_1(b) = q_2(b) = 0$, therefore the resultant of q_1 and q_2 must be zero.

∎

Let $\mathcal{S}^{m+1}$ be the set $\{ H \in \mathcal{H}_{m+1} \mid$ all critical points of the compactified Hamiltonian flow associated to H are hyperbolic $\}$. We will show that $\mathcal{H}^{m+1} \setminus \mathcal{S}^{m+1}$ is contained in an algebraic hypersurface.

THEOREM 4.2. *If $H \in \mathcal{H}^{m+1} \setminus \mathcal{S}^{m+1}$, then there exists a polynomial function $\widetilde{\mathcal{R}} : \mathbb{R}^N \to \mathbb{R}$ such that $\widetilde{\mathcal{R}}(\phi(H)) = 0$.*

Proof: Let us consider the equations which define the critical points of the compactified Hamiltonian flow associated to H in the local chart (U_1, ϕ_1) with coordinates $(z_1, z_2, \ldots, z_n)$. Let $q(\zeta)$ be the characteristic polynomial at a critical point, and $q_1(\zeta)$, $q_2(\zeta)$ be as in Lemma 4.1, then if $H \in \mathcal{H}^{m+1} \setminus \mathcal{S}^{m+1}$ the set of equations

$$
\begin{aligned}
z_1' &= Z_1(z_1, z_2, \ldots, z_n) &&= 0, \\
z_2' &= Z_2(z_1, z_2, \ldots, z_n) &&= 0, \\
&\;\;\vdots \\
z_n' &= Z_n(z_1, z_2, \ldots, z_n) &&= 0, \\
&\ q_1(z_1, z_2, \ldots, z_n, \zeta) &&= 0, \\
&\ q_2(z_1, z_2, \ldots, z_n, \zeta) &&= 0,
\end{aligned}
$$

has a common real solution (use again the above Lemma 4.1). These equations are polynomials in the coefficients of the original Hamiltonian function H. Using a similar argument as in Theorem 4.1 it follows that there exist a polynomial function of the coefficients such that $\widetilde{\mathcal{R}}_1 = 0$. Proceeding similarly with the rest of the charts we get $\widetilde{\mathcal{R}}_j$, for $j = 1, 2, \ldots, n+1$. Then defining $\widetilde{\mathcal{R}} = \Pi_{j=1}^{n+1} \widetilde{\mathcal{R}}_j$, we have that if $H \in \mathcal{H}^{m+1} \setminus \mathcal{S}^{m+1}$ then $\widetilde{\mathcal{R}} = 0$. This completes the proof.

∎

Let us observe that in fact, both Theorems 4.1 and 4.2 were proved in local coordinates.

5. Application to some particular cases. In this section we will analyze two simple cases of polynomial Hamiltonians. In the first case we will consider one degree of freedom but arbitrary degree in the Hamiltonian. In the second case we will study the case of a single term (monomial) in the homogeneous part of highest degree. In both cases a fairly complete description of the behavior of the vector field near or at infinity can be given. Although the last case is non generic it appears frequently in applications to Celestial Mechanics.

5.1. Behavior near infinity in the planar case. Let $H^*(y_1, y_2, y_3)$ be the generating function of a polynomial Hamiltonian with one degree of freedom and degree $m+1$. In the Poincaré sphere S^2 where it is defined, we will take the cylindrical projection (θ, y_3) where $y_1 = \sqrt{1 - y_3^2}\cos\theta$, $y_2 = \sqrt{1 - y_3^2}\sin\theta$. Let

$$R(\theta, y_3) = H^*\left(\sqrt{1 - y_3^2}\cos\theta, \sqrt{1 - y_3^2}\sin\theta, y_3\right).$$

Then, from the general equations for $\widetilde{X_H}$ (see Proposition 2.1), we obtain the equations associated to the compactified vector field in these coordinates. Indeed,

$$\theta' = \frac{y_1 y_2' - y_2 y_1'}{y_1^2 + y_2^2} = -\frac{1}{1 - y_3^2}\left(y_1\frac{\partial H^*}{\partial y_1} + y_2\frac{\partial H^*}{\partial y_2}\right).$$

By the chain rule,

$$\frac{\partial R}{\partial y_3} = -\frac{y_1 y_3}{1 - y_3^2}\frac{\partial H^*}{\partial y_1} - \frac{y_1 y_3}{1 - y_3^2}\frac{\partial H^*}{\partial y_2} + \frac{\partial H^*}{\partial y_3}.$$

We compute the following expression and use Euler's theorem on the right hand side, since H^* is homogeneous, getting

$$y_3\frac{\partial R}{\partial y_3} - (m+1)R = -\frac{1}{1 - y_3^2}\left(y_1\frac{\partial H^*}{\partial y_1} + \frac{\partial H^*}{\partial y_2}\right).$$

Hence, we have

$$\theta' = -(m+1)R(\theta, y_3) + y_3\frac{\partial R}{\partial y_3},$$

$$y_3' = -y_3\frac{\partial R}{\partial \theta}.$$

We have the following result for this case.

PROPOSITION 5.1. *The critical points at infinity for one degree of freedom coincides with E^∞. Moreover,*
(a) *The critical points are given by the roots of the 2π-periodic function $R(\theta, 0)$.*
(b) *If $(\theta^*, 0)$ is a critical point then $(\theta^* + \pi, 0)$ is also a critical point. Hence the critical points appear by pairs.*
(c) *If θ^* is a simple root of $R(\theta, 0)$ then the critical point is hyperbolic, in fact a node. In this case we have two possibilities.*
   (1) *If $m + 1$ is even and $(\theta^*, 0)$ is a stable node then $(\theta^* + \pi, 0)$ is also a stable node.*
   (2) *If $m + 1$ is odd then there exist at least two critical points. If $(\theta^*, 0)$ is a stable node then $(\theta^* + \pi, 0)$ is an unstable node and viceversa.*

Proof: Since E^∞ is invariant and has dimension zero it must consist of critical points. Since E^∞ has $y_3 = 0$, then the critical points are given by the equation $R(\theta, 0) = 0$. This proves (a). Notice that by homogeneity of H^*, we have

$$H^*(y_1, y_2, 0) = (-1)^{m+1} H^*(-y_1, -y_2, 0),$$

and therefore $R(\theta, 0) = (-1)^{m+1} R(\theta + \pi, 0)$. This gives (b).

In order to prove (c) we linearize at a critical point $(\theta^*, 0)$. The jacobian matrix is

$$\begin{pmatrix} -(m+1)\frac{\partial R}{\partial \theta}(\theta^*, 0) & * \\ 0 & -\frac{\partial R}{\partial \theta}(\theta^*, 0) \end{pmatrix}.$$

Then the eigenvalues are $-(m+1)\partial R/\partial\theta(\theta^*, 0)$ and $-\partial R/\partial\theta(\theta^*, 0)$. Therefore if θ^* is a simple root of $R(\theta, 0)$ then the eigenvalues have the same sign, and the critical point is a stable or unstable node.

From the relation $R(\theta, 0) = (-1)^{m+1} R(\theta + \pi, 0)$, we get

$$\frac{\partial R}{\partial \theta}(\theta^*, 0) = (-1)^{m+1} \frac{\partial R}{\partial \theta}(\theta^* + \pi, 0).$$

Therefore the stability at the symmetric critical points is the same if $m+1$ is even or is reversed if $m+1$ is odd. This completes the proof. ∎

In fact, the structurally stable plane Hamiltonian polynomial vector fields have been characterized in [JL].

5.2. Behavior at infinity in the monomial case. Given the polynomial Hamiltonian H, suppose its highest homogeneous part has the form

$$H_{m+1}(y_1, \ldots, y_n) = y_1^{a_1} y_2^{a_2} \cdots y_n^{a_n},$$

where $a_j \geq 0$ for $j = 1, 2, \ldots, n$, and $a_1 + a_2 + \ldots + a_n = m + 1$. We will give a complete description of the topological structure of E^∞ and the invariant flow there. First, let

$$S_j^{n-2} = \{(y_1, \ldots, y_n) \in S^{n-1} \mid y_j = 0\},$$

then we can state:

PROPOSITION 5.2. *Suppose H_{m+1} is a single monomial, then*
(a) $E^\infty = \bigcup_{a_j \geq 1} S_j^{n-2}$.
(b) *If $a_j \geq 2$ then S_j^{n-2} consists entirely of critical points.*
(c) *If $a_j, a_k \geq 1$ for $j \neq k$ and $n \geq 4$, then $S_j^{n-2} \cap S_k^{n-2}$ is a $(n-3)$-sphere of critical points.*
(d) *If $a_j = 1$ then on S_j^{n-2} there are no other critical points than those described in (c). If $a_{j+d} \geq 1$ then any solution starts in some sphere of critical points $S_j^{n-2} \cap S_k^{n-2}$ and ends in $S_j^{n-2} \cap S_{j+d}^{n-2}$. If $a_{j+d} = 0$, then any solution starts and ends in some sphere of critical points $S_j^{n-2} \cap S_k^{n-2}$.*

Proof: The set E^∞ is defined by the equations $H_{m+1}(y_1, \ldots, y_n) = 0$ and $y_1^2 + \cdots + y_n^2 = 1$. Since H_{m+1} contains explicitly only factors y_j for $a_j > 0$, part (a) follows easily. Suppose $a_j \geq 2$. Then in the computation of $\partial H_{m+1}/\partial y_k$ there exists some positive power of y_j, therefore, on setting $y_j = 0$ in the global equations

$$\dot{y}_i = \frac{\partial H_{m+1}}{\partial y_{i+d}} + \lambda y_i,$$

$$\dot{y}_{i+d} = -\frac{\partial H_{m+1}}{\partial y_i} + \lambda y_{i+d},$$

for $i = 1, 2, \ldots, d$, all right hand sides vanish. This means S_j^{n-2} consists of critical points, and consequently (b) is proved. For proving (c), a similar argument applies. We omit the details. Now suppose $a_j = 1$ and, without loss of generality, that $1 \leq j \leq d$; then it is readily seen that S_j^{n-2} is invariant, and the global equations restricted to S_j^{n-2} reduce to

$$\dot{y}_{j+d} = (-1 + y_{j+d}^2) \frac{\partial H_{m+1}}{\partial y_j},$$

$$\dot{y}_k = y_{j+d} y_k \frac{\partial H_{m+1}}{\partial y_j}, \quad \text{for } k \neq j + d.$$

For any initial condition not contained in the sphere of critical points $S_j^{n-2} \cap S_k^{n-2}$, we can reparametrize the solution by the common factor $(\partial H_{m+1}/\partial y_j)^{-1}$. Then the system becomes

$$y'_{j+d} = -1 + y_{j+d}^2,$$
$$y'_k = y_{j+d} y_k, \quad \text{for } k \neq j + d,$$

where the prime means derivative with respect to the new time. From the first equation above, we have that y_j decreases monotonically from $y_{j+d} = 1$ to $y_{j+d} = -1$, passing through $y_{j+d} = 0$. In the interval of values $-1 < y_{j+d} < 0$, no other coordinate y_k vanishes, therefore the solutions dies in $y_{j+d} = 0$ if the exponent $a_{j+d} \geq 1$, or passes through and dies in $y_k = 0$ if $a_{j+d} = 0$. This proves (d).

REFERENCES

[B]   S. BARNETT, *Matrices in Control Theory*, Krieger Publ., Florida, 1984.

[CL]   A. CIMA AND A. J. LLIBRE, *Bounded polynomial vector fields,* Trans. Amer. Math. Soc. **318** (1990), 557-579.

[DLLP1] J. DELGADO, E.A. LACOMBA, J. LLIBRE AND E. PÉREZ, *Poincaré compactification of the collinear three body problem.* To appear in the Proceedings of the International Symposium on Hamiltonian Systems and Celestial Mechanics, Guanajuato, México, 1991.

[DLLP2] J. DELGADO, E.A. LACOMBA, J. LLIBRE AND E. PÉREZ, *Poincaré compactification of the Kepler and the collinear three body problem.* To appear in the Proceedings of the 2nd Workshop on Dynamical Systems of the Euler Institute, Saint Petersburg, Russia, 1991.

[GP]   V. GUILLEMIN AND A. POLLACK, *Differential topology.* Prentice-Hall, 1974.

[Heg] D. C. HEGGIE, *A global regularization of the gravitational n-body problem,* Cel. Mech. **10** (1974), 217-241.

[JL]   X. JARQUE AND J. LLIBRE, Structural stability of planar Hamiltonian polynomial vector fields, preprint.

TRANSVERSE HOMOCLINIC CONNECTIONS FOR GEODESIC FLOWS

VICTOR J. DONNAY*

Abstract. Given a two dimensional Riemannian manifold for which the geodesic flow has a homoclinic (heteroclinic) connection, we show how to make a C^2 small perturbation of the metric for which the connection becomes transverse. We apply this result to several examples.

1. Introduction. The existence of a transverse heteroclinic or homoclinic connection is one of the basic mechanisms by which complicated motion is produced in a dynamical system. The discovery of such connections and their effect on dynamics dates to Poincaré. In modern language, a transverse heteroclinic connection implies the existence of a horseshoe, which in turn implies that on a subset of the phase space, the dynamics are equivalent to a symbol shift on N symbols, has infinitely many periodic orbits and has positive topological entropy [S],[M].

Connections arise in the following way. Associated to a hyperbolic periodic orbit are stable and unstable manifolds. When the unstable manifold of one periodic orbit intersects the stable manifold of another periodic orbit, we have a heteroclinic connection between the periodic orbits. If the manifolds belong to the same periodic orbit, then the connection is called homoclinic. In some cases, the stable and unstable manifolds may actually be equal to one another. Thus they intersect, but not transversely. They are said to form a separatrix. This situation is very special, arising for instance in certain integrable systems. One expects that a "typical" perturbation to the system will cause the intersection to become transverse. Much work has been done proving this result for various classes of systems.

In this note, we examine geodesic flows on surfaces and show

THEOREM 1.1. *Given a smooth two-dimensional Riemannian manifold whose geodesic flow has a homoclinic (heteroclinic) connection, one can construct a smooth metric, C^2 close to the original metric, for which the connection is transverse.*

Our construction involves a very explicit, localized perturbation to the metric. Applying this theorem to the case of an ellipsoid with three distinct axis, we can find surfaces with positive Gaussian curvature whose geodesic flow has positive topological entropy (see Example 1). This result was first obtained by Knieper and Weiss [KW] (see also [We]) who used a version of the Melnikov method. They made a perturbation of the ellipsoid which produced a topological intersection of stable and unstable manifolds with a two sided crossing. This type of intersection is weaker than a transverse crossing but still implies positive topological entropy.

---

* Department of Mathematics, Bryn Mawr College, Bryn Mawr, PA 19010. Partially supported by NSF grant DMS 9123856.

In §4, we give several examples to which our construction applies.

2. Background. Given a surface M with Riemannian metric g, the geodesic flow ϕ^t occurs in the unit tangent bundle $X = SM$. A point $z = (p, v) \in X$ has two components: a base point $p = \pi(z) \in M$, where π is the canonical projection, and unit vector $v \in S_p M$. We denote by γ_z the associated geodesic on the surface: $\gamma_z(t) = \pi(\phi^t z)$. A point $z \in X$ is periodic if $\phi^t z = z$ for some $t > 0$. A periodic point is hyperbolic if, when one takes a surface of section through this point and transverse to the flow, the induced return map is hyperbolic at the fixed point.

The stable and unstable manifolds of a periodic hyperbolic point $z_a \in X$ are defined by

$$W^s(z_a) \;=\; \{z \in X : \lim_{t \to +\infty} \text{distance}(\phi^t z, \phi^t z_a) \to 0\}$$

$$W^u(z_a) \;=\; \{z \in X : \lim_{t \to -\infty} \text{distance}(\phi^t z, \phi^t z_a) \to 0\}.$$

If $z \in W^s(z_a)$ then the vector $\xi^s(z) \in T_z X$ tangent to $W^s(z_a)$ is called the stable vector at z. This vector has the property that $\lim_{t \to +\infty} \|D\phi^t \xi^s(z)\| = 0$. Along an orbit, the stable vectors form an invariant family of directions: i.e. $\phi^t z \in W^s(\phi^t z_a)$ and $\xi^s(\phi^t z) = D\phi^t \xi^s(z)$. Analogous results hold for the unstable vector $\xi^u(z)$ with the change that $\lim_{t \to -\infty} \|D\phi^t \xi^u(z)\| = 0$. The weak stable and weak unstable manifolds of z_a are defined by

$$W^{ws}(z_a) = \bigcup_{-\infty < t < \infty} \phi^t W^s(z_a), \qquad W^{wu}(z_a) = \bigcup_{-\infty < t < \infty} \phi^t W^u(z_a).$$

We define the weak stable and weak unstable manifolds of a periodic orbit to be equal to $W^{ws}(z_a)$ and $W^{wu}(z_a)$ for any point z_a in the orbit. Note that by flow invariance, this definition is independent of the choice of point.

The geodesic flow has a heteroclinic (homoclinic) connection if the weak unstable manifold of a hyperbolic periodic point z_a intersects the weak stable manifold of a hyperbolic periodic point z_b, $z_a \neq z_b$ ($z_a = z_b$). If the two manifolds are equal, then for $z \in W^{wu}(z_a) = W^{ws}(z_b)$, we have that $\xi^u(z) = \xi^s(z)$. We will show that the manifolds cross transversely by finding a point $z_0 \in W^{wu}(z_a) \cap W^{ws}(z_b)$ for which $\xi^u(z_0) \neq \xi^s(z_0)$.

We use Jacobi fields and the Jacobi equation to determine the evolution of stable and unstable vectors. For each $z \in X$, there is an orthonormal, right handed frame $\{\xi_v, \xi_{v\perp}, \xi_\psi\}$ for the tangent space $T_z X$ [E]. The vector $\xi_v = \frac{d}{dt}|_{t=0}\phi^t z$ lies in the flow direction, $\xi_{v\perp}$ lies in the horizontal subspace, and ξ_ψ lies in the fiber direction. The perpendicular subspace spanned by $\{\xi_{v\perp}, \xi_\psi\}$ is invariant: if a vector $\xi(z)$ lies in the perpendicular subspace of $T_z X$ then $D\phi^t \xi$ lies in the perpendicular subspace at $T_{\phi^t z} X$. If we write such a vector as

$$\xi = J\xi_{v\perp} + J'\xi_\psi$$

then its image vector is given by

$$D\phi^t \xi = J(t)\xi_{v\perp} + J'(t)\xi_\psi$$

where $(J(t), J'(t))$ is the solution of the Jacobi equation

(2.1) $$J''(t) + K(t)J(t) = 0$$

with initial conditions $(J(0) = J, J'(0) = J')$, where $K(t) = K(\gamma_z(t))$ is the Gaussian curvature along the geodesic γ_z. We call $(J(t), J'(t))$ the Jacobi field associated to the vector ξ.

The stable and unstable vectors lie in the perpendicular subspace and their coordinates relative to the basis $\{\xi_{v\perp}, \xi_\psi\}$ are denoted by $(J_s(z), J_s'(z))$, $(J_u(z), J_u'(z))$. The stable and unstable Jacobi fields are defined to be the solutions of (2.1) for which $\lim_{t\to+\infty} J(t) = 0$ and $\lim_{t\to-\infty} J(t) = 0$.

The stable and unstable vectors are only determined up to projective coordinate; $u_s = J_s'/J_s, u_u = J_u'/J_u$. Thus when we say $\xi^s(z) = \xi^u(z)$, we mean their projective coordinates are equal. The evolution of this projective coordinate is given by the Riccati equation

(2.2) $$u'(t) = -K(t) - u^2(t)$$

where $K(t) = K(\gamma_z(t))$ is the curvature along the geodesic γ_z. When $J(t^*) = 0$, the Riccati solution has a singularity but one can continue past this singularity by taking the solution of (2.2) which has $\lim_{t\to t^*-} u(t) = -\infty$, $\lim_{t\to t^*+} u(t) = +\infty$.

3. Main result. Henceforth, we assume that $z_a, z_b \in X$ are hyperbolic periodic points whose weak unstable and weak stable manifolds intersect with $W^{wu}(z_a) = W^{ws}(z_b)$.

To make the intersection transverse, we will perturb the metric in a neighbourhood U of one point $\pi(z_0)$ where z_0 lies in $W^{wu}(z_a) \cap W^{ws}(z_b)$. The base point $\pi(z_0)$ should be away from the periodic geodesics $\gamma_{z_a}, \gamma_{z_b}$, and the geodesic γ_{z_0} should pass through U only once. Our perturbation will keep this geodesic unchanged but will increase the curvature along the geodesic. As a result, z_0 will lie in $\widetilde{W}^{wu}(z_a) \cap \widetilde{W}^{ws}(z_b)$ for the perturbed system but the manifolds will intersect transversely there.

LEMMA 3.1. *There exists a point $z_0 \in W^{wu}(z_a) \cap W^{ws}(z_b)$ and an open neighbourhood $U(\pi(z_0) \subset M$ such that*
i. $\gamma_{z_a}, \gamma_{z_b} \subset M \setminus U$
ii. γ_{z_0} passes through U only once; i.e. there exists times $t_1 < 0 < t_2$ such that $\gamma_{z_0}(t_1) \in \partial U, \gamma_{z_0}(t_2) \in \partial U$ and $\gamma_{z_0}(t) \in U$ only for $t \in (t_1, t_2)$.
iii. The stable vector at $\phi^t z_0$ satisfies $J_s(\phi^t z_0) \neq 0, t \in (t_1, t_2)$.

Proof. In most specific examples, this result would be geometrically obvious. For the general case, we argue as follows. When a geodesic $\gamma \subset M$ intersects itself, it must do so transversely. After crossing itself, the geodesic can not immediately intersect itself a second time: it must travel at least a distance equal to the injectivity radius of M. Thus a geodesic can have at most countably many self-intersections.

If a geodesic γ is forward asymptotic to γ_b and backward asymptotic to γ_a, then all but a finite part of γ will lie within ϵ—tubes about γ_a and γ_b. Choosing ϵ sufficiently small, we can find a segment of γ outside these tubular neighbourhoods, and in this segment γ has at most finitely many self-intersections. We choose our point z_0 so that $\pi(z_0)$ lies in this segment of γ, avoids these finitely many points and has $J_s(z_0) \neq 0$. This last condition can be satisfied since the zeros of a non-trivial solution of the Jacobi equation (2.1) are isolated. A suitably small neighbourhood $U(\pi(z_0))$ will now satisfy our requirements. $\square$

LEMMA 3.2. *There exist smooth C^2 small perturbations $\widetilde{g}$ of the metric g such that*
i. $\widetilde{g} = g$ on $M \setminus U$.
ii. γ_{z_0} *remains a geodesic for $\widetilde{g}$.*
iii. $\widetilde{K}(\gamma_{z_0}(t)) \geq K(\gamma_{z_0}(t))$ *for $t \in (t_1, t_2)$, t_1, t_2 given in Lemma 3.1, and the inequality is strict at $t = 0$.*
iv. $\sup |\widetilde{K} - K|$ *can be made arbitrarily small.*

Proof. We use Fermi coordinates as described in ([K], p.108). We choose an orthonormal frame $\{E_0, E_1\}$ in $S_{\pi(z_0)}M$ with $E_0 = \dot{\gamma}_{z_0}(0)$ and denote by $\{E_0(t), E_1(t)\}$ the orthonormal frame in $S_{\gamma_{z_0}(t)}M$, $t \in [-t_0, t_0]$, obtained by parallel translation along γ_{z_0}. Fermi coordinates are defined by the map $\Phi : [-t_0, t_0] \times R \to M$ with

$$\Phi(t; x_1) \to exp_{\gamma_{z_0}(t)}(x_1 E_1(t))$$

which gives coordinates in a sufficiently small neighborhood of the origin $(0, 0) \subset (-t_0, t_0) \times R$.

In these coordinates, the metric g satisfies

$$g_{00}(t; 0) = 1, \quad g_{10}(t; x_1) = 0, \quad g_{11}(t; x_1) = 1,$$
$$\frac{\partial g_{00}(t;0)}{\partial x_1} = 0, \quad \frac{\partial^2 g_{00}(t;0)}{\partial x_1 \partial x_1} = -K(t; 0).$$

For a perturbed metric $\widetilde{g}$ of the form

$$\widetilde{g}_{00}(t; x_1) = g_{00}(t; x_1) + \alpha(t; x_1)x_1^2$$
$$\widetilde{g}_{01}(t; x_1) = g_{01}(t; x_1)$$
$$\widetilde{g}_{11}(t; x_1) = g_{11}(t; x_1),$$

γ_{z_0} remains a geodesic; the curve $(x_0(\tau) = \tau, x_1(\tau) = 0)$ satisfies the equation for a geodesic in the $\widetilde{g}$ metric

$$\ddot{x}_i(\tau) + \sum_{k,l} \widetilde{\Gamma}^i_{kl}(x(\tau))\, \dot{x}_k \dot{x}_l = 0$$

because $\widetilde{\Gamma}^i_{00}(t; 0) = 0, i = 0, 1$. The curves $(x_0(\tau) \equiv \text{constant}, x_1(\tau) = \tau)$ also satisfy these geodesic equations. Hence the coordinates (t, x_1) are still

Fermi coordinates for the perturbed metric. Therefore the curvature $\widetilde{K}$ of the $\widetilde{g}$ metric along the geodesic γ_{z_0} is given by

$$\widetilde{K}(t;0) = -\frac{\partial^2 \widetilde{g}_{00}(t;0)}{\partial x_1 \partial x_1} = K(t;0) - 2\alpha(t;0).$$

To insure that $\widetilde{K}(t;0) > K(t;0)$, we need only choose $\alpha(t;0) < 0$.

One simple way to do this, as well as to insure that α has support inside U, is to make α a smooth radially symmetric bump function: $\alpha(t;x_1) = f(\sqrt{t^2 + x_1^2})$ where f satisfies $f(0) < 0$, $\frac{\partial f}{\partial r}(r) > 0$, $r \in (0, R)$, $f(r) \equiv 0$, $r \geq R$, and $\frac{\partial^k f}{\partial r^k}(0) = \frac{\partial^k f}{\partial r^k}(R) = 0$, $k > 0$. Choose R so that $\{\Phi(t; x_1) : t^2 + x_1^2 \leq R^2\} \subset U$.

It remains to show that we can make $\|\widetilde{g} - g\|$ arbitrarily small in the C^2 norm which will prove (iv). To do this, we define a one-parameter family of metrics $\widetilde{g}_c$ using bump functions $f_c(r) = \frac{1}{c^2} f(cr)$. The support of f_c lies in the interval $[0, R/c)$. For these metrics, one can show that $\Delta g_c = \widetilde{g}_c - g = f_c(r)x_1^2$ satisfies

$$\lim_{c \to \infty} \|\Delta g_c\|_{C^2} = 0.$$

As an example of the type of calculation involved, we find that

$$
\begin{aligned}
\frac{\partial^2 \Delta g_c}{\partial x_1^2} &= \frac{\partial^2 f_c}{\partial r^2}\left(\frac{\partial r}{\partial x_1}\right)^2 x_1^2 + \frac{\partial f_c}{\partial r}\frac{\partial^2 r}{\partial x_1^2}x_1^2 + 4x_1\frac{\partial f_c}{\partial r}\frac{\partial r}{\partial x_1} + 2f_c \\
&= \frac{\partial^2 f}{\partial r^2}\left(\frac{\partial r}{\partial x_1}\right)^2 x_1^2 + \frac{1}{c}\frac{\partial f}{\partial r}\frac{\partial^2 r}{\partial x_1^2}x_1^2 + 4x_1\frac{1}{c}\frac{\partial f}{\partial r}\frac{\partial r}{\partial x_1} + \frac{2}{c^2}f.
\end{aligned}
$$

Since $|x_1| < R/c$, we have that

$$\lim_{c \to \infty} \frac{\partial^2 \Delta g_c}{\partial x_1^2} = 0$$

uniformly on U. $\square$

Using these lemmas, we can now prove our main result.

THEOREM 3.1. *The points z_a, z_b are hyperbolic and periodic for the perturbed flow $\widetilde{\phi}^t$ and $\widetilde{W}^{wu}(z_a) \cap \widetilde{W}^{ws}(z_b)$ intersect transversely at z_0.*

Proof. Since the perturbation was localized away from $\gamma_{z_a}, \gamma_{z_b}$ these geodesics are unchanged. Since the geodesic γ_{z_0} is also unchanged, it remains forward asymptotic to γ_{z_b} and backwards asymptotic to γ_{z_a}. Hence $z_0 \in \widetilde{W}^{wu}(z_a) \cap \widetilde{W}^{ws}(z_b)$.

To show that the manifolds intersect transversely at z_0, it is enough to show that the stable and unstable vectors $\widetilde{\xi}^s(z_0), \widetilde{\xi}^u(z_0)$ are not equal. By the flow invariance of the stable and unstable vector fields, this is equivalent to showing that the unstable vector $\widetilde{\xi}^u(z_1)$ at the point $z_1 = \phi^{t_1}z_0$ does not get sent to the stable vector $\widetilde{\xi}^s(z_2)$ at the point $z_2 = \phi^{t_2}z_0$.

The coordinates $(\tilde{J}_u(z_1),\ \tilde{J}'_u(z_1))$ of the unstable vector $\tilde{\xi}^u(z_1)$ are uniquely determined by the value of the curvature $\tilde{K}(\gamma_{z_1}(t))$, $t < 0$. Since this curvature is unchanged by the perturbation, $(\tilde{J}_u(z_1), \tilde{J}'_u(z_1)) = (J_u(z_1), J'_u(z_1))$. Similarly, the stable Jacobi field at z_2 is determined by the curvature $\tilde{K}(\gamma_{z_2}(t))$, $t > 0$ which is unchanged by the perturbation so that $(\tilde{J}_s(z_1), \tilde{J}'_s(z_2)) = (J_s(z_1), J'_s(z_2))$.

For the unperturbed flow, we know that

$$D\phi^{t_2-t_1}\xi^u(z_1) = \xi^s(z_2).$$

Expressed in terms of the Riccati equation, the solution of (2.2) that starts at z_1 with initial condition $u(0) = J'_u(z_1)/J_u(z_1)$ will have value $u(t_2-t_1) = J'_s(z_2)/J_s(z_2)$ when it evolves to z_2.

We claim that, since $\tilde{K}(\gamma_{z_1}(t)) \geq K(\gamma_{z_1}(t)), t \in (0, t_2 - t_1)$ with strict inequality at some points, the solution to

$$\tilde{u}'(t) = -\tilde{K}(\gamma_{z_1}(t)) - \tilde{u}^2(t)$$

with initial condition $\tilde{u}(0) = \tilde{J}'_u(z_1)/\tilde{J}_u(z_1) = J'_u(z_1)/J_u(z_1)$ will have final value $\tilde{u}(t_2 - t_1) \neq u(t_2 - t_1)$, and hence $D\phi^{t_2-t_1}\tilde{\xi}^u(z_1) \neq \tilde{\xi}^s(t_2)$.

To see this, we define $\Delta u(t) = \tilde{u}(t) - u(t)$ and $\Delta K(t) = \tilde{K}(\gamma_{z_1}(t)) - K(\gamma_{z_1}(t))$, then we have

$$(3.1) \qquad \Delta u'(t) = -\Delta u(t)\big(\tilde{u}(t) + u(t)\big) - \Delta K(t).$$

Note that Lemma 3.1 (iii) implies that $u(t)$ is finite for $t \in (0, t_2 - t_1)$ and hence, for $\tilde{g}$ sufficiently close to g, the same is also true for $\tilde{u}(t)$. The solution to equation (3.1) with initial condition $\Delta u(t) = 0$ is

$$\Delta u(t) = \exp\Big(-\int_0^t \tilde{u}(s) + u(s)\,ds\Big)\int_0^t -\Delta K(s)\exp\Big(\int_0^s \tilde{u}(v) + u(v)\,dv\Big)\,ds.$$

Hence $\Delta u(t_2 - t_1) < 0$ since $\Delta K(s) > 0$. $\square$

4. Examples

1. The geodesic flow on an ellipsoid with three distinct axes is an integrable system. Each axes furnishes a simple closed geodesic; the middle length one is hyperbolic. Knieper and Weiss [KW] noted that this geodesic furnishes a heteroclinic connection. Specifically if γ_{z_a} is this geodesic traversed in one direction and γ_{z_b} this geodesic traversed in the opposite direction, then geodesics that are forward asymptotic to γ_b will be backwards asymptotic to γ_a. Lifting this statement to the unit tangent bundle, we get that $W^{wu}(z_a) = W^{ws}(z_b)$. Applying our theorem to this case, we get a metric $\tilde{g}$ of positive Gaussian curvature for which the heteroclinic connection is transverse. The standard techniques [M] imply positive topological entropy.

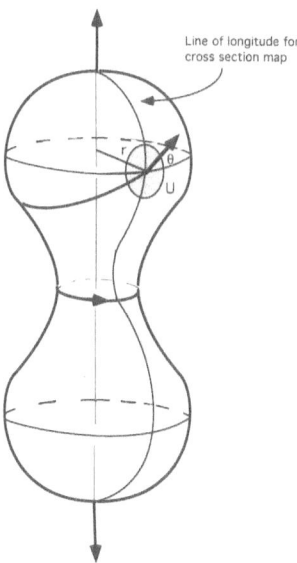

Line of longitude for
cross section map

FIG. 4.1. *Rotationally symmetric sphere with hyperbolic closed geodesic.*

2. Take a surface of revolution (Fig. 4.1) which is pinched in the middle giving a hyperbolic closed orbit. On a surface of revolution, the geodesic flow is integrable with an integral of motion given by the Clairaut Integral: $r \cos \theta$ stays constant along trajectories, where r is the radial distance of the point on the surface from the axis of revolution and θ is the angle the geodesic makes with the horizontal curve [Do]. If the hyperbolic closed geodesic is at a distance r_1 from the axis, then any orbit whose Clairaut Integral equals r_1 will be both forward and backward asymptotic to the close orbit. In the unit tangent bundle, the lifts of these geodesics lie in both the weak stable and weak unstable manifold of the periodic orbit: we have a homoclinic connection.

In (Fig. 4.2), we draw the return map to a cross section for the geodesic flow. Our cross section consists of points in phase space whose basepoints lie on a given line of longitude and whose direction vector makes an angle $\theta \in (-\pi/2, \pi/2)$ with the horizontal. The periodic orbit induces a fixed point for this return map. The stable and unstable manifolds of this fixed point are equal thereby producing the homoclinic connection. Note that there are actually two homoclinic connections: one for the geodesics that are in the top half of the sphere and one for geodesics in the bottom half.

If we now apply our perturbation in a neighborhood U of a point in the upper half of the sphere, we can produce a transverse crossing in the upper connection (Fig. 4.3). The lower connection is unaffected by this perturbation.

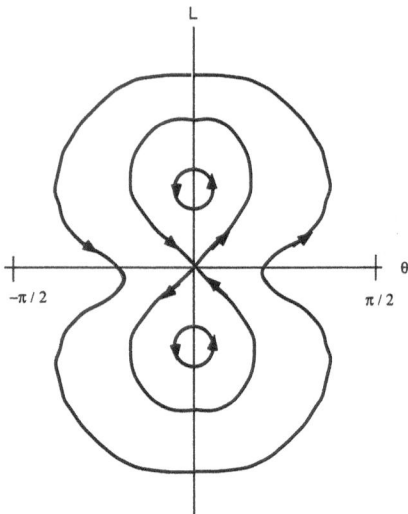

FIG. 4.2. *Homoclinic connection for integrable system.*

An interesting feature of our perturbed connection is that, although the stable and unstable manifolds cross transversely at some points, at other points the two manifolds are equal. This phenomena occurs because the perturbation was localized and hence there are points in the stable and unstable manifolds of the fixed point that never enter the perturbed region. To these points, it is as if the perturbation did not occur, and so their stable and unstable vectors remain equal.

3. In [D1], the author gave an example of a smooth metric on the two-torus, the "light bulb", (Fig. 4.4) for which the geodesic flow has both an ergodic component of positive measure on which the measure theoretic entropy is positive (i.e. positive Lyapunov exponents on a set of positive measure) and an integrable component on which the Lyapunov exponents are zero.

Taking a cross section map (Fig. 4.5) as in example 2, we see that this system has an elliptic periodic orbit corresponding to the closed geodesic going around the thick part of the light bulb. Surrounding the elliptic orbit is an elliptic island completely foliated by invariant curves. This is the integrable region of phase space. The boundary of this set consists of a homoclinic connection formed by the stable and unstable manifolds of the hyperbolic periodic orbit at the neck of the light bulb. These manifolds consist of orbits on the light bulb that are both forward and backwards asymptotic to the closed geodesic. Finally outside the connection are trajectories that enter the light bulb from the flat region and then leave the light bulb. These trajectories define an invariant region of phase space and

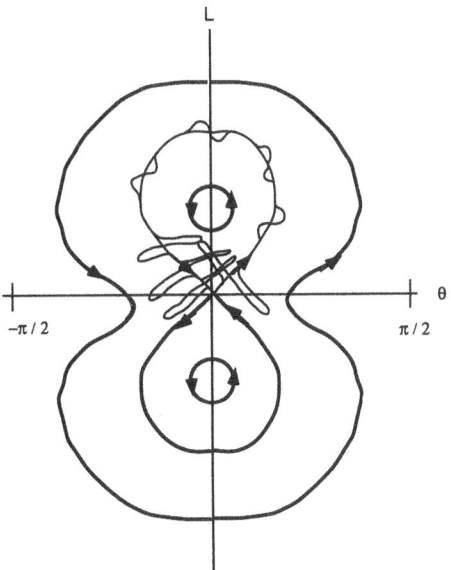

FIG. 4.3. *Transverse homoclinic connection.*

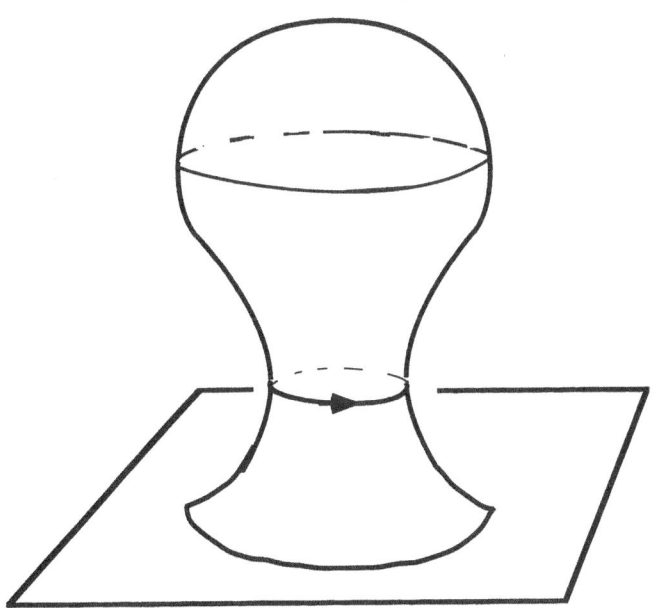

FIG. 4.4. *Flat torus with light bulb.*

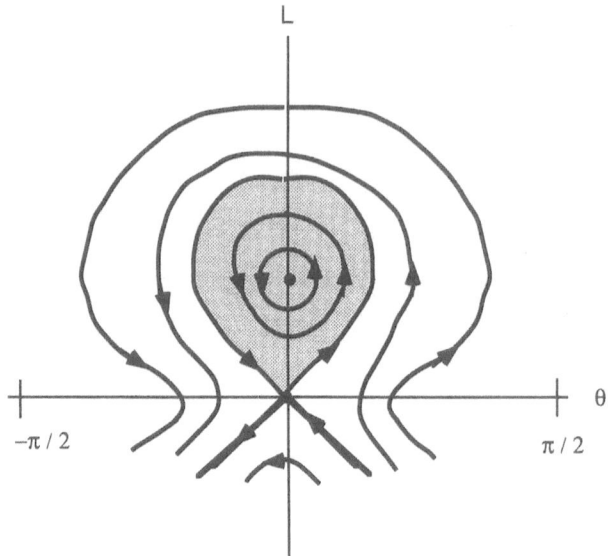

FIG. 4.5. *Homoclinic connection separating ergodic and integrable regions.*

in this region the dynamics are ergodic with positive Lyapunov exponent almost everywhere [D1] [D2] [BG]. Do not be deceived by (Fig. 4.5) which seems to indicate that the trajectories outside the homoclinic connection also lie on invariant curves. These curves are only invariant as long as the trajectory remains in the bulb. After leaving the bulb, passing through the flat region and then returning to the bulb, a trajectory will lie on a different one of these curves.

This system is one of only a very few Hamiltonian systems known (see also [P],[W]) to exhibit coexistence; that is the existence of both regions with ergodic motion and regions with stable motion. In these examples, the reason coexistence can be proven is because of the existence of a separatrix, formed by a homoclinic or heteroclinic connection, that separates the ergodic region from the stable region. In a more typical example, such as the standard map, one finds chaotic regions and stable regions intertwined in a very complicated way. At present, no one has been able to prove that such systems have ergodic components of positive measure or positive measure entropy.

If we use our perturbation technique on the light bulb, we can break the separatrix and thereby arrive at a new candidate for coexistence. This system would now have some intertwining of the stable and unstable region. Note that by choosing the region U of the perturbation away from the elliptic orbit, we can keep an integrable component of phase space. Intuitively, one would think that such a small perturbation should not destroy the positive measure entropy. However this remains to be proven and is probably a very difficult problem.

5. Acknowledgements. The author thanks Victor Bangert, Keith Burns, Gerhard Knieper and Howie Weiss for useful discussions and Keith Burns for a critical reading of the manuscript. This paper was written while the author was a visitor at the Institute for Advanced Study.

Addendum: G. Paternain has given examples of real analytic metrics on S^2 that have positive topological entropy [Pa]. His examples use the dynamics of rigid bodies. D. Petroll [Pe] has proven Theorem 1.1 independently. His proof, which uses methods similar to ours, also holds for geodesic flows on higher dimensional manifolds.

REFERENCES

[BG] K. BURNS AND M. GERBER, *Continuous invariant cone families and ergodicity of flows in dimension three*, Ergod. Th. & Dynam. Sys. 9 (1989), pp. 19–25.

[D1] V.J. DONNAY, *Geodesic flow on the two-sphere, Part I: Positive measure entropy*, Ergod. Th. & Dynam. Sys. 8 (1988), pp. 531–553.

[D2] V.J. DONNAY, *Geodesic flow on the two-sphere, Part II: Ergodicity, Dynamical Systems*, Springer Lecture Notes in Math., Vol. 1342 (1988), pp. 112–153.

[Do] M.P. DoCARMO, *Differential Geometry of Curves and Surfaces*, Prentice-Hall: New York, 1976.

[E] P. EBERLEIN, *When is a geodesic flow of Anosov type?* , J. Differential Geometry 8 (1973), pp. 437–463.

[K] W. KLINGENBERG, *Lectures on Closed Geodesics*, Grundlehren Math. Wiss. 230, Springer-Verlag, New York-Heidelberg-Berlin, 1982.

[KW] G. KNIEPER AND H. WEISS, *A surface with positive curvature and positive topological entropy*, to appear, J. Diff. Geom.

[M] J. MOSER, *Stable and Random Motions in Dynamical Systems*, Annals of Mathematics Studies 77, Princeton University Press, 1973.

[Pa] G.P. PATERNAIN, *Real analytic convex surfaces with positive topological entropy and rigid body dynamics*, Manuscripta Mathematica vol 78 (1993), pp. 397–402.

[Pe] D. PETROLL, *Transversale heterokline Orbits beim Geodatischen Flus*, preprint.

[P] F. PRZYTYCKI, *Examples of conservative diffeomorphisms of the two-dimensional torus with coexistence of elliptic and stochastic behaviour*, Ergod. Th. & Dynam. Sys. 2 (1982) no. 3-4, pp. 439–463 (1983).

[S] S. SMALE, *Diffeomorphisms with many periodic points, Differential and Combinatorial Topology*, (edited by S. S. Cairns) Princeton University Press, pp. 63–80, 1965.

[We] H. WEISS, *Genericity of symplectic diffeomorphisms and metrics on S^2 with positive topological entropy*, preprint.

[W] M. WOJTKOWSKI, *Linked twist mappings have the K-property*, Annals of the New York Academy of Sciences Vol. 357, Nonlinear Dynamics (1980), pp. 65–76.

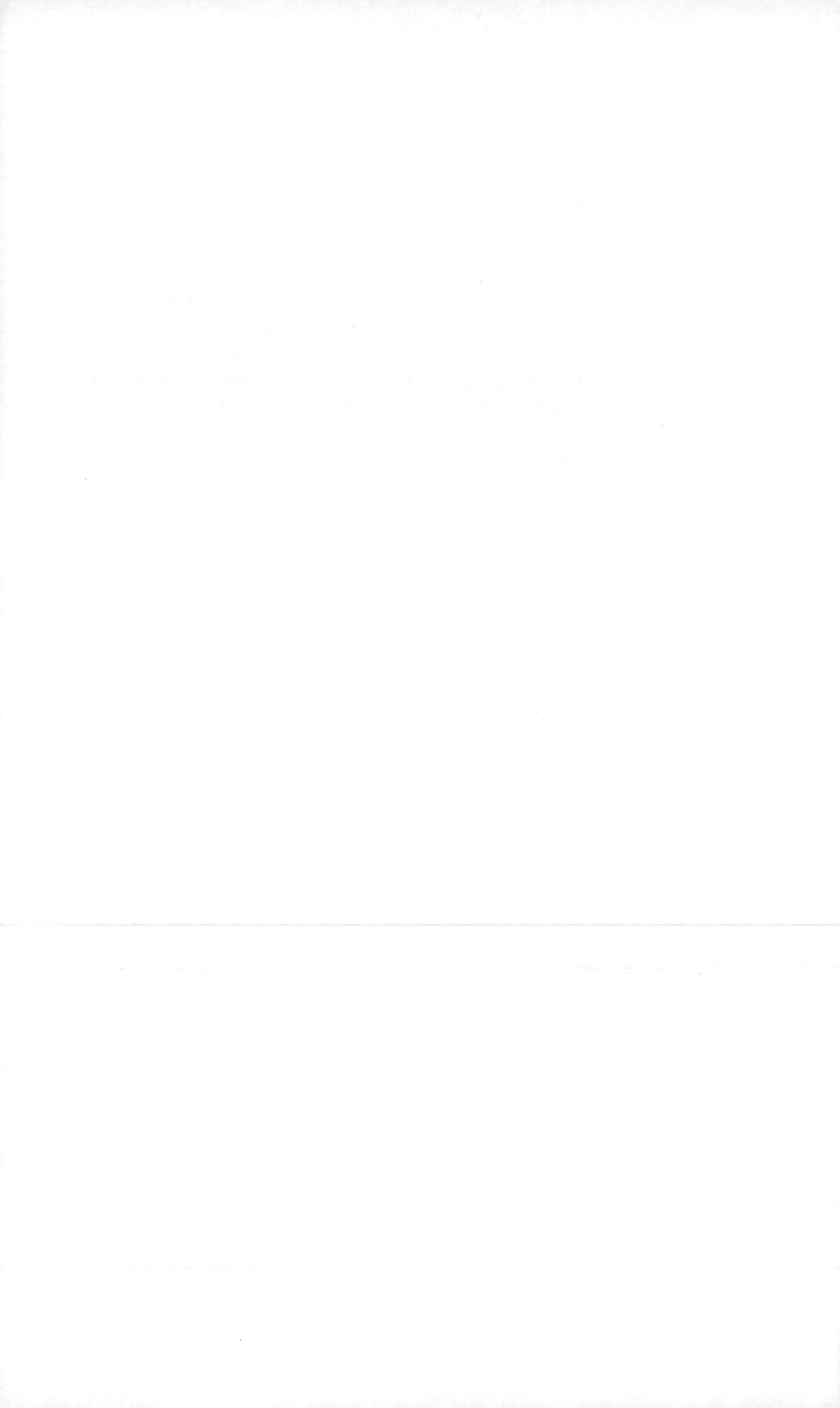

A NEW PROOF OF ANOSOV'S AVERAGING THEOREM

H. SCOTT DUMAS*

1. Introduction. I would like to describe and sketch the proof of a result in averaging theory I obtained recently with François Golse and Pierre Lochak [6]. Our result comprises both a new proof, as well as a generalization, of D.V. Anosov's general multiphase averaging theorem [1] for systems of ODEs with slow variables evolving in $\mathbf{R}^m$ and fast variables evolving on a smooth compact immersed manifold. We extend Anosov's work by allowing the fast variables to belong to an arbitrary smooth compact Riemannian manifold, and the vector field to have only Sobolev regularity. This is accomplished using normal form techniques adapted to a slightly generalized version of the DiPerna-Lions theory of generalized flows for ODEs [5]. By specializing to the case of Hamiltonian vector fields, we obtain an interesting and somewhat surprising result for Hamiltonians of low regularity, as well as a reason for including this article in these proceedings.

I do not intend here to give a complete proof of the result just described; the interested reader will find full details in [6]. Instead, I will first place it in context by briefly recalling the main developments of averaging theory, then I will state Anosov's theorem and indicate how we have generalized it. Finally, I will outline our method of proof as it might be applied to a less general result like Anosov's (i.e., without worrying about optimal smoothness or the more general geometric setting). This provides a chance to see the proof unencumbered by the technicalities required by the DiPerna-Lions theory and its generalization to flows on Riemannian manifolds, and also shows our method to be a simple means of obtaining Anosov's original theorem.

2. Brief description of averaging theory. Averaging methods go back at least as far as Laplace and Lagrange, who, in celestial mechanics, saw the advantages of replacing the periodic perturbations due to orbiting planets by constant perturbations obtained by smearing the planets out over their orbits. Since then, the subject has come to embrace almost any method in which one seeks to approximate the solutions of a given set of differential equations by solutions of a simpler set obtained by "averaging" the original vector field.

The method has a clear intuitive appeal, and various perturbation approximations involving averaging techniques were used in celestial mechan-

* Department of Mathematical Sciences, University of Cincinnati, Cincinnati, Ohio 45221-0025 USA. This work was supported in part by the National Science Foundation under Grant DMS-9106812. I would like to thank the mathematics departments of the Universities of Picardy (Amiens), and Paris VII (Jussieu) for their hospitality while much of this work was carried out, and especially Ken Meyer for organizing this conference while I enjoyed that hospitality.

127

ics throughout the 19th century. Yet rigorous results did not appear until
Fatou's work in the early part of this century, and a systematic theory
of averaging—at least in the simpler, single-phase case—was not widely
known until the appearance of the book by Bogoliubov and Mitropolski
[4] (see [14], Appendix 1, for a detailed account of the history of single-
phase averaging). The more complex multiphase averaging theory was also
brought to light mostly by Russian mathematicians, notably Anosov [1],
Arnold [2], and Neishtadt [11,12,13].

The basic idea of averaging is very simple, and is perhaps best first
approached by considering the ODE

$$(2.1) \qquad \frac{dx}{dt} = \varepsilon f(x,t,\varepsilon)$$

where ε $(0 \leq \varepsilon \ll 1)$ is a small parameter, and $f : U \times \mathbf{R} \times [0, \varepsilon_0) \to \mathbf{R}^m$
$(U \subset \mathbf{R}^m)$ is T-periodic in t and sufficiently smooth. The so-called averaged
system associated to 2.1 is

$$(2.2) \qquad \frac{dy}{dt} = \varepsilon \overline{f}(y) \equiv \varepsilon \frac{1}{T} \int_0^T f(y,t,0) dt,$$

and it is easy to see why solutions of 2.1 with initial condition $x(0) = x_0$
should be approximated to $O(\varepsilon)$ over time intervals of length $O(\varepsilon^{-1})$ by
solutions of 2.2 with the same initial condition. In fact, writing

$$(2.3) \qquad x(t) - y(t) = \varepsilon \int_0^t \big(f(x(\tau), \tau, \varepsilon) - \overline{f}(y(\tau))\big) d\tau$$

and denoting by $\widetilde{f}(z,t,\varepsilon) = f(z,t,\varepsilon) - \overline{f}(z)$ the oscillating part of f (which
has zero mean; i.e., vanishing t-integral over any interval of length T with
z and ε fixed), we see that, since x and y are slowly varying in t, it is
reasonable to expect the integrand of 2.3 to be of the form $\widetilde{f}(x_0, \tau) +$
$\varepsilon g(x_0, \tau, \varepsilon)$ for some smooth, bounded function g. In this case $|x(t) - y(t)| \leq$
$\varepsilon M T + \varepsilon^2 N t$ (M and N bound $\widetilde{f}$ and g), so that $y(t)$ is indeed $O(\varepsilon)$-close
to $x(t)$ on time intervals of $O(\varepsilon^{-1})$. (A rigorous proof along these lines of
this simplest of averaging results may be found in Remark (3) following
Theorem 1 of [8]).

To find higher order averaging approximations (i.e., autonomous sys-
tems with solutions that remain $O(\varepsilon^N)$-close to solutions of 2.1 over $O(\varepsilon^{-1})$
t-intervals), one introduces a near-identity transformation

$$(2.4) \qquad x(t) = y(t) + \varepsilon P(y(t), t, \varepsilon)$$

which takes Eq. 2.1 to Eq. 2.2 (or, more generally, to some other suitable
autonomous system which agrees to leading order with 2.2) up to a t-
dependent remainder of order $O(\varepsilon^N)$. Differentiating 2.4 along solutions of

2.1 and demanding that y satisfy 2.2 up to order $O(\varepsilon^N)$ leads to a system of $N - 1$ so-called homological equations for P, the first of which is

$$(2.5) \qquad\qquad \partial_t P(y, t, 0) = \tilde{f}(y, t, 0).$$

Some freedom in choosing functions of integration is available at higher orders, giving rise to different schemes for higher order averaging (cf. [8], [10], or [14] for discussions of various schemes and references, as well as the smoothness required of f).

Because of the possibility of "resonances," it requires more work to show that averaging approximations are valid for *multiphase* systems of ODEs, one of the simplest examples of which is the so-called standard form:

$$(2.6a) \qquad\qquad \frac{dI}{dt} = \varepsilon f(I, \theta, \varepsilon)$$

$$(2.6b) \qquad\qquad \frac{d\theta}{dt} = \omega(I) + \varepsilon g(I, \theta, \varepsilon),$$

where $I \in U \subset \mathbf{R}^m$, $\theta \in \mathbf{T}^n \equiv \mathbf{R}^n / \mathbf{Z}^n$, and f, g, and ω are sufficiently smooth in all arguments. In this case one seeks to show that the evolution of the slow variables I is well approximated, on an appropriate $O(\varepsilon^{-1})$-time interval, by solutions of the associated (slow) averaged system

$$(2.7) \qquad\qquad \frac{dJ}{dt} = \varepsilon \overline{f}(J) \equiv \varepsilon \int_{\mathbf{T}^n} f(J, \theta, 0) d\theta.$$

Increasingly refined results for systems of this type were obtained by Arnold and Neishtadt. Arnold first showed that, in the case $n = 2$ and under certain (rather strong) nonresonance conditions on ω, the deviation of solutions of 2.7 from those of 2.6 is no greater than $O(\varepsilon^{1/2}(\log\varepsilon)^2)$ on $O(\varepsilon^{-1})$-time intervals [2]. The optimal error estimate for $n > 1$ (and the natural way of expressing it) was later announced by Neishtadt as the culmination of a series of papers [11,12,13]. Using methods descended from Kasuga's work on adiabatic ergodic theorems [9], Neishtadt shows that if ω is mildly nondegenerate, then except for a set of (nearly resonant) initial conditions of small measure, the average separation between solutions of 2.6 and 2.7 with the same initial condition is $O(\varepsilon^{1/2})$ over an $O(\varepsilon^{-1})$-time interval. He also estimates the measure of the exceptional set of initial conditions to be $O(\varepsilon^{1/2})$, and proves that these estimates are optimal.

A still more general sort of multiphase averaging theory was considered by Anosov in the early 1960's [1], and it is on this type that I will focus here. Anosov allows the fast variables to evolve on an arbitrary smooth compact immersed manifold, and allows the frequency map ω to depend on both the fast and slow variables. In this very general case, the notion of resonance (and the recourse to small divisor techniques) is lost, and the nonresonance or nondegeneracy conditions on ω must be replaced by a hypothesis of uniform distribution (e.g., ergodicity or mixing) on the unperturbed flow. Under the weaker hypothesis of ergodicity, it is apparently no longer possible to formulate order estimates of the precision with which solutions of the averaged system approximate those of the perturbed system. Anosov instead proves a theorem of the type described in the next section.

3. Anosov's theorem and its generalization. Let $\mathcal{M}$ be a smooth compact immersed manifold of dimension n. For small $\varepsilon \geq 0$, we consider the system of ODEs on $\mathbf{R}^m \times \mathcal{M}$

$$(3.1a) \qquad \frac{dX}{dt} = \varepsilon f(X, Y)$$

$$(3.1b) \qquad \frac{dY}{dt} = \omega(X, Y) + \varepsilon g(X, Y)$$

associated to the perturbation $(\varepsilon f, \omega + \varepsilon g)$ of the vector field $(0, \omega)$. Here X and Y denote elements of $\mathbf{R}^m$ and $\mathcal{M}$, respectively, and $d\mu_Y$ denotes Lebesgue measure on $\mathcal{M}$ normalized to satisfy the relation $\int_{\mathcal{M}} d\mu_Y = 1$. We group the assumptions on 3.1 and its associated averaged system into four hypotheses, the first three of which are:

(i) f, g, and ω are C^1 in all their arguments
(ii) $\mathrm{div}_Y \omega = \mathrm{div}(0, \omega) = 0$
(iii) $f(X, Y) = 0$ for all $X \notin B_R(0)$, the open ball of radius $R > 0$ centered on 0 in $\mathbf{R}^m$

As described above, the goal of averaging theory is to approximate the X-components of solutions of 3.1 by solutions of the associated (slow) averaged system in $\mathbf{R}^m$:

$$(3.2) \qquad \frac{dW}{dt} = \varepsilon \overline{f}(W) \equiv \varepsilon \int_{\mathcal{M}} f(W, Y) d\mu_Y .$$

Standard ODE theory shows that, under hypotheses (i)–(iii), the systems 3.1 and 3.2 generate global flows which leave the respective domains $D_R \equiv \overline{B_R}(0) \times \mathcal{M}$ and $\overline{B_R}(0)$ invariant (here $\overline{B_R}(0)$ is the closed ball of radius R centered on the origin in $\mathbf{R}^m$). We shall denote these flows by $(X_t^\varepsilon, Y_t^\varepsilon) \equiv (X_t^\varepsilon, Y_t^\varepsilon)(x, y)$ and $W_t^\varepsilon \equiv W_t^\varepsilon(x)$ respectively. This notation means that $t \mapsto (X_t^\varepsilon, Y_t^\varepsilon) \equiv (X_t^\varepsilon, Y_t^\varepsilon)(x, y)$ is the integral curve of the vector field $(\varepsilon f, \omega + \varepsilon g)$ passing through the initial condition (x, y) at time $t = 0$. It is perhaps worth pointing out that the unperturbed flow is consistently denoted (X_t^0, Y_t^0).

Using this notation, the last assumption on the system 3.1 is the hypothesis of ergodicity, expressed as:

(iv) Any function $F \in L^2(D_R)$ such that $F \circ (X_t^0, Y_t^0) = F$ for all $t \in \mathbf{R}$ is $dx d\mu_y$-almost everywhere equal to a function of X only.

Finally, we introduce the pointwise error E_t^ε between solutions of the perturbed system 3.1 and its associated averaged system 3.2, defined as

$$E_t^\varepsilon(x, y) = X_t^\varepsilon(x, y) - W_t^\varepsilon(x).$$

We may now state

THEOREM 3.1. **Anosov's averaging theorem.** *Assume hypotheses (i) through (iv) pertaining to systems 3.1 and 3.2, and fix $T > 0$. Then for any p, $1 \le p < \infty$, the average error*

$$\int_{D_R} \sup_{0 \le t \le T/\varepsilon} |E_t^\varepsilon(x, y)|^p \, dx \, d\mu_y \to 0 \quad \text{as} \quad \varepsilon \to 0.$$

In particular, for any $r > 0$, we have

$$\text{meas}\{(x, y) \in D_R | \sup_{0 \le t \le T/\varepsilon} |E_t^\varepsilon(x, y)| > r\} \to 0 \quad \text{as} \quad \varepsilon \to 0.$$

Remarks 3.1.
(1) Hypotheses (ii) and (iii) may be weakened somewhat, although doing so has the principal effect of making the proof messier. Hypothesis (ii) means that the invariant measure on $\mathcal{M}$ of the unperturbed flow is simply the ordinary measure $d\mu_Y$, while hypothesis (iii) simplifies technical aspects of the proof in an obvious way.
(2) It is not hard to describe the generalization of this theorem which we prove in [6]. In fact, our theorem may be obtained from the one stated above as follows. First, the geometric setting is generalized by allowing $\mathcal{M}$ to be an arbitrary smooth compact Riemannian manifold of dimension n. Second, we remove the smoothness hypothesis (i) above, and replace it with the following (considerably weaker) hypotheses:
   (a) $f, g \in L_{loc}^\infty \cap W_{loc}^{1,1}(\mathbf{R}^m \times \mathcal{M})$ and $\omega \in L_{loc}^\infty \cap W_{loc}^{1,2}(\mathbf{R}^m \times \mathcal{M})$
   (b) $\bar{f} \in W_{loc}^{1,\infty}(\mathbf{R}^m)$
   (c) $\text{div}_X f + \text{div}_Y g = \text{div}(f, g) \in L_{loc}^\infty(\mathbf{R}^m \times \mathcal{M})$
   (Recall that in standard Sobolev space notation, $f \in W_{loc}^{1,p}$ means that f together with all its first-order derivatives belong to L_{loc}^p.)
(3) The dispassionate reader may well wonder if weakening hypothesis (i) as just described does not also have the principal effect of making the proof messier. To this natural query I have two responses: First; yes, it does make the proof messier, but the result is quite strong and worth the additional mess (cf. [6]). Second; yes, it does make the proof messier, and it is precisely the point of this article to describe the new proof under the older, stronger hypotheses, in which case the new proof is more accessible.

4. Sketch of the proof. In essence, the proof outlined below shows that the functional analysis methods pioneered by Kasuga [9] and later used by Neishtadt [13] may be extended to the more general result of Anosov [1]. A few remarks on the way this differs from Anosov's original approach may be found in the concluding remarks, below.

The proof sketched here consists of two parts. In the first part, we derive a "weak" homological equation by attempting to conjugate 3.1a and 3.2 via a near identity transformation. In the second part, we use functional analysis techniques to approximately solve the weak homological equation.

Sketch of Proof. **Part 1).** In order to examine the relation between 3.1a and 3.2, we follow Neishtadt [13] by introducing the smooth auxiliary function $S \in C_0^\infty(\mathbf{R}^m \times \mathcal{M}; \mathbf{R}^m)$ (to be determined later) and the near-identity transformation $P = X + \varepsilon S(X, Y)$ (analogous to 2.4 , above).

Now integrating the difference $\frac{d}{dt}(X - W) = \frac{d}{dt}(P - W - \varepsilon S)$ from 0 to t leads to

$$E_t^\varepsilon(x, y) = X_t^\varepsilon(x, y) - W_t^\varepsilon(x) = \varepsilon \int_0^t \left(\overline{f}(X_u^\varepsilon(x, y)) - \overline{f}(W_u^\varepsilon(x))\right) du$$

(4.1)

$$+ \varepsilon(S(X_t^\varepsilon(x, y), Y_t^\varepsilon(x, y)) - S(x, y)) + \varepsilon \int_0^t R^\varepsilon(X_u^\varepsilon(x, y), Y_u^\varepsilon(x, y)) du$$

where

$$(4.2) \qquad R = f - \overline{f} - \varepsilon f \cdot \mathrm{grad}_x S - (\omega + \varepsilon g) \cdot \mathrm{grad}_y S .$$

This implies that for all $t > 0$,

$$|E_t^\varepsilon(x, y)| \leq L\varepsilon \int_0^t |X_u^\varepsilon - W_u^\varepsilon|(x, y) du \; + \; \varepsilon|S((X_t^\varepsilon, Y_t^\varepsilon)(x, y))|$$

$$(4.3) \qquad + \varepsilon|S(x, y)| + \varepsilon \int_0^t |R^\varepsilon((X_u^\varepsilon, Y_u^\varepsilon)(x, y))| du$$

where L is the Lipschitz constant for $\overline{f}$.

Again following Neishtadt, we set $Z(t) = \int_{D_R} |E_t^\varepsilon(x, y)| dx d\mu_y$ (recall $D_R = \overline{B_R}(0) \times \mathcal{M}$). Then integrating 4.3 over all initial conditions $(x, y) \in D_R$ (using Fubini's Theorem to reverse the order of integration), we obtain

$$Z(t) \leq L\varepsilon \int_0^t Z(u) du + \varepsilon \int_{D_R} |S(X_t^\varepsilon, Y_t^\varepsilon)| dx d\mu_y + \varepsilon\|S\|_{L^1}$$

(4.4)

$$+ \varepsilon \int_0^t \int_{D_R} |R^\varepsilon(X_u^\varepsilon, Y_u^\varepsilon)| dx d\mu_y du .$$

In view of assumption (ii), it is easy to verify that the Jacobian of the reverse flow $(X_{-t}^{\varepsilon}, Y_{-t}^{\varepsilon})$ is $O(1)$ over $O(\varepsilon^{-1})$-time intervals. We use this fact to ensure that the dilation or compression of the image measure $(X_t^{\varepsilon}, Y_t^{\varepsilon})_*$ $(dxd\mu_y)$ is $O(1)$, bounded away from 0, on $O(\varepsilon^{-1})$-time intervals. More precisely, we deduce the existence of constants $C_2 \geq C_1 > 0$ such that, given $T > 0$, for all $\varepsilon > 0$, the inequality

$$C_1 dx d\mu_y \leq (X_t^{\varepsilon}, Y_t^{\varepsilon})_*(dxd\mu_y) \leq C_2 dx d\mu_y$$

holds for all $0 \leq t \leq T/\varepsilon$. We may therefore write

$$Z(t) \leq L\varepsilon \int_0^t Z(u)du + c\left(\varepsilon\|S\|_{L^1} + T\|R\|_{L^1}\right)$$

for some positive constant c and for all $t \in [0, T/\varepsilon]$. A standard Gronwall estimate then leads to

(4.5) $$Z(t) \leq c\left(\varepsilon\|S\|_{L^1} + T\|R\|_{L^1}\right)\exp(\varepsilon L t).$$

But we know (cf. 4.2) that

(4.6) $\|R\|_{L^1} \leq \left\|\widetilde{f} - \omega \cdot \mathrm{grad}_y S\right\|_{L^1} + \varepsilon\left\|f \cdot \mathrm{grad}_x S + g \cdot \mathrm{grad}_y S\right\|_{L^1}$

(here $\widetilde{f} = f - \overline{f}$), so if we can choose the auxiliary function S so as to make the difference

(4.7) $\left\|\widetilde{f} - \omega \cdot \mathrm{grad}_y S\right\|_{L^1}$

sufficiently small with ε, we obtain

(4.8) $\displaystyle \sup_{t \in [0, T/\varepsilon]} Z(t) = \sup_{t \in [0, T/\varepsilon]} \int_{D_R} |E_t^{\varepsilon}(x, y)| dx d\mu_y \to 0 \quad$ as $\varepsilon \to 0$

by taking the supremum of 4.5 over $[0, T/\varepsilon]$. Observe that demanding that 4.7 be small constitutes a kind of weak homological equation for S. However, the convergence stated in 4.8 is weaker than that announced in Anosov's Averaging Theorem, since only the following inequality is true:

$$\sup_{t \in [0, T/\varepsilon]} \int_{D_R} |E_t^{\varepsilon}(x, y)| dx d\mu_y \leq \int_{D_R} \sup_{t \in [0, T/\varepsilon]} |E_t^{\varepsilon}(x, y)| dx d\mu_y,$$

and not (at least in general) the equality, as claimed in [13]. The stronger statement in Anosov's Averaging Theorem is instead obtained after a not-entirely trivial calculation which we give in Section 4 of [6]. (The same calculation also shows why p in Anosov's theorem may range between 1 and ∞.)

If this latter calculation is admitted, the proof of Anosov's Averaging Theorem reduces to approximately solving the homological equation (i.e, to finding a smooth S so that 4.7 is sufficiently small), which we set out to do below in Part 2).

Part 2). In the spirit of Kasuga [9], we use functional analysis methods to approximately solve (i.e., to demonstrate the existence of an approximate weak solution of) the homological equation

$$(4.9) \qquad \tilde{f} = \omega \cdot \mathrm{grad}_y S \,.$$

To this end, we recall that the unperturbed flow of 3.1 is denoted by (X_t^0, Y_t^0). To this flow is associated the evolution operator U_t on $L^2(D_R)$ defined by

$$(4.10) \qquad U_t(F) = F \circ (X_t^0, Y_t^0)$$

for any $F \in L^2(D_R)$. It is easy to verify that U is a strongly continuous group of isometries in $L^2(D_R)$ with infinitesimal generator A defined by

$$(4.11) \qquad AF = \omega \cdot \mathrm{grad}_y F$$

and with domain $\mathcal{D}(A) = \{F \in L^2(D_R) \mid \omega \cdot \mathrm{grad}_y F \in L^2(D_R)\}$. From Stone's Theorem, we conclude that A is skew-adjoint, from which it follows that the closure of the range of A is the orthogonal complement of the null space of A:

$$\overline{\mathcal{R}(A)} = \mathcal{N}(A)^\perp \,.$$

We will now use the ergodicity hypothesis (iv) to show that $\mathcal{N}(A) = \{F(X)\}$, the set of functions depending a.e. only on X.

It is clear that $\mathcal{N}(A) \supset \{F(X)\}$. To see that the reverse inclusion also holds, let $F \in \mathcal{N}(A)$, and $t \neq 0$. Since $\mathcal{N}(A) \subset \mathcal{D}(A)$, we may write $U_t F - F = \int_0^t A(U_s F)\,ds$, and since A commutes with the group it generates, this last integral becomes $\int_0^t U_s(AF)\,ds = 0$. F is therefore invariant under U, and by (iv), we must have $F \in \{F(X)\}$.

Since $\tilde{f} = f - \bar{f}$ has zero mean over $\mathcal{M}$, it follows that $\tilde{f} \in \{F(X)\}^\perp = \mathcal{N}(A)^\perp$, which in turn leads to $\tilde{f} \in \overline{\mathcal{R}(A)}$. Therefore, given any $\eta > 0$, we may choose a transformation $S \in \mathcal{D}(A)$ with

$$(4.12) \qquad \|\tilde{f} - AS\|_{L^2} < \eta.$$

The procedure just described approximately solves the homological equation, but only in the weak sense—that is, the S appearing in 4.12 need not be a smooth function. To conclude, we must show that we can regularize S while maintaining the estimate 4.12.

For this purpose we use a family of mollifiers defined as follows. Let $\rho \equiv \rho(x)$ be a function in $C^\infty(\mathbf{R}^d)$ (d sufficiently large) satisfying

$$(4.13) \qquad \rho > 0 \text{ on } B_1(0), \quad \mathrm{supp}\rho = \overline{B}_1(0), \quad \int \rho(x)\,dx = 1 \,.$$

We then scale this bump function so that its support is a ball of size ε by setting $\rho_\varepsilon(x) = \varepsilon^{-d}\rho(x/\varepsilon)$. Functions may now be regularized by convolution with ρ_ε; the family of mollifiers R_ε is defined by $R_\varepsilon f = \rho_\varepsilon * f$. In Lemma 2 of [6], we also show that the commutator $[A, R_\varepsilon]$ of A and R_ε converges to zero in the following sense:

$$(4.14) \qquad \|[A, R_\varepsilon]f\|_{L^1(B_R)} \to 0 \quad \text{as} \quad \varepsilon \to 0$$

for all $f \in L^2_{loc}(\mathbf{R}^d)$.

Using the family of regularized auxiliary functions $S_\varepsilon = R_\varepsilon S$, we have

$$\|\tilde{f} - AS_\varepsilon\|_{L^1(D_R)}$$

$$\leq (\text{meas } D_R)^{1/2}\|\tilde{f} - AS\|_{L^2(D_R)} + \|R_\varepsilon(AS) - AS\|_{L^1(D_R)} + \|[A, R_\varepsilon]S\|_{L^1(D_R)}$$
(4.15)
since

$$AS_\varepsilon - AS = AR_\varepsilon S - AS = R_\varepsilon(AS) - AS + [A, R_\varepsilon]S.$$

Now let η be an arbitrary positive number. According to 4.12, we can choose $S \in \mathcal{D}(A)$ so that the first term on the right side of 4.15 is less than η. Since $S \in \mathcal{D}(A)$ and $\text{meas}(D_R) < \infty$, $AS \in L^1(D_R)$ and therefore the second term on the right side of 4.15 can be made smaller than η by choosing $0 < \varepsilon < \varepsilon_0(\eta)$. As for the third term on the right side of 4.15, Eq. 4.14 shows that it can also be made less than η by choosing $0 < \varepsilon < \varepsilon_1(\eta)$. Therefore, for $0 < \varepsilon < \min(\varepsilon_0(\eta), \varepsilon_1(\eta))$, the left side of 4.15 is less than 3η.

This shows that 4.12 can be made as small as desired by choosing ε sufficiently small. In view of the remarks at the end of Part 1), the outline of the proof of Anosovs' Averaging Theorem is complete.   □

5. Concluding remarks. To appreciate the simplicity of the proof outlined above, it suffices to look at Anosov's original proof [1] (which, unfortunately, has never been translated; see [3] or [10] for discussions in English). Very roughly speaking, Anosov's approach involves the partition of the $O(\varepsilon^{-1})$-time interval of evolution into two types of subintervals: those on which the ergodicity of the unperturbed flow is rapidly realized, and those on which it is not. It is only during subintervals of the latter type that appreciable separation between averaged and exact solutions can occur. Careful estimate of the total measure of such subintervals for various initial conditions yields the conclusions of Anosov's theorem.

A comparison between Anosov's proof of his theorem and the proof sketched here shows an interesting parallel with the relationship between Birkhoff's and Von Neuman's proofs of their respective ergodic theorems. In both cases the second proof uses basic functional analysis methods to replace careful ad-hoc estimates.

The proof outlined here also dovetails nicely with the DiPerna-Lions theory of generalized flows for ODEs having vector fields with only Sobolev regularity [5]. In [6], we first verify that the DiPerna-Lions theory carries over to flows on Riemannian manifolds. Extending the averaging theorem to generalized flows in this more general geometric setting involves constructing the operators U_t of Eq. 4.10 and the mollifiers R_ε of Eq. 4.14 above, as well as verifying the estimates leading to the homological equation 4.9.

Finally, I should mention that we recently obtained results showing that if the hypothesis of "uniform distribution" on the unperturbed flow is strengthened from ergodicity to mixing, it is possible to obtain order estimates of the precision with which averaged solutions approximate exact solutions; this will be the subject of a future publication [7].

REFERENCES

[1] D. V. ANOSOV, *Averaging in systems of ODEs with rapidly oscillating solutions*, Izv. Akad. Nauk. SSSR 24 (1960), 721–742 (untranslated).

[2] V. I. ARNOLD, *Conditions for the applicability, and estimate of the error, of an averaging method for systems which pass through states of resonance in the course of their evolution*, Dokl. Akad. Nauk. 161 (1) (1965), 331–334. English translation: Soviet Math. Doklady 6 (2) (1965), 581–585.

[3] V. I. ARNOLD, (ed.) *Dynamical Systems III*, Encyclopedia of Mathematical Sciences III, Springer-Verlag, Berlin 1988.

[4] N. N. BOGOLIUBOV, YU. A. MITROPOLSKI, *Asymptotic Methods in the Theory of Nonlinear Oscillations*, Gordon and Breach, New York 1961.

[5] R. DIPERNA, P.-L. LIONS, *Ordinary differential equations, transport theory and Sobolev spaces*, Inv. Math. 98 (1989), 511–548.

[6] H. S. DUMAS, F. GOLSE, P. LOCHAK, *Multiphase averaging for generalized flows on manifolds*, Erg. Th. Dynam. Sys. 14 (1994), 53–68.

[7] H. S. DUMAS, F. GOLSE, *The averaging method for mixing flows*, (to appear).

[8] J. A. ELLISON, A. W. SÁENZ, H. S. DUMAS, *Improved N th order averaging theory for periodic sytems*, J. Diff. Equations 84 (2) (1990), 383–403.

[9] T. KASUGA, *On the adiabatic theorem for the Hamiltonian system of differential equations in classical mechanics (parts I, II, and III)*, Proc. Acad. Japan 37 (1961), 366–382.

[10] P. LOCHAK, C. MEUNIER, *Multiphase Averaging for Classical Systems*, Applied Mathematical Sciences 72, Springer-Verlag, New York, 1988.

[11] A. I. NEISHTADT, *Passage through resonance in a two-frequency problem*, Dokl. Akad. Nauk. SSSR Mechanics 221 (2) (1975), 301–304. English translation: Soviet Phys. Doklady 20 (3) (1975), 189–191.

[12] A. I. NEISHTADT, *Averaging in multi-frequency systems*, Dokl. Akad. Nauk. SSSR Mechanics 223 (2) (1975), 314–317. English translation: Soviet Phys. Doklady 20 (7) (1975), 492–494.

[13] A. I. NEISHTADT, *Averaging in multi-frequency systems II*, Dokl. Akad. Nauk. SSSR Mechanics 226 (6) (1976), 1295–1298. English translation: Soviet Phys. Doklady 21 (2) (1976), 80–82.

[14] J. A. SANDERS, F. VERHULST, *Averaging Methods in Nonlinear Dynamical Systems*, Applied Mathematical Sciences 59, Springer-Verlag, New York, 1985.

BIFURCATIONS IN THE GENERALIZED VAN DER WAALS INTERACTION: THE POLAR CASE ($M = 0$)

ANTONIO ELIPE* AND SEBASTIÁN FERRER*

Abstract. In the regimen for bounded motions and considering a weak perturbation, we study the classical dynamics of the averaged system of the generalized van der Waals interaction for $m = 0$, whose orbit manifold is a 2-dimensional sphere. Complementing the work of Alhassid et al. and Ganesan and Lakshmanan, we show that the global flow is characterized by three parametric bifurcations of butterfly type corresponding to the dynamical symmetries of the problem.

Key words. Perturbed Hydrogen atom, Zeeman effect, van der Waals, normalization, bifurcations.

PACS numbers. 31.10.+z, 03.65.Fd, 31.60.+b, 32.60.+i

1. Introduction. The sensitivity of Rydberg atoms to perturbations and the wealth of experimental information they can yield has led to the revival of interest in the study of atoms in external fields (for a review see Kleppener et al. 1983). In those studies on Rydberg atoms, as well as other large families of non linear dynamical systems, a number of different roads have been taken: the semiclassical quantum mechanics approach searching for adiabatic invariants and dynamical symmetries, has been followed by Alhassid et al. (1987); and more recently the Painlevé singularity analysis by Ganesan and Lakshmanan (1990).

Also recently another approach is emerging which consists of on the qualitative analysis of the normalized phase space (see for instance Deprit 1983, David et al. 1989a, 1989b, Uzer et al. 1991). This method proves particularly useful when secular perturbations are insufficient and higher order terms have to be taken into account (Miller 1991). For a large family of systems whose normalized orbital space is a sphere, the pattern of the energy levels is obtained readily from the analysis of the extrema on the sphere. Adding color graphics to this technique, specially when several parameters are involved, increases considerably the possibilities in searching for the equilibria on the whole domain of parameters and integrals. More details on this "picturing of the Hamiltonian" is given in Coffey et al. (1991), Healy and Deprit (1991); see also Uzer et al. (1991).

In this paper we take the approach of the first paragraph for the study of a dynamical system proposed by Alhassid et al. 1987, which applies to the hydrogen atom subjected to a certain class of perturbations, including the van der Waals and the diamagnetic interactions. After presenting the three integrable cases (for $\beta = \frac{1}{2}, 1, 2$) obtained by Ganesan and Lakshmanan 1990, we make a normalization by a Lie transform which leads us in particular to the "hidden" dynamical constant of motion first computed

---

* Universidad de Zaragoza, Grupo de Mecánica Espacial, 50009 Zaragoza, Spain.

by Alhassid et al., which is an extension of the one found by Solov'ev (1982) from the secular perturbation analysis from the Zeeman effect.

Within a given hydrogenic manifold of principal quantum number— which we denote here by L—we study the the global flow of classical dynamics of the system. We take advantage of the fact that a perturbed Keplerian system with an axial symmetry, the orbit manifold is a 2-dimensional sphere for each principal quantum number. With a set of coordinates proposed by Deprit (1983) we get the equilibria, showing the stability domains as a function of the parameter β, and find three butterfly bifurcations (see David et al.). The values of the parameters for which the bifurcations occur are precisely the ones found previously, when they defined the dynamical symmetries, that is, the integrable cases. The flow for each domain is obtained and shows the disappearance of separatrices for the integrable cases. In this way we recapture the conclusions of the study of Ganesan and Lakshmanan (1990), about the connection between the existence of separatrices and their disappearance with what these authors call chaos-order-chaos, without having to rely on the Poincaré surface of section study done by these authors.

For the integrable case $\beta = 1$, the flow reduces to a pure rotation around the z axis. In relation to the analysis of the global flow we show that there are two different regimes: for values $\beta \in (\frac{1}{2}, 2)$ to which the van der Waals case belongs, the circular orbits are stable. When $\beta \notin (\frac{1}{2}, 2)$, the Zeeman problem being one of the important physical examples, all the trajectories go to the origin. Finally the pattern of the energy levels in general are explained in terms of its values at the equilibria as functions of the parameters.

For more details on this problem as well as the study of the general case of the magnetic quantum number $m \neq 0$, see Elipe and Ferrer (1992).

2. The Hamiltonian, integrals and normalization. This paper deals with the hydrogen atom in a generalized van der Waals potential whose Hamiltonian may be expressed as $\mathcal{H} = \mathcal{H}_0 + \mathcal{P}$, with

$$(2.1) \qquad \mathcal{H}_0 = \tfrac{1}{2} \boldsymbol{X} \cdot \boldsymbol{X} - \frac{\mu}{\|\boldsymbol{x}\|}, \qquad \mathcal{P} = \gamma(x^2 + y^2 + \beta^2 z^2)$$

where $[\gamma] = 1/T^2$, considered as a small parameter, either positive or negative, is a square of a frequency and β is dimensionless. This Hamiltonian, studied by Alhassid et al. (1987) defines a three degrees of freedom system and represents several cases of physical interest in solid-state physics and physical chemistry. They are:

- When $\gamma = 0$, we obtain the standard hydrogen– atom problem or Coulomb problem.
- When $\beta = 0$ (and $\gamma = -\omega^2/2$; where ω is the electron cyclotron frequency) we have the quadratic Zeeman effect in moderately strong magnetic fields.

- If $\beta = 1$, there results the spherical quadratic Zeeman effect (integrable system).
- For $\beta = \sqrt{2}$ (and $\gamma = -1/(16d^3)$), where d is the distance from atom to the surface), we have the instantaneous van der Waals potential (see Baym 1969).

Considering the system (2.1), a large body of information is now available for nonzero values of γ and $\beta = 0$ (see, for instance, Gay, 1986; Hasegawa et al., 1989). However, the hydrogen–atom problem involving the generalized van der Waals force is yet to be studied in detail both by classical and quantum mechanical methods. In this paper, by means of analytical methods, we are interested in understanding the classical dynamics of the system defined by (2.1). We will get this global picture of the dynamics from the search for equilibria and their bifurcations in the normalized Hamiltonian.

System (2.1) possesses cylindrical symmetry. If we introduce cylindrical coordinates (namely, $x = \rho\cos\phi$, $y = \rho\sin\phi$, and $z = z$) in (2.1) we have

$$(2.2) \qquad \mathcal{H} = \tfrac{1}{2}(P_\rho^2 + P_z^2) + \frac{P_\phi^2}{2\rho^2} + \gamma(\rho^2 + \beta^2 z^2) - \frac{\mu}{\sqrt{\rho^2 + z^2}}$$

where P_ρ, P_ϕ, and P_z are the canonical momenta conjugate to the coordinates ρ, ϕ, and z, respectively. This Hamiltonian defines a three degrees of freedom system in which ϕ is a cyclic variable and so the momenta P_ϕ is conserved. Since P_ϕ is the z component of the angular momentum, it can be quantized as $P_\phi = m\hbar$, where m is the magnetic quantum number. We will refer to this quantity under different notations; in this the paper either $H = 0$ or $\sigma = 0$ will mean $m = 0$.

2.1. Integrals in the polar case. Limiting to $P_\phi = m = 0$ Ganesan and Lakshmanan have discussed this problem. Introducing semiparabolic coordinates so as to avoid the Coulomb singularity problem, $\rho = uv$, $z = \tfrac{1}{2}(u^2 - v^2)$, then the classical dimensionless Hamiltonian $H = E$ of the problem becomes that of a set of two coupled sextic anharmonic oscillator (the perturbation is a polynomial of degree six) with the Hamiltonian

$$\overline{\mathcal{H}} = \tfrac{1}{2}(P_u^2 + P_v^2) - \tfrac{1}{2}(u^2 + v^2) + A(u^6 + v^6) + B(u^4 v^2 + u^2 v^4),$$

where $A = (\gamma\beta^2)/4$, $B = \gamma(4 - \beta^2)/4$. Here, the true energy occurs as a parameter and the physical trajectories evolve in an effective potential whose pseudoenergy is always equal to 2.

By applying Painlevé singularity analysis to the equations of motion of the above coupled oscillator system they have located at least three integrable cases, giving the integrals which explain the dynamical symmetries: When $B = 0$ ($\beta = 2$) we have two independent sextic oscillators. For $B = 3A$ ($\beta = 1$) we have a second integral of motion $\mathcal{I}_2 = (uP_v - vP_u)^2$, which is the spherically symmetric case in the original variables. For $B = 15A$ $(\beta = \frac{1}{2})$ we have $\mathcal{I}_2 = P_u + P_v + uv + 6A(u^4 + v^4)uv + 20Au^3v^3$.

Ganesan and and Lakshmanan showed by the Poincaré surface of sections the connection of this integrable cases with the general case by what they call chaos-order-chaos. In this paper we show that the normalized system gives this picture immediately. Finally let us point out that using the same technique, a generalization of this coupled oscillators has been recently studied by Deprit and Ferrer (1990b).

2.2. Averaged Hamiltonian and generalized Solov'ev invariant: Dynamical symmetries. As we have said, the generalized van der Waals interaction for $m = 0$ defines a biparametric dynamical system with two degrees of freedom. Considering a weak perturbation, we focus now ion the regimen of bounded motions. Ew will study the classical dynamics of the reduced averaged system by means of a Delaunay normalization (Deprit 1983). In fact the secular perturbation will be sufficient for our analysis. Thus, we make first a Delaunay transformation $\Phi : (\boldsymbol{x}, \boldsymbol{X}) \rightarrow (L, G, H, \ell, g, h)$, (see for instance Goldstein, 1980, p. 483). Then, by a Lie transform

$$\Psi : (L', G', H', \ell', g', h'; \gamma) \rightarrow (L, G, H, \ell, g, h),$$

a first order term of the normalized Hamiltonian is obtained straightforward converting $\mathcal{H}$ into its average over the mean anomaly. At the first order

$$\mathcal{H}_1' = \langle \mathcal{H} \rangle_\ell = \frac{1}{2\pi} \int_0^{2\pi} \mathcal{H}_1 \, d\ell.$$

From $x^2 + y^2 = r^2 - z^2$ we write $\mathcal{P} = r^2 - (1 - \beta^2)z^2$. An easy calculation based on properties of the Delaunay transformation supplies the averages of r^2 and z^2 over the mean anomaly. Indeed, with the semi-major axis $a = L^2/\mu$ and the *eccentric* anomaly defined by Kepler's equation $E - e \sin E = \ell$, it is readily seen that

$$r = a(1 - e \cos E), \quad r \cos f = a(\cos E - e), \quad r \sin f = b \sin E,$$

where f is the true anomaly, $e^2 = 1 - G^2/L^2$ and $b = a\sqrt{1 - e^2}$. It follows from Kepler's equation that $\partial E/\partial \ell = a/r$. We will use this formula to change the independent variable from the mean anomaly to the eccentric anomaly in the quadratures.

From the relation $z = r \sin \theta \sin I$ together with $\theta = g + f$ and introducing the eccentric anomaly E, after some elementary manipulations and computations, the first order normalized Hamiltonian is

$$\langle \mathcal{H} \rangle = -\frac{\mu^2}{2L^2}$$

$$(2.3) \qquad + \frac{1}{2}\gamma a^2 \left[2 + 3e^2 + (1 - \beta^2)\left(-\eta^2 + \sigma^2 - 5e^2 s^2 \sin^2 g\right) \right] + \mathcal{O}(\gamma^2),$$

where $\eta = G/L$, $\sigma = H/L$, $c = H/G$ and $s^2 = 1 - c^2$ and where we follow the usual rule of dropping the primes in the averaged Delaunay variables in order to simplify the notation. Denoting by

$$\mathcal{H}_e = -\frac{\mu^2}{2L^2} + \frac{1}{2}\gamma a^2[5 + \sigma^2(1 - \beta^2)]$$

we have

(2.4) $\Lambda = \langle\mathcal{H}\rangle - \mathcal{H}_e = \frac{1}{2}\gamma a^2 \left[-3\eta^2 - (1 - \beta^2)\left(\eta^2 + 5e^2s^2\sin^2 g\right)\right],$

which is precisely the generalized Solov'ev integral (see Braun and Solov'ev, 1984) given by Alhassid et al., (1987)

(2.5) $\Lambda = (4 - \beta^2)\|\boldsymbol{A}\|^2 + 5(\beta^2 - 1)A_z^2,$

where $\boldsymbol{A}$ is the Laplace–Runge–Lenz vector and A_z is the z component of $\boldsymbol{A}$.

Alhassid et al. reported the dynamical symmetries of this problem using the quantum mechanics approach. Related to (2.5) called by them as "universal adiabatic invariant," they found in particular dynamical symmetries for the three special choices $\beta = 2, 1$ and $\frac{1}{2}$. As we will see, our analysis of the phase space will give us special features of the system for these three particular cases.

3. Global flow on a two-dimensional sphere: Equilibria and bifurcations. The orbital space of bounded Keplerian systems is given by $\mathrm{R}^+ \times S^2 \times S^2$ (see for instance Moser, 1970). Besides, if we have a system with an axial symmetry Meyer (1973) showed that we can make another reduction in the orbital space. In other words, after the double reduction our orbital space for each pair (L, H) is given by a 2-dimensional sphere S^2. This is the case for our system (2.1). As the result of the normalization the angle ℓ is ignorable and the momenta L is an integral. Then, the rotational symmetry maps it onto a conservative system with only one degree of freedom, on the 2-dimensional sphere S^2 of radius L.

Addressing the reader to the explanations of Cushman (1983) and Coffey et al. (1986) for details on what the reduction stands for, we will use a set of coordinates proposed by Deprit (1983) and already used in the analysis of the polar Zeeman case (see Deprit and Ferrer 1990a). They are given by

(3.1) $\zeta_1 = Le\cos g,$ $\zeta_2 = Le\sin g,$ $\zeta_3 = G,$

which satisfy the equation $\zeta_1^2 + \zeta_2^2 + \zeta_3^2 = L^2$, and their mutual Poisson brackets satisfy the relation

$$\{\zeta_i, \zeta_j\} = \delta_{ijk}\zeta_k.$$

where δ_{ijk} is the Levi-Civita symbol. Graphing the phase orbits of such a system amounts to drawing the level contours of its Hamiltonian. For the Zeeman effect, which has been studied extensively (for references see Deprit and Ferrer 1990a), some authors have chosen a representation on cylinders (G, g).

TABLE 3.1
Equilibria in polar case

	E_1, E_3	E_2, E_4	E_0, E_5
$0 \leq \beta < 1/2$	Stable	Stable	Unstable
$\beta = 1/2$	Stable	Degeneracy	Degeneracy
$1/2 < \beta < 1$	Stable	Unstable	Stable
$\beta = 1$	Degeneracy	Degeneracy	Stable
$1 < \beta < 2$	Unstable	Stable	Stable
$\beta = 2$	Degeneracy	Stable	Degeneracy
$\beta > 2$	Stable	Stable	Unstable

Such representations are somewhat misleading, for the orbital space of $\mathcal{H}'_1$ is not made of cylinders, but of spheres. Maps of $\mathcal{H}'_1$ in planes (G, g) are local, not global: they fail to represent faithfully the phase space in the neighborhood of the circular orbits $(e = 0)$. Yet these are precisely the areas in the phase space where significant events take place.

The points such that $\zeta_3 > 0$ which comprise what we shall call here the northern hemisphere of the sphere $S^2(L)$ stand for ellipses traveled in the direct sense while those for which $\zeta_3 < 0$ in the southern hemisphere represent ellipses traveled in the retrograde sense. Any point on the equatorial circle $\zeta_3 = 0$ corresponds to a segment along a straight line passing through the origin having precisely the origin as its midpoint. The north pole $(\zeta_1 = \zeta_2 = 0, \quad \zeta_3 = L)$, is a circle traveled in the direct sense, and the south pole $(\zeta_1 = \zeta_2 = 0, \quad \zeta_3 = -L)$, the same circle but traveled in the retrograde sense.

From (2.5), the normalized perturbation in these variables is given by

$$(3.2) \qquad \mathcal{P} = \frac{\gamma a^2}{2L^2} \left[(4 - \beta^2)\zeta_1^2 + (4\beta^2 - 1)\zeta_2^2 \right],$$

and on account of Liouville's theorem we obtain the equations of the motion for this system:

$$(3.3) \qquad \begin{aligned} \dot{\zeta_1} &= \frac{\gamma a^2}{L^2}(4\beta^2 - 1)\zeta_2\zeta_3, \\ \dot{\zeta_2} &= \frac{\gamma a^2}{L^2}(\beta^2 - 4)\zeta_3\zeta_1, \\ \dot{\zeta_3} &= 5\frac{\gamma a^2}{L^2}(1 - \beta^2)\zeta_1\zeta_2. \end{aligned}$$

Clearly, from these equations (3.3), we observe that in general there are six critical points, placed on the intersection of the sphere with the coordinate

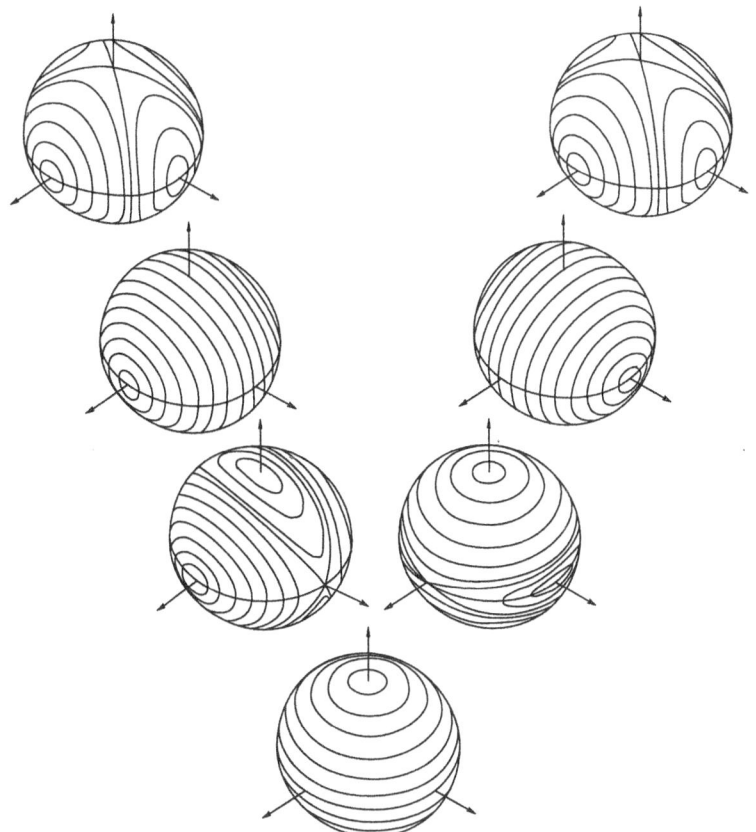

FIG. 3.1. *Phase flow in polar case through butterfly bifurcations*

axes $\zeta_1, \zeta_2, \zeta_3$ $(E_1, E_3, E_2, E_4, E_0, E_5)$. For specific values of the parameter: $\beta = 1/2$, $\beta = 1$ and $\beta = 2$, we obtain the great circles $\zeta_1 = 0$, $\zeta_3 = 0$ and $\zeta_2 = 0$ as set of non isolated equilibrium points, that is to say, degeneracies.

Since the Hamiltonian (3.2) is equivalent to one previously studied by the authors (see Ferrer *et al.*, 1990 and Deprit and Elipe, 1991), we refer to these papers for details. Let us to say only that the phase flow and bifurcations in our case is explained by means of *butterfly-* bifurcations (Fig. 3.1), and the evolution is given in Table 3.1. The evolution of the energy at the equilibria points is represented in Fig. 3.2. Finally we would like to point out that we could make use here of the set of coordinates introduced by Coffey *et al.*, (1986). Details on how the dynamics on the orbital space looks like with these coordinates will be given elsewhere.

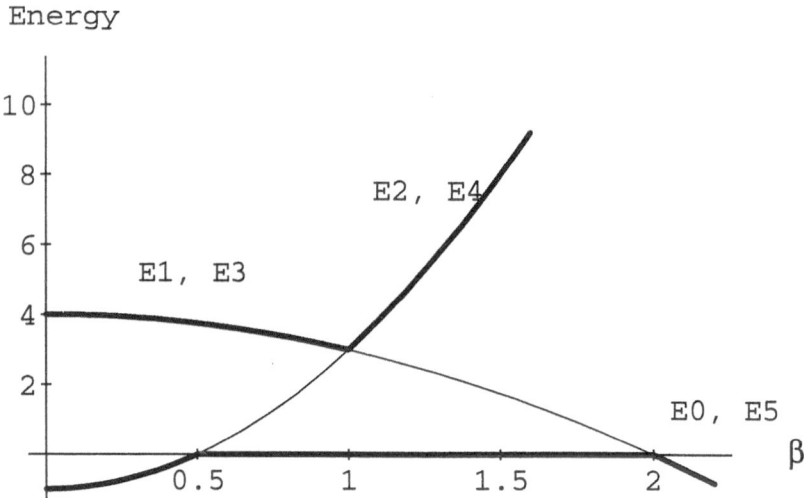

FIG. 3.2. *Evolution of the energy at the equilibrium points as function of β. Thicker lines indicate where an equilibrium is stable, thinner lines indicate instability.*

Acknowledgements

We are extremely grateful to Dr. Deprit at the U. S. National Institute of Standards and Technology who introduced us to these techniques. We are also thankful to the anonymous referee, whose comments have improved the final version of the article. Partial support comes from the Comisión Interministerial Científica y Técnica (CICYT PB90–0921 and ESP91-0919). Authors appear in alphabetical order.

REFERENCES

[1] Alhassid, Y., Hinds, E. A. and Meschede, D.: 1987, *Phys. Rev. Lett.* **59**, 1545–1548.
[2] Baym, G., Ed.: 1969, *Lectures on Quantum Mechanics*, Benjamin, Inc, New York, Ch. 11, 234–237.
[3] Coffey, S., Deprit, A., and Miller B.R.: 1986, *Celest. Mech.* **39**, 365–406.
[4] Coffey, S., Deprit, A., Deprit, E., and Healy, L.: 1990, *Science* **247**, 833.
[5] Cushman, R.: 1983, *Celest. Mech.*, **31**, 401–429.
[6] David, D., Holm, D. D., and Tratnik, M. V. 1989a, *Physics Lett. A*. **137**, 355–364.
[7] David, D., Holm, D. D., and Tratnik, M. V. 1989b, *Physics Lett. A*. **138**, 29–36.
[8] Deprit A.: 1983, *Celest. Mech.* **29**, 229–247.
[9] Deprit, A., and Elipe, A.: 1991,*Celest. Mech. & Dyn. Astron.* **51**, 227–250.
[10] Deprit, A. and Ferrer, S.: 1990a, *Rev Acad. Ciencias. Zaragoza* **45**, 111–125.
[11] Deprit, A. and Ferrer, S.: 1990b, *Phys. Lett. A*. **148**, 412–414.
[12] Elipe, A. and Ferrer, S.: 1992, submitted to *Phys. Rev. A*.
[13] Ferrer, S., Marco, M.J., Osásar, C., and Palacián, J.F: 1990, *Physics Lett A*. **146**, 411–420.
[14] Ganesan, K. and Lakshmanan, M.: 1989, *Phys. Rev. Lett.* **62**, 232.

[15] Ganesan, K. and Lakshmanan, M.: 1990, *Phys. Rev. A.* **42**, 3940–3947.

[16] Gay, J. C.: 1986, *Atoms in unusual situations*, Ed. J.P. Briand, Proceedings NATO, ASI Series B, Physics, Plenum, New York, v. 143, p. 1073.

[17] Goldstein, H.: 1980, *Classical Mechanics*, Addison-Wesley, Reading, Mass.

[18] Hasegawa, H., Robnik, M., and Wunner, G.: 1989, *Prog. Theor. Phys. Suppl.* **98**, 198.

[19] Healy, L. and Deprit, E.: 1991, *Computers in Physics* **5**, 491–496.

[20] Kleppener, D., Littman, M.G., and Zimmerman, M.L.: 1983, in *Rydberg States of Atoms and Molecules*, Ed. R.F. Stebbings and F.B. Dunning, Cambridge Univ. Press, New York, p. 73.

[21] Marsden, J. and Weinstein, A.: 1974, *Rep. Math. Phys.* **5**, 121–130.

[22] Meyer, K.: 1973, in *Dynamical Systems*, Ed. M.M. Peixoto, Academic Press, New York, pp. 259–272.

[23] Miller, B.R.: 1991, *Celest. Mech. & Dyn. Astron.* **51**, 251–270.

[24] Moser, K.: 1970, *Comm. Pure Appl. Math.* **23**, 609–636.

[25] Solov'ev, E. A.: 1982, *Sov. Phys. JETP* **55**, 1017.

[26] Uzer, T., Farrelly, D., Milligan, J., Raines, P. and Skelton, J.: 1991, *Science* **253**, 42–48.

ENERGY EQUIPARTITION AND NEKHOROSHEV-TYPE ESTIMATES FOR LARGE SYSTEMS

ANTONIO GIORGILLI*

Abstract. The relevance of Nekhoroshev-like results for systems with a large number of degrees of freedom is discussed. It is shown that numerical experiments suggest that such results could be extended to systems of interest in classical statistical mechanics. A possible connection with the problem of equipartition of energy is also indicated.

1. Overview. In recent decades the quantitative methods of perturbation theory have produced very beautiful and significant results. I refer here on the one hand to the celebrated KAM theorem (Kolmogorov, Moser, and Arnold) [17,19,1], and, on the other hand, to the less famous, but for some purposes more relevant, Nekhoroshev's theorem [20,21]; see also [4]. The analytical aspect of both of these theorems is the subject of other contributions to the present volume. So, I will consider some problems related to the relevance of such results for physical systems. In particular, I will discuss the possible interest of Nekhoroshev-like results for systems with a large number of degrees of freedom.

Let me briefly illustrate the problem in some detail. In its most naive formulation, perturbation theory tries to prove that a near-to-integrable Hamiltonian system is in fact integrable. That such an attempt generally fails was shown by Poincaré: he proved that the series expansions of perturbation theory are generally non convergent. But Poincaré himself made substantial contributions to the development of perturbation methods. In particular, he also considered some problems which are solved either by the KAM theorem or by Nekhoroshev's theorem.

i. The connection with KAM theory can be found in his discussion about the "Divergence des séries de M. Lindstedt" (chapter XIII of the *Méthodes Nouvelles*, in particular sect. 149). Poincaré realized that the series could converge for particular nonresonant values of the frequencies. However, after a quite detailed discussion, he concluded: "Les raisonnements de ce Chapitre ne me permettent pas d'affirmer que ce fait ne se présentera pas. Tout ce qu'il m'est permis de dire, c'est qu'il est fort invraisemblable." After the works of Siegel, Kolmogorov, Moser and Arnold we know that actually there are cases in which the Lindstedt series do converge.

ii. The case of Nekhoroshev's theorem is even more interesting. The fact that the series of perturbation theory, although generally divergent, are very useful in astronomical computation was, of course, well known to Poincaré. But he also raised the problem of finding quantitative estimates of the validity of the formal expansions.

* Dipartimento di Matematica dell' Universitá di Milano, Via Saldini 50, 20133 Milano, Italy.

Furthermore, he realized that there could be a strict connection between the series of perturbation theory and the asymptotic series, like Stirling's series. After Nekhoroshev's work, we know that this is exactly the case.

There are some aspects common to the KAM and Nekhoroshev's approach: essentially, in both cases we get a strong mathematical result by relaxing our requirements. Let me elucidate this point. If the perturbation series were convergent, they would typically give solutions valid *(i)* for all times and *(ii)* for initial data in an open region of phase space. According to Poincaré's results, it is generally impossible to satisfy both these requirements. The KAM approach looks for solutions valid for all times, but relinquishes all possible initial conditions. In contrast, Nekhoroshev's approach looks for approximate solutions over finite time intervals, but for initial conditions in open regions of phase space.

In the case of KAM theory the result is that for most initial conditions the motion is quasiperiodic with rationally independent frequencies. The corresponding orbits lie on invariant tori which appear as deformations of the corresponding unperturbed tori with the same frequencies. However, the set of invariant tori is a nowhere dense set, and moreover its complement is connected for systems with more than two degrees of freedom.

In the case of Nekhoroshev's theory the result is that the variation in time of the unperturbed actions of the system is small for a finite, but large time. More precisely, denoting by I the action variables and by ε a perturbation parameter, one has an estimate of the form

$$(1.1) \qquad |I(t) - I(0)| < \varepsilon^b \quad \text{for } |t| \leq T_0 \exp\left[(\varepsilon_*/\varepsilon)^a\right] \;,$$

where T_0, ε_*, a and b are suitable constants, depending in particular on the number n of degrees of freedom. The interesting fact is the exponential dependence of the time on the inverse of the perturbation parameter.[1] The common characteristic of both theorems is that one has to pay something, and even a lot, in order to get a significant result.

Let me now skip the discussion about the possible interest of the KAM theorem from a probabilistic viewpoint. I will instead concentrate my attention on Nekhoroshev's theorem. This in view of the fact that the possi-

---

[1] It should be stressed that such a statement cannot be deduced from the continuity of solutions with respect to the initial data. The following example illustrates the difference. Consider an equilibrium point, and try to solve the following problem: given a starting point $P(0)$ at distance ϱ from the equilibrium, find a time $T(\varrho)$ such that the distance of $P(t)$ from the equilibrium is less than, say, 2ϱ for $|t| \leq T(\varrho)$. The continuity with respect to initial data cannot give in general a result better than $T(\varrho) = \text{const}$; this is easily seen by considering the equation $\dot{x} = x$. Nekhoroshev's theorem gives instead $T(\varrho) \sim \exp(\varrho^{-a})$. Of course, additional hypotheses on the system are necessary. In our case, for instance, the Hamiltonian character of the system plays a fundamental role, together with the facts that the system is near to integrable and that the unperturbed system satisfies some steepness conditions.

bility of choosing the initial condition in an open set is clearly compatible with experimental errors in the initial data.

It is quite obvious that a Nekhoroshev-like result could be very relevant for the problem of stability of orbits in systems with a small number of degrees of freedom, e.g., the solar system. That it can be effectively applied is a less obvious fact. Indeed, Nekhoroshev's theorem, like the KAM theorem, states that the result holds true provided the perturbation is small enough. For instance, in the case of the solar system this means that the Jupiter's mass should be small. The trouble is that, according to the general estimates available, "small enough" could mean, e.g., "smaller than the mass of a proton". Of course, the estimates could be improved, but it is not at all evident that Nekhoroshev's results are directly applicable to a realistic model.[2]

Another field where the results of perturbation theory could be very relevant is the ergodic problem in statistical mechanics. In this framework one typically has to deal with the computation of time averages of some interesting quantities. In order to deal with notions with a well defined mathematical sense, one usually looks for time averages over an infinite time. But Nekhoroshev's theorem shows that it can well happen that some relevant quantities are essentially invariant ("frozen") up to very large times, possibly exceeding any physically realistic time scale for the system at hand. The idea is not new: the fact that one has to take into account the relaxation time to (some sort of) equilibrium was already pointed out by Boltzmann and Jeans [9,15,16]; see also [3,12]. However, they were not able to give their idea strong mathematical support. Such support could perhaps be found in the framework of a Nekhoroshev–like theory.

The main trouble is the following. According to the existing general formulations, the constants T_0, ε_* and a in the exponential estimate 1.1 present a bad dependence on the number n of degrees of freedom. If such a dependence is optimal, then Nekhoroshev's theorem is deprived of any interest for very large systems. One could imagine two possible ways out of this trouble:

(i) improving the estimates in order to remove the dependence on n, and/or

(ii) trying a further relaxation of the requirements.

Let me elucidate the latter point.

In a naive approach one could think that the badness of the estimates could be simply due to some bad setting or approximation in the theory;

---

[2] An attempt to produce realistic estimates has been made in the case of the Lagrangian equilibria of the restricted spatial problem of three bodies. The final result is that there is a neighborhood of about 10^4 to 10^5 Km which is effectively stable, in the sense that orbits starting in that neighbourhood cannot leave a neighbourhood of twice that size up to a time of the order of the estimated age of the universe [10,24]. Such a result is clearly realistic, but still not enough to prove the stability of the orbits of the Trojan asteroids.

if this were the case, then one should be able to remove the dependence on n from the estimates. But some recent results indicate that in general such an improvement cannot be made (see for instance [18,23]). Thus, one has to rely on some more or less strange properties of some specific system in order to get a further improvement of the estimates. This is a hard mathematical problem.

A more interesting question is based on the following considerations. Nekhoroshev's result has been paid for by relinquishing information over an infinite time. By the way, we rely on the fact that the age of the universe is a time large enough for physical purposes. There remains a bad dependence on the number of degrees of freedom. Indeed, even if we admit that the number of atoms in the universe is finite, nevertheless all the constants go to zero too rapidly for our purposes. By analogy, we observe that in Nekhoroshev's approach one takes into account the *actions* of the system, and tries to prove that *all of them* are essentially frozen. But we can also remark that the freezing of *all* the actions is still a quite strong request. Indeed, we know very well that a resonance typically causes the actions to vary quite rapidly. The question is: can we relax our request, so that the dependence on n can be removed, or at least weakened? More explicitly, the suggestion is to look for quantities eligible as adiabatic invariants of a particular system, without making reference to the actions.

A possible answer takes into account two facts. The first one is the stabilizing effect of the resonances;[3] the second fact is that the frequency

---

[3] Resonance has been often considered as a typical source of instability. This is perhaps due to the fact that the resonant behavior of a near-to-integrable system is quite different from the behavior of the corresponding unperturbed system. For instance, even at purely formal level, perturbation theory indicates that resonance causes a loss of approximate first integrals (or adiabatic invariants). In fact, things are more complicated. Resonance can typically be responsible for the locking of some angles, thus giving rise to libration phenomena. This fact has been well known to astronomers for a long time (consider, for instance, the 5/2 resonance between Jupiter and Saturn). Numerical simulations clearly indicate that the regions of libration associated to a resonance often survive even when all the ordered behavior in the nonresonant regions has been destroyed by the perturbation: it is very easy to observe this phenomenon in the phase portrait of a system with two degrees of freedom. Furthermore, a clever quantitative approach to perturbation theory reveals that the regions of libration are, in some sense, the most stable ones. Roughly, one can say that the loss of some approximate first integrals is compensated by a gain in robustness of the surviving ones. Illustrating the technical mechanism requires a short discussion about the role of the small divisors in the series generated via perturbation methods. A small divisor is an expression of the form $k \cdot \omega$, where $k \in \mathbf{Z}^n$ is a vector of integers, and $\omega \in \mathbf{R}^n$ is the vector of frequencies. If the components $\omega_1, \dots, \omega_n$ of ω are independent over the rationals, then $k \cdot \omega$ never vanishes for $k \neq 0$, but it can assume arbitrarily small values. A typical bound from below has the form $|k \cdot \omega| \geq \gamma |k|^{-\tau}$, with suitable constants $\gamma > 0$ and $\tau > n - 1$, and with $|k| = |k_1| + \dots + |k_n|$. The exponent $\tau > n-1$ is the source of the worst dependence on n in the exponential estimate 1.1: the estimated value of the constant a is $1/(\tau + 1)$ in the best case. If however the frequencies are completely resonant, i.e., if the relation $k \cdot \omega = 0$ holds for $n - 1$ independent integer vectors, then $|k \cdot \omega|$ either is zero, or is larger than some positive constant. The existence of $n - 1$ resonant k's causes a loss of

can play the role of a perturbation parameter. These facts have been pointed out by Benettin, Galgani and the author in the context of research concerned with systems of interest in statistical mechanics [4,5,6,7,8]. A general scheme is the following. One considers a mechanical system composed of two separate subsystems: the first one is a system of identical harmonic oscillators characterized by a fixed frequency, ω say; the second system is a generic one, the only prescription being that it is characterized by some typical time τ, which is very large with respect to the period $2\pi/\omega$ of the first system. The subsystems are coupled by a term which is at least proportional to the coordinates of the oscillators. As a concrete example, one can consider a system of identical diatomic molecules interacting via a short range smooth potential: the internal vibrations represent the first subsystem; the translations and the rotations represent the second subsystem; the interaction potential contributes the coupling between the two subsystems. Under such conditions the sharing of energy internal to each subsystem cannot generally be bounded: the oscillators form a completely resonant system, and the exchange of energy between identical oscillators is completely free; the internal behavior of the second subsystem is completely unknown. However, it turns out that *(i)* the coupling between the subsystems is proportional to the inverse $1/\omega$ of the frequency, and *(ii)* the total harmonic energy of the high frequency subsystem is a good adiabatic invariant. With quantitative estimates in the spirit of Nekhoroshev's theory one proves that no significant exchange of energy between the subsystems can take place up to a time of order $T_0 \exp(\omega/\omega_*)$, with constants T_0 and ω_* possibly still dependent on n. Strictly speaking, the problem of removing the dependence on n has not been solved, but at least the worst dependence (the exponent a in 1.1) has disappeared. A complete elimination of the dependence on n is a hard mathematical task: one should probably complement the perturbation techniques with statistical considerations in order to get a satisfactory result.

Since the mathematical aspects of the problems illustrated here seemed to be too hard, with some collaborators I decided to perform a numerical exploration. The goal of our computation was to make a numerical check of the dependence of the parameters entering the exponential estimates on the number n of degrees of freedom. Of course, we are well aware that a numerical experiment cannot replace a mathematical proof. So, we consider our results only as indications. We hope to be able, at some future time, to support our numerical results with mathematical proofs.

My aim is now to give a brief report on our numerical results. The interested reader will find more details in [13].

$n - 1$ approximate integrals, so that only one approximate integral survives, in general independent of the Hamiltonian. The reward is that the bound of the small denominator with a constant removes the worst dependence on n in the exponential estimate: one finds $a = 1$.

2. The model problems. In the spirit of the discussion in the previous section, we tried to answer the following two questions:

(i) Is there a numerical indication that the general Nekhoroshev estimates
   are essentially optimal, at least concerning the dependence on n?

(ii) Is there an indication that a result weaker than Nekhoroshev's can be
   possibly extended to the thermodynamic limit?

By "general Nekhoroshev estimates" I mean here estimates concerning the conservation of *all* the action variables of the system.

To address the first question, we investigated the celebrated model of Fermi, Pasta and Ulam [11], usually referred to as the FPU model. The reason of our choice was that it is representative of the general problem of the stability of an elliptic equilibrium point. Concerning the second question, we considered a variant of the FPU model which is often used in solid state physics. Although the models are quite well known, let me briefly recall the relevant information.

. The FPU model is a chain of $n + 2$ equal mass points on a line, connected with identical nonlinear springs; the end points of the chain are fixed, so that the system has n degrees of freedom. If $x_0, \ldots, x_{n+1}$ denote the displacements of the mass points from their equilibrium positions, and $y_1, \ldots, y_n$ denote the corresponding momenta, the Hamiltonian of the system takes the form

$$H(x, y) = \frac{1}{2} \sum_{j=0}^{n} [y_j^2 + (x_{j+1} - x_j)^2] + \frac{\alpha}{3} \sum_{j=0}^{n} (x_{j+1} - x_j)^3 + \frac{\beta}{4} \sum_{j=0}^{n} (x_{j+1} - x_j)^4.$$

(2.1)

The condition of fixed ends is $x_0 = x_{n+1} = 0$. Here, as in the original paper by Fermi, Pasta and Ulam, the nonlinearity is due to the cubic and quartic terms of the potential; the parameters α and β control the size of the interaction. For $\alpha = \beta = 0$ the system is well known to be integrable, and is equivalent to a system of n harmonic oscillators with frequencies

(2.2) $$\omega_j = 2 \sin \frac{j\pi}{2(n+1)} , \quad 1 \leq j \leq n .$$

The canonical transformation

(2.3)
$$q_l = \sqrt{\frac{2}{\omega_l(n+1)}} \sum_{j=1}^{n} x_j \sin \frac{jl\pi}{n+1} ,$$

$$p_l = \sqrt{\frac{2\omega_l}{n+1}} \sum_{j=1}^{n} y_j \sin \frac{jl\pi}{n+1} , \quad 1 \leq l \leq n ,$$

introduces the normal modes of the system. The transformed Hamiltonian reads

(2.4) $$H(p, q) = \frac{1}{2} \sum_{l=1}^{n} \omega_l (q_l^2 + p_l^2) + H_3 + H_4 ,$$

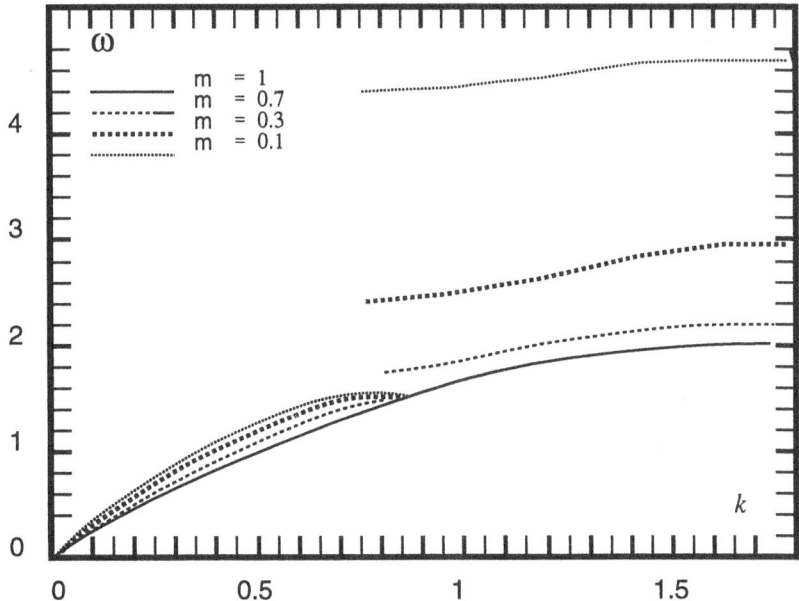

FIG. 2.1. *Frequency spectrum ω vs. k for the modified FPU model, for different values of the mass parameter. The figure clearly exhibits the separation of the acoustic and the optical branches.*

where H_3 and H_4 are the cubic and quartic parts respectively of the Hamiltonian; the explicit form of these polynomials is not relevant here. It is evident that the Hamiltonian has the form of a system of nonlinearly coupled harmonic oscillators.

The modified FPU model differs from the original one in that n is even and the points have alternating light and heavy masses. Considering only a quartic nonlinear term, the Hamiltonian of the model can be written as

$$(2.5) \quad H(x,y) = \frac{1}{2} \sum_{j=0}^{n} \left[\frac{y_j^2}{m_j} + (x_{j+1} - x_j)^2 \right] + \frac{\beta}{4} \sum_{j=0}^{n} (x_{j+1} - x_j)^4 ,$$

where

$$(2.6) \qquad m_j = 1 \text{ for } j \text{ odd} , \; m_j = m < 1 \text{ for } j \text{ even} .$$

Here too the system with $\beta = 0$ is integrable, and is equivalent to a system of harmonic oscillators with frequencies

$$(2.7) \qquad (\omega_l^{\pm})^2 = \frac{1 + m \pm \sqrt{1 + m^2 + 2m \cos 2k_l}}{m} ,$$

where

(2.8)
$$k_l = \frac{l\pi}{n+1} \ , \quad 1 \le l \le \frac{n}{2} \ .$$

Here, the signs $-$ and $+$ refer to the two possible choices of the sign in the formula; the set of frequencies ω^- characterizes the so-called acoustic modes, while the frequencies ω^+ characterize the so-called optical modes. The reason for these names turns out to be evident if one looks at the plot of the frequency spectrum in fig. 2.1: it is clearly seen that for $m \ll 1$ the spectrum splits into two well separated components. The low frequency part of the spectrum—the acoustic modes—exhibits essentially the same pattern as the spectrum of the FPU system. In contrast, the high frequency part of the spectrum becomes flatter and flatter with decreasing m. To make this information quantitative, let me introduce the following relevant quantities:

(2.9)
$$\omega^-_{\text{max}} = \sqrt{2} \ , \quad \omega^+_{\text{min}} = \sqrt{\frac{2}{m}} \ , \quad \omega^+_{\text{max}} = \sqrt{\frac{2(1+m)}{m}} \ ;$$

the meaning of these quantities is quite obvious. Let me also introduce the average frequency ω of the optical modes, defined as

(2.10)
$$\omega = \sqrt{\frac{{\omega^+_{\text{min}}}^2 + {\omega^+_{\text{max}}}^2}{2}} = \sqrt{\frac{2+m}{m}} \ .$$

For $m \to 0$ it is clearly seen that: *(i)* the average optical frequency ω goes to infinity as $\sqrt{2/m}$, *(ii)* the spreading $\delta\omega = \omega^+_{\text{max}} - \omega^+_{\text{min}}$ goes to zero as $\sqrt{m/2}$, and *(iii)* the ratio $\omega/\omega^-_{\text{max}}$ between the average optical frequency and the maximal acoustic frequency goes to infinity as $\sqrt{1/m}$. Thus, for m small enough we can represent the whole system as composed of two separate subsystems—the acoustic and the optical modes respectively—characterized by well separated frequencies. With a canonical transformation to normal modes, and with a suitable rearrangement, the Hamiltonian can be given the form

(2.11)
$$H(p, x, \pi, \xi) = \hat{h}(p, x) + h_\omega(\pi, \xi) + f(p, x, \pi, \xi) \ .$$

Here, p, x are canonical coordinates describing the acoustic modes, while π, ξ describe the optical modes. The rearrangement of the various terms of the transformed Hamiltonian is done as follows. All the terms independent of π, ξ are collected in $\hat{h}(p, x)$; this is a system of nonlinear oscillators. The term $h_\omega(\pi, \xi)$ has the form

(2.12)
$$h_\omega(\pi, \xi) = \frac{1}{2} \sum_l (p_l^2 + \omega^2 q_l^2) \ ,$$

where ω is the average optical frequency defined by 2.10; this describes a set of identical harmonic oscillators. Finally, all the remaining terms are collected in $f(p, x, \pi, \xi)$; this contains some small quadratic terms in π, ξ (because the optical frequencies are not exactly equal), and the quartic part of the Hamiltonian depending at least linearly on ξ. Thus we have a system of the type described in section 1, to which the results of Benettin et al. [8] can be applied.

3. Results for the FPU model. This study is concerned with the first question at the beginning of sect. 2. As a theoretical basis for the model we used the results of Giorgilli [14]. I give here a short summary. Making reference to the variables p, q of the normal modes, consider the harmonic actions I_l of the oscillators defined by

$$(3.1) \qquad I_l = \frac{1}{2}\left(p_l^2 + \omega_l^2 q_l^2\right), \quad 1 \le l \le n;$$

consider now a neighborhood of the origin in $\mathbf{R}^{2n}$ of the form

$$(3.2) \qquad \mathcal{D}_\varrho = \left\{(p, q) \in \mathbf{R}^{2n} : \sqrt{I_l} \le \varrho,\, 1 \le l \le n\right\}.$$

Then there are constants a, c, d, μ, ϱ_0 depending on the number n of degrees of freedom such that:

i. for $\varrho < \varrho_0$ there exist n approximate first integrals $\Phi^{(1)}, \ldots, \Phi^{(n)}$ of the Hamiltonian 2.1 which are polynomials of a suitable degree $r(\varrho)$, namely of the form

$$(3.3) \qquad \Phi^{(l)} = I_l + \Phi_3^{(l)} + \ldots + \Phi_r^{(l)},$$

with $\Phi_s^{(l)}$ a homogeneous polynomial of degree s in p, q.

ii. for every $(p, q) \in \mathcal{D}_\varrho$ these approximate first integrals satisfy the estimates (uniform with respect to l)

$$(3.4) \qquad \begin{aligned} \left|\Phi^{(l)}(p, q) - I_l(p, q)\right| &< \delta(\varrho) := d\varrho^3, \\ \left|\dot{\Phi}^{(l)}(p, q)\right| &\le b(\varrho) := a \exp\left(-\frac{\mu}{\varrho^c}\right). \end{aligned}$$

As to the behavior of a, c, μ and ϱ_0 with respect to n, we have the theoretical estimates

$$(3.5) \qquad c \sim \frac{1}{n}, \quad \mu \sim n, \quad \ln a \sim n, \quad \varrho_0 \sim \frac{1}{3^n n^{5/2}}.$$

Our aim was to check the estimates 3.4, and in particular the dependence on n of a, μ and c as given by 3.5. Our computational scheme was the following.

1. For fixed n, take several initial conditions at different total energies, and integrate numerically Hamilton's equations for the system 2.1; the relation between the total energy E and the parameter ϱ of the domain is

(3.6)
$$\varrho = \sqrt{\frac{2E}{\omega_1}} \; .$$

2. For every orbit, compute the quantity

(3.7)
$$\tilde{b}(\varrho) = \max_l \sup_t \left| \frac{I_l(t + \Delta t) - I_l(t)}{\Delta t} \right| \; ;$$

this quantity should be identified with $b(\varrho)$ in 3.4. We checked that the result does not significantly depend on Δt provided we take $10T_1 < \Delta t < 50T_1$, where $T_1 = 2\pi/\omega_1$ is the maximal harmonic period.[4]

3. Evaluate the constants a, μ and c in the expression 3.4 of $b(\varrho)$ via a best fit with the computed values of $\tilde{b}(\varrho)$.

4. Do the steps 1, 2, and 3 above for different values of n, and plot the estimated values of a, μ and c vs. n. This provides an opportunity to compare with the theoretically expected behavior 3.5.

I report in fig. 3.1 only the results for c; similar results were obtained for a and μ (for a complete report, see [13]). At first sight, they look quite far from the theoretical behavior $c \sim 1/n$. However, one should notice that all the theoretical estimates are concerned with asymptotic behavior. In this respect, the relevant fact is that for $n > 30$ the numerical data clearly indicate that c is a decreasing function of n. Our data are not enough, and in particular not accurate enough, to exclude the possibility that c goes, say, as $1/\sqrt{n}$. But, as far as the question of whether the dependence on n can be removed or not, our data give at least a strong indication that such a dependence cannot be removed. Thus, the numerical investigations suggest that the general Nekhoroshev estimates for this model are essentially optimal.

---

[4] The problem here is that one should evaluate the time derivative of a function $\Phi^{(l)}$ which is unknown. Thus, all we can do is to use the action I_l, which is a first approximation of $\Phi^{(l)}$, and to remove the quasiperiodic components in the variation of I_l by averaging over a suitable time interval. For a more detailed discussion of this point see [13]

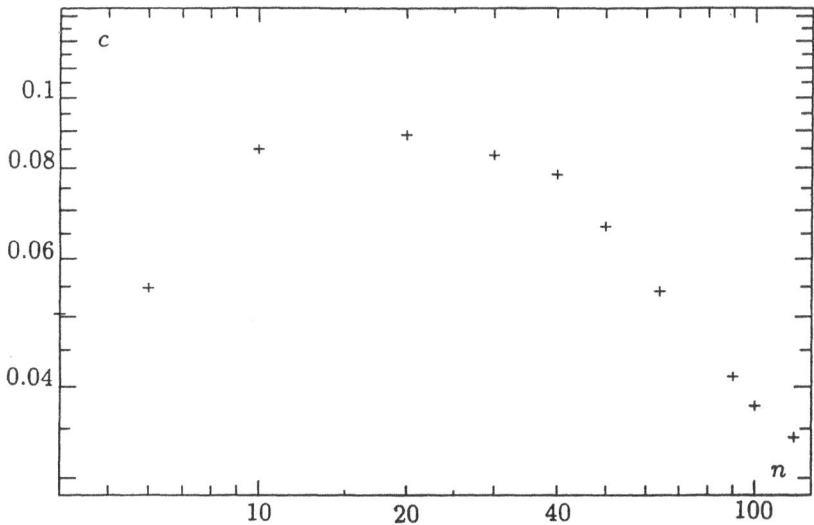

FIG. 3.1. *Numerical estimate of the quantity c in 3.5 vs. n, to be compared with the theoretical estimate 2.10. Although the expected behavior $c \sim 1/n$ cannot be considered as fully confirmed, it seems nevertheless reasonable to conclude that c decreases to zero with increasing n.*

4. The modified FPU model. I come now to consider the second question at the beginning of sect. 2. We considered the modified FPU model. As already remarked, the theoretical basis is furnished by the estimates in [8]. Here too I summarize the results. Make reference to the form 2.11 of the Hamiltonian, and introduce the parameter

$$(4.1) \qquad \lambda = \frac{\omega}{\omega_{\max}^{-}} ,$$

where $\omega_{\max}^{-}$ and ω are defined by 2.9 and 2.10. Denote now by E the total energy of the system. Then there are constants A and C, depending on n and E, such that:

i. for $\lambda > CE$ there exists an approximate first integral of the form $\Phi = h_\omega + \ldots$;

ii. for every p, x, π, ξ satisfying $H(p, x, \pi, \xi) \le E$ one has the estimate

$$(4.2) \qquad \lambda|\dot{\Phi}(p, x, \pi, \xi)| \le A\exp(-B\lambda) , \quad B = \frac{e}{CE} .$$

Here too we are interested in the dependence of the constants A and C on n. Having in mind the possible extension of the results to the thermodynamic limit, we consider as a parameter the specific energy E/n instead of the total energy E. In this case the theoretical estimates give

$$(4.3) \qquad\qquad A \sim n^6 , \quad C \sim n^8 .$$

The numerical scheme is as follows.

1. Fix both the specific energy E/n and the number n of degrees of freedom; for different values of λ choose several different initial conditions (e.g., change only the phases of the oscillators) and integrate numerically Hamilton's equations for the system 2.5.

2. Compute the quantity

$$(4.4) \qquad\qquad \dot{E}^+ = \sup_t \langle |\dot{h}_\omega(t)| \rangle ;$$

this quantity should be considered as a bona fide estimate of $|\dot{\Phi}|$.

3. Plot $\lambda \dot{E}^+$ vs. λ on a log scale: if the exponential law 4.2 holds, then the data should be aligned on a straight line; the slope of the line estimates the value of B, while the intercept with the origin estimates the value of A.

4. Keeping the specific energy E/n fixed, repeat steps 1, 2 and 3 above for different values of n, and compare all the graphs.

The results are illustrated in figs. 4.1 and 4.2. Fig. 4.1 shows that the data corresponding to different values of n are quite well aligned on straight lines, and that the slope of the lines is independent of n; this means that the constant B in the exponential law 4.2 is independent of n. There remains possibly a weak dependence on n in the constant A. In fig. 4.2 the same data are used to plot $\lambda \dot{E}^+ / \ln n$ vs. λ: it is evident that the intercepts of the straight lines are now very close, indicating that A should grow at most like $\ln n$. This shows that the theoretical estimates in 4.3 are very pessimistic: an improvement of the theory should be possible.[5]

---

[5] After publication of these results, and after my talk at the Conference in Cincinnati, D. Bambusi succeeded in proving that the constants A and B do not depend on n in the case of fixed *total* energy. The extension of the results to the case of fixed *specific* energy seems to require statistical considerations on the initial conditions. The reason is as follows: the possibility that all the energy concentrates on a single oscillator for a long time cannot be excluded on the basis of dynamical considerations, but is clearly improbable from a statistical viewpoint. Taking into account such unlikely situations requires the condition of fixed total energy. The results will be published elsewhere.

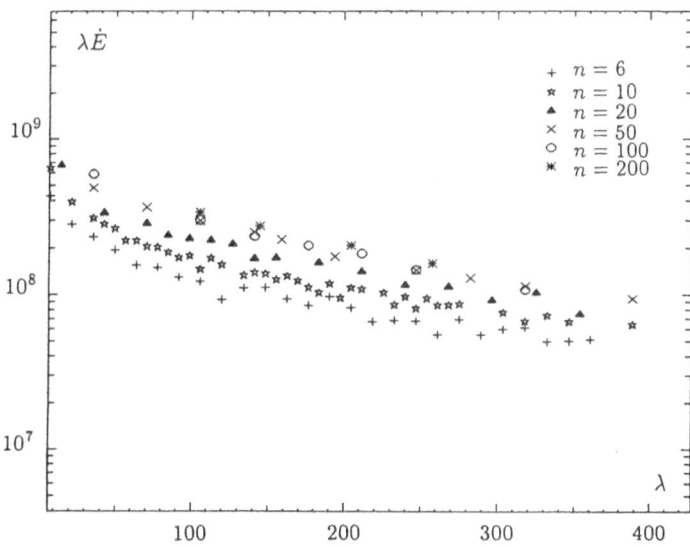

FIG. 4.1. *Numerical estimate of* $\lambda\dot{E}^+$ *vs.* λ, *to be compared with the theoretical esti-mate 4.2. The data corresponding to different values of* n *are quite well aligned on straight lines, in agreement with the purely exponential estimate. The slope of the straight lines appears to be independent of* n, *which suggests* $B = $ const; *the intercepts of the straight lines at the origin exhibit instead a small dependence on* n, *which suggests a small dependence on* n *in* A.

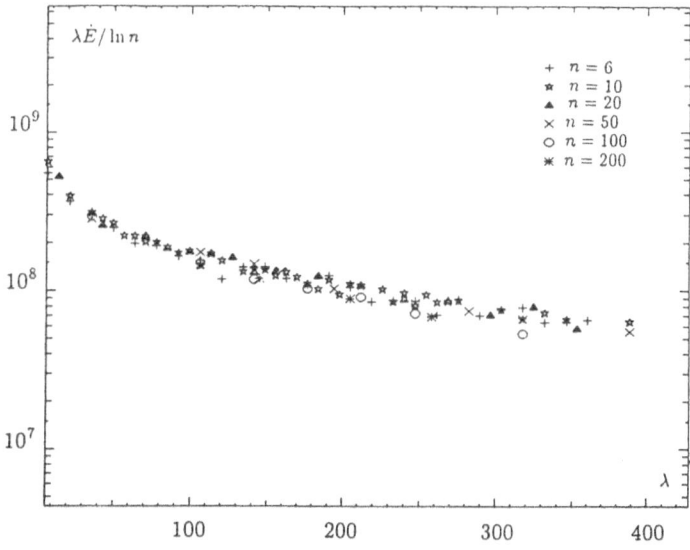

FIG. 4.2. *Same as figure 4.1, with* $\lambda\dot{E}$ *replaced by* $\lambda\dot{E}/\ln n$, *in order to check the* ansatz $A \sim \ln n$. *The figure shows that the increase of* A *with* n *should be less than* $\ln n$.

5. Conclusions. The numerical investigation of systems of FPU type suggest that Nekhoroshev's estimates concerning the actions of a-near-to integrable Hamiltonian system cannot be extended to the case of systems with a large number of degrees of freedom. This fact seems to support the common belief that for statistical systems there are no adiabatic invariants, and that equipartition of energy should hold.

However, there exists a relevant class of systems for which the relaxation time to equipartition could be very large. These systems are characterized by a splitting of the whole system into two different subsystems characterized by widely different proper frequencies: in such a case the relaxation time grows as fast as the exponential of the ratio between the high frequency and the low one. The validity of these results for large systems is however still an open problem, because the available theoretical estimates contain a dependence on the number n of degrees of freedom which could render the result invalid in the thermodynamic limit. In this case the numerical investigations suggest that the estimates could be improved, and made significant also for large n, e.g., on the order of Avogadro's number.

REFERENCES

[1] V.I. ARNOLD, *Small denominators II. Proof of a theorem of A. N. Kolmogorov on the invariance of quasi-periodic motions under small perturbations of the hamiltonian*, Uspekhi Mat. Nauk 18 (1963), 13–39.

[2] V.I. ARNOLD, *Small denominators and problems of stability of motion in classical and celestial mechanics*, Uspekhi Mat. Nauk 18 (1963), 91–179.

[3] G. BENETTIN, L. GALGANI, A. GIORGILLI, *Boltzmann's ultraviolet cutoff and Nekhoroshev's theorem on arnold diffusion*, Nature 311 (1984), 344.

[4] G. BENETTIN, L. GALGANI, A. GIORGILLI, *A proof of Nekhoroshev's theorem for the stability times in nearly integrable hamiltonian systems*, Celest. Mech. 37 (1985), 1–25.

[5] G. BENETTIN, L. GALGANI, A. GIORGILLI, *Classical perturbation theory for a system of weakly coupled rotators*, Nuovo Cim. B 89 (1985), 89–102.

[6] G. BENETTIN, L. GALGANI, A. GIORGILLI, *Numerical investigations on a chain of weakly coupled rotators in the light of classical perturbation theory*, Nuovo Cim. B 89 (1985), 103–119.

[7] G. BENETTIN, L. GALGANI, A. GIORGILLI, *Exponential estimate of the stability time among translational and vibrational degrees of freedom*, Phys. Lett. A 120 (1987), 23–27.

[8] G. BENETTIN, L. GALGANI, A. GIORGILLI, *Realization of holonomic constraints and freezing of high frequency degrees of freedom in the light of classical perturbation theory, Part II*, Comm. Math. Phys. 121 (1989), 557–601.

[9] L. BOLTZMANN, Nature 51 (1895), 413–415.

[10] A. CELLETTI, A. GIORGILLI, *On the stability of the lagrangian points in the spatial restricted problem of the three bodies*, Celest. Mech. 50 (1991), 31–58.

[11] E. FERMI, J. PASTA, S. ULAM, *Los Alamos report No. LA-1940*, later published in Lect. Appl. Math. 15 (1955), 143.

[12] L. GALGANI, *Relaxation times and the foundations of classical statistical mechanics in the light of modern perturbation theory*, Nonlinear Evolution and Chaotic Phenomena, Nato ASI Series B 176, G.Gallavotti, P.F.Zweifel, eds. 1987, 147–159.

[13] L. GALGANI, A. GIORGILLI, A. MARTINOLI, S. VANZINI, On the problem of energy equipartition for large systems of the Fermi-Pasta-Ulam type: analytical and numerical estimates, Physica D 59 (1992), 334–348.

[14] A. GIORGILLI, Rigorous results on the power expansions for the integrals of a hamiltonian system near an elliptic equilibrium point, A. Inst. H. Poincaré 48 (1988), 423–439.

[15] J.H. JEANS, On the vibrations set up in molecules by collisions, Phil. Magazine 6 (1903), 279.

[16] J.H. JEANS, On the partition of energy between matter and aether, Phil. Magazine 10 (1905), 91.

[17] A.N. KOLMOGOROV, The conservation of conditionally periodic motions with a small variation in the hamiltonian, Dokl. Akad. Nauk SSSR 98 (1954), 527–530.

[18] P. LOCHAK, Stabilité en temps exponentiels des systèmes hamiltoniens proches de systèmes intégrables: résonances et orbites fermées, preprint (1989).

[19] J. MOSER, On invariant curves of area-preserving mappings on an annulus, Nachr. Akad. Wiss. Göttingen Math. Phys. Kl. II (1962), 1–20.

[20] N.N. NEKHOROSHEV, Exponential estimates of the stability time of near-integrable hamiltonian systems, Russ. Math. Surveys 32 (1977), 1-65.

[21] N.N. NEKHOROSHEV, Exponential estimates of the stability time of near-integrable hamiltonian systems, 2, Trudy Sem. Petrovs. 5 (1979), 5–50.

[22] H. POINCARÉ, Les méthodes nouvelles de la mécanique céleste, Gauthier-Villars, Paris 1892.

[23] J. PÖSCHEL, On Nekhoroshev's estimate for quasi-convex hamiltonians, preprint ETH, Zürich 1991.

[24] C. SIMÓ, Estabilitat de sistemes hamiltonians, Memorias de la Real Academia de Ciencias y Artes de Barcelona 48 (1989), 303–348.

SUSPENSION OF SYMPLECTIC TWIST MAPS BY HAMILTONIANS

CHRISTOPHE GOLÉ*

Abstract. We extend some results of Moser [17], Bialy and Polterovitch [1], on the suspension of symplectic twist maps by Hamiltonian flows.

1. Introduction. At this conference, the author talked about symplectic twist maps in the cotangent bundle T^*M of a general compact manifold M, and their use to prove theorems of existence of periodic orbits for Hamiltonians in T^*M. The crucial observation was that, for large classes of (time dependent) Hamiltonian systems on T^*M, the time-1 map can be decomposed into a finite product of symplectic twist maps. These symplectic twist maps each come equipped with a generating function which translates the problem of finding periodic orbits into one of finding critical points of a function W on a finite dimensional manifold (W is a certain sum of the generating functions at successive points.) This can be seen as a generalization of the classical method of broken geodesics.

There has been a considerable development of similar finite dimensional, or "discrete" techniques these last few years in symplectic geometry [2,3], [13], [14,15], [18], [19,20], [5-10]. All can be given the geometric interpretation of "generating phases" for Lagrangian manifolds. For instance, dW is a generating phase for the graph of the time-1 of our Hamiltonian. These techniques, when they apply, are usually much simpler (because finite dimensional) than the more analytical techniques of calculus of variation. The main stumbling block at this point is to find a correct way to apply the technique of generating phases to compact symplectic manifolds.

Here, we go somewhat against this trend, as instead of transforming a continuous (Hamiltonian) system into a discrete one, we present a result that makes a small contribution to the question: when is a symplectic twist map the time-1 of a Hamiltonian?

J. Moser [17] showed that any C^∞ twist map of the (bounded) annulus can be suspended by a Hamiltonian flow. M. Bialy and L. Polterovitch [1] remarked that his method could be extended to certain maps of $T^*\mathbb{T}^n$. The Hamiltonian is constructed in such a way that it is "optical" (H_{pp} definite positive). However, a certain condition of symmetry in the second derivative of the generating function is necessary for Moser's method to work.

The method we use here does not require this symmetry condition. But we pay the price of not being able (as to now) to make the Hamiltonian optical, or time periodic.

---

* Institute for Mathematical Sciences, SUNY at Stony Brook, Stony Brook, NY 11794-3660. Partially supported by an NSF postdoctoral grant DMS 91-07950.

We start with some definitions and lemmas about symplectic twist maps. In the third section, we prove the theorem of suspension. Finally, in the last section, we state a few results for Hamiltonian systems in cotangent bundles.

2. Symplectic twist maps and their generating functions. Since our main result is about maps of $T^*\mathbb{T}^n$, we will concentrate on this space. We will let the reader consult [5–10] for a natural generalization of symplectic twist maps on cotangents of arbitrary manifolds.

Let $\mathbb{T}^n = \mathbb{R}^n/\mathbb{Z}^n$ be the n–dimensional torus. Its cotangent bundle $T^*\mathbb{T}^n \xrightarrow{\pi} \mathbb{T}^n$ is trivial: $T^*\mathbb{T}^n = \mathbb{T}^n \times \mathbb{R}^n$, the cartesian product of n cylinders. We give it the coordinates (q, p) in which the symplectic structure is

$$\Omega = dq \wedge dp = \sum_{k=1}^{n} dq_k \wedge dp_k.$$

As in any cotangent bundle, Ω is exact: $\Omega = -d\lambda$, where $\lambda = p\,dq$.

It is useful to work in the covering space $\mathbb{R}^{2n} = \tilde{\mathbb{T}}^n \times \mathbb{R}^n$ of $T^*\mathbb{T}^n$, with projection $pr : \mathbb{R}^{2n} \to \mathbb{T}^n \times \mathbb{R}^n$.

We will fix the lift of a map F once and for all, remembering that two lifts only differ by a composition with a translation of some integer vector m.

DEFINITION 2.1. *A map F of $T^*\mathbb{T}^n$ is called a* **symplectic twist map** *if*

(1) F *is homotopic to Id.*

(2) F *is* **exact symplectic:** $F^*p\,dq - p\,dq = dh$ *for some* $h : T^*\mathbb{T}^n \to \mathbb{R}$.

(3) **(Twist Condition)** *If* $\tilde{F}(q, p) = (Q, P)$ *is a lift of F then the map $p \to Q(q_0, p)$ is a diffeomorphism of $\mathbb{R}^n$ for all q_0, and thus the map*

$$\psi : (q, p) \to (q, Q)$$

is a diffeomorphism (change of coordinates) of $\mathbb{R}^{2n}$.

The usual twist maps of the annulus $\mathbb{S}^1 \times \mathbb{R}$ are easily seen to be symplectic twist maps. In this case, *(3)* is equivalent to the usual monotone twist condition:

$$\frac{\partial Q}{\partial p} > a > 0.$$

The following results are also helpful to construct symplectic twist maps. Their proofs can be found in [10] (see also [11] for Corollary 2.7.)

PROPOSITION 2.1. *There is a homeomorphism between the set of lifts $\tilde{F}$ of C^1 symplectic twist maps of $T^*\mathbb{T}^n$ and the set of C^2 real valued functions S on $\mathbb{R}^{2n}$ satisfying the following:*

(a) $S(q + m, Q + m) = S(q, Q), \quad \forall m \in \mathbb{Z}^n$

(b) The maps: $q \to \partial_2 S(q, Q_0)$ and $Q \to \partial_1 S(q_0, Q)$ are diffeomorphisms of $\mathbb{R}^n$ for any Q_0 and q_0 respectively.

(c) $S(0, 0) = 0.$

This correspondence is given by:

$$(2.1) \qquad \tilde{F}(q, p) = (Q, P) \Leftrightarrow \begin{cases} p = & -\partial_1 S(q, Q) \\ P = & \partial_2 S(q, Q). \end{cases}$$

LEMMA 2.1.*Let $f : \mathbb{R}^N \to \mathbb{R}^N$ be a local diffeomorphism at each point, such that:*

$$\sup_{x \in \mathbb{R}^N} \left\| (Df_x)^{-1} \right\| = K < \infty.$$

Then f is a diffeomorphism of $\mathbb{R}^N$.

COROLLARY 2.1.*Let $S : \mathbb{R}^{2n} \to \mathbb{R}$ be a C^2 function satisfying:*

$$(2.2) \qquad \begin{aligned} &\det \partial_{12} S \neq 0 \\ &\sup_{(q, Q) \in \mathbb{R}^{2n}} \left\| \partial_{12} S(q, Q))^{-1} \right\| = K < \infty. \end{aligned}$$

Then the maps: $q \to \partial_2 S(q, Q_0)$ and $Q \to \partial_1 S(q_0, Q)$ are diffeomorphisms of $\mathbb{R}^n$ for any Q_0 and q_0 respectively, and thus S generates an exact symplectic map of $\mathbb{R}^{2n}$.

Thus, if S satisfies (2.2), as well as the periodicity condition *(a)* of Proposition 2.1, it generates a symplectic twist map.

COROLLARY 2.2. *Let F be an exact symplectic map of $T^* \mathbb{T}^n$, homotopic to Id. Let $\tilde{F}(q, p) = (Q, P)$ be a lift of F. Suppose that*

$$(2.3) \qquad \sup_{z \in \mathbb{R}^{2n}} \left\| \left(\frac{\partial Q}{\partial p} \right)^{-1}_z \right\| < \infty.$$

Then $\tilde{F}$ is a symplectic twist map.

3. Suspension of symplectic twist maps by Hamiltonian flows

Let $\tilde{F}(q, p) = (Q, P)$ be the lift of a symplectic twist map of $T^* \mathbb{T}^n$, and $S(q, Q)$ its generating function. In this section, we impose the:

Convexity Condition 3.1. There is a positive a such that:

$$\langle \partial_{12} S(q, Q) . v, v \rangle \leq -a \left\| v \right\|^2 .$$

uniformly in (q, Q).

*Remark 3.1.*Note that:

$$\frac{\partial Q}{\partial p}(q, p) = - (\partial_{12} S(q, Q))^{-1} ,$$

as can easily be derived by implicit differentiation of $p = -\partial_1 S(q, Q)$. The convexity condition 3.1 thus translates to:

$$(3.1) \qquad \left\langle \left(\frac{\partial Q}{\partial p}\right)^{-1} v, v \right\rangle \geq a \left\|v\right\|^2, \quad \forall v \in \mathbb{R}^n.$$

uniformly in (q, p). This means that F has bounded twist. MacKay, Meiss and Stark [16] imposed this condition on their definition of symplectic twist maps, a terminology that we have taken from them.

In [1], Bialy and Polterovitch make the remark that, if the map F satisfies the convexity condition, and if moreover $\partial_{12} S(q, Q)$ is a symmetric matrix, then Moser's method of suspending twist maps by optical Hamiltonian flows goes through. Here, we present a theorem of suspension that does not require $\partial_{12} S$ to be symmetric, at the cost of opticity, and time periodicity of the Hamiltonian:

THEOREM 3.1. Let $F(q, p) = (Q, P)$ be a symplectic twist map of $T^* \mathbb{T}^n$ which satisfies the convexity condition 3.1. Then F is the time 1 map of a (time dependent) Hamiltonian H.

Proof. Let $S(q, Q)$ be the generating function of F. Condition 3.1 can be rewritten:

$$(3.2) \qquad \inf_{(q,Q) \in \mathbb{R}^{2n}} \langle -\partial_{12} S(q, Q) v, v \rangle > a \left\|v\right\|^2, \quad a > 0, \forall v \neq 0 \in \mathbb{R}^n.$$

$\square$

The following lemma, whose proof is left to the reader shows that this inequality implies (2.2). Hence whenever we have a function on $\mathbb{R}^{2n}$ which is suitably periodic and satisfies (3.2), it is the generating function for some symplectic twist map.

LEMMA 3.1. Let $\{A_x\}_{x \in \Lambda}$ be a family of $n \times n$ real matrices satisfying:

$$\sup_{x \in \Lambda} \langle A_x v, v \rangle > a \|v\|^2, \quad \forall v \neq 0 \in \mathbb{R}^n.$$

Then :

$$\sup_{x \in \Lambda} \left\| A_x^{-1} \right\| < a^{-1}.$$

We construct a differentiable family S_t of generating functions, with $S_1 = S$, and then show how to make a Hamiltonian vector field out of it, whose time 1 map is F. Let

$$S_t(q, Q) = \begin{cases} \frac{1}{2} a f(t) \|Q - q\|^2 & \text{for } 0 < t \leq \frac{1}{2} \\ \frac{1}{2} a f(t) \|Q - q\|^2 + (1 - f(t)) S(q, Q) & \text{for } \frac{1}{2} \leq t \leq 1. \end{cases}$$

where f is a smooth positive functions, $f(1) = f'(1/2) = 0$, $f(1/2) = 1$, $\lim_{t \to 0+} f(t) = +\infty$. We will ask also that $1/f(t)$, which can be continued to $1/f(0) = 0$ be differentiable at 0. The choice of f has been made so that S_t is differentiable with respect to t, for $t \in (0, 1]$. Furthermore, it is easy to verify that:

$$\sup_{(q,Q) \in \mathbb{R}^{2n}} \langle -\partial_{12} S_t(q, Q)v, v \rangle > a\, ||v||^2, \quad a > 0, \forall v \neq 0 \in \mathbb{R}^n, t \in (0, 1].$$

Hence S_t generates a smooth family F_t, $t \in (0, 1]$ of symplectic twist maps, and in fact $F_t(q, p) = (q + (af(t))^{-1}p, p)$, $t \leq 1/2$), so that $\lim_{t \to 0^+} F_t = Id$, in the C^2 topology (on compact sets.) Let us write

$$\tilde{S}_t(q, p) = S_t \circ \psi_t(q, p),$$

where ψ_t is the change of coordinates given by the fact that F_t is twist. It is not hard to verify that $\psi_t(q, p) = (q, q - (af(t))^{-1}p)$, $t \leq 1/2$. so that:

$$\tilde{S}_t(q, p) = \frac{1}{2}(af(t))^{-2}||p||^2$$

In particular, by our assumption on $1/f(t)$, $\tilde{S}_t$ can be differentiably continued for all $t \in [0, 1]$, with $S_0 \equiv 0$. Hence, in the q, p coordinates, we can write:

$$F_t^* p dq - p dq = d\tilde{S}_t, \quad t \in [0, 1].$$

A family of maps that satisfies this with $\tilde{S}_t$ differentiable in (q, p, t) is called an *exact symplectic isotopy*.

The proof of the theorem derives from the standard:

LEMMA 3.2. *Let g_t be an exact symplectic isotopy of $T^*\mathbb{T}^n$ (or T^*M, in general.) Then g_t is a Hamiltonian isotopy.*

Proof. Let g_t be an exact symplectic isotopy:

$$g_t^* p dq - p dq = dS_t$$

for some S_t differentiable in all of (q, p, t). We claim that the (time dependent) vector field:

$$X_t(z) = \frac{dg_t}{dt}(g_t^{-1}(z))$$

whose time t is g_t, is Hamiltonian. To see this, we compute:

$$\frac{d}{dt}(d\tilde{S}_t) = \frac{d}{dt} g_t^* p dq = g_t^* L_{X_t} p dq = g_t^* \left(i_{X_t} d(p dq) - d(i_{X_t} p dq) \right),$$

from which we get

$$i_{X_t} dq \wedge dp = dH_t$$

with

$$H_t = \left((g_t^{-1})^* \frac{dS_t}{dt} - i_{X_t} p dq \right),$$

which exactly means that X_t is Hamiltonian.   □

4. Some recent results for Hamiltonian systems. We want to state here some results about the use of symplectic twist maps for Hamiltonian systems. They are proven in [9]. The results announced at this conference about periodic orbits on *compact* generalized annuli in the cotangent of an arbitrary manifold have been presented in [6,8].

The following proposition can, in some sense, be seen as a converse to Theorem 3.1.

PROPOSITION 4.1. *Let h^ϵ be the time ϵ of a Hamiltonian flow for a C^2 Hamiltonian function $H(q, p, t) = H_t(z)$ on $T^* \mathbb{T}^n \times \mathbb{R}$ (or $T^* M \times \mathbb{R}$, where $\tilde{M} = \mathbb{R}^n$) and satisfying the following:*

(1) $\sup \|\nabla^2 H_t\| < K$

(2) The matrices $H_{pp}(z, t)$ are positive definite and $C < \|H_{pp}\| < C^{-1}$.

Then, for all sufficiently small ϵ, h^ϵ is a symplectic twist map of $T^* \mathbb{T}^n$. Moreover, h^ϵ satisfies the convexity condition 3.1.

Using this proposition, one can prove, with the Aubry type variational setting, the following theorem:

THEOREM 4.1. *Let $H(q, p, t)$ be a Hamiltonian function on $T^* \mathbb{T}^n \times \mathbb{R}$ as in Proposition 4.1. Then the time 1 map h^1 of the associated Hamiltonian flow has at least $n + 1$ periodic orbits of type m, d, for each prime m, d, and 2^n when they are all non degenerate (i.e. generically.)*

(A periodic orbit of type m, d for the lift $\tilde{F}$ of a map F is one such that $\tilde{F}^d(q, p) = (q + m, p)$. One fixes a lift once and for all.)

We also have a corresponding result on $T^* M$ where M supports a metric of negative curvature (in particular $\tilde{M} = \mathbb{R}^n$). There, m represents a free homotopy class of loops in M. On such spaces, one only gets a lower bound of 2 orbits of type m, d in general, unless m is the class of the trivial loops, in which case one gets the usual lower bounds $cl(M)$ and $sb(M)$.

REFERENCES

[1] M. BIALY, L. POLTEROVITCH, *Hamiltonian systems, lagrangian tori and birkhoff's theorem*, Math. Ann. 292 (1992), 619–627.

[2] M. CHAPERON, *Une idée du type "géodésiques brisées" pour les systèmes hamiltoniens*, C.R. Acad. Sc. 298 Série I (13), Paris (1984) 293–296.

[3] M. CHAPERON, *Recent results in symplectic geometry*, Dyn. Sys. and Erg. Th. 23 Banach Center Publications, Warsaw (1989).

[4] C.C. CONLEY, E. ZEHNDER, *The birkhoff-lewis fixed point theorem and a conjecture of v.i. arnold*, Invent. Math. 73 (1983).

[5] C. GOLÉ, *Monotone maps and their periodic orbits*, in The geometry of Hamiltonian systems, T. Ratiu (ed.) Springer-Verlag 1991.

[6] C. GOLÉ, *Periodic orbits for hamiltonian systems in cotangent bundles*, IMS preprint, SUNY at Stony Brook 1991.

[7] C. GOLÉ, *Ghost circles for twist maps*, Jour. of Diff. Eq. 97 (1) (1992), 140–173.

[8] C. GOLÉ, *Symplectic twist maps and the theorem of conley-zehnder for general cotangent bundles*, in Mathematical Physics X, proceedings of

the Xth congress, K.Schmüdgen (ed.) Springer-Verlag 1991.

[9] C. GOLÉ, *Optical hamiltonians and symplectic twist maps*, Physica D 71 (1994), 185–195.

[10] C. GOLÉ, *Symplectic twist maps*, World Scientific Publishers (to appear).

[11] M.R. HERMAN, *Inegalités a priori pour des tores invariants par des difféomorphismes symplectiques*, Publ. Math. I.H.E.S. 70 (1989), 47–101.

[12] F.W. JOSELLIS, *Global periodic orbits for hamiltonian systems on* $\mathbb{T}^n \times \mathbb{R}^n$, Ph.D. Thesis Nr. 9518, ETH Zürich 1991.

[13] F. LAUDENBACH, J.C. SIKORAV, *Persistence d'intersection avec la section nulle...*, Invent. Math. 82 (1985).

[14] P. LECALVEZ, *Existence d'orbites de birkhoff généralisées pour les difféomorphismes conservatifs de l'anneau*, preprint, Université Paris-Sud, Orsay 1989.

[15] P. LECALVEZ, *Un théorème de translation pour les difféomorphismes du tore ayant des points fixes*, preprint, Université Paris-Sud, Orsay 1991.

[16] R.S. MACKAY, J.D. MEISS, J. STARK, *Converse KAM theory for symplectic twist maps*, Nonlinearity 2 (1989), 555–570.

[17] J. MOSER, *Monotone twist mappings and the calculus of variations*, Ergod. Th. and Dyn. Sys. 6 (1986), 401–413.

[18] J.C. SIKORAV, *Sur les immersions lagrangiennes dans un fibré cotangent admettant une phase generatrice globale*, C.R. Acad. Sc. Paris, t. 302 Série I (3) (1986).

[19] C. VITERBO, *Intersection de sous variétés lagrangiennes, fonctionnelles d'action et indice de systèmes hamiltoniens*, Bull. Soc. math. France 115 (1987), 361–390.

[20] C. VITERBO, *Symplectic topology as the geometry of generating functions*, Math. Ann. 292 (1992), 685–711.

GLOBAL STRUCTURAL STABILITY OF PLANAR HAMILTONIAN VECTOR FIELDS[*]

XAVIER JARQUE[†] AND JAUME LLIBRE[‡]

Abstract. We characterize all structurally stable C^r planar Hamiltonian vector fields with respect to perturbations in the set of all C^r planar vector fields. Also, we extend this characterization to C^r planar integrable vector fields, and we show that the usual requirement that the homeomorphism lies near the identity in the definition of structural stability is redundant for the C^r planar Hamiltonian vector fields.

1. Introduction and statement of the results. The global structural stability of C^r Hamiltonian vector fields with n degree of freedom is a hard open problem. In this paper we will characterize the C^r planar Hamiltonian vector fields (i.e., with 1 degree of freedom) which are structurally stable with respect to perturbations in the set of all C^r planar vector fields. Furthermore, we extend this characterization to C^r planar integrable vector fields.

In [JL] we classify the structural stability of a planar Hamiltonian polynomial vector field with respect to perturbations, first in the set of all C^r planar vector fields, second in the set of all planar polynomial vector fields, and finally in the set of all planar Hamiltonian polynomial vector fields. A summary of these results is given in the Appendix.

In dimension 2 the study of the structural stability of vector fields goes back to Andronov and Pontrjagin [AP] who in 1937 studied the structural stability of analytic vector fields on the closed 2-dimensional disk. Peixoto [P] in 1962 characterized the C^1 vector fields defined on the orientable differentiable compact connected 2-manifolds without boundary which are structurally stable. In 1982 Kotus, Krych and Nitecki [KKN] gave sufficient conditions in order to a C^r vector field on an orientable differentiable open connected 2-manifold without boundary S to be structurally stable (furthermore they proved that these conditions are necessary if $S = \mathbb{R}^2$). Other papers related with the structural stability of some classes of planar vector fields on are due to Sotomayor [So2] and Shafer [S1],[S2].

In order to state our results we will need some definitions and notation. Since a C^r planar vector field X can be regarded as a C^r map from $\mathbb{R}^2$ to $\mathbb{R}^2$, we can endow the space of all C^r vector fields with the *strong C^r-topology or Whitney C^r-topology*. This topology permits (in contrast with the *compact-open topology*) the control of the behaviour of perturbations of $X(p)$ when p goes to infinity. We denote by $\mathcal{X}^r, r \geq 1$, all the C^r vector fields on $\mathbb{R}^2$

---

[*] The authors are partially supported by DGICYT grant PB90-0695.

[†] Departament d'Economia i d'Historia Econòmica, Universitat Autònoma de Barcelona, Bellaterra, 08193 Barcelona, Spain.

[‡] Departament de Matemàtiques, Universitat Autònoma de Barcelona, Bellaterra, 08193 Barcelona, Spain.

with the *strong C^r-topology*. We denote by $\mathcal{H}_{\mathbb{R}^2}$ the homeomorphisms of $\mathbb{R}^2$ endowed with the compact-open topology (for more details on these topologies see [Hi]).

Let $H : \mathbb{R}^2 \to \mathbb{R}^2$ be a C^{r+1} function with $r \geq 1$. Then the vector field $X : \mathbb{R}^2 \to \mathbb{R}^2$ defined by

$$X(x,y) = \left(\frac{\partial H}{\partial y}(x,y), -\frac{\partial H}{\partial x}(x,y) \right) = \left(H_y(x,y), -H_x(x,y) \right)$$

is called the C^r *Hamiltonian vector field associate to H*. The set of all planar Hamiltonian vector fields with the induced C^r-topology is denoted by $\mathcal{H}^r$.

Let $X \in \mathcal{X}^r$. We say that X is *structurally stable with respect to perturbations in $\mathcal{X}^r$* if there exists a neighborhood N of X in $\mathcal{X}^r$ such that $Y \in N$ implies X and Y are topologically equivalent; that is, there exists a homeomorphism of $\mathcal{H}_{\mathbb{R}^2}$ near the identity carrying orbits of the flow induced by X onto orbits of the flow induced by Y, preserving sense but not necessarily parametrization.

Let X be a C^1 planar vector field. For each $p \in \mathbb{R}^2$ let $(I^-(p), I^+(p))$ be the maximal open interval where the solution $\varphi_t(p)$ of X such that $\varphi_0(p) = p$ is defined. Then we denote by $O^+(p)$ (respectively $O^-(p)$) the *positive* (respectively *negative*) *semi-orbit* of p; i.e., $O^+(p) = \{\varphi_t(p) : 0 \leq t < I^+(p)\}$ and $O^-(p) = \{\varphi_t(p) : I^-(p) < t \leq 0\}$. The *orbit of p*, denoted by $O(p)$, is defined by $O(p) = O^-(p) \cup O^+(p)$.

A positive semi-orbit $O^+(p)$ of X is *bounded* if it is contained in a compact set; *escapes to infinity* if for every compact set K there exist $q \in O^+(p)$ such that $O^+(q) \cap K = \emptyset$; and *oscillates* if it is neither bounded nor escapes to infinity. In a similar way we define bounded, escapes to infinity and oscillates for a negative semi-orbit $O^-(p)$.

A *saddle connection at infinity* (SAI) of X is a pair $(O^+(p), O^-(q))$ of semi-orbits each escaping to infinity such that there is a sequence $p_n \to p$ in $\mathbb{R}^2$ and $t_n \to I^+(p)$ such that $\varphi_{t_n}(p_n) \to q$ in $\mathbb{R}^2$. Then $O^+(p)$ (respectively $O^-(q)$) is called the *stable* (respectively *unstable*) *separatrix* of the SAI. A *saddle connection* is an orbit $O(p)$ such that $O^+(p)$ is a stable separatrix of a saddle or a SAI, and $O^-(p)$ is an unstable separatrix of a saddle or a SAI. We denote by $W^s(X)$ (respectively $W^u(X)$) the union of all orbits formed by stable (respectively unstable) separatrices of a saddle or a SAI of X.

THEOREM 1.1. *A vector field $X \in \mathcal{H}^r$ is structurally stable with respect to perturbations in $\mathcal{X}^r$ if and only if*
(a) all the critical points of X are hyperbolic saddles; and
(b) $Cl(W^u(X)) \cap Cl(W^s(X))$ consist only of saddle points.

As usual here $Cl(A)$ denotes the closure of a subset A of $\mathbb{R}^2$. The proof of Theorem 1.1 is given in Section 2 and uses essentially the properties of the planar Hamiltonian vector fields and a theorem of Kotus, Krych and

Nitecki [KKN] on the structural stability of vector fields $X \in \mathcal{X}^r$ with respect to perturbations in $\mathcal{X}^r$.

We say that a C^r planar vector field $X = (f, g) : \mathbb{R}^2 \to \mathbb{R}^2$ is C^r *integrable* if it has a C^{r+1} first integral (i.e., there exist a non-constant C^{r+1} function $H : \mathbb{R}^2 \to \mathbb{R}^2$ which is constant on each solution of X) and at least one of the two following quadratures

$$\int_0^t \frac{1}{\frac{\partial H}{\partial y}(x(s), y(s))} f(x(s), y(s)) ds, \quad \int_0^t \frac{1}{\frac{\partial H}{\partial x}(x(s), y(s))} g(x(s), y(s)) ds$$

it is well-defined on each solution of the differential system defined by X.

COROLLARY 1.1. *A C^r integrable vector field X on $\mathbb{R}^2$ is structurally stable with respect to perturbations in $\mathcal{X}^r$ if and only if satisfies conditions (a) and (b) of Theorem 1.1.*

We will prove Corollary 1.1 in Section 2.

The requirement that the homeomorphism lies near the identity (in a suitable topology) in the definition of structurally stable C^r planar vector fields is studied for many authors. In fact Peixoto [P] proved that for orientable differentiable compact connected 2-manifolds without boundary the requirement is redundant. Kotus, Krych and Nitecky [KKN] showed, with an example, that for orientable differentiable open 2-manifold without boundary when only C^4-perturbations are allowed the requirement is not redundant. More details can be find in [DS]. But as the following theorem shows the requirement that the homeomorphism lies near the identity in the definition of structurally stable C^r planar Hamiltonian vector fields is redundant.

THEOREM 1.2. *In the definition of structural stability of elements of $\mathcal{H}^r$ with respect to perturbations in $\mathcal{X}^r$ the requirement that the equivalence homeomorphism be closed to the identity is redundant.*

Theorem 1.2 will be proved at the end of Section 2. From Corollary 1.1 and Theorem 1.2 we get the following result.

COROLLARY 1.2. *In the definition of structural stability of C^r integrable vector fields respect to perturbations in $\mathcal{X}^r$ the requirement that the equivalence homeomorphism be closed to the identity is redundant.*

2. The proofs. We start by proving some basic results about the planar Hamiltonian vector fields which will be necessary for studying their structural stability.

A critical point p of X is *non-degenerate* if the determinant of the linear part of X at p is non-zero; i.e., $\det(DX(p)) \neq 0$. A critical point p of X is *hyperbolic* if the real part of the eigenvalues of $DX(p)$ are non-zero.

PROPOSITION 2.1. *Let $X \in \mathcal{H}^r$ and let p be a critical point of X. Then the following statements hold.*
(a) The divergence of X is identically zero.
(b) The flow φ_t defined by X is area preserving.

(c) If p is non-degenerate, then p is a saddle or a center.

(d) If p is hyperbolic, then p is a saddle.

Proof . Since $X = (H_y, -H_x)$ for some C^{r+1} function $H : \mathbb{R}^2 \to \mathbb{R}^2$ with $r \geq 1$, from the Schwartz Lemma $\mathrm{div}(X) = H_{yx} - H_{xy} = 0$. Hence (a) is proved.

From ([AA], pp 22) if $A(t)$ denotes the Euclidean area of the image of a domain D under φ_t, then

$$\frac{dA}{dt} = \int_{\varphi_t(D)} \mathrm{div}(X) \ dx \wedge dy,$$

where $dx \wedge dy$ is the Euclidean area-element of $\mathbb{R}^2$. Therefore, from (a) it follows that $dA/dt = 0$. Consequently we get (b).

Clearly

$$DX(p) = \left(\begin{array}{cc} H_{yx}(p) & H_{yy}(p) \\ -H_{xx}(p) & -H_{xy}(p) \end{array} \right).$$

Then, $\mathrm{trace}(DX(p)) = 0$ and $\det(DX(p)) = H_{xx}(p)H_{yy}(p) - H_{xy}^2(p)$. If p is non-degenerate, then $\det(DX(p)) \neq 0$. If $\det(DX(p)) < 0$ then p is a saddle (see [ALGM], pp 183). Since $\mathrm{trace}(DX(p)) = 0$ if $\det(DX(p)) > 0$ then p is a center focus or a centrofocus (see [ALGM], pp 160). But from (b) it follows that p cannot be a focus or a centrofocus. Therefore, we have proved (c).

If p is hyperbolic, then p is non-degenerate and the real part of the eigenvalues of $DX(p)$ are non-zero. Then, from (c) and ([ALGM], pp 183) it follows that p is a saddle, and (d) is proved.

∎

PROPOSITION 2.2. *Let $X \in \mathcal{H}^r$. Suppose that all the critical points of X are saddles. Then the following statements hold.*

(a) X has no periodic orbits.

(b) X has no oscillating orbits.

Proof . Suppose that $O(p)$ is a periodic orbit of X. From ([Pr], pp 279) the sum of the indices of the critical points contained in the bounded region of $\mathbb{R}^2 \setminus O(p)$ is 1. But if all the critical points of X are saddles every sum of indices is always negative. Hence, (a) follows.

Let $O(p)$ be an orbit of X which oscillates in forward time. If it oscillates in backward time the proof is similar. Clearly we can assume that $O(p)$ is neither a periodic orbit nor a critical point. Now, as in the proof of the Poincaré-Bendixson Theorem (see for instance [So1]), the semi-orbit $O^+(p)$ accumulates to a "separatrix cycle S having points at infinity"; i.e. S is formed by many (at least one) curves starting and ending at infinity each one formed by separatrices of saddles or SAI's and saddles (the number of saddles can be zero). Therefore, as in the proof of ([Pr], pp 279) the sum of the indices of the critical points of X contained into the component of $\mathbb{R}^2 \setminus S$ which contains $O(p)$ must be 1. In contradiction with the fact that all the critical points of X are saddles, and consequently every sum of indices is always negative. Hence, (b) follows.

∎

We say that the periodic orbit $(x(t), y(t))$ of X is *hyperbolic* if

$$\int_{\gamma(t)} \operatorname{div} X(x(t), y(t)) dt \neq 0.$$

The next result is proved in [KKN].

THEOREM 2.1. *A vector field $X \in \mathcal{X}^r$ is structurally stable with respect to perturbations in $\mathcal{X}^r$ if and only if*
(a) X has no oscillating orbits;
(b) all the critical points and periodic orbits are hyperbolic ; and
(c) $Cl(W^u(X)) \cap Cl(W^s(X))$ consist only of saddle points.

Proof of Theorem 1.1. Clearly it will be sufficient to show that statements (a) and (b) of Theorem 1.1 are equivalent to statements (a)-(c) of Theorem 2.1 when $X \in \mathcal{H}^r$.

Suppose that X satisfies (a) and (b) of Theorem 1.1. By statement (a) of Theorem 1.1 and statement (b) of Proposition 2.2, statement (a) of Theorem 2.1 holds. From statement (a) of Theorem 1.1 and (a) of Proposition 2.2 it follows that X has no periodic orbits. Then, since all critical points are hyperbolic, we get statement (b) of Theorem 2.1. Finally, from statement (b) of Theorem 1.1 it follows immediately statement (c) of Theorem 2.1.

Now, we assume that the Hamiltonian vector field X satisfies (a)-(c) of Theorem 2.1. Then, from statement (b) of Theorem 2.1 and statement (d) of Proposition 2.1, we get statement (a) of Theorem 1.1. Finally statement (c) of Theorem 2.1 and statement (b) of Theorem 1.1 are the same.

∎

Proof of Corollary 1.1. It is sufficient to show that every C^r integrable vector field on $\mathbb{R}^2$ can be written as a C^r Hamiltonian vector field after a change on the time variable. This idea goes back to Andronov, Vitt and Khaikin (see [AVK], Chapter II, §7.2, see also [FC]).

Suppose that $X = (f, g)$, where f and g are C^r functions. Then X define the differential system

(2.1)
$$\frac{dx}{dt} = f(x, y), \qquad \frac{dy}{dt} = g(x, y).$$

Let $H : \mathbb{R}^2 \to \mathbb{R}^2$ be the C^{r+1} first integral of X. We introduce a new time θ by

$$\frac{dx}{d\theta} = \frac{\partial H}{\partial y},$$

more precisely

$$d\theta = \frac{1}{\frac{\partial H}{\partial y}} \frac{dx}{dt} dt = \frac{1}{\frac{\partial H}{\partial y}} f(x, y) dt,$$

if

(2.2)
$$\int_0^t \frac{1}{\frac{\partial H}{\partial y}(x(s), y(s))} f(x(s), y(s)) ds$$

is well-defined; or by

$$\frac{dy}{d\theta} = -\frac{\partial H}{\partial x},$$

more precisely

$$d\theta = -\frac{1}{\frac{\partial H}{\partial x}} \frac{dy}{dt} dt = -\frac{1}{\frac{\partial H}{\partial x}} g(x, y) dt,$$

if (2.2) is not well-defined and if

$$\int_0^t \frac{1}{\frac{\partial H}{\partial x}(x(s), y(s))} g(x(s), y(s)) ds$$

is well-defined.

Consequently, for each solution $(x(t), y(t))$ of system (2.1) the relation between θ and t is reduced to a well-defined quadrature. Now, system (2.1) goes over to

$$\frac{dx}{d\theta} = \frac{\partial H}{\partial y}, \quad \frac{dy}{d\theta} = -\frac{\partial H}{\partial x};$$

because

$$\frac{\partial H}{\partial x} \frac{dx}{d\theta} + \frac{\partial H}{\partial y} \frac{dy}{d\theta} = 0.$$

∎

Proof of Theorem 1.2. It is sufficient to show that if X is topologically equivalent (via a homeomorphism not necessarily near the identity) to all of its neighbours then satisfies hypoteses (a) and (b) of Theorem 1.1.

As it shows in Section 8 of [KKN] condition (b) of Theorem 1.1 never uses the distance between the homeomorphism h and id. Then we only need to prove that all the critical points of X are hyperbolic saddles.

If X has a curve of critical points then for an arbitrarily small C^r-perturbations there exist vector fields with all the critical points isolated and we get a contradiction with the fact that X is topologically equivalent to all of its neighbours.

Assume that all the critical points of X are isolated and that at least one of them (say p) is not hyperbolic. Then, from Proposition 2.1(c), either p is a center or $\det(DX(p)) = 0$. Consider U_p a neighborhood of p which do not contain any other critical point of X and construct a C^∞ bump function $\rho(x,y)$ with support contained in U_p. If $X = (X_1, X_2)$ we define $X_\epsilon = (X_1 + \epsilon\rho, X_2)$. Clearly X_ϵ is C^r close to X for small enough ϵ. Furthermore it is equal to X on the complement of U_p and has a critical point p_ϵ such that $p_\epsilon \to p$ as $\epsilon \to 0$. Now, for a suitable ρ the critical point p_ϵ of X_ϵ has local behaviour distinct to p but the same behaviuor outside of U_p.

∎

3. Appendix. As we said in Section 1 we proved in [JL] some theorems about the structural stability of a planar Hamiltonian polynomial vector field with respect to perturbations, first in $\mathcal{X}^r$, second in the set of all planar polynomial vector fields and finally in the set of all planar Hamiltonian polynomial vector fields. In fact, when we consider only polynomial perturbations we can think the vector fields either on $\mathbb{R}^2$ or on $\mathbb{S}^2$ (Poincaré sphere) but we will prove that the set of all structurally stable planar Hamiltonian polynomial vector fields is the same in both cases. Here we present these results without proof.

We denote by $\mathcal{P}_n(\mathbb{R}^2)$ the polynomial vector fields on $\mathbb{R}^2$ of the form

$$X(x,y) = \big(P(x,y), Q(x,y)\big),$$

where P and Q are polynomials in the variables x and y of degree at most n (with $n \in \mathbb{N}$). Any such X is uniquely specified by the $(n+1)(n+2)$ coefficients of P and Q, hence it may be identified with a unique point in $\mathbb{R}^{(n+1)(n+2)}$. The *coefficient topology* of $\mathcal{P}_n(\mathbb{R}^2)$ is the induced topology through the above identification from the usual Euclidean topology of $\mathbb{R}^{(n+1)(n+2)}$. Let $\mathcal{H}P_n(\mathbb{R}^2)$ be the subset of $\mathcal{P}_n(\mathbb{R}^2)$ formed by the Hamiltonian polynomial vector fields of the form

$$X(x,y) = \big(H_y(x,y), -H_x(x,y)\big),$$

where $H(x,y)$ is a polynomial of degree $n+1$ in the variables x and y.

Let $\mathcal{D}$ be $\mathcal{X}^r$, $\mathcal{P}_n(\mathbb{R}^2)$ or $\mathcal{H}P_n(\mathbb{R}^2)$. A vector field $X \in \mathcal{X}^r$ is *structurally stable with respect to perturbations in* $\mathcal{D}$ (when these perturbatiosn have meaning) if there exits a neighborhood N of X in $\mathcal{D}$ such that $Y \in N$ implies X and Y are topologically equivalent; that is, there exists a homeomorphism $h : \mathbb{R}^2 \to \mathbb{R}^2$ near the identity carrying orbits of the flow induced by X onto orbits of the flow induced by Y, preserving sense but not necessarily parametrization.

For $X \in \mathcal{P}_n(\mathbb{R}^2)$ we can consider the Poincaré compactified vector field $p(X)$ on $\mathbb{S}^2 = \{(x,y,z) \in \mathbb{R}^3 : x^2 + y^2 + z^2 = 1\}$ (see [Go]) which is an analytical vector field on $\mathbb{S}^2$. We denote by $\mathcal{P}_n(\mathcal{S}^2)$ all the vector fields, $p(X)$ on $\mathbb{S}^2$ defined as above and endowed with the coefficient topology. Also we denote by $\mathcal{H}P_n(\mathbb{S}^2)$ the subset of $\mathcal{P}_n(\mathbb{S}^2)$ which comes from the Hamiltonian polynomial vector fields on $\mathbb{R}^2$ and we endow it with the relative topology of $\mathcal{P}_n(\mathbb{S}^2)$. So, if $X \in \mathcal{P}_n(\mathbb{R}^2)$ then $p(X) \in \mathcal{P}_n(\mathbb{S}^2)$ and if $X \in \mathcal{H}P_n(\mathbb{R}^2)$ then

$p(X) \in \mathcal{HP}_n(\mathbb{S}^2)$. A critical point p of the analytic vector field $p(X)$ on $\mathbb{S}^2$ is called *infinite* (respectively *finite*) if $p \in \mathbb{S}^1 = \{(x, y, z) \in \mathbb{S}^2 : z = 0\}$ (respectively $p \in \mathbb{S}^2 \setminus \mathbb{S}^1$).

Let $\mathcal{D}$ be $\mathcal{P}_n(\mathbb{S}^2)$ or $\mathcal{HP}_n(\mathbb{S}^2)$. A vector field $p(X) \in \mathcal{P}_n(\mathbb{S}^2)$ is said to be *structurally stable with respect to perturbation in* $\mathcal{D}$ (when these perturbations have meaning) if there exists a neighborhood N of $p(X)$ in $\mathcal{D}$ such that $p(Y) \in N$ implies that $p(X)$ and $p(Y)$ are topologically equivalent; that is, there exists a homeomorphism of $\mathbb{S}^2$ carrying orbits of the flow induced by $p(X)$ onto orbits of the flow induced by $p(Y)$, preserving sense but not necessarily parametrization.

THEOREM 3.1. *The vector field $X \in \mathcal{HP}_n(\mathbb{S}^2)$ is structurally stable with respect to perturbations in $\mathcal{X}^r$ if and only if*
(a) all the finite critical points of X are hyperbolic saddles; and
(b) X has no saddle connections (where separatrices of SAI's are taken into account).

THEOREM 3.2. *The vector field $X \in \mathcal{HP}_n(\mathbb{S}^2)$ is structurally stable with respect to perturbations in $\mathcal{P}_n(\mathbb{S}^2)$ if and only if*
(a) all the finite critical points of X are hyperbolic saddles;
(b) all the infinite critical points of X are non-degenerate nodes with different eigenvalues; and
(c) X has no saddle connections.

We denote by $\Sigma \subset \mathcal{HP}_n(\mathbb{R}^2)$ the subset of all structurally stable vector fields with respect to perturbations in $P_n(\mathbb{R}^2)$ and by $\Sigma_{\mathbb{S}^2} \subset \mathcal{HP}_n(\mathbb{S}^2)$ the subset of all structurally stable vector fields with respect to perturbations in $\mathcal{P}_n(\mathbb{S}^2)$.

THEOREM 3.3. $X \in \Sigma$ *if and only if* $p(X) \in \Sigma_{\mathbb{S}^2}$.

Let $X \in \mathcal{HP}_n$ and let p be a non-degenerate saddle of X. Suppose that X has a saddle connection between two separatrices of p. If the other two separatrices do not connect between them we will say that X has a *saddle loop* or simply a *loop* (see Figure 1(a)). If the other two separatrices of p connect as in Figure 1(b) we will say that X has a *simple two-loop*, and if they connect as in Figure 1(c) we will say that X has a *nested two-loop*. We will say that X has a *two-loop* when X has either a simple two-loop or a nested two-loop.

THEOREM 3.4. *The vector field $X \in \mathcal{HP}_n(\mathbb{S}^2)$ is structurally stable with respect to perturbations in $\mathcal{HP}_n(\mathbb{S}^2)$ if and only if:*
(a) all its finite critical points are either hyperbolic saddles or non-degenerate centers;
(b) all its infinite critical points of X are non-degenerate nodes; and
(c) all its saddle connections are either saddle loops or simple two-loops or nested two-loops.

We denote by $\Sigma' \subset \mathcal{HP}_n(\mathbb{R}^2)$ the subset of all structurally stable vector fields with respect to perturbations in $\mathcal{HP}_n(\mathbb{R}^2)$ and by $\Sigma'_{\mathbb{S}^2} \subset \mathcal{HP}_n$ the subset of all structurally stable vector fields with respect to perturbations in $\mathcal{HP}_n(\mathbb{S}^2)$.

THEOREM 3.5. *X* $\in \Sigma'$ *if and only if* $p(X) \in \Sigma'(\mathbb{S}^2)$

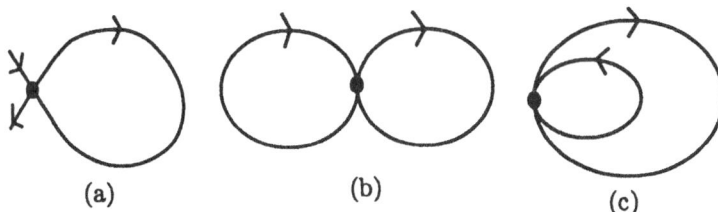

(a)   (b)   (c)

Figure 1. Loops and two-loops.

Further we consider the interesting question in the definition of structural stability of whether or not it is important that the equivalence homeomorphism between the structurally stable vector field and its neighbours be close to the identity map.

REFERENCES

[ALGM] A.A. ANDRONOV, E.A. LEONTOVICH, I.I. GORDON AND A.G. MAIER, "Qualitative Theory of Second Order Dynamical Systems," translated by John Wiley & Sons, New York, 1973, 1966.

[AP] A.A. ANDRONOV AND L. PONTRJAGIN, *Systemes grossiers*, Acad. Sci. URRS **14** (1937), pp. 247–250.

[AVK] A.A. ANDRONOV, A.A. VITT AND S.E. KHAIKIN, "Theory of oscillations," Dover Edition, 1987, 1937.

[AA] D.V. ANOSOV AND V.I. ARNOLD, "Dynamical Systems Vol. 1," Encyclopedia of Mathematical Science, Springer-Verlag, New-York, 1988.

[DS] F. DUMORTIER AND D.S. SHAFER, *Restrictions on the equivalence homeomorphism in stability of polynomial vector fields*, J. London Math. Soc. **41** (1990), pp. 100–108.

[FC] M.P. FEIX AND L. CAIRÓ, *Hamiltonian form for a general two-dimensional ODE system. Application to the 2d Lotka-Volterra equations*, preprint, Univ. Orléans (1991).

[Go] E.A. GONZÁLEZ, *Generic properties of polynomial vector fields at infinity*, Trans. Amer. Math. Soc. **143** (1969), pp. 201–222.

[Hi]   M.W. HIRSCH, "Differential Topology," Springer-Verlag, New York, 1976.

[JL]   X. JARQUE AND J. LLIBRE, *Structural stability of planar Hamiltonian poly-nomial vector fields*, Proc. London Math. Soc. **68** (1994), pp. 617–640.

[KKN] J. KOTUS, M. KRYCH AND Z. NITECKI, "Global structural stability of flows on open surfaces," Memoirs Amer. Math. Soc. **261** (1982).

[P]   M.M. PEIXOTO, *Structural stability on two-dimensional manifolds*, Topology **1** (1962), pp. 101–120.

[Pr]   L. PERKO, "Differential equations and Dynamical Systems," Springer-Verlag, New York, 1991.

[S1]   D.S. SHAFER, *Structural stability and generic properties of planar polynomial vector fields*, Rev. Mat. Hisp.-Amer. **3** (1987), pp. 337–355.

[S2]   D.S. SHAFER, *Structure and stability of gradient polynomial vector fields*, J. London Math. Soc. **41** (1990), pp. 109–121.

[So1] J. SOTOMAYOR, "Lições de equações diferenciais ordinárias," Instituto de Matemática Pura e Aplicada, Rio de Janeiro, 1979.

[So2] J. SOTOMAYOR, *Stable planar polynomial vector fields*, Rev. Math. Iber. **1** (1985), pp. 15–23.

ANALYTIC TORSION, FLOWS AND FOLIATIONS

STEPHANE LAEDERICH*

Abstract. We present an overview of the known results in Lefschetz formulas for flows, that is, on the problem of relating the topology of a manifold to the number and nature of periodic orbits of a vector field.

1. Introduction. It is well known that the question of the existence of periodic orbits of a given vectorfield on a compact manifold is of central importance in the theory of dynamical systems and other fields. One may, for example, consider an energy surface of a given Hamiltonian system and, according to the topology of this surface, decide whether (and which) periodic orbits do occur. To give but a few applications, periodic orbits can be used to understand stability for large times as in P. Lochak's [Lo] proof of Nekhoroshev's theorem or appear in the theory of semiclassical approximation, in the Gutzwiller trace formula (see Guillemin and Uribe [GU]).

Unfortunately, there is no general formula which counts periodic orbits of a given vector field on a given compact manifold. Looking at the standard Lefschetz formula on compact manifolds suggests the form of a formula counting periodic orbits; that is, a formula relating a topological invariant of the manifold to closed orbits of vector fields via a sum or a product over dynamical quantities. These dynamical quantities should, for example, involve the eigenvalues of the Poincaré map around the orbit or the length or period of the orbit. The aim of this paper is to present a survey of what is now known as Lefschetz formulas for flows as well as the techniques needed for their proofs and to present some new ideas and open problems in this field.

This paper is divided into four parts. We start by presenting the topological invariant used in these formulas as well as a few of its properties. In the second section, we present results of Ruelle, Selberg and Fried on the geodesic flow on manifolds of negative curvature. We then present some of our results in the case of fibrations, for which one can consider suspension vector fields. In the last section, we study the more general case of arbitrary foliations of compact manifolds, of which a special case is given by nonsingular flows.

2. Analytic torsion. We start with the topological invariant. In the twenties, Reidemeister introduced an invariant, now called Reidemeister torsion, which was used to distinguish between different lens spaces. This invariant was defined topologically in terms of the twisted cochain complex for acyclic bundles E (i.e. without cohomology). Ray and Singer [RS1]

---

* Department of Mathematics, University of Michigan, Ann Arbor, Michigan 48109-1003.

suggested a new definition in terms of traces of heat kernels on the manifold. The equivalence was then proved independently by Cheeger and Müller [Ch1], [Mu].

In the following, M will always denote a compact real or complex manifold and $E \to M$ will denote a flat vector bundle over M constructed from a representation of $\pi_1(M)$ into either $SO(n)$, or $U(n)$ in the complex case.

The analytic definition of the torsion is based on the following concept: Let $\mathcal{E}^q$, $q = 0, \ldots, N$, be vector bundles over M and let E^q, $q = 0, \ldots, N$ be the spaces of sections of the bundles $\mathcal{E}^q$. We will consider two cases, the twisted deRham complex, where $\mathcal{E}^q = \Lambda^q T^* M \otimes E$ or the twisted Dolbeault complex, where $\mathcal{E}^q = \Lambda^{p,q} T^* M \otimes E$. Let d^q denote either the deRham d operator or the $\bar{\partial}$ operator.

$$d^q : E^q \longrightarrow E^{q+1}, \ d^{q+1} d^q = 0.$$

That is, (E, d) is an elliptic complex. (See Atiyah-Bott [AB1&2], Gilkey [Gi] or Wells [We]). Taking the adjoints $(d^q)^*$ with respect to the natural inner products on E^q arising from a metric g on M, we consider the corresponding Laplacians

$$(2.1) \qquad \triangle^q = (d^q)^* d^q + d^{q-1} (d^{q-1})^* : E^q \longrightarrow E^q.$$

Following Seeley [Se], one can define a zeta function associated to the Laplacians.

DEFINITION 2.1. For $s \in \mathbb{C}$ with $Re(s)$ sufficiently large, define

$$(2.2) \qquad \zeta^q(s) = \frac{1}{\Gamma(s)} \int_0^\infty t^{s-1} tr(e^{-t\triangle^q} - P_q) dt$$

where P_q is the projection on Ker $\triangle^q$.

In his paper, Seeley showed that $\zeta^q(s)$ is analytic for $Re(s)$ sufficiently large and possesses a meromorphic extension to $\mathbb{C}$ which is regular at $s = 0$. Using the zeta functions one can define a zeta–regularized determinant of $\triangle^q$ by

DEFINITION 2.2.

$$(2.3) \qquad \det_\zeta \triangle^q = \exp - \frac{d}{ds}\bigg|_{s=0} \zeta^q(s).$$

Formally, the zeta function defined by (2.2) is equal to $\sum \lambda^{-s}$ where the sum runs over the non zero eigenvalues of the Laplacian. Thus, by taking a derivative at $s = 0$ we obtain $- \sum \log(\lambda)$. Taking the exponential, we formally obtain a product over the nonzero eigenvalues, which motivates the definition (2.3).

We can now define the torsion of the elliptic complex (E, d).

DEFINITION 2.3. The torsion of the elliptic complex (E, d) is defined by

$$(2.4) \qquad T(E) = \prod_q (\det_\zeta \Delta^q)^{(-1)^{q+1}q} = \frac{(\det \Delta^1)}{(\det \Delta^2)^2} \frac{(\det \Delta^3)^3}{\cdots} \cdots$$

That is, the analytic torsion is an alternating ratio of regularised determinants of Laplacians. Note that in the complex case, one obtains $\dim_{\mathbb{C}} M$ analytic torsions. In both the case of the deRham and of the Dolbeault complex, these definitions are the square of the ones given by Ray and Singer [RS1&2]. In the real case, this is thus the square of the Reidemeister torsion.

We will state a few properties of the torsion. For further properties and the proof of the following theorem, we refer to [RS1&2].

THEOREM 2.1. (i) Let M be a real compact manifold. Let $E \to M$ be a flat acyclic bundle over M (i.e. there is no deRham cohomology with coefficients in E). Then $T(E)$ is independent of the choice of the metric g on M.

(ii) Let $E \to M$ be as above. If M is even dimensional, then $T(E) = 1$, that is the torsion is trivial.

(iii) Let M be a compact complex manifold of complex dimension n. Let $E_1 \to M$ and $E_2 \to M$ be two flat hermitian bundles of the same rank over M. Then if the Dolbeault cohomology of M with coefficients in E_i, $i = 1, 2$ vanishes, i.e. if $H^{p,q}(M, E_{1,2}) = 0$, the ratios

$$\frac{T_p(E_1)}{T_p(E_2)}, \quad p = 0, \ldots, \dim_{\mathbb{C}} M$$

is independent of the choice of metric on M.

3. Results. The known results are of three kinds. The first kind concerns the geodesic flow on manifolds of negative curvature. The second kind concerns fibrations. On a fibration, one can consider the flow of a suspension vector field to obtain dynamically relevant results. Finally, the last kind concerns more general foliations than the ones arising from the flow of a vector field.

3.1. Geodesic flow. We let $\mathbb{X}$ be a compact hyperbolic manifold of dimension $d > 2$ given by

$$(3.1) \qquad\qquad\qquad \mathbb{X} = \Gamma \backslash \mathbb{H}$$

where $\mathbb{H}$ denotes the Poincaré upper half space and where Γ is a group of isometries of $\mathbb{H}$. We will consider the geodesic flow on the unit tangent bundle $S\mathbb{X}$. To that end, we let

$$(3.2) \qquad \rho : \pi_1(S\mathbb{X}) \to SO(m)$$

be a representation of $\pi_1(S\mathbb{X})$ such that the corresponding flat bundle $E \to S\mathbb{X}$ is acyclic, that is, such that

$$H^*(S\mathbb{X}, E) = 0.$$

With this data, one can construct a so called Ruelle Zeta function (see D. Fried [Fr1...5] and Ruelle [Ru]). For $s \in \mathbb{C}$, it is defined formally by

$$(3.3) \qquad R_\rho(s) = \prod_\gamma det(I - \rho(\gamma)e^{-sl(\gamma)})$$

where the product runs over all primitive closed geodesics, where $l(\gamma)$ denotes the length of γ, where $\rho(\gamma) \in SO(m)$ is the image under ρ of the homotopy class of γ. The following theorem, due to D. Fried [Fr4] justifies this definition.

THEOREM 3.1. $R_\rho(s)$ *is analytic for* $Re(s) > d-1$, *has a meromorphic extension to the whole complex plane which is regular at* $s = 0$ *and*

$$(3.4) \qquad |R_\rho(0)^{(-1)^{d-1}}| = T(\mathbb{X}, E).$$

One should note that the theorem remains valid for $d = 2$ even though several technical difficulties do arise. This theorem is a perfect example of a Lefschetz formula for a flow. It relates the torsion to the closed orbits of a vector field on the manifold as described in the introduction. It should be noted that Selberg [Se] and Ray and Singer [RS1] obtained results in this direction.

The standard idea for such computations is the following: One first starts by computing the heat kernel on $\mathbb{X}$ in terms of the one on $\mathbb{H}$,

$$k_{\mathbb{X}}(t, x, y) = \sum_{g \in \Gamma} \rho(g) k_{\mathbb{H}}(t, \tilde{x}, g\tilde{y})$$

where $\tilde{x}, \tilde{y}$ are points above x, y. To obtain the trace one considers these kernels on the diagonal

$$(3.5) \qquad k_{\mathbb{X}}(t, x, x) = k_{\mathbb{H}}(t, \tilde{x}, \tilde{x}) + \sum_\gamma \sum_{g \in \gamma} \rho(g) k_{\mathbb{H}}(t, \tilde{x}, g\tilde{x})$$

where the sum runs over nontrivial conjugacy classes γ in Γ. The first term represents the heat flow around contractible loops while the second term gives the heat flow around the free homotopy class of γ.

In this case, to compute the terms arising in (3.5) one has an efficient algebraic tool. Following [Fr4] we consider $\mathbb{H}$ as one sheet of the hyperboloid $-x_0^2 + x_1^2 + ... + x_d^2 = 0$. One can then rewrite $\mathbb{X}$ as

$$(3.6) \qquad \mathbb{X} = \Gamma \backslash G / K$$

where $G = SO(1,d)$ and $K = SO(d)$ is the isotropy group of $(1,0,...,0)$. Now a complex valued form ω on G/K pulls back as a form $\tilde{\omega}$ on G which can be written as

$$\tilde{\omega} = \sum f_{i_1,...,i_j} \omega_{i_1} \wedge ... \wedge \omega_{i_j}$$

where the components f give a map

$$f \; : \; G \rightarrow \Lambda^j \mathbb{C}^d, \; \; f(gk) = \xi_j(k^{-1}) f(g).$$

The ξ_j are the usual representations of $SO(d)$ into $\Lambda^j \mathbb{C}^d$. If we let M be the rotations fixing the vertical axis, i.e. $M \cong SO(d-1)$, one easily obtains that $\xi_j|_M = \sigma_j \oplus \sigma_{j-1}$, where σ_j is the action of M on $\Lambda^j \mathbb{C}^d$. Now the Selberg trace formula (see [Fr3..5]) computes the trace of the heat kernel in terms of these representations of $SO(d)$ into $\Lambda^j \mathbb{C}^d$. Before we state this formula, we need some additional notation. Given a closed geodesic γ on $\mathbb{X}$, the Poincaré map around γ (of length $l(\gamma)$) is the direct sum of

$$A^u(\gamma) = m_\gamma e^{l(\gamma)}, \; A^s(\gamma) = m_\gamma e^{-l(\gamma)}$$

where m_γ is the holonomy around γ obtained by the transport of orthogonal vectors. Let

$$(3.7) \qquad \Delta(\gamma) = det(I - A^s(\gamma)).$$

The Selberg trace formula states that

$$(3.8) \qquad \text{Trace } (e^{-t\Delta_j}) = I_t(\sigma_j) + I_t(\sigma_{j-1}) + G_t(\sigma_j) + G_t(\sigma_{j-1})$$

where

$$(3.9) \qquad I_t(\sigma_j) = a_1 \int_{-\infty}^{\infty} e^{-t(v^2+c^2)} P_{\sigma_j}(v) dv$$

where P_{σ_j} is a smooth Plancherel density and where

$$(3.10) \qquad G_t(\sigma_j) = \sum_\gamma a_\gamma \frac{1}{\sqrt{4\pi t}} e^{-l(\gamma)/4t} e^{-tc^2} e^{-(d-1)l(\gamma)/2}.$$

The sum in (3.10) runs over closed geodesics in $\mathbb{X}$, c, a_1 are constants depending on the manifold, d and j and where

$$a_\gamma = tr(\rho(\gamma)) tr(\sigma_j(m_\gamma)) l(\gamma_0)/\Delta(\gamma)$$

where γ_0 is the primitive geodesic corresponding to γ. Perhaps the best way to think of this formula is to consider (3.6). The terms in I correspond to the heat flow around null homotopic geodesics while the G terms correspond to the heat flow around the free homotopy class of γ. For details of the (algebraic) proof of the formula we refer to [Fr4]. To obtain the main result, one has to compute the contribution of each of the terms in (3.8) to the torsion.

We note that this result has been extended to the case of the geodesic flow on complex hyperbolic manifolds (see [Fr5]) as well as the case of locally symmetric spaces (see [Fr3]).

3.2. Fibrations. We consider fibrations

(3.11)
$$
\begin{array}{ccc}
F & \longrightarrow & M \\
& & \downarrow{\scriptstyle \pi} \\
& & B
\end{array}
$$

where M, F and B are either real or complex manifolds. In the complex case we will require the projection map π to be holomorphic. From a dynamical point of view, we will be specially interested in the cases where $B = S^1$ or $B = \mathbb{C}/\Gamma$, $\Gamma = \mathbb{Z} + \tau\mathbb{Z}$, $Im \ \tau > 0$. In the real case, given a fibration over S^1, one can consider suspension vectorfields of the form $\pi^* 1$.

In the complex case, fibrations over the torus are more general. When M is Kähler, the existence of a nonsingular holomorphic vectorfield v on M guarantees the existence of a fibration of the form (3.11) (see Crew and Fried [CF]). In general, the projection will only be C^∞. However, we restrict ourselves to the holomorphic case.

To obtain a Lefschetz formula for the flows, one first has to obtain a formula relating the torsion of M in terms of the torsion of the base and the cohomology of the fiber, seen as a bundle over B.

In the real case, $B = S^1$. We let $\mathcal{E} \to S^1$ be a flat vector bundle such that

(3.12) $H^*(M, \pi^*\mathcal{E}) = 0.$

We let χ denote the holonomy of $\mathcal{E}$ around S^1. Denoting by f the monodromy of the fiber bundle, we let $H^*(f)$ denote the holonomy of the bundle $H^*(F)$, viewed as a vector bundle over S^1. This map f can be thought of as being the time one map of the lift of the vectorfield 1 on S^1 onto M. One has

THEOREM 3.2. *Under the above conditions,*

(3.13) $T(M, \pi^*\mathcal{E}) = \prod_q det(I - \chi \otimes H^q(f))^{(-1)^q}$

where the product runs over $q = 0, ..., \dim F$.

In the complex case, since the topological invariant is the ratio of Ray–Singer torsions, one has to consider two flat bundles $\mathcal{E}_i \to B$, $i = 1, 2$ satisfying

(3.14) $\qquad H^{p,q}(M, \pi^* \mathcal{E}_i) = 0, \; i = 1, 2, \; p, q = 0, ..., \dim_{\mathbb{C}} M.$

We then have the

THEOREM 3.3. *Under the condition (3.3) and for* $p = 1, \ldots, \dim_{\mathbb{C}} M$,

(3.15) $\qquad \dfrac{T_p(\pi^* \mathcal{E}_1)}{T_p(\pi^* \mathcal{E}_2)} = \displaystyle\prod_{p_1 + p_2 = p} \prod_{j=0}^{\dim_{\mathbb{C}} F} \left[\dfrac{T_{p_1}(\mathcal{E}_1 \otimes H^{p_2, j} F)}{T_{p_1}(\mathcal{E}_2 \otimes H^{p_2, j} F)} \right]^{(-1)^j}$

where $T_p(\pi^* \mathcal{E}_i)$ *denotes the* $\overline{\partial}$*-torsion defined by (2.4) and where* $T_p(\mathcal{E}_i \otimes H^{p_2, j} F)$ *is the* $\overline{\partial}$*-torsion for the Laplacian on* B *with coefficients in* $\mathcal{E}_i \otimes H^{p_2, j} F$.

For a proof of Theorem 3.2, we refer either to D. Fried [Fr3] or X. Dai [Da1,2] and for a proof of Theorem 3.3, see [La1]. We will nevertheless sketch the idea of the proof as well as the techniques which have to be used. We will consider a simpler real case. We let $\tilde{M} = \mathbb{R} \times F$ and let $M = \tilde{M}/(x, y) \sim (x + n, f^n(y))$, f being an isometry of the fiber. One chooses a direct product metric $g_M = g_{S^1} + g_F$ on M. By a standard argument, one can thus write the heat kernel $k_M(t, u_1, u_2)$ on q–forms over M in terms of the one on $\mathbb{R}$ and the one on F. One has

(3.16) $\quad k_M^q(t, u_1, u_2) = \displaystyle\sum_{i=0,1} \sum_{n \in \mathbb{Z}} \chi^n k_{\mathbb{R}}^i(t, x_1, x_2 + n) \otimes k_F^{k-i}(t, y_1, f^n(y_2))$

where $u_i = (x_i, y_i)$. Taking (3.16) on the diagonal and taking the local trace yields

$\quad tr(k_M^q(t, u, u)) = \displaystyle\sum_{i=1,2} \sum_{n \in \mathbb{Z}} tr(\chi^n) tr(k_{\mathbb{R}}^i(t, x, x+n)) tr(k_F^{k-i}(t, y, f^n(y))).$

(3.17)

Integration on M will give the total trace. By a standard argument due to Kotake, [Ko], the fiber part only contributes to the trace at fixed points of f^n. This is intuitively clear from the fact that if $f^n(y) \neq y$, the kernel is smooth and thus does not contribute to the total trace. In fact, one has that

(3.18) $\quad \displaystyle\sum_q (-1)^q \int_F tr(k_F^q(t, y, f^n(y))) dvol_F = \sum_q (-1)^q tr H^q(f^n).$

Thus, using the definition of the torsion and the standard algebraic identity

$$(3.19) \qquad \log(det(I - A)) = \sum_n \frac{(-1)^n}{n} tr(A^n)$$

one obtains Theorem 3.2. When the monodromy of the fibration, that is f, is not an isometry of the fiber, one cannot choose a product metric on M. This prevents one from computing the trace of the heat kernel directly. One thus applies a trick introduced by Witten [Wi], the so–called adiabatic limit. One chooses a metric on M of the form

$$(3.20) \qquad g_M^\varepsilon = \pi^* \varepsilon^{-2} g_B + g_F$$

where π denotes the projection on the base. Note that g_F will in general vary from fiber to fiber. One then lets $\varepsilon \to 0$. In the limit, one obtains a product metric. The analysis was made possible by Melrose's theory [Me] of pseudo differential operators on manifolds with boundaries. The main result, due to X. Dai [Da1] can be summarized as follows. Denote the eigenvalues of the Laplacian Δ_ε for the metric given by (3.20) by λ_ε. The discrete set of eigenvalues ($\varepsilon > 0$) can be divided in two parts: The ones which remain bounded away from zero as $\varepsilon \to 0$ and the ones tending to zero asymptotically. When $\lambda_\varepsilon \to 0$ as $\varepsilon \to 0$, one has an asymptotic expansion of λ_ε of the form

$$(3.21) \qquad \lambda_\varepsilon = \varepsilon^2 \mu + O(\varepsilon^3)$$

where μ is an eigenvalue of the Laplacian for the g_B metric on B. One then uses standard finite propagation speed techniques due to Cheeger, Gromov and Taylor [CGT] to show that asymptotically large eigenvalues do not contribute and that the contribution of the others yields the desired result.

From these formulas, it is now possible to compute the torsion in terms of the dynamics. We start with the real case, following D. Fried's paper [Fr3]. Let $H^*(f)$ be the holonomy of the bundle $H^*(F)$, seen as a bundle over S^1. Let χ denote the holonomy of the vector bundle $\mathcal{E}$ over S^1. The torsion of S^1 with coefficients in $\mathcal{E} \otimes H^q(F)$ is given by

$$(3.22) \qquad T(S^1, \mathcal{E} \otimes H^*(F)) = det(I - \chi \otimes H^q(f)).$$

Taking the logarithm of the torsion of M as given by (3.13) and making use of (3.19) we obtain

$$(3.23) \qquad \log(T(M, \pi^* E)) = \sum_n \sum_q tr(\chi \otimes H^q(f^n)).$$

Supposing that F has isolated periodic orbits, one can then decompose (3.23) into terms arising from the same periodic orbits and thus obtain an expression of the torsion in terms of indices of primitive periodic orbits of f.

In the complex case one proceeds along the same lines. One difficulty arises from the fact that we use Atiyah–Bott fixed point indices [AB1&2]. These are defined as follows: We let f be a holomorphic map $f : F \longrightarrow F$. The index of a fixed point is then given by

$$(3.24) \qquad \operatorname{ind}^p(x, f) = \sum_{q=0}^{\dim_{\mathbb{C}} F} (-1)^q \frac{\operatorname{tr}\Lambda^{p,q}\partial f(x)}{\det_{\mathbb{C}}(I - \partial f(x))}$$

for $p = 0, \ldots, \dim_{\mathbb{C}} F$. Suppose one is now given a nonsingular holomorphic vectorfield v on the fibration M with base a torus $B = \mathbb{C}/\Gamma$, $\Gamma = \mathbb{Z} + \tau\mathbb{Z}$, $\operatorname{Im} \tau > 0$. We assume that v is transverse to the fibration. This naturally gives rise to a holomorphic foliation $\mathcal{F}$ of the manifold (spanned by the vector field) and a natural $\mathbb{Z}^2$ action (the Poincaré map)

$$(3.25) \qquad \varphi_{n,m} : F \longrightarrow F, \ n, \ m \in \mathbb{Z}.$$

Due to the nature of the Atiyah–Bott indices, we will have to require that

$$(3.26) \qquad \text{for all } n, m \in \mathbb{Z}, \varphi_{n,m} \text{ has isolated fixed points}$$

We will also need a nondegeneracy condition on the fixed points. This condition is rather technical and arises from the presence of small divisors (see (3.24)). We require that there exists a constant $C > 0$ and an positive integer $r > 2$ such that for any n, $m \in \mathbb{Z}$, $n^2 + m^2 > 0$, given a fixed point x of $\varphi_{n,m}^z$, the eigenvalues λ of $\partial\varphi_{n,m}^z(x)$ satisfy

$$(3.27) \qquad |1 - \lambda^k| \geq \frac{C}{|k|^r}.$$

When $\lambda = e^{2\pi\theta}$, (3.27) is just a usual diophantine condition on θ. It is known (see Hardy and Wright [HW]) that for fixed C the set of θ not satisfying this condition has measure of order C. It is thus easy to see that (3.27) is not satisfied on a set of measure vanishing as C^2.

With these rather restrictive conditions, we can mimic the construction in the real case. For each closed leaf $\mathcal{L}$ of the foliation (i.e. for each closed orbit) of the vector field, we have functions $\Theta^p(\mathcal{L}, \pi^*\mathcal{E}_i)$ which can be computed from the Atiyah–Bott indices of $\varphi_{n,m}$ along the leaf and from the transition functions of the bundles $\mathcal{E}_i$. One then obtains the

THEOREM 3.4. *Under the assumptions of Theorem 3.3,*

$$(3.28) \qquad \frac{T_p(\pi^*\mathcal{E}_1)}{T_p(\pi^*\mathcal{E}_2)} = \prod_{\mathcal{L} \text{ compact}} \frac{\Theta^p(\mathcal{L}, \pi^*\mathcal{E}_1)\Theta^{p-1}(\mathcal{L}, \pi^*\mathcal{E}_1)}{\Theta^p(\mathcal{L}, \pi^*\mathcal{E}_2)\Theta^{p-1}(\mathcal{L}, \pi^*\mathcal{E}_2)}.$$

In the simple complex analogue of the twisted product, we consider complex line bundles generated by a representation of $\pi_1(B)$ into $SU(1)$.

As shown by Fay, [Fa], one can then consider the torsion as a function of $\mathbb{C}^2$ into $\mathbb{C}$. In that case, the dynamical functions used in (3.28) are actually classical theta functions. In a sense, by varying the variables, one can retrieve part of the dynamics. For details, we refer to [La2].

3.3. General foliations. As was seen in the previous section, one of the drawbacks of the analytical techniques as well as the use of either the Lefschetz or the Atiyah–Bott indices is that one requires leaves of the foliations (i.e. closed orbits) to be isolated and that the eigenvalues of the Poincaré map around such a leaf satisfy some Diophantine condition. One way around this problem is to consider what is now know as noncommutative geometry. This theory, introduced by A. Connes (see [Co1&2]) allows one to define pseudodifferential operators on the leaves of a arbitrary smooth foliation $\mathcal{F}$ of dimension p on a compact manifold M of dimension n. We consider the holonomy groupoid of the foliation, denoted by $G = G(M, \mathcal{F})$. It is given as the collection of all triples $u = (x, y, [\gamma])$ where x and y are in the same leaf, where γ is a path from x to y and where $[\gamma]$ is the holonomy equivalence class of γ. The groupoid comes with range and source mappings $r, s : G \to M$, $r(x, y, [\gamma]) = x$, $s(x, y, [\gamma]) = y$. We let $G^x = \{s(u) = x\}$, $G_y = \{r(u) = y\}$ and $G_y^x = G^x \cap G_y$.

We introduce measures on M as follows. Let us denote by $\lambda = \{\lambda^x\}$ a family of tangential measures on $\{G^x\}$, obtained from a Riemannian metric on the leaves. In the presence of an invariant transverse measure ν, G can be made into a measured groupoid $(G, \tilde{\mu})$, where

$$(3.29) \qquad \tilde{\mu}(E) = \int_M \lambda^x(E \cap G^x) d\mu(x), \ \mu = \int d\nu.$$

The basic trick is now to consider smooth functions on this groupoid. Denote by $C_c^{\infty,0}(G)$ the space of tangentially smooth functions on G with compact support. The reduced $C^\star$ algebra of $G(M, \mathcal{F})$ is the completion of $C_c^{\infty,0}(G)$ with respect to the norm $\|\phi\| = \sup_{x \in M} \|\pi_x(\phi)\|$ where π_x is the natural representation of $C_c^{\infty,0}(G)$ into the algebra of bounded operators in $L^2(G^x, \lambda^x)$ given by

$$(3.30) \qquad ((\pi_x \phi)\psi)u = \int_{G^x} \phi(u^{-1}v)\psi(v)d\lambda^x(v), \ \psi \in L^2(G^x, \lambda^x).$$

Here, we assumed that G was Hausdorff. For further details, we refer to [Co1] and [MS]. The existence of an invariant transverse measure ν guarantees the existence of a canonical faithful, semifinite normal trace Φ_ν on the von Neumann algebra $W^\star(G)$.

With these tools at hand, it is now possible to define differential and pseudo differential operators on the leaves of the foliation. To simplify to the extreme, they are defined as follows. Consider now two flat bundles E_1 and E_2 on M. A tangential operator is a family $(P_x)_{x \in M}$ of left G_x^x-

invariant operators

$$(3.31) \qquad P_x : C^\infty(G^x, s^*E_1) \to C^\infty(G^x, s^*E_2).$$

The left invariance implies that the vector–valued distribution kernel $k(u, v)$ on G is of the form $k(u^{-1}v)$, i.e., for every $\xi \in C_c^{\infty,0}(s^*E_1)$,

$$(3.32) \qquad P_x(\xi)(u) = \int_{G^x} k(u^{-1}v)\xi(v)d\lambda^x(v).$$

This approach has two advantages. It allows one to use the tools of the theory of pseudo–differential operators even though the leaves are in general noncompact. In particular, the heat kernel of the Laplacian in the tangential direction will behave as the classical one. This allows one, for example, to compute the index of elliptic operators or to define the extension of the analytic torsion by just using (2.4), where the usual trace is now replaced by Φ_ν. The other advantage is that one can apply the theory in the case of group actions on a manifold. Suppose Γ is a group acting from the right on a manifold M. Dynamically, think of the iterates of a diffeomorphism. The groupoid is defined by the identification of points on the same orbit. The tangential measures now reduce to Haar measures on Γ and the rest of the definitions carry over without changes. There is great hope that this will provide new insights about diffeomorphisms on compact manifolds.

It is to be noted that the torsion will depend on the transverse measure chosen for the construction. However, in many cases, one has a natural invariant transverse measure, as in the case of Hamiltonian flows. The main and somewhat astonishing result for the dynamics is

THEOREM 3.5. *Let $E \to M$ be a flat bundle over M. Let $T(E)$ denote the generalised analytic torsion. Then, $T(E) \neq 1$ if and only if the measure of closed leaves without holonomy is nonzero.*

That is, this version of the torsion fails to catch isolated leaves or periodic orbits. There is hope that one can improve this result by designing other analytic invariants or by simply considering another differential operator. It is to be noted that a different version of the torsion has been introduced by Cohen and Connes [CC].

4. Conclusion. As seen in the previous sections, the scope of the Lefschetz formulas for flows is rather limited. In fact, as was pointed out by D. Fried [Fr3], there is no hope to find as general a result as the standard Lefschetz Formula. Nevertheless, there is hope to improve the known results. Several new approaches, specially by R. Forman [Fo], have allowed one to relax the hypothesis on the flow of the vector fields. In fact, it should be possible to consider the case of almost fibrations (where the nature of the fiber may vary).

In the noncommutative context, much remains to be done. Can one design analytical invariants like the index (see [Co1,2]) which have a dynamical significance? The tools at hand in this case also allow one to study discrete group actions. In the case of the action of a diffeomorphism, very little is known about the relation between such invariants and the dynamics even though such actions are well understood from a noncommutative geometrical point of view. (see Baum and Connes [BC]).

REFERENCES

[AB1] M. ATIYAH AND R. BOTT, *A Lefschetz point formula for elliptic complexes*, Ann. of Math., 86 (1967), pp. 374–407.

[AB2] M. ATIYAH AND R. BOTT, *A Lefschetz fixed point formula for elliptic complexes II*, Ann. of Math., 88 (1968), pp. 451–491.

[BC] P. BAUM AND A. CONNES, *Chern character for discrete groups*, A fete in Topology, Academic Press (1988).

[Ch] J. CHEEGER, *Analytic torsion and the heat equation*, Annals of Math. 109, pp. 259–322.

[CGT] J. CHEEGER, M. GROMOV AND M. TAYLOR, *Finite propagation speed, kernel estimates for functions of the Laplace Operator and the geometry of complete Riemannian manifolds*, J. Diff. Geom., 17 (1982), pp. 15–53.

[CC] P. COHEN AND A. CONNES, *On the noncommutative torus of dimension two*, Springer Proc. in Physics, 47 (1990), pp. 151–156.

[Co1] A. CONNES, *A survey of foliations and operator algebras*, Proc. Symp. Pure Math., 38 (1982), pp. 521–628.

[Co2] A. CONNES, *Noncommutative differential geometry*, Publications Mathematiques de l'IHES (1985).

[CF] R. CREW AND D. FRIED, *Nonsingular holomorphic flows*, Topology 25 No 4 (1986), pp. 471–473.

[D1] X. DAI, *Adiabatic limits non multiplicativity of signature and the Leray spectral sequence*, J. Amer. Math. Soc., vol 4, 2 (1991), pp. 265–321.

[D2] X. DAI, *Geometric invariants and their adiabatic limits*, MIT Preprint (1990).

[Fa] J. FAY, *Analytic torsion and Prym differentials*, Ann. of Math. Studies, Princeton University Press (1981).

[Fo] R. FORMAN, *Personal communication*.

[Fr1] D. FRIED, *Zeta functions of Ruelle and Selberg I*, Ann. Sci. Ec. Norm. sup., 19 (1986), pp. 491–517.

[Fr2] D. FRIED, *Zeta functions of Ruelle and Selberg II*, In preparation.

[Fr3] D. FRIED, *Lefschetz Formulas for flows*, Contemp. Math., 58 III (1987), pp. 19–69.

[Fr4] D. FRIED, *Analytic torsion and closed geodesics on hyperbolic manifolds*, Invent. Math., 84 (1986), pp. 523–540.

[Fr5] D. FRIED, *Torsion and closed geodesics on complex hyperbolic manifolds*, Invent. Math., 91 (1988), pp. 31–51.

[Fr6] D. FRIED, *Periodic points of holomorphic maps*, Topology 25 No 4 (1986), pp. 429–441.

[Gi] P. B. GILKEY, *Invariance theory, the heat equation and the Atiyah–Singer index theorem*, Publish or Perish 11 (1984).

[GU] V. GUILLEMIN AND A. URIBE, *Circular symmetry and the trace formula*, Invent. Math., 96 (1989), pp. 385–423.

[HW] G. HARDY AND E. WRIGHT, *An introduction to the theory of numbers*, Oxford Univ. Press (1978).

[Ko] T. KOTAKE, *The fixed point theorem of Atiyah–Bott via parabolic operators*, Comm. Pure and Appl. Math., 23 (1969), pp. 789–806.

[La1] S. LAEDERICH, $\bar{\partial}$-Torsion for complex manifolds and the adiabatic limit, Preprint, to be published in Comm. Math. Phys. (1992).

[La2] S. LAEDERICH, $\bar{\partial}$-Torsion, foliations and holomorphic vector fields, Preprint, to be published in Comm. Math. Phys. (1992).

[Lo] P. LOCHAK, Personal communication.

[Me] R. B. MELROSE, The Atiyah Patodi Singer index theorem, MIT Preprint (1991).

[Mu] W. MÜLLER, Analytic torsion and Reidemeister torsion of Riemannian manifolds, Adv. in Math., 28, 3 (1978), pp. 233–305.

[MS] C. MOORE AND C. SCHOCHET, Global analysis on foliated spaces, MSRI Publ. 9, Springer-Verlag (1988).

[RS1] D. RAY AND I. SINGER, R-torsion and the Laplacian on Riemannian Manifolds, Adv. in Math., 7 (1971), pp. 145–210.

[RS2] D. RAY AND J. SINGER, Analytic torsion for complex manifolds, Ann. of Math., 108 (1978), pp. 1–39.

[Ru] D. RUELLE, Zeta functions for expanding maps and Anosov flows, Invent. Math., 34 (1976), pp. 231–242.

[Se] R. T. SEELEY, Complex powers of an elliptic operator, Proc. Symp. Pure Math., 10 (1967), pp. 288–307.

[We] R. O. WELLS, Differential analysis on complex manifolds, Springer-Verlag, GTM 65 (1980).

[Wi] E. WITTEN, Supersymmetry and Morse theory, Adv. in Math., 4 (1970), pp. 109–126.

LINEARIZED DYNAMICS OF SYMMETRIC LAGRANGIAN SYSTEMS

DEBRA LEWIS*

1. Introduction. The structure induced by the Lagrangian or Hamiltonian character of a mechanical system and its symmetries can make explicit analyses of the stability and bifurcation behavior tractable even for moderately large systems. The variational characterization of relative equilibria, i.e. steady motions generated by elements of the symmetry group, greatly simplifies many of the necessary calculations. Local minima modulo symmetries of an appropriate energy-momentum functional correspond to nonlinearly orbitally stable motions; if the second variation of the energy-momentum functional is positive (negative) definite on an appropriate subspace, the relative equilibrium is said to be formally stable. For finite dimensional systems, formal stability typically implies nonlinear orbital stability.

The relationship between formal and linear stability of steady motions appears to be quite complicated in many cases. 'Gyroscopic' effects may linearly stabilize a steady motion which fails to be formally stable. At symmetric relative equilibria, the relationship between formal and linear stability appears to be particularly interesting; in several mechanical systems, the loss of formal and linear stability coincides with the termination of a 'window' of bifurcations from the symmetric state. (See Lewis [1991] and Lewis et al. [1992] for examples of this behavior.) The block structure of the linearized equations derived here and used in Bloch et al. [1992] is used to identify conditions under which small dissipation destabilizes a linearly stable, but formally unstable relative equilibrium.

The formal and nonlinear stability of relative equilibria has been the subject of numerous treatments. We consider here the extension of one such treatment, the reduced energy momentum method of Simo et al. [1991a, b] and Lewis [1992], to the analysis of linear stability. The reduced energy momentum method specifies a decomposition of an appropriate subspace of the tangent space to the phase space with respect to which the second variation of the energy-momentum functional block diagonalizes. The tangent space $\mathcal{G} \cdot q_e$ to the orbit of the symmetry group through the equilibrium configuration is central to the block diagonalization procedure. Two (typically) distinct complements to $\mathcal{G} \cdot q_e$ figure prominently in the analysis: the orthogonal complements $\mathcal{V}_{int}$ and $(\mathcal{G} \cdot q_e)^\perp$ with respect to the configuration and velocity stability forms. The full stability form block diagonalizes with

* Mathematics Board, University of California at Santa Cruz, Santa Cruz, CA 95064. I would like to thank Tudor Ratiu and J.C. Simo for many valuable discussions. This work was partially supported by NSF grant DMS-9122708 and Faculty Research Grant 503128-09523.

respect to $\mathcal{G} \cdot q_e$, $\mathcal{V}_{\text{int}}$, and $(\mathcal{G} \cdot q_e)^{\perp}$.

The linearized dynamics do not block diagonalize with respect to the spaces discussed above; however, there is still substantial blocking structure — three of the nine blocks determined by the decomposition are equal to zero. The analysis of the linearized equations suggests a further decomposition beyond that used in the formal stability analysis. Variations lying in the intersection $\mathcal{V}_S$ of $\mathcal{V}_{\text{int}}$ and $(\mathcal{G} \cdot q_e)^{\perp}$ are not directly coupled by the linearized dynamics to the space $\mathcal{V}_{\text{rig}}$ of admissible variations lying in $\mathcal{G} \cdot q_e$ (modulo $\mathcal{G}_\mu \cdot q_e$). In some cases, the linearized dynamics can be factored into two blocks: one acting on three spaces isomorphic to $\mathcal{V}_{\text{rig}}$, the other acting on two copies of $\mathcal{V}_S$.

We consider the linearized equations on a subspace of the tangent space at the equilibrium. The subspace is determined by the linearized momentum constraint. Conservation of momentum is guaranteed by Noether's theorem for symmetric Lagrangian systems. The collection of all such conserved momenta can be viewed as a map from the phase space to the dual of the Lie algebra of the symmetry group, known as the momentum map. (Linear and angular momentum are the best known such maps.) We restrict our attention to the kernel of the linearized momentum map at the relative equilibrium and consider the linearized equations on a complement $\mathcal{S}$ to the space of infinitesimal group motions lying in the kernel. A geometric motivation for this restriction is symplectic reduction. Given a canonical Hamiltonian system with momentum map $\boldsymbol{J}$, the quotient $\boldsymbol{J}^{-1}(\mu)/G_\mu$ is a symplectic manifold for regular values of $\mu \in \mathcal{G}^*$. Here $\mathcal{G}^*$ denotes the dual of the Lie algebra of the symmetry group G and G_μ denotes the isotropy subgroup of μ. (See Abraham and Marsden [1978] or Meyer [1973] for discussions of symplectic reduction.) The subspace discussed above models the tangent space to the reduced symplectic manifold at an asymmetric relative equilibrium.

The momentum associated to a symmetric relative is not a regular value of the momentum map; the level set $\boldsymbol{J}^{-1}(\mu)$ is a cone, rather than a manifold. (See Arms et al. [1981] for the proof of this result.) Hence the space $\mathcal{S}$ need not be the tangent space of a manifold. However, we believe that even in the symmetric case the space $\mathcal{S}$ is a natural choice for study, given that both the kernel of the linearization of the momentum map and the tangent space to the isotropy subgroup orbit are invariant under the linearized dynamics. Results of Montaldi et al. [1990] and Lewis et al. [1992] suggest that in most cases the linearized dynamics off the subspace $\mathcal{S}$ are relatively simple.

As an illustration of the decomposition technique and as an example of a system possessing the finer block structure discussed above, we consider the linearized dynamics of rigidly rotating pseudo-rigid bodies. Pseudo-rigid bodies are a mechanical system on the Lie group $GL(3)^+$ which models homogeneous hyperelasticity. In this example, the components of the decomposition appear to be relevant to the mechanical properties of the system.

2. Block diagonalization and the linearized dynamics. The variational approach to the nonlinear stability analysis of an equilibrium of a conservative system is widely known — equilibria correspond to critical points of an appropriate energy functional and stable equilibria correspond to local extrema of this functional. In the presence of symmetry, this variational approach can often be extended to the analysis of simple steady motions generated by the action of the symmetry group. In this case, steady motions correspond to critical points of an appropriate energy-momentum functional; again local extrema correspond to stable motions. The construction of a variational formulation on the configuration manifold for the identification of relative equilibria of simple mechanical systems, i.e. systems with a free group action and energy of the form 'kinetic plus potential', is due to Smale [1970a, b]. An extension of Smale's construction to systems other than simple mechanical systems can be found in Lewis [1992].

The general technique for the location of relative equilibria can be briefly described as follows: Consider a Lagrangian system with Lagrangian L and symmetry group G. The energy of the system is given by the function $E : TQ \to \mathbb{R}$ defined by

$$(2.1) \qquad\qquad E(v) := \langle \boldsymbol{FL}\, v, v \rangle - L(v).$$

Here $\langle\,,\,\rangle$ denotes the dual pairing between T^*Q and TQ and $\boldsymbol{FL} : TQ \to T^*Q$ denotes the Legendre transformation

$$(2.2) \qquad\qquad \langle \boldsymbol{FL}\, v, w \rangle := \tfrac{d}{d\epsilon}\big|_{\epsilon=0} L(v + \epsilon\, w).$$

The equations of motion are determined by the Lagrangian and the two form ω_L on the tangent bundle TQ constructed by pulling the canonical symplectic two form ω on the cotangent bundle T^*Q back to TQ by means of the Legendre transformation, i.e.

$$(2.3) \qquad\qquad \omega_L(v)(\delta v, \Delta v) := \omega(\boldsymbol{FL}(v))(T_v\, \boldsymbol{FL}\, \delta v, T_v\, \boldsymbol{FL}\, \Delta v)$$

for δv and $\Delta v \in T_v TQ$. The equations of motion associated to the Lagrangian L are $\dot{v} = X_E(v)$, where the vector field $X_E : TQ \to TTQ$ is determined by the relationship $\iota_{X_E}\omega_L = DE$. If L is regular, then X_E is guaranteed to exist.

Given a configuration $q_e \in Q$ and element ξ of the Lie algebra $\mathcal{G}$ of G, we wish to determine conditions on q_e and ξ under which the configuration q_e will undergo a steady group motion with group velocity ξ, i.e., that the evolution of q_e according to the dynamics of the Lagrangian system will be given by

$$(2.4) \qquad\qquad q_t = \exp(t\, \xi) \cdot q_e,$$

with velocity

$$(2.5) \qquad \dot{q}_t = \xi_Q(q_t) := \tfrac{\partial}{\partial t} \exp(t\,\xi) \cdot q_e.$$

Such steady motions are known as *relative equilibria*. For example, if $G = SO(3)$, the group of rigid rotations in $\mathbb{R}^3$, then an element $\xi \in \mathcal{G}$ can be identified with a vector in $\mathbb{R}^3$ and the motion q_t consists of a steady rotation about the axis ξ. Given $\xi \in \mathcal{G}$, to identify pairs (q_e, ξ) satisfying (2.4), we define the *locked Lagrangian* $L_\xi := L \circ \xi_Q$. If L is regular, then $\xi_Q(q)$ is a relative equilibrium if and only if q is a critical point of L_ξ.

At a symmetric relative equilibrium v_e the generator ξ is not unique; ξ may be replaced by $\xi + \zeta$ for any element ζ of the isotropy algebra of v_e. To obtain sharp formal stability conditions for a symmetric relative equilibrium, it is necessary to consider all possible choices of generator. (See Lewis [1991, 1992] for detailed discussions of this issue.)

2.1. Formal stability. We first consider the formal stability of relative equilibria. The essential features of the reduced energy momentum (REM) method are sketched below; readers unfamiliar with the method should consult Simo et al. [1991a] (for the simple mechanical case) or Lewis [1992] (for the symmetric and general Lagrangian cases) and references therein.

The conservation laws given by Noether's theorem are central to the stability analysis of conservative systems with symmetry. These laws can be summarized as follows: Given $\xi \in \mathcal{G}$, define $J_\xi : TQ \to \mathbb{R}$ by

$$(2.1.1) \qquad J_\xi(v) := \langle FL\,v, \xi_Q(\pi(v)) \rangle = \tfrac{d}{d\epsilon}\big|_{\epsilon=0} L(v + \epsilon\,\xi_Q(\pi(v))),$$

where $\pi : TQ \to Q$ is the canonical projection, and define the *momentum map* $J_L : TQ \to \mathcal{G}^*$ by $J_L(v) \cdot \xi := J_\xi(v)$. Noether's theorem states that the momentum map is conserved by the dynamics generated by a G invariant Lagrangian L.

We shall say that a relative equilibrium $v_e = \xi_Q(q_e)$ is *formally stable* if the second variation of the energy-momentum functional $E_\xi := E - J_\xi$ at v_e is positive semi-definite on $\ker[DJ_L(v_e)]$ with kernel $\mathcal{G}_\mu \cdot v_e$, where $\mathcal{G}_\mu := \{\zeta \in \mathcal{G} : \mathrm{ad}_\zeta^*\mu = 0\}$ is the isotropy subalgebra of $\mu := J_\xi(v_e)$. For asymmetric relative equilibria of finite dimensional systems, formal stability implies nonlinear stability modulo the isotropy group $G_\mu = \{g \in G : \mathrm{Ad}_g^*\mu = \mu\}$. For symmetric and infinite dimensional systems, additional criteria must be imposed to obtain a nonlinear stability result. The reduced energy-momentum method specifies an identification of $\ker[DJ_L(v_e)]$ with subspaces of $T_{q_e}Q$ (and, in the Hamiltonian treatment, $T_{q_e}^*Q$). To construct this identification, we make use of the *locked momentum map* $\mathbb{I}_\xi := J_L \circ \xi_Q$ mapping Q into $\mathcal{G}^*$ and the *linearized momentum tensor* $\mathbb{I}_e : \mathcal{L}(\mathcal{G}, \mathcal{G}^*)$ given by

$$(2.1.2) \qquad \mathbb{I}_e\boldsymbol{\eta} := \tfrac{d}{d\epsilon}\big|_{\epsilon=0} \mathbb{I}_{(\xi+\epsilon\eta)}(q_e) = \langle\!\langle \xi_Q(q_e), \eta_Q(q_e) \rangle\!\rangle,$$

where $\langle\!\langle\ ,\ \rangle\!\rangle : T_{q_e}Q \times T_{q_e}Q \rightarrow \mathbb{R}$ is the bilinear form associated to the *linearized Legendre transformation* $\mathbf{m} : T_{q_e}Q \rightarrow T_{q_e}^*Q$ given by

(2.1.3) $\langle \mathbf{m}\,\delta\mathrm{v}, \Delta\mathrm{v} \rangle := \langle\!\langle \delta\mathrm{v}, \Delta\mathrm{v} \rangle\!\rangle := \frac{d}{d\epsilon}\big|_{\epsilon=0} \frac{d}{d\tau}\big|_{\tau=0} L(v_e + \epsilon\,\delta\mathrm{v} + \tau\,\Delta\mathrm{v}).$

Note that we use the symbols $\delta\mathrm{v}$ and $\Delta\mathrm{v}$ to denote variations within the fiber $T_{q_e}Q$ and the symbols δv and Δv to denote full tangent vector variations, i.e. elements of $T_{v_e}TQ$. (This notation is intended to suggest local coordinates $v = (q, \mathrm{v})$.) G invariance of the Lagrangian implies that

(2.1.4) $$DL_\xi(q) \cdot \eta_Q(q) = \mathrm{ad}_\xi^* \mathbb{I}_\xi(q) \cdot \eta$$

for all $\eta \in \mathcal{G}$. Equation (2.1.4) yields the 'rigid equilibrium condition' $\mathrm{ad}_\xi^* \mathbb{I}_\xi(q_e) = 0$, i.e. $\xi \in \mathcal{G}_\mu$.

It is clear from (2.1.2) that

(2.1.5) $$\ker[\mathbb{I}_e] \supseteq \mathcal{G}_{q_e} := \{\zeta \in \mathcal{G} : \zeta_Q(q_e) = 0\}.$$

If the restriction of $\langle\!\langle\ ,\ \rangle\!\rangle$ to $\mathcal{G} \cdot q_e$ is non-degenerate, then $\ker[\mathbb{I}_e] = \mathcal{G}_{q_e}$ and there exists a local map $\Xi : \mathcal{U} \rightarrow \mathcal{G}$ from a neighborhood $\mathcal{U}$ of q_e in Q satisfying $\mathbb{I}_{\Xi(q)}(q) = \mu$, i.e.

(2.1.6) $$\mathbf{v}_\mu(q) := (\Xi(q))_Q(q) \in \mathbf{J}_L^{-1}(\mu).$$

We introduce the notation $[\mathbb{I}_e]^{-1} : \mathrm{range}[\mathbb{I}_e] \times \mathrm{range}[\mathbb{I}_e] \rightarrow \mathbb{R}$ to denote the pairing induced by the pseudo-inverse of $\mathbb{I}_e$; even though the preimage of an element of $\mathrm{range}[\mathbb{I}_e]$ is only defined modulo $\ker[\mathbb{I}_e]$, the pairing itself is well-defined. If $\ker[\mathbb{I}_e] = \mathcal{G}_{q_e}$, then the map $\gamma : \mathrm{range}[\mathbb{I}_e] \rightarrow T_{q_e}^*Q$ determined by the relationship

(2.1.7) $$\langle \gamma\,\nu, \delta q \rangle = \langle\!\langle \eta_Q(q_e), \delta q \rangle\!\rangle$$

for $\eta \in [\mathbb{I}_e]^{-1}(\nu)$ and $\delta q \in T_{q_e}Q$ is well-defined.

The space $\mathcal{Q}$ of admissible configuration variations is defined by

(2.1.8) $$\mathcal{Q} := \left\{ \delta q \in T_{q_e}Q : D\mathbb{I}_\xi(q_e) \cdot \delta q \in \mathcal{G}_{q_e}^A \right\},$$

where $\mathcal{G}_{q_e}^A$ denotes the annihilator of the isotropy algebra of q_e. Let $(\mathcal{G} \cdot q_e)^\perp$ denote the $\langle\!\langle\ ,\ \rangle\!\rangle$-orthogonal complement to the tangent $\mathcal{G} \cdot q_e$ to the group orbit and $\mathrm{vert}(\ ) : T_{q_e}Q \rightarrow T_{v_e}TQ$ denote the vertical lift

(2.1.9) $$\mathrm{vert}(\delta\mathrm{v}) := \frac{d}{d\epsilon}\big|_{\epsilon=0}(v_e + \epsilon\,\delta\mathrm{v}).$$

If we define $\mathcal{K}_\xi : \mathcal{Q} \times (\mathcal{G} \cdot q_e)^\perp \rightarrow \ker[D\mathbf{J}_L(v_e)]$ by

(2.1.10) $\begin{aligned} \mathcal{K}_\xi(\delta q, \delta\mathrm{v}) &:= T_{q_e}\mathbf{v}_\mu \cdot \delta q + \mathrm{vert}(\delta\mathrm{v}) \\ &= T_{q_e}\xi_Q \cdot \delta q + \mathrm{vert}\left(\delta\mathrm{v} - \mathbf{m}^{-1}\gamma\left(D\mathbb{I}_\xi(q_e) \cdot \delta q\right)\right), \end{aligned}$

then $\ker[D\boldsymbol{J}_L(v_e)] = \mathrm{range}\,[\mathcal{K}_\xi]$. Note that $\mathcal{K}_\xi(\zeta_Q(q_e), 0) = \zeta_{TQ}(v_e)$ for $\zeta \in \mathcal{G}_\mu$, where

$$(2.1.11) \qquad \xi_{TQ}(v_e) = \tfrac{d}{d\epsilon}\big|_{\epsilon=0} \exp(\epsilon\,\xi) \cdot v_e$$

is the infinitesimal generator associated to the induced action of G on TQ.

We can pull back $D^2 E_\xi(q_e)$ from $\ker[D\boldsymbol{J}_L(v_e)]$ to $\mathcal{Q} \times (\mathcal{G} \cdot q_e)^\perp$ using $\mathcal{K}_\xi$. The *stability two form* $\mathcal{B}_e : \mathcal{Q} \times \mathcal{Q} \to \mathbb{R}$ defined by

$$(2.1.12) \qquad
\begin{aligned}
\mathcal{B}_e(\delta q, \Delta q) \;&:=\; [\mathbb{I}_e]^{-1}\left(D\mathbb{I}_\xi(q_e) \cdot \delta q, D\mathbb{I}_\xi(q_e) \cdot \Delta q\right) \\
&\quad - D^2 L_\xi(q_e)(\delta q, \Delta q)
\end{aligned}$$

satisfies

$$(2.1.13) \quad D^2 E_\xi(q_e)(\mathcal{K}_\xi(\delta q, \delta \mathrm{v}), \mathcal{K}_\xi(\Delta q, \Delta \mathrm{v})) = \mathcal{B}_e(\delta q, \Delta q) + \langle\!\langle \delta \mathrm{v}, \Delta \mathrm{v} \rangle\!\rangle$$

for $\delta q,\ \Delta q \in \mathcal{Q}$ and $\delta \mathrm{v},\ \Delta \mathrm{v} \in (\mathcal{G} \cdot q_e)^\perp$. The relative equilibrium v_e is formally stable if $\langle\!\langle\ ,\ \rangle\!\rangle$ is positive (negative) definite on $(\mathcal{G} \cdot q_e)^\perp$ and $\mathcal{B}_e$ is positive (negative) semi-definite with kernel $\mathcal{G}_\mu \cdot q_e$.

Remarks. In the simple mechanical case, the symmetry group G acts freely; hence the map $\Xi : Q \to \mathcal{G}$ and vector field $\mathbf{v}_\mu : Q \to \boldsymbol{J}_L^{-1}(\mu)$ can be constructed globally. In this case, we can construct the *amended potential* $V_\mu : Q \to \mathbb{R}$, given by

$$(2.1.14) \qquad V_\mu(q) := E(\mathbf{v}_\mu(q)).$$

The amended potential, which is due to Smale [1970a, b], is a unique globally defined function which captures both the potential energy and the kinetic energy associated to the momentum constraint. Critical points q_e of V_μ correspond to relative equilibria $\mathbf{v}_\mu(q_e)$; local minima (modulo symmetries of the equilibrium momentum μ) of V_μ correspond to (formally) stable relative equilibria.

2.2. Block diagonalization: rigid and internal variations.

The block diagonalization procedure specifies a decomposition of the tangent space $T_{q_e}Q$ which substantially simplifies the stability analysis. This decomposition consists of 'rigid' variations, which lie in a subspace of the tangent space $\mathcal{G} \cdot q_e$ to the group orbit $G \cdot q_e$, and 'internal' variations, which satisfy the linearized group equilibrium condition. The space $\mathcal{Q}_{\mathrm{rig}}$ of rigid variations is defined as the intersection of $\mathcal{G} \cdot q_e$ and the space $\mathcal{Q}$ of admissible variations. The variation of the locked momentum map $\mathbb{I}_\xi$ satisfies

$$(2.2.1) \qquad D\mathbb{I}_\xi(q_e) \cdot \eta_Q(q_e) = -\left(\mathbb{I}_e(\mathrm{ad}_\eta \xi) + \mathrm{ad}_\eta^* \mu\right)$$

for all $\eta \in \mathcal{G}$. Thus

$$(2.2.2) \qquad
\begin{aligned}
\mathcal{Q}_{\mathrm{rig}} \;&:=\; \mathcal{G} \cdot q_e \cap \mathcal{Q} = \tilde{\mathcal{G}} \cdot q_e \qquad \text{for} \\
\tilde{\mathcal{G}} \;&:=\; \left\{ \eta \in \mathcal{G} : \mathrm{ad}_\eta^* \mu \in \mathrm{range}\,[\mathbb{I}_e] \right\}.
\end{aligned}$$

The G invariance of the Lagrangian L implies that the stability two form $\mathcal{B}_e$ takes a particularly simple form on $\mathcal{Q}_{\text{rig}}$. We define the *generalized Arnold form* $\mathcal{A} : \widetilde{\mathcal{G}} \times \widetilde{\mathcal{G}} \to \mathbb{R}$ by

$$
\begin{aligned}
(2.2.3) \qquad \mathcal{A}(\zeta, \eta) \quad &:= \quad \mathcal{B}_e(\eta_Q(q_e), \zeta_Q(q_e)) \\
&= \quad [\mathbb{I}_e]^{-1}\left(\operatorname{ad}_\zeta^* \mu, \operatorname{ad}_\eta^* \mu\right) + \operatorname{ad}_\zeta^* \mu \cdot \operatorname{ad}_\eta \xi.
\end{aligned}
$$

The space $\mathcal{Q}_{\text{int}}$ of internal variations is defined as the subspace $\mathcal{B}_e$ orthogonal to $\mathcal{Q}_{\text{rig}}$, i.e. $\mathcal{Q}_{\text{int}} := \ker[\beta]$, where the map $\beta : \mathcal{Q} \to \widetilde{\mathcal{G}}^*$ is defined by

$$(2.2.4) \quad (\beta\,\delta q) \cdot \eta := \mathcal{B}_e(\delta q, \eta_Q(q_e)) = -[\mathbb{I}_e]^{-1}\left(\operatorname{ad}_\eta^* \mu, D\mathbb{I}_\xi(q_e) \cdot \delta q\right).$$

The space $\mathcal{Q}_{\text{int}}$ can also be specified as the space of variations satisfying the linearized version of the 'rigid equilibrium condition' $\operatorname{ad}_\zeta^* \mathbb{I}_\xi(q) = 0$, i.e.

$$(2.2.5) \quad \mathcal{Q}_{\text{int}} = \{\delta q \in \mathcal{Q} : D\mathbb{I}_\xi(q_e) \cdot \delta q = \mathbb{I}_e \zeta \qquad \text{for some } \zeta \in \mathcal{G}_\mu\}.$$

The space $\mathcal{Q}$ of admissible variations can be decomposed as

$$(2.2.6) \quad \mathcal{Q} = \mathcal{Q}_{\text{rig}} + \mathcal{Q}_{\text{int}}, \qquad \text{with} \qquad \mathcal{Q}_{\text{rig}} \cap \mathcal{Q}_{\text{int}} = \ker[\mathcal{A}] \cdot q_e.$$

The choice of a complement $\mathcal{S}$ to $\mathcal{G}_\mu \cdot v_e$ in $\ker[DJ_L(v_e)]$ determines a complement $\mathcal{V}$ to $\mathcal{G}_\mu \cdot q_e$ in $\mathcal{Q}$ (and vice versa) by means of the relationship $\mathcal{S} = \mathcal{K}_\xi \left[\mathcal{V} \times (\mathcal{G} \cdot q_e)^\perp\right]$. Given $\mathcal{V}$, we define

$$(2.2.7) \qquad \mathcal{V}_{\text{rig}} := \mathcal{Q}_{\text{rig}} \cap \mathcal{V} \qquad \text{and} \qquad \mathcal{V}_{\text{int}} := \mathcal{Q}_{\text{int}} \cap \mathcal{V}.$$

If $\ker[\mathcal{A}] = \mathcal{G}_\mu + \mathcal{G}_{q_e}$, then

$$(2.2.8) \qquad\qquad\qquad \mathcal{V}_{\text{rig}} \cap \mathcal{V}_{\text{int}} = \{0\}$$

and

$$(2.2.9) \quad \mathcal{S} = \mathcal{K}_\xi \cdot (\mathcal{V}_{\text{rig}} x\{0\}) \oplus \mathcal{K}_\xi \cdot (\mathcal{V}_{\text{int}} x\{0\}) \oplus \operatorname{vert}\left[(\mathcal{G} \cdot q_e)^\perp\right].$$

In this case, the restriction of the second variation $D^2 E_\xi(v_e)$ to $\mathcal{S}$ block diagonalizes to

$$(2.2.10) \qquad\qquad D^2 E_\xi(v_e) = \operatorname{diag}\left[\mathcal{A}, \mathcal{B}_{\text{int}}, \langle\!\langle\ ,\ \rangle\!\rangle\right]$$

with respect to the decomposition (2.2.9) where $\mathcal{B}_{\text{int}} := \mathcal{B}_e|\mathcal{V}_{\text{int}}$.

If (2.2.8) holds, we can construct the projection $\mathbb{P}_{\text{rig}}$ of $\mathcal{V}$ onto the 'rigid' component $\mathcal{V}_{\text{rig}}$ using the Arnold form and the map β as follows: Let $\mathcal{G}_\mathcal{A}^\perp$ be a complement to $\mathcal{G}_\mu + \mathcal{G}_{q_e}$ in $\widetilde{\mathcal{G}}$ such that $\mathcal{V}_{\text{rig}} = \mathcal{G}_\mathcal{A}^\perp \cdot q_e$. The condition $\ker[\mathcal{A}] = \mathcal{G}_\mu + \mathcal{G}_{q_e}$ implies that for every $\delta q \in \mathcal{Q}$ there exists $\kappa_{\delta q} \in \mathcal{G}_\mathcal{A}^\perp$ satisfying

$$(2.2.11) \qquad\qquad\qquad \mathcal{A}(\kappa_{\delta q}, \cdot) = \beta\delta q.$$

Hence, if we define the projection $\mathbb{P}_{\text{rig}} : \mathcal{V} \to \mathcal{V}_{\text{rig}}$ by

$$(2.2.12) \qquad\qquad\qquad \mathbb{P}_{\text{rig}}\,\delta q := (\kappa_{\delta q})_Q(q_e),$$

then $(\mathbf{1} - \mathbb{P}_{\text{rig}})$ maps $\mathcal{V}$ into $\mathcal{V}_{\text{int}}$.

2.3. Linearized equations of motion. When considering relative equilibria it is often convenient to write the equations of motion in the form

$$(2.3.1) \qquad \iota_{(X_E - \xi_{TQ})} \omega_L = DE_\xi,$$

which makes use of the identity $DJ_\xi = \iota_{\xi_{TQ}} \omega_L$. If we let L denote the linearization of the vector field $X_E - \xi_{TQ}$ at a relative equilibrium $v_e = \xi_Q(q_e)$, then (2.3.1) implies that

$$(2.3.2) \qquad \iota_{(L\,\delta v)} \omega_L(v_e) = \iota_{\delta v} D^2 E_\xi(v_e)$$

for all $\delta v \in T_{v_e} TQ$. As was previously mentioned, at a symmetric relative equilibrium v_e the generator ξ is not unique. The choice of generator will typically be reflected in the expression of the linearization of the vector field $X_E - (\xi + \zeta)_{TQ}$ which describes the dynamics relative to a background motion (e.g., a rotating frame). The equilibrium does not 'see' the isotropy algebra element, but the nearby asymmetric states are effected by that choice.

The restriction of the linearized equations to a complement S to the tangent $\mathcal{G}_\mu \cdot v_e$ to the isotropy subgroup orbit in the kernel of the linearized momentum map is easily justified in the asymmetric case by identifying S with the tangent space to the symplectically reduced manifold given by the reduction theorem. However, the space S is a natural object of study even in the symmetric case. We shall show that the subspaces $\mathcal{G}_\mu \cdot v_e$ and $\ker[DJ_L(v_e)]$ are invariant under the linearized equations of motion. Hence, given an arbitrary complement $\mathcal{U}$ to $\ker[DJ_L(v_e)]$ in $T_{v_e} TQ$, the linearized equations have a block upper triangular form with respect to the decomposition $\mathcal{G}_\mu \cdot v_e \oplus S \oplus \mathcal{U}$. The fact that $L[\ker[DJ_L(v_e)]] \subseteq \ker[DJ_L(v_e)]$ is a consequence of the identities (2.3.2), $\iota_{\eta_{TQ}} \omega_L = DJ_\eta$, and

$$(2.3.3) \qquad D^2 E_\xi(v_e)(\delta v, \eta_{TQ}(v_e)) = DJ_{\mathrm{ad}_\xi \eta}(v_e) \cdot \delta v$$

for all $\delta v \in T_{v_e} TQ$ and $\eta \in \mathcal{G}$, which imply that

$$
\begin{aligned}
DJ_\eta(v_e) \cdot L\,\delta v &= \omega_L(v_e)(\eta_{TQ}(v_e), L\,\delta v) \\
&= -D^2 E_\xi(v_e)(\eta_{TQ}(v_e), \delta v) \\
&= DJ_{\mathrm{ad}_\xi \eta}(v_e) \cdot \delta v \\
(2.3.4) \qquad\qquad &= 0
\end{aligned}
$$

for all $\delta v \in \ker[DJ_L(v_e)]$ and $\eta \in \mathcal{G}$. The same identities can be used to show that

$$(2.3.5) \qquad
\begin{aligned}
\omega_L(v_e)(L\,\eta_{TQ}(v_e), \delta v) &= DJ_{\mathrm{ad}_\xi \eta}(v_e) \cdot \delta v \\
&= \omega_L(v_e)((\mathrm{ad}_\xi \eta)_{TQ}(v_e), \delta v)
\end{aligned}
$$

for all $\eta \in \mathcal{G}$ and $\delta v \in T_{v_e} TQ$. Thus $\boldsymbol{L}\,\eta_{TQ}(v_e) = (\mathrm{ad}_\xi \eta)_{TQ}(v_e)$. Since ξ is an element of the isotropy subalgebra $\mathcal{G}_\mu$, we see that $\mathcal{G}_\mu \cdot v_e$ is $\boldsymbol{L}$ invariant. For a detailed analysis of the full linearized equations, see Montaldi et al. [1990] or Lewis et al. [1992].

The Lagrangian symplectic two form ω_L and the map $\mathcal{K}_\xi : \mathcal{Q} \times (\mathcal{G} \cdot q_e)^\perp \to \ker\left[D\boldsymbol{J}_L(v_e)\right]$ induce a two form $\widetilde{\omega}_e$ on $\mathcal{Q} \times (\mathcal{G} \cdot q_e)^\perp$ by

$$
\begin{aligned}
\widetilde{\omega}_e((\delta q, \delta v),(\Delta q, \Delta v)) &:= \omega_L(v_e)(\mathcal{K}_\xi(\delta q, \delta v)\mathcal{K}_\xi(\Delta q, \Delta v)) \\
&= \langle\!\langle \Delta v, \delta q \rangle\!\rangle - \langle\!\langle \delta v, \Delta q \rangle\!\rangle - \mathrm{d}\alpha_\mu(\delta q, \Delta q),
\end{aligned}
$$
(2.3.6)

where $\alpha_\mu := \boldsymbol{FL} \circ \mathbf{v}_\mu$ is a locally defined one form. The two form $\mathrm{d}\alpha_\mu$ in (2.3.6) can be expressed in terms of $\mathrm{d}\alpha_\xi$ and $D\mathbb{II}_\xi(q_e)$ as

$$
\begin{aligned}
\mathrm{d}\alpha_\mu(\delta q, \Delta q) &= \mathrm{d}\alpha_\xi(\delta q, \Delta q) - \langle \gamma(D\mathbb{II}_\xi(q_e) \cdot \delta q), \Delta q \rangle \\
&\quad + \langle \gamma(D\mathbb{II}_\xi(q_e) \cdot \Delta q), \delta q \rangle
\end{aligned}
$$
(2.3.7)

for $\delta q, \Delta q \in \mathcal{Q}$.

The expression (2.3.7) takes a particularly simple form when one of the arguments lies in the tangent to the group orbit. G equivariance of the Legendre transformation and the identity $\mathcal{L}_{\eta_Q} \xi_Q = (\mathrm{ad}_\xi \eta)_Q$ imply that

$$
\iota_{\eta_Q(q_e)} \mathrm{d}\alpha_\xi = \mathbf{m}\,(\mathrm{ad}_\xi \eta)_Q(q_e) - \eta \cdot D\mathbb{II}_\xi(q_e).
$$
(2.3.8)

Thus (2.3.7) and (2.2.1) imply that

$$
\begin{aligned}
\widetilde{\omega}_e((\eta_Q(q_e),0),(\delta q, \delta v)) &= \mathrm{d}\alpha_\mu(\delta q, \eta_Q(q_e)) \\
&= \mathrm{d}\alpha_\xi(\delta q, \eta_Q(q_e)) - (D\mathbb{II}_\xi(q_e) \cdot \delta q) \cdot \eta \\
&\quad + \langle \gamma(D\mathbb{II}_\xi(q_e) \cdot \eta_Q(q_e)), \delta q \rangle \\
&= -\langle \gamma(\mathrm{ad}_\eta^* \mu), \delta q \rangle
\end{aligned}
$$
(2.3.9)

for all $\eta \in \widetilde{\mathcal{G}}$, $\delta q \in \mathcal{Q}$, and $\delta v \in (\mathcal{G} \cdot q_e)^\perp$.

We can solve for the $\mathcal{S}$ component of the linearized dynamics as follows: If we identify the pair $(\widetilde{\delta q}, \widetilde{\delta v}) \in \mathcal{V} \times (\mathcal{G} \cdot q_e)^\perp$ with the $\mathcal{S}$ component of $\boldsymbol{L}\,\delta v$, then (2.3.2) implies that

$$
\begin{aligned}
&\langle\!\langle \Delta v, \widetilde{\delta q} \rangle\!\rangle - \langle\!\langle \widetilde{\delta v}, \Delta q \rangle\!\rangle - \mathrm{d}\alpha_\mu(\widetilde{\delta q}, \Delta q) \\
&\quad = \mathcal{B}_e(\delta q, \Delta q) + \langle\!\langle \delta v, \Delta v \rangle\!\rangle
\end{aligned}
$$
(2.3.10)

for all $\Delta q \in \mathcal{Q}$ and $\Delta v \in (\mathcal{G} \cdot q_e)^\perp$. If $\mathcal{Q} + \mathcal{G} \cdot q_e = T_{q_e} Q$, then (2.3.10) determines $\widetilde{\delta q}$ modulo $\mathcal{G}_\mu \cdot q_e$ and completely determines $\widetilde{\delta v}$. An immediate consequence of (2.3.10) is that $\widetilde{\delta q} = \delta v$ modulo $\mathcal{G} \cdot q_e$, i.e.

$$
\widetilde{\delta q} = \delta v + \zeta_Q(q_e)
$$
(2.3.11)

for some $\zeta \in \widetilde{\mathcal{G}}$. Substituting (2.3.11) into (2.3.10) and applying (2.3.7) and (2.3.8) yields the equation

$$(2.3.12) \quad \left\langle \mathbf{m}\,\widetilde{\delta v} + \iota_{\delta v}d\alpha_\xi + \gamma(\mathrm{ad}_\zeta^*\mu - D\mathbb{I}_\xi(q_e)\cdot\delta v) + \iota_{\delta q}\mathcal{B}_e, \Delta q \right\rangle = 0$$

for $\widetilde{\delta v}$ and ζ.

We solve (2.3.12) for ζ by setting $\Delta q = \eta_Q(q_e)$ for some $\eta \in \widetilde{\mathcal{G}}$. Applying (2.3.8), we see that

$$(2.3.13) \qquad \mathcal{B}_e(\delta q, \eta_Q(q_e)) = -\mathrm{ad}_\zeta^*\mu\cdot\eta.$$

Thus ζ satisfies

$$(2.3.14) \qquad -\mathrm{ad}_\zeta^*\mu = J(\iota_{\delta q}\mathcal{B}_e),$$

where the cotangent bundle momentum map $J : T^*Q \to \mathcal{G}^*$ is determined by the relationship $J(z)\cdot\eta = \langle z, \eta_Q(q)\rangle$ for all $\eta \in \mathcal{G}$.

We now solve (2.3.12) for $\widetilde{\delta v}$. We note that (2.3.8) implies that

$$(2.3.15) \qquad J(\iota_{\delta v}d\alpha_\xi) = D\mathbb{I}_\xi(q_e)\cdot\delta v$$

for all $\delta v \in (\mathcal{G}\cdot q_e)^\perp$. Hence, if we define $\mathbb{P}_A : T_{q_e}^*Q \to (\mathcal{G}\cdot q_e)^A$ by $\mathbb{P}_A := 1 - \gamma \circ J$, then (2.3.12) implies that

$$(2.3.16) \qquad \left\langle \mathbf{m}\,\widetilde{\delta v} + \mathbb{P}_A\left(\iota_{\delta v}d\alpha_\xi + \iota_{\delta q}\mathcal{B}_e\right), \Delta q \right\rangle = 0$$

for all $\Delta q \in \mathcal{Q}$. Thus if $\mathcal{Q} + \mathcal{G}\cdot q_e = T_{q_e}Q$, then (2.3.16) completely determines $\widetilde{\delta v}$. If we define $\mathcal{L} : \mathcal{V}\times(\mathcal{G}\cdot q_e)^\perp \to \mathcal{V}\times(\mathcal{G}\cdot q_e)^\perp$ by

$$(2.3.17) \quad \mathcal{L}(\delta q, \delta v) = \left(\delta v + \zeta_Q(q_e), -\mathbf{m}^{-1}\,\mathbb{P}_A\left(\iota_{\delta v}d\alpha_\xi + \iota_{\delta q}\mathcal{B}_e\right)\right),$$

where $\zeta \in \widetilde{\mathcal{G}}$ is determined by (2.3.14), then

$$(2.3.18) \qquad \mathcal{K}_\xi \circ \mathcal{L} = L \circ \mathcal{K}_\xi \qquad \text{modulo} \quad \mathcal{G}_\mu\cdot v_e.$$

The linearized equations on $\mathcal{S}$ are obtained by composing the projection of $\ker[DJ_L(v_e)]$ onto $\mathcal{S}$ with L.

2.4. Block structure of the linearized equations. If $\ker[\mathcal{A}] = \mathcal{G}_\mu + \mathcal{G}_{q_e}$, and hence the projection $\mathbb{P}_{\mathrm{rig}}$ is well-defined, then we can decompose (2.3.17) in accordance with the decomposition (3.4.2). If $\eta_Q(q_e) = \mathbb{P}_{\mathrm{rig}}\,\delta q$ is the $\mathcal{V}_{\mathrm{rig}}$ component of δq, then equation (2.3.14) implies that ζ satisfies

$$(2.4.1) \qquad -\mathrm{ad}_\zeta^*\mu = J(\iota_{\delta q}\mathcal{B}_e) = \mathcal{A}(\eta, \cdot),$$

i.e. $\zeta_Q(q_e) = \Gamma\,\eta_Q(q_e)$, where $\Gamma : \mathcal{V}_{\mathrm{rig}} \to \mathcal{V}_{\mathrm{rig}}$ satisfies

$$(2.4.2) \quad \Gamma\,\eta_Q(q_e) := (\mathrm{ad}_\eta\xi)_Q(q_e) + \mathbf{m}^{-1}\gamma(\mathrm{ad}_\eta^*\mu) \qquad \text{modulo } \mathcal{G}_\mu\cdot q_e.$$

Let $\mathcal{G}_{\widetilde{\mathcal{A}}}^{\perp}$ be the complement to $\mathcal{G}_\mu + \mathcal{G}_{q_e}$ in $\widetilde{\mathcal{G}}$ determined by $\mathcal{V}_{\text{rig}}$. We can decompose $\widetilde{\delta q}$ into its $\mathcal{V}_{\text{rig}}$ and $\mathcal{V}_{\text{int}}$ components:

$$(2.4.3) \qquad \widetilde{\delta q} = \underbrace{\Gamma\,\mathbb{P}_{\text{rig}}\,\delta q + \mathbb{P}_{\text{rig}}\,\delta v}_{\mathcal{V}_{\text{rig}}} + \underbrace{(1 - \mathbb{P}_{\text{rig}})\,\delta v}_{\mathcal{V}_{\text{int}}}\,.$$

If $\mathcal{G}_{q_e}^A = \text{range}\,[\mathbb{I}_e]$ and $\mathcal{Q} + \mathcal{G} \cdot q_e = T_{q_e}\mathcal{Q}$, we define $\Upsilon : \mathcal{V}_{\text{rig}} \to (\mathcal{G} \cdot q_e)^\perp$ by

$$(2.4.4) \qquad \Upsilon\,\eta_Q(q_e) := -\mathbf{m}^{-1}(\mathbb{P}_A\,\iota_{\eta_Q(q_e)}\mathcal{B}_e)\,.$$

The map Υ satisfies

$$(2.4.5) \qquad \langle\!\langle \Upsilon\,\eta_Q(q_e), \delta q \rangle\!\rangle = (\beta\,\delta q) \cdot \eta - \langle \gamma(\iota_\eta \mathcal{A}), \delta q \rangle$$

for all $\delta q \in \mathcal{Q}$. Thus the linear map $\mathcal{L}$ can be decomposed with respect to (3.4.2) as

$$(2.4.6) \qquad \mathcal{L} = \begin{pmatrix} \Gamma & 0 & \mathbb{P}_{\text{rig}} \\ 0 & 0 & 1 - \mathbb{P}_{\text{rig}} \\ \Upsilon & -\mathbf{m}^{-1} \circ \mathcal{B}_e & \Phi \end{pmatrix},$$

where $\Phi : (\mathcal{G} \cdot q_e)^\perp \to (\mathcal{G} \cdot q_e)^\perp$ is defined as

$$(2.4.7) \qquad \Phi := \mathbf{m}^{-1} \circ (d\alpha_\xi - \gamma \circ D\mathbb{I}_\xi(q_e))\,.$$

The Hamiltonian form of (2.4.6) is given by

$$(2.4.8) \qquad \mathcal{L}_H = \text{diag}\,[1, 1, \mathbf{m}] \circ \mathcal{L} \circ \text{diag}\,[1, 1, \mathbf{m}^{-1}]\,,$$

where the orthogonal subspace $(\mathcal{G} \cdot q_e)^\perp \subseteq T_{q_e}\mathcal{Q}$ is replaced by the annihilator $(\mathcal{G} \cdot q_e)^A \subseteq T_{q_e}^*\mathcal{Q}$ in (2.2.9). See Simo et al. [1991a] for the Hamiltonian treatment of the reduced energy momentum method and the identifications used there.

Additional block structure may arise in the linearized equations (2.4.6). There are two bilinear forms of $\mathcal{Q}$ which are clearly relevant to the block structure of the equations: the stability two form $\mathcal{B}_e$ and the linearized Legendre transformation $\langle\!\langle\,,\,\rangle\!\rangle$. Equation (2.3.8) implies that $D\mathbb{I}_\xi(q_e)$ maps the metric orthogonal subspace $(\mathcal{G} \cdot q_e)^\perp$ into $\mathcal{G}_{q_e}^A$. Hence, if we assume that $\mathcal{G}_{q_e}^A = \text{range}\,[\mathbb{I}_e]$, then $(\mathcal{G} \cdot q_e)^\perp \subseteq \mathcal{Q}$. The spaces $\mathcal{V}_{\text{int}}$ and $(\mathcal{G} \cdot q_e)^\perp$ are the orthogonal complements in $\mathcal{Q}$ to $\widetilde{\mathcal{G}} \cdot q_e$ with respect to these forms. We let $\mathcal{V}_S$ denote the intersection of $\mathcal{V}_{\text{int}}$ and $(\mathcal{G} \cdot q_e)^\perp$ and $\mathcal{V}_\chi$ (respectively $\mathcal{V}_M$) denote the metric orthogonal complement to $\mathcal{V}_S$ in $\mathcal{V}_{\text{int}}$ (respectively $(\mathcal{G} \cdot q_e)^\perp$). We note that $\mathcal{V}_S \subseteq \ker\,[\mathbb{P}_{\text{rig}}]$,

$$(2.4.9) \qquad \begin{aligned} \mathcal{V}_M &= \mathbf{m}^{-1}\left[(\mathcal{G} \cdot q_e + \mathcal{V}_S)^A\right] = \Upsilon\,[\mathcal{V}_{\text{rig}}] \quad \text{and} \\ \mathcal{V}_\chi &= (1 - \mathbb{P}_{\text{rig}})\,[\mathcal{V}_M]\,. \end{aligned}$$

The spaces $\mathcal{V}_{\text{rig}}$, $\mathcal{V}_M$, and $\mathcal{V}_\chi$ all have dimension $\dim \widetilde{\mathcal{G}} - \dim (\mathcal{G}_\mu + \mathcal{G}_{q_e})$.
  If we introduce the further decomposition

(2.4.10) $\mathcal{V}_{\text{int}} \approx \mathcal{V}_\chi + \mathcal{V}_S$ and $(\mathcal{G} \cdot q_e)^\perp \approx \mathcal{V}_M + \mathcal{V}_S$,

then three of the blocks in (2.4.6) block diagonalize. Specifically, we have

(2.4.11) $\quad \begin{pmatrix} \Upsilon \\ 0 \end{pmatrix} \qquad\qquad (\; \mathbb{P}_{\text{rig}} \quad 0 \;) \qquad\qquad \begin{pmatrix} 1 - \mathbb{P}_{\text{rig}} & 0 \\ 0 & 1 \end{pmatrix}.$
$\qquad\qquad \mathcal{V}_{\text{rig}}-(\mathcal{G} \cdot q_e)^\perp \qquad\quad (\mathcal{G} \cdot q_e)^\perp-\mathcal{V}_{\text{rig}} \qquad\quad \mathcal{V}_{\text{rig}}-\mathcal{V}_{\text{int}}$

Thus $\mathcal{V}_{\text{rig}}$ and the two copies of $\mathcal{V}_S$ are not directly coupled by $\mathcal{L}$. If

(2.4.12) $\mathcal{B}_e [\mathcal{V}_M] + d\alpha_\xi [\mathcal{V}_M] \subseteq \mathcal{V}_S^A$,

then $\mathcal{L} = \text{diag} [L_M, L_S]$, where

(2.4.13)
$$L_M \;=\; \begin{pmatrix} \Gamma & 0 & \mathbb{P}_{\text{rig}} \\ 0 & 0 & 1 - \mathbb{P}_{\text{rig}} \\ \Upsilon & -\mathbf{m}^{-1} \circ \mathcal{B}_\chi & \Phi_M \end{pmatrix} \qquad \text{and}$$

$$L_S \;=\; \begin{pmatrix} 0 & 1 \\ -\mathbf{m}^{-1} \circ \mathcal{B}_S & \Phi_S \end{pmatrix}.$$

Here $\mathcal{B}_\chi$, etc. denotes the restriction of the appropriate map to the specified subspace. The condition $\mathcal{B}_e [\mathcal{V}_M] \subseteq \mathcal{V}_S^A$ is equivalent to $\mathcal{B}_e [\mathcal{V}_\chi] \subseteq \mathcal{V}_S^A$, since for any $\delta q \in \mathcal{V}_{\text{int}}$ and $\Delta q \in \mathcal{V}_M$,

(2.4.14) $\mathcal{B}_e(\mathbb{P}_{\text{rig}} \Delta q, \delta q) = \beta(\delta q) \cdot \kappa_{\Delta q} = 0$

for $\kappa_{\Delta q}$ given by (2.2.11). Similarly, $\Phi [\mathcal{V}_M] \subseteq \mathcal{V}_S^A$ if and only if $d\alpha_\xi [\mathcal{V}_M] \subseteq \mathcal{V}_S^A$, since γ maps range $[\mathbb{I}_e]$ into the annihilator of $(\mathcal{G} \cdot q_e)^\perp$.
  Condition (2.4.12) can be interpreted as stipulating that the orthogonal complements to $\mathcal{V}_S$ in $(\mathcal{G} \cdot q_e)^\perp$ with respect to $\langle\!\langle \, , \, \rangle\!\rangle$, $\mathcal{B}_e$, and $d\alpha_\xi$ all coincide. This condition certainly need not be satisfied; however, there are at least some non-trivial systems for which it is satisfied. (We shall examine one such system in the next section.) We hope to eventually develop some insight into the geometric and mechanical implications of this additional structure.

3. Pseudo-rigid bodies

3.1. Introduction.
We shall illustrate the constructions defined in the previous sections by means of a simple, but non-trivial example. We take as our example the equations for the evolution of a pseudo-rigid body. Pseudo-rigid bodies model homogeneous hyperelasticity, i.e. they model a deformable body undergoing only *affine* deformations. We assume in addition that the body is fixed at its center of mass. While the assumption of homogeneity is obviously an extreme simplification of nonlinear elasticity,

this model still captures many interesting features of the full infinite dimensional system; it also provides an example of the additional block structure discussed in §2.4. Pseudo-rigid bodies have been studied by Slawianowski [1974] and Cohen and Muncaster [1988], among others. The special case of an incompressible pseudo-rigid body with the stored energy of a perfect fluid is the classical Riemann [1861] problem. The results summarized in §3.a–c are derived in Lewis and Simo [1990] and Simo et al. [1991].

The configuration manifold for a pseudo-rigid body is $Q = C$, the Lie group of orientation-preserving invertible linear transformations of $\mathbb{R}^3$, i.e. the manifold of three by three matrices with positive determinant. A tangent vector to $F \in Q$ is of the form (F, V) for some $V \in L(3)$, the space of general 3×3 matrices. Elements of the cotangent bundle T^*Q are likewise pairs (F, P) for $P \in L(3)$. The dual pairing between $(F, P) \in T_F^*Q$ and $(F, V) \in T_F Q$ is given by

$$\langle P, V \rangle := \mathrm{tr}\left(P^T V\right).$$

The elastic properties of the material are captured by the stored energy function W. We shall express the stored energy as a function of the right Cauchy-Green tensor $C = F^T F$. In the absence of external forces, a pseudo-rigid body evolves according to the dynamics

$$(3.1.1) \qquad \ddot{F} = -2\, F\, \partial_C W(F^T F)\, E^{-1},$$

where E is the inertia tensor

$$(3.1.2) \qquad E := \int_{\mathcal{B}} \rho_{\mathrm{ref}}(X)\, X \otimes X\, dX$$

of the reference configuration $\mathcal{B}$. The equations of motion (3.1.1) are Lagrange's equations associated to the Lagrangian

$$
\begin{aligned}
(3.1.3) \qquad L(F, V) \quad &:= \quad \tfrac{1}{2} \langle VE, V \rangle - W(F^T F) \\
&= \quad \int_{\mathcal{B}} \rho_{\mathrm{ref}}(X) \left(\tfrac{1}{2} |VX|^2 - \widetilde{W}(C) \right) dX,
\end{aligned}
$$

where $\widetilde{W}$ is the pointwise stored energy function for nonhomogeneous elasticity.

The rotation group $SO(3)$ acts on Q by multiplication on the left and right. We shall be particularly concerned with the left action of $SO(3)$. We denote by ξ_Q an infinitesimal spatial rotation about ξ, i.e.

$$(3.1.4) \qquad \xi_Q(F) := \tfrac{d}{d\epsilon}\big|_{\epsilon=0} \exp(\epsilon\,\xi)F = \hat{\xi}\, F,$$

where exp denotes the exponential map from the Lie algebra $so(3) \approx \mathbb{R}^3$ to $SO(3)$ and $\hat{\xi}$ denotes the skew-symmetric matrix satisfying $\hat{\xi}x = \xi \times x$ for all $x \in \mathbb{R}^3$. We shall abuse notation by using the symbol $\xi_Q(F_e)$ to denote both the configuration-velocity pair $(F, \hat{\xi}F) \in TQ$ and the velocity matrix $\hat{\xi}F \in L(3)$; it will be clear in context which interpretation is meant. The group orbit $G \cdot F = \{RF : R \in SO(3)\}$ through a configuration F has tangent space $\mathcal{G} \cdot F = \{\xi_Q(F) : \xi \in \mathbb{R}^3\}$. The momentum map $J_L : TQ \to \mathbb{R}^3$ associated to the left action of $SO(3)$ on TQ is angular momentum, which is given by

$$(3.1.5) \quad J_L(F, V) \cdot \xi = \tfrac{d}{d\epsilon}\big|_{\epsilon=0} L(F, V + \epsilon\,\xi_Q(F)) = \left\langle VEF^T, \hat{\xi} \right\rangle.$$

Remarks. If the reference configuration is symmetric, i.e. if E commutes with some subgroup of $SO(3)$, then the Lagrangian is invariant under the right (body) action of that subgroup of $SO(3)$, as well as the left (spatial) action of $SO(3)$. In this case more complicated steady motions than rigid rotations may be analysed by the reduced energy momentum method. We defer the analysis of such motions to a future paper.

As discussed in §2, relative equilibria are critical points of the locked Lagrangian $L_\xi = L \circ \xi_Q$. Equation (3.1.3) for the Lagrangian implies that

$$(3.1.6) \qquad\qquad L_\xi(F) = \tfrac{1}{2}\mathbb{I}_\xi(F) \cdot \xi - W(F^T F),$$

where the locked momentum map $\mathbb{I}_\xi$ is given by

$$(3.1.7) \qquad\qquad \mathbb{I}_\xi(F) = \left(\operatorname{tr}\left[FEF^T\right]\mathbf{1} - FEF^T\right)\xi,$$

with linearization

$$(3.1.8) \qquad \mathbb{I}_e = \operatorname{tr}[e]\,\mathbf{1} - e, \qquad \text{where} \qquad e := F_e E F_e^T.$$

The 'rigid' equilibrium condition $\operatorname{ad}_\xi^*\mathbb{I}_\xi(F_e) = 0$ implies that the angular velocity ξ at a relative equilibrium F_e is an eigenvector of e. Thus, without loss of generality, we can assume that $e = \operatorname{diag}[e_1, e_2, e_3]$ for some positive constants e_i, and that ξ is parallel to $k = (0, 0, 1)$. Given these assumptions, the relative equilibrium conditions are

$$(3.1.9) \qquad 2\,F_e\,\partial_C W(F_e^T F_e)\,F_e^T = |\xi|^2\,\operatorname{diag}[e_1, e_2, 0].$$

3.2. Formal stability of relative equilibria. As a first step in analyzing the stability of a relative equilibrium $\xi_Q(F_e)$, we construct the subspaces specified by the block diagonalization algorithm. The group $SO(3)$ acts freely on Q; thus the space $\mathcal{Q}$ of admissible variations is the full tangent space, i.e. $\mathcal{Q} = T_{F_e}Q$. Given $M \in L(3)$, define the map $M_\tau : GL(3) \to L(3)$ and the matrix $M_{\tau_e} \in L(3)$ by

$$(3.2.1) \qquad M_\tau(F) := MF^{-T}E^{-1} \qquad \text{and} \qquad M_{\tau_e} := M_\tau(F_e).$$

The metric orthogonal to $\mathcal{G} \cdot q_e$ satisfies $(\mathcal{G} \cdot F_e)^{\perp} = \text{range}[\mathbf{v}_{\perp}]$, where $\mathbf{v}_{\perp} : \mathbb{R}^3 \times \mathbb{R}^3 \to (\mathcal{G} \cdot F_e)^{\perp}$ is defined by

(3.2.2)

$$\mathbf{v}_{\perp}(\mathbf{x}, \mathbf{y}) := S(\mathbf{x}, \mathbf{y})_{\tau_e} \quad \text{for}$$

$$S(\mathbf{x}, \mathbf{y}) := \begin{pmatrix} y_1 + y_2 & x_3 & x_2 \\ x_3 & y_2 - y_1 & x_1 \\ x_2 & x_1 & y_3 \end{pmatrix}.$$

The locked momentum map associated to ξ has first variation

(3.2.3) $D\mathbb{II}_{\xi}(F_e) \cdot (\delta f \, F_e) = 2 \left(\text{tr} \left[\delta f \, e \right] \mathbf{1} - \text{sym} \left[\delta f \, e \right] \right) \xi.$

The linearized locked momentum tensor $\mathbb{II}_e$ satisfies

(3.2.4) $\mathbb{II}_e = \text{diag} \left[I_1, I_2, I_3 \right] := \text{diag} \left[e_2 + e_3, e_1 + e_3, e_1 + e_2 \right].$

The linearized locked momentum tensor (3.2.4) is clearly invertible. Hence the pairing $[\mathbb{II}_e]^{-1} : \text{range}[\mathbb{II}_e] \times \text{range}[\mathbb{II}_e] \to \mathbb{R}$ is simply

(3.2.5) $[\mathbb{II}_e]^{-1}(\nu, \tilde{\nu}) = \nu \cdot \text{diag} \left[I_1, I_2, I_3 \right]^{-1} \tilde{\nu}.$

In order to determine the space of 'internal' variations, we note that $\mathcal{G}_{\mu} = \text{span} \{k\}$. Substituting (3.2.3) into (2.2.5), we find that the space $\mathcal{Q}_{\text{int}}$ of internal variations is

(3.2.6) $\mathcal{Q}_{\text{int}} = \{ \delta f \, F_e : \text{sym}[\delta f \, e] \xi = \lambda \, \xi \quad \text{for some } \lambda \in \mathbb{R} \}.$

We select the metric orthogonal subspace as the complement $\mathcal{V}$ to $\mathcal{G}_{\mu} \dot{} F_e$ in $\mathcal{Q} = T_{F_e} \mathcal{Q}$. This yields the 'rigid-internal' decomposition $\mathcal{V} = \mathcal{V}_{\text{rig}} \oplus \mathcal{V}_{\text{int}}$ for

(3.2.7) $\mathcal{V}_{\text{rig}} = \{ \alpha_Q(F_e) = \hat{\alpha} \, F_e : \alpha \cdot \xi = 0 \}$

and

$$\mathcal{V}_{\text{int}} = \left\{ \mathbf{v}_{\text{int}}(\sigma, \delta) := \widehat{(\sigma_1, \sigma_2, 0)}_{\tau_e} + \mathbf{v}_{\perp}((0, 0, \sigma_3), \delta) : \sigma, \, \delta \in \mathbb{R}^3 \right\}.$$
(3.2.8)

Note that $\mathbf{v}_{\text{int}}((0, 0, \sigma_3), \delta) = \mathbf{v}_{\perp}((0, 0, \sigma_3), \delta)$; thus

(3.2.9) $\mathcal{V}_S = \mathcal{V}_{\text{int}} \cap (\mathcal{G} \cdot F_e)^{\perp} = \left\{ \mathbf{v}_{\perp}((0, 0, \sigma_3), \delta) : \sigma_3 \in \mathbb{R}, \, \delta \in \mathbb{R}^3 \right\},$

with complement

(3.2.10) $\mathcal{V}_M = \left\{ \mathbf{v}_{\perp}((\sigma_1, \sigma_2, 0), 0) : \sigma_1, \, \sigma_2 \in \mathbb{R}^3 \right\}$

in $(\mathcal{G} \cdot F_e)^{\perp}$.

We next turn to the computation of the Arnold form and the stability form $\mathcal{B}_e$ at a critical point F_e of the locked Lagrangian L_{ξ}. Equations

(2.2.3), (3.2.4), and (3.2.5) imply that the Arnold form $\mathcal{A} : \mathbb{R}^3 \times \mathbb{R}^3 \to \mathbb{R}$ satisfies

$$
\begin{aligned}
\mathcal{A}(\eta, \zeta) &= [\mathbb{I}_e]^{-1} (\mu \times \eta, \mu \times \zeta) - (\mu \times \eta) \cdot (\xi \times \zeta) \\
&= [\mathbb{I}_e]^{-1} (I_3 \xi \times \eta, (I_3 \mathbf{1} - \mathbb{I}_e)(\xi \times \zeta)) \\
&= |\xi|^2 I_3 \, \eta \cdot \text{diag} \left[\frac{I_3 - I_2}{I_2}, \frac{I_3 - I_1}{I_1}, 0 \right] \zeta.
\end{aligned}
$$

(3.2.11)

The construction of the stability two form $\mathcal{B}_e$ involves the computation of the second variation $D^2 L_\xi(F_e)$ of the locked Lagrangian. We simply state that result here and refer the reader to Simo et al. [1991a] for the derivation of the second variation.

$$
(3.2.12) \quad D^2 L_\xi(F_e) \cdot (\delta f \, F_e, \Delta f \, F_e) = \begin{aligned} &- \langle \text{sym} \, [\delta f], \mathbf{C} \, [\text{sym} \, [\Delta f]] \rangle \\ &- G(\Delta f, \delta f), \end{aligned}
$$

where $\mathbf{C}$ denotes the spatial elasticity tensor

$$
(3.2.13) \qquad \mathbf{C}^{ijkl} = 4 \, F^i{}_I F^j{}_J F^k{}_K F^l{}_L \frac{\partial^2 W(C)}{\partial C_{IJ} \partial C_{KL}}
$$

and the 'geometric' term $G(\Delta f, \delta f)$ satisfies

$$
(3.2.14) \qquad G(\Delta f, \delta f) = I_3(\delta f \, \xi) \cdot (\Delta f \, \xi) - (\delta f^T \xi) \cdot \mathbb{I}_e(\Delta f^T \xi).
$$

The 'fine' decomposition (2.4.10) appears to be relevant to the stability form $\mathcal{B}_e$. Specializing (3.2.14) to internal variations yields

$$
(3.2.15) \quad G(\mathbf{v}_{\text{int}}(\sigma, \delta), \mathbf{v}_{\text{int}}(\tilde{\sigma}, \tilde{\delta})) = (\xi \times \sigma) \cdot \left(e^{-1} - e_3^{-1} \mathbf{1} \right) (\xi \times \tilde{\sigma}).
$$

Thus the geometric term depends only on variations lying in $\mathcal{V}_\chi$. Equations (3.2.3) and (3.2.5) imply that the momentum correction term satisfies

$$
\begin{aligned}
&[\mathbb{I}_e]^{-1} \left(D\mathbb{I}_\xi(F_e) \cdot \delta f \, F_e, D\mathbb{I}_\xi(F_e) \cdot \Delta f \, F_e \right) \\
&\qquad = \frac{4 \, |\xi|^2}{I_3} \left(\langle \delta f, e \rangle - \mathbf{k} \cdot \delta f \, e \, \mathbf{k} \right) \left(\langle \Delta f, e \rangle - \mathbf{k} \cdot \Delta f \, e \, \mathbf{k} \right).
\end{aligned}
$$

(3.2.16)

Specializing (3.2.16) to internal variations yields

$$
(3.2.17) \quad [\mathbb{I}_e]^{-1} \left(D\mathbb{I}_\xi(F_e) \cdot \mathbf{v}_{\text{int}}(\sigma, \delta), D\mathbb{I}_\xi(F_e) \cdot \mathbf{v}_{\text{int}}(\tilde{\sigma}, \tilde{\delta}) \right) = \frac{4 \, |\xi|^2}{I_3} \delta_2 \, \tilde{\delta}_2.
$$

Thus the momentum correction term depends only on the in-plane, 'symmetric' variation in $\mathcal{V}_S$.

3.3. Isotropic materials. We now consider a special class of materials with additional symmetry properties beyond frame invariance. A homogeneous elastic material is *isotropic* if the stored energy function is invariant under both the left and the *right SO*(3) action. In this case, there exists a function $\Phi : \mathbb{R}^3 \to \mathbb{R}$ such that $W(C) = \Phi(\mathrm{I}, \mathrm{II}, \mathrm{III})$, where $\mathrm{I} = \mathrm{tr}\,[C]$, $\mathrm{II} = \frac{1}{2}\left(\mathrm{tr}\,[C]^2 - \mathrm{tr}\,[C^2]\right)$, and $\mathrm{III} = \det[C]$ are the principal invariants of C. Note that the Lagrangian (3.1.3) is right $SO(3)$ invariant only if the reference inertia tensor E is a multiple of the identity, e.g. if the reference configuration is equivalent to a sphere. However, we shall see that the equations associated to isotropic materials possess additional structure even when the Lagrangian is not right invariant.

The assumption of isotropy implies that at equilibrium the left Cauchy-Green tensor and spatial inertia matrix share a common eigenbasis. The equilibrium conditions for a diagonal equilibrium deformation of an isotropic material may be expressed as follows: Define $\kappa_1 := E_1(\Lambda_2 - \Lambda_3)$, $\kappa_2 := E_2(\Lambda_1 - \Lambda_3)$, and

$$(3.3.1) \qquad \kappa_\xi := \frac{\xi^2}{2\,(\Lambda_1 - \Lambda_2)\,(\Lambda_1 - \Lambda_3)\,(\Lambda_2 - \Lambda_3)},$$

A diagonal matrix F_e is a relative equilibrium if

$$(3.3.2) \qquad \left.\begin{array}{rcl} \Phi_{\mathrm{I}} & = & \kappa_\xi\left(\Lambda_1^2 \kappa_1 - \Lambda_2^2 \kappa_2\right) \\ \Phi_{\mathrm{II}} & = & -\kappa_\xi\left(\Lambda_1 \kappa_1 - \Lambda_2 \kappa_2\right) \\ \Phi_{\mathrm{III}} & = & \kappa_\xi\left(\kappa_1 - \kappa_2\right). \end{array}\right\}$$

Let Ψ express the stored energy as a function of the squares of the principal stretches, i.e. define $\Psi : \mathbb{R}^3 \to \mathbb{R}$ by $\Psi(\Lambda) := W\left(\mathrm{diag}\,[\Lambda]\right)$, Then the second variation takes the form

$$(3.3.3) \qquad \mathcal{B}_e(\delta F, \Delta F) = |\xi|^2\,\sigma \cdot \chi\widetilde{\sigma} + 4\delta \cdot \Delta\widetilde{\delta},$$

where

$$(3.3.4)\ \chi = \mathrm{diag}\left[\frac{\Lambda_2 \Lambda_3 (I_3 - I_2)(E_2 - E_3)}{I_2(\Lambda_2 - \Lambda_3)},\ \frac{\Lambda_1 \Lambda_3 (I_3 - I_1)(E_1 - E_3)}{I_1(\Lambda_1 - \Lambda_3)},\ \frac{I_3^2(E_1 - E_2)}{E_1 E_2(\Lambda_1 - \Lambda_2)}\right]$$

and Δ is a symmetric 3×3 matrix with entries

$$(3.3.5) \qquad \Delta_{ij} = \frac{\partial^2 \Psi(\Lambda_e)}{\partial \Lambda_i \partial \Lambda_j}\frac{1}{E_i E_j} + |\xi|^2\,R_{ij},$$

where

$$R_{11} = \frac{1}{I_3} \qquad R_{12} = \frac{1}{I_3} - \frac{(E_1 - E_2)}{2E_1 E_2(\Lambda_1 - \Lambda_2)} \qquad R_{22} = \frac{1}{I_3}$$

$$R_{13} = \frac{1}{2E_3(\Lambda_3 - \Lambda_1)} \qquad R_{23} = \frac{1}{2E_3(\Lambda_3 - \Lambda_2)} \qquad R_{33} = 0.$$

$(3.3.6)$

If $\Lambda_1 = \Lambda_2$ and $E_1 = E_2$, then

$$(3.3.7) \qquad \chi_3 = 4 \left(\frac{|\xi|^2 \Lambda_3 e_1}{\Lambda_1 - \Lambda_3} - 2\Lambda_1 \, \Phi_{\mathrm{II}}(\Lambda_1 - \Lambda_3) \right).$$

A rigidly rotating diagonal relative equilibrium of an isotropic material is nonlinearly stable modulo G_{μ_e} if

    (i) The body is in rotation about the axis of maximal inertia of the equilibrium configuration, i.e. $I_3 > I_1$ and $I_3 > I_2$

    (ii) the ordering of the principal stretches and the principal axes of the reference configuration agree, i.e.

$$(3.3.8) \qquad \frac{E_i - E_j}{\Lambda_i - \Lambda_j} > 0 \qquad\qquad \text{for } i < j$$

    (iii) Δ is positive definite.

For a rotating relative equilibrium, the first two sets of stability conditions imply that the Baker-Ericksen inequalities, which require agreement of the ordering of the principal stretches and the force per current area, are satisfied.

For two relatively simple isotropic materials, the Ciarlet-Geymonat material and the St. Venant-Kirchhoff material, any relative equilibrium rotating about its axis of maximal inertia is nonlinearly stable. For a Mooney-Rivlin material, both stable and unstable relative equilibria may exist. See Lewis and Simo [1990] for the details of the results regarding these materials.

3.4. The linearized equations. We now apply the results of §2.3 to the pseudo-rigid body equations (3.1.1) for an isotropic material. The space S is represented by

$$(3.4.1) \qquad \begin{aligned} S \approx \Big\{ (\delta F, \delta V) &= ((\alpha_1, \alpha_2, 0)_Q(F_e) + \mathbf{v}_{\mathrm{int}}(\sigma, \delta), \mathbf{v}_\perp(\varphi, \psi)) \\ &: \alpha_1, \ \alpha_2 \in \mathbb{R}, \ \sigma, \ \delta, \ \phi, \ \psi \in \mathbb{R}^3 \Big\}. \end{aligned}$$

The variation $(\delta F, \delta V)$ is identified with the vector

$$(3.4.2) \qquad (\underbrace{(\alpha_1, \alpha_2)}_{v_{\mathrm{rig}}}, \underbrace{(\sigma_1, \sigma_2)}_{v_X}, \underbrace{(\varphi_1, \varphi_2)}_{v_M}, \underbrace{\sigma_3, (\delta_1, \delta_2, \delta_3)}_{v_S}, \underbrace{\varphi_3, (\psi_1, \psi_2, \psi_3)}_{v_S}).$$

With respect to the decomposition (3.4.2) the linearized equations on S take the form $\mathcal{L} = \mathrm{diag}\,[L_1, L_2]$, where

$$L_1 = \begin{pmatrix} \mathrm{diag} \left[\dfrac{K_2}{I_1}, \dfrac{K_1}{I_2} \right] \rho & \mathbf{O} & 2\,\mathrm{diag}\,[K_1, K_2]^{-1} \\[2ex] \mathbf{O} & \mathbf{O} & \mathrm{diag} \left[\dfrac{-K_1}{I_1}, \dfrac{K_2}{I_2} \right] \\[2ex] \dfrac{2\,|\xi|^2 I_3 e_3}{I_1 I_2}\,\mathrm{diag}\,[e_2, -e_1] & -\mathrm{diag}\,[\tilde{\chi}_1, \tilde{\chi}_2] & 2\,\mathrm{diag} \left[\dfrac{e_2}{I_1}, \dfrac{e_1}{I_2} \right] \rho \end{pmatrix}$$

$$(3.4.3)$$

and

$$(3.4.4) \quad L_2 = \begin{pmatrix} 0 & 0 & 1 & 0 \\ 0 & \mathbf{O} & 0 & 1 \\ -\tilde{\chi}_3 & 0 & 0 & -2\,|\xi|\,(1, I_3^{-1}K_3, 0) \\ 0 & -e\Delta & 2\,|\xi|\,(1,0,0)^T & \mathbf{O} \end{pmatrix},$$

where

$$(3.4.5) \quad K := (I_3 - I_1, I_3 - I_2, I_1 - I_2) = (e_2 - e_3, e_1 - e_3, e_2 - e_1),$$

$$(3.4.6) \quad \rho := \begin{pmatrix} 0 & -|\xi| \\ |\xi| & 0 \end{pmatrix}, \quad \text{and} \quad \tilde{\chi} := |\xi|^2 \,\mathbf{w}^{-1}\chi$$

for χ given by (3.3.4), Δ given by (3.3.5), and

$$(3.4.7) \quad \mathbf{w} = \text{diag}\left[-\frac{K_1}{e_2 e_3}, \frac{K_2}{e_1 e_3}, \frac{I_3}{e_1 e_2}\right].$$

We now make a few general observations regarding the linearized equations. The L_1 block depends primarily on the current configuration; the deformation from the reference to current configuration enters only through the terms χ_1 and χ_2. There is no explicit dependence on the stored energy function. The characteristic polynomial of L_1 has the form

$$
\begin{aligned}
(3.4.8) \quad \text{char}\,[L_1] \;=\; & \lambda^6 + \lambda^4\,|\xi|^2\,(c_{40} + c_{41}\,\chi_1 + c_{42}\,\chi_2) \\
& + \lambda^2\,|\xi|^4\,(c_{20} + c_{21}\,\chi_1 + c_{22}\,\chi_2 + c_{23}\,\chi_1\,\chi_2) \\
& + |\xi|^6\,c_{03}\,\chi_1\,\chi_2,
\end{aligned}
$$

where the coefficients c_{ij} are rational functions of the e_i. By rescaling λ, we can eliminate the dependence of char $[L_1]$ on ξ. Thus the stability conditions arising from (3.4.8) impose conditions on the 'in-plane' deformation information captured in χ_1 and χ_2 which are independent of the material properties of the pseudo-rigid body and the angular velocity. We hope to explore the significance of these conditions in future work. The characteristic polynomial of the L_2 block explicitly involves information regarding the current configuration, reference configuration, angular velocity, and stored energy function.

3.5. Derivation of the linearized equations. The $\mathcal{V}_{\text{rig}}$–$\mathcal{V}_{\text{rig}}$ term:
The map Γ satisfies $\Gamma\,\eta_Q(F_e) = \zeta_Q(F_e)$, where ζ is determined by the relationships

$$(3.5.1) \quad \zeta \times \mu = \iota_\eta \mathcal{A} \quad \text{and} \quad \zeta \cdot \xi = 0.$$

Equations (3.2.11) and (3.5.1) imply that

$$(3.5.2) \quad \zeta = \text{diag}\left[\frac{I_3 - I_1}{I_1}, \frac{I_3 - I_2}{I_2}, 0\right](\xi \times \eta).$$

The $\mathcal{V}_{\text{rig}}$–$\mathcal{V}_M$ and $\mathcal{V}_\chi$–$\mathcal{V}_M$ terms: We first show that the map κ defined by (2.2.11) satisfies

$$(3.5.3) \quad \kappa_{\mathbf{v}_\perp(\varphi,\psi)} = \check{\kappa}(\varphi_1,\varphi_2) := 2\big((e_2 - e_3)^{-1}\varphi_1, (e_3 - e_1)^{-1}\varphi_2, 0\big),$$

where $(\varphi_1, \varphi_2, \varphi_3 = \varphi$. Equation (3.2.11) implies that

$$(3.5.4)\, \mathcal{A}(\eta, \kappa_{\mathbf{v}_\perp}(\varphi,\psi)) = [\mathbb{I}_e]^{-1}\left(\mu \times \eta, ((I_3\mathbf{1} - \mathbb{I}_e)(\xi \times \kappa_{\mathbf{v}_\perp}(\varphi,\psi)))\right).$$

The righthand side of equation (2.2.11) is given by

$$(3.5.5)\quad\begin{aligned} &- [\mathbb{I}_e]^{-1}\left(\mu \times \eta, D\mathbb{I}_\xi(F_e) \cdot \mathbf{v}_\perp(\varphi,\psi)\right) \\ &= -2\,[\mathbb{I}_e]^{-1}\left(\mu \times \eta, (2\,\psi_2 + \psi_3)\xi - S(\varphi,\psi)\xi\right) \\ &= -2\xi\,[\mathbb{I}_e]^{-1}\left(\mu \times \eta, (\varphi_2, \varphi_1, 2\,\psi_2)\right). \end{aligned}$$

Equating (3.5.4) and (3.5.6) and imposing the constraint $\kappa_{\mathbf{v}_\perp}(\varphi,\psi) \cdot \xi = 0$ given by (3.2.7) yields (3.5.3).

Given the formula (3.5.3) for the map $\check{\kappa}$, we see that the projections $\mathbb{P}_{\text{rig}}$ and $(1 - \mathbb{P}_{\text{rig}})$ satisfy

$$(3.5.6) \quad \mathbb{P}_{\text{rig}}\,\mathbf{v}_\perp((\varphi_1,\varphi_2,0),0) = (\check{\kappa}(\varphi_1,\varphi_2))_Q(F_e)$$

and

$$(3.5.7)\quad\begin{aligned} &(1 - \mathbb{P}_{\text{rig}})(\mathbf{v}_\perp((\varphi_1,\varphi_2,0),0)) \\ &= \mathbf{v}_\perp((\varphi_1,\varphi_2,0),0) - (\check{\kappa}(\varphi_1,\varphi_2))_Q(F_e) \\ &= \mathbf{v}_{\text{int}}\left((-a_2^{-1}\varphi_1, a_1^{-1}\varphi_2, 0),0\right). \end{aligned}$$

The $\mathcal{V}_{\text{rig}}$–$(\mathcal{G}\cdot F_e)^\perp$ term: Since $T_{F_e}Q = \mathcal{V}_{\text{int}} + \mathcal{G}\cdot F_e$, $\mathbf{v}_\perp(\dot\varphi,\dot\psi) \in (\mathcal{G}\cdot F_e)^\perp$ is determined by its pairing with an arbitrary element of $\mathcal{V}_{\text{int}}$. For $\dot\varphi$, $\dot\psi$, σ, and $\delta \in \mathbb{R}^3$,

$$(3.5.8) \quad \left\langle\!\left\langle \mathbf{v}_\perp(\dot\varphi,\dot\psi), \mathbf{v}_{\text{int}}(\sigma,\delta) \right\rangle\!\right\rangle = \dot\varphi \cdot \mathbf{w}\sigma + \dot\psi \cdot \mathbf{e}^{-1}\delta,$$

where $\mathbf{w}$ is given by (3.4.7). Given $\nu \in \mathcal{G}^* \approx \mathbb{R}^3$, the pairing of $\gamma(\nu)$ with an internal variation $\mathbf{v}_{\text{int}}(\sigma,\delta)$ is given by

$$(3.5.9)\quad\begin{aligned} \gamma(\nu) \cdot \mathbf{v}_{\text{int}}(\sigma,\delta) &= \left\langle\!\left\langle (\mathbb{I}_e^{-1}\nu)_Q(F_e), \mathbf{v}_{\text{int}}(\sigma,\delta)\right\rangle\!\right\rangle \\ &= 2\nu \cdot \text{diag}\left[I_1^{-1}, I_2^{-1}, 0\right]\sigma. \end{aligned}$$

The identities (3.2.11), (3.5.8), and (3.5.9) imply that

$$(3.5.10)\quad\begin{aligned} \gamma(\iota_\eta\mathcal{A}) \cdot \mathbf{v}_{\text{int}}(\sigma,\delta) &= |\xi|^2 I_3\,\sigma \cdot \text{diag}\left[\frac{I_3 - I_2}{I_2}, \frac{I_3 - I_1}{I_1}, 0\right]\eta \\ &= \frac{2\,|\xi|^2 I_3\,e_3}{I_1\,I_2}\,\langle \mathbf{v}_\perp(-e_2\eta_1, e_1\eta_2, 0),0), \mathbf{v}_{\text{int}}(\sigma,\delta)\rangle \end{aligned}$$

for all σ, $\delta \in \mathbb{R}^3$. Since the internal variations lie in the kernel of β, this determines the $\mathcal{V}_{\text{rig}}$-$(\mathcal{G} \cdot F_e)^\perp$ term.

The $(\mathcal{G} \cdot q_e)^\perp$–$(\mathcal{G} \cdot q_e)^\perp$ term: We first determine the contribution from $d\alpha_\xi$. The differential of the one-form α_ξ is computed using the general formula

$$(3.5.11) \quad d\alpha_\xi(v, w) = D(\alpha_\xi \cdot w) \cdot v - D(\alpha_\xi \cdot v) \cdot w - \langle \alpha_\xi, \mathcal{L}_v w \rangle$$

for vector fields v and w. We note that for a fixed matrix M the pairing $\langle \alpha_\xi(F), M_\tau(F) \rangle = \left\langle M, \hat{\xi} \right\rangle$ is constant. Thus

$$(3.5.12) \qquad\qquad D(\alpha_\xi \cdot M_\tau)(F) = 0.$$

The Lie bracket of two vector fields of the form M_τ satisfies

$$(3.5.13) \qquad\qquad \mathcal{L}_{M_\tau} N_\tau(F_e) = 2 \, \text{skew} \left[Me^{-1}N^T \right]_{\tau_e}.$$

Hence combining (3.5.11), (3.5.12) and (3.5.13) yields

$$d\alpha_\xi(M_{\tau_e}, N_{\tau_e}) = -2 \left\langle \alpha_\xi(F_e), \text{skew} \left[Me^{-1}N^T \right]_{\tau_e} \right\rangle = -2 \left\langle Me^{-1}N^T, \hat{\xi} \right\rangle.$$
$$(3.5.14)$$

In particular, for φ, ψ, σ, and $\delta \in \mathbb{R}^3$,

$$
\begin{aligned}
d\alpha_\xi(\mathbf{v}_\perp(\varphi, \psi), \mathbf{v}_{\text{int}}(\sigma, \delta)) &= -2 \left\langle S(\varphi, \psi) F_e^{-T} \, \mathbf{v}_{\text{int}}(\sigma, \delta)^T, \hat{\xi} \right\rangle \\
&= -2\xi \Big((\varphi_2 e_3^{-1}, \varphi_1 e_3^{-1}, \psi_2 e_2^{-1} - \psi_1 e_1^{-1}) \cdot \sigma \\
&\qquad + \varphi_3 \left(e_1^{-1}, -e_2^{-1}, 0 \right) \cdot \delta \Big).
\end{aligned}
$$
$$(3.5.15)$$

Equations (3.2.3) and (3.5.9) imply that

$$(3.5.16)
\begin{aligned}
&\langle \gamma \left(D\mathbb{II}_\xi(F_e) \cdot \mathbf{v}_\perp(\varphi, \psi) \right), \mathbf{v}_{\text{int}}(\sigma, \delta) \rangle \\
&\quad = 2 \, \langle \gamma \left(\text{tr} \left[S(\varphi, \psi) \right] \mathbf{1} - S(\varphi, \psi) \right) \xi \rangle, \mathbf{v}_{\text{int}}(\sigma, \delta) \rangle \\
&\quad = -4\xi \left(\tfrac{\varphi_2}{I_1}, \tfrac{\varphi_1}{I_2}, 0 \right) \cdot \sigma.
\end{aligned}
$$

Combining (3.5.15) and (3.5.16) yields

$$(3.5.17)
\begin{aligned}
&\langle -\gamma \left(D\mathbb{II}_\xi(F_e) \cdot \mathbf{v}_\perp(\varphi, \psi) \right) - \iota_{\mathbf{v}_\perp(\varphi, \psi)} d\alpha_\xi, \mathbf{v}_{\text{int}}(\sigma, \delta) \rangle \\
&\quad = 2\xi((a_2 e_3^{-1}\varphi_2, a_1 e_3^{-1}\varphi_1, e_1^{-1}\psi_1 - e_2^{-1}\psi_2) \cdot \sigma \\
&\qquad + \varphi_3 \left(e_1^{-1}, -e_2^{-1}, 0 \right) \cdot \delta).
\end{aligned}
$$

The final step in the computation of the $(\mathcal{G} \cdot F_e)^\perp$–$(\mathcal{G} \cdot F_e)^\perp$ term consists of solving for the actual element of $(\mathcal{G} \cdot F_e)^\perp$ determined by the pairing (3.5.17) using (3.5.8).

The $\mathcal{V}_{\text{int}}-(\mathcal{G}\cdot F_e)^{\perp}$ term: Using (2.4.13), (3.3.3), and (3.5.8), we find that the contribution of an internal variation $\mathbf{v}_{\text{int}}(\sigma, \delta)$ to the $(\mathcal{G}\cdot F_e)^{\perp}$ term is

$$(3.5.18) \qquad -\left(\mathbf{w}^{-1}\chi\,\sigma + \mathbf{e}\Delta\,\delta\right).$$

REFERENCES

[1] R. ABRAHAM, J.E. MARSDEN, *Foundations of mechanics* (1978), second edition, Benjamin/Cummings Publishing Company, Reading, MA.

[2] J.M. ARMS, V. MONCRIEF, J.M. MARSDEN, *Symmetry and bifurcation of momentum maps*, Comm. Math. Phys. 78 (1981), 455–478.

[3] H. COHEN, R.G. MUNCASTER, *The theory of pseudo-rigid bodies* (1988), Springer-Verlag, New York.

[4] D. LEWIS, *Bifurcation of liquid drops*, Nonlinearity 6 (1993), 491–522.

[5] D. LEWIS, *Lagrangian block diagonalization*, J. Dyn. and Diff. Eqs. 4 No.1 (1992), 1–41.

[6] D. LEWIS, T.S. RATIU, *Linearized stability of relative equilibria*, (tentative title) (1994), manuscript in preparation.

[7] D. LEWIS, T.S. RATIU, J.C. SIMO, J.E. MARSDEN, *The heavy top: a geometric treatment*, Nonlinearity 5 (1992), 1–48.

[8] D. LEWIS, J.C. SIMO, *Nonlinear stability of rotating pseudo-rigid bodies*, Proc. R. Soc. Lond. A 427 (1990), 281–319.

[9] A. BLOCH, P.S. KRISHNAPRASAD, J.E. MARSDEN, T.S. RATIU, *Dissipation induced instabilities*, Ann. Inst. Henri Poincaré, Analyse non linéare 11 (1) (1994), 37–90.

[10] K. MEYER, *Symmetries and integrals in mechanics*, Dynamical Systems (1973), 259–273, ed. M. Peixoto, Academic Press, New York.

[11] J. MONTALDI, M. ROBERTS, I. STEWART, *Stability of normal modes of symmetric Hamiltonian systems*, Nonlinearity 3 No.3 (1990), 731–772.

[12] B. RIEMANN, *Ein beitrag zu den untersuchungen über die bewegung eines flüssigen gleichartigen ellipsoides*, Abh. d. Königl. Gesell. der Wis. zu Göttingen 9 (1861), 3–36.

[13] J.C. SIMO, D. LEWIS, J.E. MARSDEN, *The stability of relative equilibria. Part I: the reduced energy-momentum method*, The Archive for Rational Mechanics and Analysis 115 (1991), 15–60.

[14] J.C. SIMO, T. POSBERGH, J.E. MARSDEN, *The stability of relative equilibria. Part II: application to nonlinear elasticity*, The Archive for Rational Mechanics and Analysis 115 (1991), 61–100.

[15] J. SLAWIANOWSKI, *Analytical mechanics of finite homogeneous strains*, Arch. Mech. 26 (1974), 569–587.

[16] S. SMALE, *Topology and Mechanics I*, Inventiones mathematica 10 (1970a), 305–331.

[17] S. SMALE, *Topology and Mechanics II*, Inventiones mathematica 11 (1970b), 45–64.

A 1 : −1 SEMISIMPLE HAMILTONIAN HOPF
BIFURCATION IN VORTEX DYNAMICS

CHJAN LIM*

The instability of the von Karman vortex streets [1] has remained an enigma in fluid mechanics for over 80 years. This is due in part to the fact that the von Karman model has successfully predicted several properties of the wake of cylinders [2], [3], and yet, it does not offer stable theoretical candidates for the vortex trails observed in experiments. In this note I will sketch a fresh approach to this old problem, which is based mainly on a systematic application of the Principle of Genericity, and Hamiltonian Bifurcation theory, [4], [5], [6], [7].

The von Karman model is a one-dimensional infinite lattice composed of 2 staggered rows of vortices (of opposite circulations) which are governed by the equations,

$$(1) \qquad \frac{d\overline{Z}_m^{\pm}}{dt} = \frac{i\Gamma}{2\pi} \left(\sum_{k \neq m} \frac{1}{Z_m^{\pm} - Z_k^{\pm}} \pm \sum_{\ell=-\infty}^{\infty} \frac{1}{Z_m^{\pm} - Z_\ell^{\mp}} \right)$$

Here $Z_m^{\pm}(t)$ denotes the location of the m-th vortex in the upper and lower rows respectively, Γ is the circulation of the upper row, and the bar denotes complex conjugation. The (relative) equilibrium von Karman vortex streets form a two parameter family given by:

$$(2) \qquad Z_m^{\pm}(t) = \left(m \pm \frac{1}{4} \right) \ell + U_s t \pm \frac{ih}{2}$$

$$U_s = \frac{\Gamma}{2\ell} \tanh \left(\frac{\pi h}{\ell} \right)$$

where h is the separation between the two parallel rows of vortices and ℓ is the separation between adjacent vortices (the parameter Γ can be absorbed into a rescaled time). From von Karman's analysis [1], it follows that the vortex street (2) is linearly stable only when the *aspect ratio* $k \equiv h/\ell$ has the special value $k^* = \frac{1}{\pi} \sinh^{-1}(1)$.

Following Kochin [8], note that spatially-periodic disturbances on the von Karman lattice (1) (with period N) correspond to finite-dimensional $2N$-body problems. Since the shortest wavelength disturbance on lattice (1), i.e., the case $N = 2$, is also the most unstable mode [3], I will focus

* This research was partially supported by the National Science Foundation under Grant DMS 9006091, Rensselaer Polytechnic Institute, Mathematical Sciences Department, Troy, NY 12180. e-mail:limc@rpi.edu

only on this case here. Let the periodic perturbations ρ_k from the relative equilibrium (2) be defined by:

$$z_k(t) = U_s t + z_{ko} + \frac{2\ell}{\pi}\rho(t)_k, \qquad k = 1, 2, 3, 4$$

(3)
$$z_{10} = \frac{\ell}{4} + \frac{ih}{2}, \qquad z_{20} = -\frac{\ell}{4} - \frac{ih}{2}$$
$$z_{30} = \frac{5\ell}{4} + \frac{ih}{2}, \qquad z_{40} = \frac{3\ell}{4} - \frac{ih}{2}$$

Substituting (3) into (1), and transforming to new variables given by

(4) $\alpha = 2(\rho_2 - \rho_1), \quad \beta = 2(\rho_2 - \rho_3)$

one obtains the equations of motion,

(5)
$$\frac{d\overline{\alpha}}{d\tau} = 4i\sin\beta \left(\frac{1}{\cos\alpha + \cos\beta} - \frac{1}{\cos\beta + \cos G} \right)$$
$$\frac{d\overline{\beta}}{d\tau} = 4i\sin\alpha \left(\frac{1}{\cos\alpha + \cos\beta} - \frac{1}{\cos\beta - \cos G} \right)$$

Note ьhat equations (5) in the two complex variables α, β have the Hamiltonian formulation

(6)
$$\frac{d\alpha_1}{d\tau} = \frac{\partial H}{\partial \beta_2}, \qquad \frac{d\beta_2}{d\tau} = -\frac{\partial H}{\partial \alpha_1}$$
$$\frac{d\beta_1}{d\tau} = \frac{\partial H}{\partial \alpha_2}, \qquad \frac{d\alpha_2}{d\tau} = -\frac{\partial H}{\partial \beta_1}$$

$$H = 4\log \left| \frac{(\cos\alpha - \cos G)(\cos\beta + \cos G)}{\cos\alpha + \cos\beta} \right|$$

Equations (6) have two degrees of freedom while the 4-body problem derived from (1) in the $N = 2$ case, has four degrees of freedom; this is because of the transformation to relative coordinates in (4), and the first integral $J \equiv \rho_1 - \rho_2 + \rho_3 - \rho_4$ which is an analogue of linear momentum. The above Hamiltonians are parameterized by $G = \pi \left(0.5 + i\frac{h}{\ell} \right)$ which is essentially the aspect ratio k of the equilibrium vortex streets (2). Furthermore, Kochin gave a Lyapunov function which implies not only the nonlinear instability of the special vortex street (2) with aspect ratio k^* but also the nonexistence of bounded solutions for (6) in this case. Thus, the isolated linearly stable vortex street turns out to be nonlinearly unstable after all.

Despite this serious difficulty, most of the research on the von Karman model continued to focus on the isolated vortex street with special k^*. Clearly, it violates the principle of genericity to use *only* the special case in a model in comparisons with observations. The point of departure of my approach to this dilemma, is to focus on all values of k that are near but unequal to k^*, in the system (6). First, I will show that Kochin's Lyapunov function is a *non-generic* property of the Hamiltonian system (6) i.e., there is no global Lyapunov function when the aspect ratio $k \neq k^*$. The nonexistence of a Lyapunov function removes the main obstacle to the existence of bounded solutions in (6). In fact, a rich variety of such bounded solutions can be established. A special subset of these, namely the near-equilibrium periodic and quasi-periodic solutions of (6), are suitable theoretical candidates for observed vortex trails because they are statically and dynamically indistinguishable from the equilibrium streets.

The nonexistence of a global Lyapunov function when the aspect ratio $k \neq k^*$ follows directly from the existence of two 1-parameter families of normal modes for the Hamiltonian system (6) at each value of $k \neq k^*$. These near-equilibrium periodic solutions of (6) bifurcate from the origin (the equilibrium vortex streets) via a $1 : -1$ semisimple resonance [4], [6] at $k = k^*$.

To see this, one notes that the Hamiltonian (6) becomes

$$(7) \qquad H(q_1, q_2, p_1, p_2; \lambda) =$$

$$\frac{1}{2}(q_1^2 + p_1^2) - \frac{1}{2}(q_2^2 + p_2^2) + \frac{1}{2}\lambda_1(q_1^2 + p_1^2) - \frac{1}{2}\lambda_2(q_2^2 + p_2^2)$$

$$+\lambda_4(q_1 p_2 - q_2 p_1) + \frac{1}{8}\left[4q_1 q_2 p_1 p_2 - q_1^2 q_2^2 - p_1^2 p_2^2 + q_1^2 p_2^2 + q_2^2 p_1^2\right]$$

in the new canonical variables defined by

$$(8) \qquad \begin{array}{ll} \alpha_1 = (q_1 + p_2)/\sqrt{2}, & \alpha_2 = (p_2 - q_1)/\sqrt{2} \\ \beta_1 = (q_2 + p_1)/\sqrt{2}, & \beta_2 = (p_1 - q_2)/\sqrt{2}. \end{array}$$

It is clear from the form of (7) and the following relations

$$(9) \qquad \lambda_1 = \lambda_2 = 2\left(\frac{g}{1+g} - \frac{1}{2}\right) \sim -\frac{1}{2}(g - g_*)^2 + \ldots$$

$$\lambda_3 \equiv 0, \quad \lambda_4 = \left(1 - \frac{2}{1+g^2}\right) \sim (g - g_*) + \ldots$$

$$g \equiv \sinh\left(\pi\frac{h}{\ell}\right),$$

that the 1-parameter family of Hamiltonians (6) has a $1 : -1$ semisimple resonance at $g = g_*$ where $g_* \equiv \sinh(\pi k^*) = 1$, corresponds to the special aspect ratio.

Now applying the semisimple Hamiltonian Hopf Bifurcation Theorem [6], one obtains

THEOREM. *For each value of the aspect ratio k near k^*, there are two 1-parameter families of near-equilibrium periodic solutions of Hamiltonian system(6).*

COROLLARY. *For each k near k^*, the Hamiltonian system (6) does not possess a global Lyapunov function.*

In conclusion, by focusing on the cases $k \neq k^*$ in the von Karman model, one finds several families of near - equilibrium periodic solution which are better theoretical counterparts for real vortex trails than the isolated equilibrium vortex street with aspect ratio k^*. Moreover, some of these normal modes are elliptic, and one expects to find subharmonic and/or torus bifurcation [9]. Once again, these near-equilibrium solutions behave like the equilibrium vortex streets. The nonlinear stability of the above normal modes can be treated by the Kolmogorov-Arnold-Moser Theorems (cf. Kummer [7]).

REFERENCES

[1] T. VON KARMAN, Gottingen Nach. Math. Phys. Kl (1911) 509-519.

[2] Y. COUDER, C. BASDEVANT, J. Fluid Mech. 173 (1986) 225.

[3] C. LIM, L. SIROVICH, Phys. Fluids 29 (12) (1986) 3910-3911.

[4] J.C. VAN DER MEER, *The Hamiltonian Hopf Bifurcation*, Lecture Notes in Math., Springer-Verlag, 1160 (1985).

[5] D. SCHMIDT, Celestial Mech. 9 (1974) 81-103.

[6] S.N. CHOW, Y. KIM, Appl. Anal. 31 (3) (1988) 163-199.

[7] M. KUMMER, Comm. Math. Phys. 58 (1978) 85-112.

[8] N. KOCHIN, I. KIBEL, N.V. ROZE, *Theoretical Hydromechanics*, Interscience (1964).

[9] C. LIM, *Subharmonic and torus bifurcation in the von Karman vortex lattice*, in preparation.

STABILITY OF HAMILTONIAN SYSTEMS
OVER EXPONENTIALLY LONG TIMES:
THE NEAR-LINEAR CASE

PIERRE LOCHAK*

1. Introduction. We shall be interested here in the classical problem of the stability of the action variables of a nearly integrable hamiltonian system, namely one governed by the following Hamiltonian:

(1.1) $H(p,q) = h(p) + \varepsilon f(p,q)$ with $(p,q) \in R^n \times T^n$, $T = R/Z$,

where (p,q) are action-angle variables of the integrable Hamiltonian h. Let us first recall informally the basic result of the theory, due to N.N. Nekhoroshev ([5], [6]; see also [7], [8]): Assume H is defined and *analytic* over some domain $\mathcal{D}$ of phase space; assume moreover that the unperturbed Hamiltonian h is *convex*, i.e. the hessian matrix $\nabla^2 h(p)$ is sign definite; then:

For any admissible initial condition $(p(0), q(0))$, one has:

$$\|p(t) - p(0)\| \le R(\varepsilon) = c \, \varepsilon^b \quad \text{for} \quad |t| \le T(\varepsilon) = \exp(c/\varepsilon^a) \,,$$

provided ε is small enough, i.e. $|\varepsilon| \le \varepsilon_0$.

This can be summarized by saying that *action variables are stable over exponentially long times if the unperturbed part of the system is analytic and convex* (this may be enlarged to the class of so-called "steep" functions; cf. [5]). We call $R(\varepsilon)$ the *radius of confinement*, $T(\varepsilon)$ the *time of stability*, and ε_0 the *threshold of stability*; (a,b) are the *stability exponents*, and they satisfy $0 < a, b \le 1$.

This result was made more explicit and improved in [2] (see also [3], [4]) under the convexity assumption on h, using a new method based on the periodic orbits of the unperturbed system. In the present paper, we shall be concerned with the important case when the unperturbed Hamiltonian is in fact *linear* (hence not convex nor steep of course), i.e. we consider perturbations of (an ensemble of non interacting) harmonic oscillators. More precisely, in section II we investigate the stability properties of Hamiltonian systems with Hamiltonians:

(1.2) $$H(p,q) = \omega_0 \cdot p + \varepsilon h_1(p) + \varepsilon^2 f(p,q).$$

We shall assume that h_1 satisfies a convexity assumption. In other words, we are looking at a "regular" degenerate problem, one in which the nonlinearity is of order ε. The main result asserts that (1.2) enjoys essentially the same stability properties as (1.1) (with convex h), *irrespective of*

---

* D.M.I., Ecole Normale Supérieure, 45 rue d'Ulm, F-75230 Paris Cedex 05, France.

the arithmetic properties of the unperturbed frequency $\omega_0 \in R^n$. This is the crucial fact, as we shall outline below. Indeed, if ω_0 is badly approximable by rationals, i.e. if it satisfies some diophantine condition, one can obtain stability properties through what we call Gevrey type estimates, via purely algebraic methods (see [8], [12], [14]). These amount to viewing the *linear* part $h_0(p) = \omega_0 \cdot p$ as the unperturbed system and controlling the growth of the coefficients of the corresponding Birkhoff series. No convexity assumption on the first order term of the perturbation is required, since the nonlinearity is not used. By contrast, here we include the first order term $\varepsilon h_1(p)$ in the unperturbed part, obtaining stability results which are independent of ω_0; this is indeed the correct extension of Nekhoroshev type results to a (simple) degenerate case. In more physical terms, it shows how "genuine" nonlinearity restores stability, even in the presence of linear resonance.

The study of a Hamiltonian system in the neighbourhood of an elliptic equilibrium point is connected with that of Hamiltonians of type (1.2), although specific properties and difficulties arise in this context. This is detailed in part III.

We note that the discrete versions of the aforementioned problems, in particular the investigation of a symplectic map near an elliptic fixed point, can be reduced to the continuous case, by means of the constructive interpolation results contained in [11].

We also point out that the results detailed below are essentially already contained in [2], and that they can most likely be also obtained using the improved version of the "traditional" approach to Nekhoroshev's theory presented in [9]. Here we adopt a semi-formal exposition style, much as in [1], which the reader is urged to consult for an overview of the theory and some background and historical information.

Lastly the author expresses his sincere thanks to the organizers for inviting him to participate in the 1992 Cincinnati Conference on Hamiltonian Dynamical Systems.

2. Perturbations of harmonic oscillators.

As already mentioned, this section is devoted to the stability properties of the action variables in systems with Hamiltonians of type (1.2). Let us first introduce the necessary pieces of notation. We assume that H is *analytic* over some domain $\mathcal{D} = \mathcal{D}(R, \rho, \sigma)$, $(\rho > 0, \sigma > 0)$ defined as follows: let B_R be the ball of radius R around the origin, then:

$$\mathcal{D} = \mathcal{D}(R, \rho, \sigma) = \left\{ (p, q) \in C^{2n}, dist(p, B_R) \leq \rho \; ; \; \Re(q) \in T^n; \; |\Im(q)| \leq \sigma \right\},$$

with $|\Im(q)| = Sup_i(|\Im(q_i)|)$. Note that the real part of $\mathcal{D}$ is nothing but $B_{R+\rho} \times T^n$.

Assume further that h_1 is *convex*. More precisely, denote $\omega_1(p) = \nabla h_1(p) \in R^n$ and $A(p) = \nabla^2 h_1(p) \in \mathcal{M}_n(R)$ the frequency vector and the hessian matrix. We assume that $A(p)$ is sign definite (say positive). Let m

be a lower bound of the spectrum of A over the real part of the domain, then:

$$\forall p \in B_{R+\rho} \subset R^n, \ \forall v \in R^n \ : \ A(p)v.v \ge m||v||^2 \ ,$$

where $m > 0$ and $u.v$ denotes the scalar product of two vectors u, v. We will denote by M the operator norm of A on $\mathcal{D}$ (the complex domain): $||A(p)v|| \le M||v||$.

Lastly, we set $\Omega = ||\omega_0||$ which is assumed to be nonzero; notice that if $\omega_0 = 0$, we are back to the nondegenerate case, modulo a scaling transformation. We collect the parameters which describe the system as $\mathcal{P} = (R, \rho, \sigma, M, m, E, \Omega)$. As in [1], in order to simplify notation, we introduce the following symbols: if α and β are any two scalar quantities, we write $\alpha \preceq \beta$ if there is a constant $c = c(\mathcal{P})$ such that $\alpha \le c\beta$; similarly, $\alpha \succeq \beta$ means $\alpha \ge c\beta$ and $\alpha \asymp \beta$ means $\alpha = c\beta$.

The main result of this section now reads as follows:

THEOREM 2.1. *For any initial condition $(p(0), q(0))$ (say with $p(0) \in B_R$ so that the trajectory is defined), one has:*

$$||p(t) - p(0)|| \preceq \varepsilon^b \quad \text{for} \quad |t| \preceq \exp(c/\varepsilon^a) \quad (c = c(\mathcal{P})),$$

provided ε is small enough; one has $a = b = \frac{1}{2n}$.

Except for minor modifications, the implied constants, including the threshold of validity, are the same as in the nondegenerate case. The pair of global stability exponents is also identical and we refer to [1] and [2] regarding the many possible local refinements. The proof of Theorem 2.1 can essentially be reduced to that of the nondegenerate case and we shall here concentrate on this reduction.

It is convenient to perform the scaling:

$$(2.1) \qquad\qquad t \to \varepsilon t, \quad H \to \varepsilon^{-1} H$$

and work with the transformed Hamiltonian (keeping the same letters)

$$(2.2) \qquad\qquad H(p,q) = \frac{\omega_0}{\varepsilon} \cdot p + h_1(p) + \varepsilon f(p,q).$$

We set $h(p) = \frac{\omega_0}{\varepsilon} \cdot p + h_1(p)$, the unperturbed part, which depends on ε, and $\omega(p) = \frac{\omega_0}{\varepsilon} + \omega_1(p)$ the unperturbed frequency.

As in the nondegenerate case, one first has to prove stability around periodic points in frequency space. To this end, *fix ε* and suppose that $\omega(0) = \frac{\omega_0}{\varepsilon} + \omega_1(0)$ is rational with period $T > 0$, which means that $T\omega(0) \in Z^n$. Here the action vector corresponding to the given rational point has been taken as the origin of the coordinates, so as to work around $p = 0$. In this context, the following holds:

LEMMA 2.1.

$$||p(0)|| \le r \Rightarrow ||p(t)|| \preceq r \quad \text{if} \quad |t| \preceq \varepsilon \ e^{cs} \quad (c = c(\mathcal{P})),$$

provided the following two conditions obtain:

$$i) \quad s \, r \, T \preceq 1, \quad ii) \quad r \succeq \sqrt{\varepsilon}.$$

To prove Lemma 2.1, just go through the proof for Hamiltonian (1.1); there are only very minor adjustments to make, including the factor ε in front of the time of validity, which stems from the scaling transformation (2.1) (scaling back (2.2) to (1.2) regains this factor).

The second step in the proof of Theorem 2.1 consists, as in the ordinary case, in approximating any point by rational ones. Again, we do not repeat the reasoning, but only point out the differences with the nondegenerate case.

First pick an arbitrary initial condition $p(0) = p^* \in B_R$, and write $\omega_1^* = \omega_1^*(p^*)$, $\omega^* = \frac{\omega_0}{\varepsilon} + \omega_1(p^*)$. By Dirichlet's theorem,

$$\forall \, Q > 1, \exists \, \omega \quad T\text{-periodic such that}: ||\omega - \omega^*|| \leq \frac{\sqrt{n}}{T Q^{\frac{1}{n}}}, \quad 1 \leq T \leq Q.$$

Set $\omega_1 = \omega - \frac{\omega_0}{\varepsilon}$; then of course:

$$||\omega_1 - \omega_1^*|| \leq \frac{\sqrt{n}}{T Q^{\frac{1}{n}}}, \quad 1 \leq T \leq Q;$$

then *define* p by $\omega_1 = \omega_1(p)$ which is possible if Q is large enough, because the map $p \rightarrow \omega_1(p)$ is locally invertible; p is rational with period $T < Q$ and the reasoning then goes through as in the ordinary case (except for a very small detail: one should rescale one of the components of ω^* to an integer, which here is "large" not 1).

The crux of the matter can in fact be simply stated as follows: although the frequency is large (of order $1/\varepsilon$), *low* frequency orbits can be used in the approximation process, and uniformly so as $\varepsilon \rightarrow 0$. This simply means that $\frac{\omega_0}{\varepsilon}$ can be shifted back near the origin, via an integer vector translation: approximation is concerned only with the unit cube.

This completes the sketch of the proof of Theorem 2.1; the scaling transformation (2.1) has been taken into account in the statement, which applies to Hamiltonian (1.2). We add a few remarks:

- Here is one situation in which Hamiltonian (1.2) arises; start from

(2.3) $H(p,q) = \omega_0 \cdot p + \varepsilon g(p,q)$

where g is a trigonometric polynomial:

$$g(p,q) = \sum_{-N \leq k \leq N} g_k(p) e^{ik \cdot q}.$$

This situation is often encountered in practice. Assume:

$$\omega_0 \cdot k = 0 \Rightarrow g_k = 0; \quad g_0(p) \text{ convex.}$$

Then, after one step of perturbation theory, Hamiltonian (2.3) will take the form (1.2), with $h_1 = g_0$.

- One can accommodate the case when h_1 is only *quasi*-convex, which here means that it is convex when restricted to the plane $\omega_0 \cdot p = 0$ or the case of a time dependent perturbation $f(p, q, t)$ which is periodic w.r.t. t, but both extensions are not permitted simultaneously.

- As mentioned in the introduction, Theorem 2.1 is the first Nekhoroshev type result in a degenerate situation and it should be compared with what we call Gevrey type estimates. These can be described as follows; consider again the Hamiltonian $H(p, q) = \omega_0 \cdot p + \varepsilon g(p, q)$ where the first order term of the perturbation need not be specified. Assume the frequency ω_0 is very irrational, specifically that it satisfies a diophantine condition:

$$\forall k \in Z^n, \quad |\omega_0 \cdot k| \geq \frac{\gamma}{|k|^{\tau-1}}, \ \gamma > 0, \ \tau \geq n.$$

Then one has the following stability estimate:

$$\|p(t) - p(0)\| \preceq \varepsilon \quad \text{for} \quad |t| \preceq \exp(c/\varepsilon^{1/\tau}), \text{ if } \varepsilon \text{ is small enough.}$$

For a proof of this result, see [8], [10], [12] and especially [14]; it is obtained in a purely algebraic way, by controlling the growth of the coefficients of the Birkhoff series, using the diophantine bound from below for the intervening small divisors.

So one obtains the stability exponents $(a, b) = (\frac{1}{\tau}, 1)$ with τ typically on the order of n. But of course, the crux of the matter is that everything is very sensitive to the arithmetic properties of ω_0; more precisely, the implied constants here depend, not only on the set of parameters $\mathcal{P}$, but also on the constants τ and γ.

- Here we give an example of a non convex situation, where a fast drift of the action variables occurs. This is an elaboration of the corresponding constructions for the nondegenerate case (see [5], Section 1.17 and [6], Section 4). Consider the Hamiltonian:

$$(2.4) \qquad H = \omega_0 \cdot p + \frac{1}{2}\varepsilon \sum_{i=1}^{n} a_i p_i^2 - \varepsilon^2 \sin(k \cdot q),$$

with $(p, q) \in R^n \times T^n, a = (a_i) \in R^n, \omega_0 \in R^n, k \in Z^n$. Assume that $\omega_0 \cdot k = 0$ and that $\sum_{i=1}^{n} a_i k_i^2 = 0$, which of course precludes convexity.

Then if $k \cdot q(0) = 0$, one checks that there are trajectories $(p(t), q(t))$ such that
$p(t) = (c + \varepsilon^2 t)k, (c \in R \text{ arbitrary})$ i.e. the action p drifts along k (belonging to the resonant plane) with speed ε^2.

The following conclusions can thus be drawn concerning Hamiltonian (1.2): - If ω_0 is very nonresonant, i.e., satisfies a diophantine condition, Gevrey type estimates obtain and stability stems from the absence of linear resonance; these estimates are thus sensitive to the arithmetic properties of the frequency vector.

- If the (first-order) nonlinear term satisfies a (quasi-) convexity assumption, a Nekhoroshev type result (Theorem 2.1) guarantees a stability of the same order as in the nondegenerate (fully nonlinear) situation, whatever the resonance properties of the linear frequency: this is the stabilization effect of nonlinearity. - In the absence of any nonresonance or convexity assumption, fast drift of the action variables may occur (it would be precluded by a finer "steepness" condition, but, apart from the convex situation, steepness cannot be read off the 2-jet of the perturbative term).

3. Neighbourhood of an elliptic equilibrium point. We turn to the problem of studying a Hamiltonian vector field in the neighbourhood of an elliptic equilibrium point, and we shall use Theorem 2.1 in order to derive a result for this situation (see [2] for a bit more details). We denote by $\pm i\alpha_j, (i = \sqrt{-1}), j = 1, ..., n$, the eigenvalues of the linearized system at the fixed point, which we take as the origin of the coordinate system $(x_j, y_j) \in R^{2n}$; we write $z = (x, y) \in R^{2n}$. We assume that the linear part can be diagonalized, and that there are no resonances of order $\leq s$ (a positive integer), which means that $\alpha \cdot k \neq 0$ $(\alpha = (\alpha_j) \in R^n)$ for $k \in Z^n, k \neq 0, |k| = |k_1| + ... + |k_n| \leq s$. We write $r_j = \frac{1}{2}(x_j^2 + y_j^2), r = (r_j) \in R_+^n$.

Assume everything is analytic and $s > 4$; then following Birkhoff, there exists a normalizing change of coordinates such that H reads:

$$(3.1) \quad H(z) = \sum_j \alpha_j r_j + \frac{1}{2} \sum_{i,j} \alpha_{ij} r_i r_j + O(||r||^3) = \alpha \cdot r + \frac{1}{2} Ar \cdot r + O(||r||^3),$$

where $A = (\alpha_{ij})$ is a symmetric matrix. This way of writing determines the sign of the α_j's; notice that if they are all of the same sign, the stability problem is immediately settled (positively) because the origin is a local maximum (or minimum) of H.

We now add the convexity assumption (which in this context, is also called "monotone twisting" that A is sign definite, say positive, with a spectrum bounded from below and above by $m > 0$ and M.

Under these assumptions, the following obtains:

THEOREM 3.1. *Let $z = z(0)$ be small enough and assume it satisfies:*

$$(3.2) \qquad r_j = \frac{1}{2}(x_j^2 + y_j^2) \succeq ||z||^{2+1/n}, \quad j = 1, ..., n,$$

(where $z = z(0), r_j = r_j(0)$, etc.); then:

$$||r_j(t) - r_j|| \preceq ||z||^{2+1/n} \quad if \quad |t| \preceq \exp(c||z||^{-1/n}).$$

Again, the implied constants can be estimated explicitly. We shall comment on condition (3.2) later. This result is a direct consequence of Theorem 2.1. First, introduce the usual polar symplectic coordinates $(r, \theta) \in R_+^n \times T^n$:

$$x_j = \sqrt{2r_j} \cos \theta_j, \quad y_j = \sqrt{2r_j} \sin \theta_j.$$

Fix $z = z(0)$ and set $\varepsilon = \frac{1}{2}||z||^2 = \sum_j r_j$. Then perform the scaling $r = \varepsilon\rho$, $H = \varepsilon K$, which multiplies the symplectic form by the factor ε and leaves the equations invariant. The new Hamiltonian reads:

$$(3.3) \qquad K(\rho, \theta) = \alpha \cdot \rho + \frac{1}{2}\varepsilon \, A\rho \cdot \rho + \varepsilon^2 f(\sqrt{\varepsilon\rho}, \theta),$$

where we write componentwise: $\sqrt{r} = (\sqrt{r_1}, ..., \sqrt{r_n}) \in R_+^n$. The function f is analytic in its two arguments and periodic w.r.t. θ.

Everything is now set up nicely to apply Theorem 2.1 (compare hamiltonians (1.2) and (3.3)), except for the analyticity problem near the coordinate planes $r_j(= \rho_j) = 0$. If one goes through the proof of Theorem 2.1, or rather the full proof in the nondegenerate case (cf. [2]) one sees that there is no need of an analyticity width (parameter ρ in Section 2) of order 1 but that it can be made to depend on ε. This can be done in order to approach the singularities and it gives rise to condition (3.2) in Theorem 3.1, which excludes cusp-shaped neighbourhoods of the coordinate planes. This leads to the statement of the theorem above.

Again, some comments may be in order:

- One can treat the case of a resonance of order $s = 4$; we have required $s > 4$ for simplicity, namely to get a remainder of order r^3 in (3.1). $s \geq 4$ is the natural condition under which KAM theory applies.

- As in the case of the perturbation of harmonic oscillators, it is enough to cast a *quasi*-convexity condition on the first order nonlinear term: here it means requiring $Ar \cdot r > 0$ *when* $\alpha \cdot r = 0$ (which never happens if the components of α are all of the same sign). One can also accommodate a time periodic term, but again, both extensions are not permitted simultaneously.

When α is very nonresonant, i.e. satisfies the diophantine condition:

$$\forall k \in Z^n, \quad |\alpha \cdot k| \geq \frac{\gamma}{|k|^{\tau-1}}, \; \gamma > 0, \; \tau \geq n,$$

one can obtain Gevrey type estimates which read as follows:

$$||r_j(t) - r_j|| \preceq ||z||^3 \quad \text{if} \quad |t| \preceq \exp(c||z||^{-1/\tau}).$$

For a proof of this result, see [12] and especially [13]; again, this is obtained in a purely algebraic way, controlling the growth of the coefficients in Birkhoff's normalizing series. One needs not require anything on the nonlinear part (the "first Birkhoff invariant" A) nor of course a condition of type (3.2) restricting the initial condition. These Gevrey type estimates thus provide a true exponential exit time estimate for the neighbourhood of an elliptic equilibrium point, but again they are "linear" in nature and the implied constants strongly depend on the arithmetics of the frequency, here on τ and γ.

- Regarding the stability exponents, one should consider r (or equivalently $||z||^2$) as the natural variable (the action variable). So the exponents in Theorem 3.1 are the same as in Theorem 2.1: $a = b = \frac{1}{2n}$. In the Gevrey estimates above, one finds the pair $(a, b) = (\frac{1}{2\tau}, \frac{1}{2})$. So the values of the first exponent, governing the time of stability, essentially agree for τ of order n (for a thorough discussion of the local value of b, see [1] and [2]). This is in sharp contrast with the harmonic oscillator case, where Theorem 2.1 yields $a = \frac{1}{2n}$, whereas Gevrey estimates produce $a = \frac{1}{\tau}$, without a factor 2. We shall briefly return to this discrepancy in the last item below.

- One should be able to show the existence of a possible instability when convexity does not obtain; this amounts to constructing an example similar to Hamiltonian (2.4) above, which should be feasible.

- The stability problem in the convex case is more delicate. Because of condition (3.2) in Theorem 3.1, we have *not* proven stability over an exponential timescale. This raises the obvious question: is condition (3.2) simply an artefact or does it reflect something real? Namely, can there occur "leaks" in the vicinity of the coordinate planes, where also ordinary KAM theory breaks down (one proves there the existence of invariant tori of smaller dimensions)?

We do not know the answer, but one can make the following remarks. Condition (3.2) stems from a lack of analyticity of Hamiltonian (3.3) near the coordinate planes. This in turn seems artificial at first sight and can be traced back to the fact that we use the polar coordinates (r, θ). By contrast, in deriving Gevrey estimates, one uses either cartesian coordinates $z = (x, y)$ or complex variables ($w = x + iy, \bar{w} = x - iy$) and no singularity occurs. But in fact these coordinates *cannot* be used in the truly nonlinear theory; they are adapted to the case when the *linear* part is taken as the unperturbed part. If one wants to take advantage of the nonlinear terms, one has to work with the action-angle coordinates for the nonlinear integrable part, which here coincide with the symplectic polar coordinates. To illustrate the fact that the choice of variables giving rise to *analytic* Hamiltonians is crucial, it may be pointed out that if one could "translate" Gevrey estimates in the harmonic oscillators case directly into such estimates for the elliptic fixed point problem, one would find $a = \frac{2}{\tau}$ in the latter case, which is simply wrong.

REFERENCES

[1] P. Lochak, Hamiltonian Perturbation Theory: Periodic Orbits, Resonances, Intermittency, Nonlinearity 6 (1993) pp. 885–906.

[2] P. Lochak, Canonical Perturbation Theory via Simultaneous Approximation, Uspekhi. Math. Nauk 47(6) (1992) pp. 59–140 (English translation in Russian Math Surveys 47(6), (1992), pp. 57–133).

[3] P. Lochak, A.I. Neistadt, Estimates in the theorem of N.N. Nekhoroshev for systems with a quasi-convex Hamiltonian, Chaos 2(4) (1992) pp. 495–499.

[4] P. Lochak, A.I. Neistadt, L. Niederman, Stability of nearly integrable convex Hamiltonian systems over exponentially long times, in Seminar on Dynamical Systems, PNDLE vol. 13, S. Kuksin, V. Lazutkin and J. Pöschel Ed. 1994, Birkhaüser Publ., pp. 15–34.

[5] N.N. Nekhoroshev, An exponential estimate of the time of stability of nearly integrable Hamiltonian systems, Russian Math Surveys 32 (1977) pp. 1–65.

[6] N.N. Nekhoroshev, An exponential estimate of the time of stability of nearly integrable Hamiltonian systems II, Trud. Sem. Petrovski 5 (1979) pp. 5–50.

[7] G. Benettin, L. Galgani, A. Giorgilli, A proof of Nekhoroshev's theorem for the stability times in nearly integrable Hamiltonian systems, Celestial Mechanics 37 (1985) pp. 1–25.

[8] G. Benettin, G. Gallavotti, Stability of motion near resonances in quasi integrable Hamiltonian systems, J. Stat. Physics 44 (1986) pp. 293–338.

[9] J. Pöschel, Nekhoroshev estimates for quasiconvex Hamiltonian systems, Math. Zeitschrift 213 (1993) pp. 187–216.

[10] A. Giorgilli and L. Galgani, Rigorous estimates for the series expansions of Hamiltonian perturbation theory, Celestial Mechanics 37 (1985) pp. 95–112.

[11] S.B. Kuksin and J. Pöschel, On the inclusion of analytic symplectic maps in analytic hamiltonian flows and its applications, in Seminar on Dynamical Systems, PNDLE vol. 13, S. Kuksin, V. Lazutkin and J. Pöschel Ed. 1994, Birkhaüser Publ.

[12] A. Giorgilli, A. Delshams, E. Fontich, L. Galgani and C. Simó, Effective stability of a Hamiltonian system near an elliptic fixed point with an application to the restricted three body problem, Journal Diff. Eq. 77 (1989) pp. 167–198.

[13] A. Giorgilli, Rigorous results on the power expansions for the integrals of a Hamiltonian system near an elliptic equilibrium point, Annales IHP Physique Théorique 48 (1989) pp. 423–439.

[14] F. Fassó, Lie series method for vector fields and Hamiltonian perturbation theory, Journal of Applied Math. and Physics (ZAMP) 41 (1990) pp. 843–864.

CONSTRAINED VARIATIONAL PRINCIPLES AND STABILITY IN HAMILTONIAN SYSTEMS

JOHN H. MADDOCKS* AND ROBERT L. SACHS†

Abstract. This article describes ways in which constrained variational principles can be used to study the stability properties of special solutions, e.g. equilibria, relative equilibria and other special motions, in Hamiltonian systems. Much of what is described is classic material, but some aspects are not well known, and others parts are of comparatively recent origins. In particular a new approach to the analysis of the second variation in constrained variational principles was recently combined with the classic Hamiltonian formulation of the Korteweg-de Vries equation to obtain a stability result for the relative equilibria or multi-soliton solutions. This new method and related issues are here described in the context of certain classic mechanical systems that are modelled by systems of ordinary differential equations.

1. Introduction. There is a long and venerable tradition connecting variational principles and the stability of special solutions in Hamiltonian systems. (For background concerning Hamiltonian systems, see e.g. Siegel and Moser, 1971, Arnol'd, 1978, or Arnol'd *et al*, 1988.) In this survey we shall primarily be concerned with the manner in which constrained variational principles can be exploited in the stability analysis of relative equilibria or steady motions. There are three observations that we wish to stress. First, in the variational principles that characterize relative equilibria in Hamiltonian systems, constrained critical points naturally arise in parametrized families (in a variety of ways). We shall describe the manner in which these parametrizations can be exploited to determine which critical points are actually constrained minima. This classification then leads to our second observation, namely that for any given constrained minimum the theory of the augmented Lagrangian can be used to construct a new variational principle in which the critical point is an unconstrained minimum. This last characterization is of interest because a (dynamic) stability is often implied. Our third observation is that the manner in which constrained critical points are embedded in families also contains considerable information about the linearized dynamics. In particular we shall describe circumstances in which we may conclude (dynamic) instability of the special motion associated with a critical point that is it not a constrained minimum.

Before discussing relative equilibria, and in order to fix ideas and set notation, we shall first review a simpler context in which variational prin-

---

* Department of Mathematics and Institute for Physical Science and Technology, University of Maryland, College Park, MD 20742.
Research partially supported by grants from the Air Force Office of Scientific Research and the Office of Naval Research.
† Department of Mathematical Sciences, George Mason University, Fairfax, VA 22030.
Research partially supported by a grant from the National Science Foundation.

ciples arise, namely in the characterization of equilibria of an autonomous Hamiltonian system of ordinary differential equations

$$(1.1) \qquad\qquad \mathbf{z}' = \mathcal{J} \nabla H(\mathbf{z}).$$

Here ∇H denotes the gradient of the Hamiltonian $H(\mathbf{z})$ with respect to the variable $\mathbf{z}$, the matrix $\mathcal{J}$ is skew-symmetric, and solutions $\mathbf{z}(t)$ of (1.1) are curves in phase space. In §2 we shall be concerned with the case in which the matrix $\mathcal{J}$ has explicit dependence on $\mathbf{z}$ and $\mathcal{J}(\mathbf{z})$ may be singular. However in the classic setting

$$(1.2) \qquad\qquad \mathcal{J} = \begin{pmatrix} 0 & 1 \\ -1 & 0 \end{pmatrix}, \quad \text{and} \quad \mathbf{z}(t) \in \Re^{2n},$$

i.e. $\mathcal{J}$ is the standard symplectic matrix and the phase space is $\Re^{2n}$, with n degrees of freedom. In particular in the standard case, we can write $\mathbf{z} = (\mathbf{q}, \mathbf{p})$ with n state variables $\mathbf{q}$ and n conjugate momenta $\mathbf{p}$. Then, because $\mathcal{J}$ is nonsingular, the equilibrium solutions of (1.1), i.e. trajectories satisfying $\dot{\mathbf{z}}_e(t) = \mathbf{0}$, are precisely the critical points of the Hamiltonian, i.e. those points in phase space satisfying

$$(1.3) \qquad\qquad \nabla H(\mathbf{z}_e) = \mathbf{0}.$$

The content of the Lagrange-Dirichlet Theorem (cf. Siegel & Moser, 1971, p. 208) is that those critical points that are actually isolated minima (or maxima) of the Hamiltonian are necessarily stable solutions of (1.1). The proof is that the Hamiltonian H is an integral of the dynamics (1.1) and so can serve as what is now usually (and in this context somewhat ironically) called a Lyapunov function. In particular, one need never address difficulties associated with marginally stable modes of the linearization of (1.1) at $\mathbf{z}_e$ that arise due to the pure imaginary spectrum of the (nonsymmetric) eigenvalue problem

$$(1.4) \qquad\qquad \rho \mathbf{g} = \mathcal{J} \nabla^2 H(\mathbf{z}_e) \mathbf{g}$$

that is obtained after separating out the time variable in the linearized dynamics. Instead in order to apply the theorem one typically examines the (symmetric) eigenvalue problem

$$(1.5) \qquad\qquad \lambda \mathbf{h} = \nabla^2 H(\mathbf{z}_e) \mathbf{h}$$

associated with the second variation of H at $\mathbf{z}_e$. If (1.5) does not have zero as an eigenvalue the critical point $\mathbf{z}_e$ is said to be nondegenerate and in particular must be isolated. If (1.5) has only positive eigenvalues, then $\mathbf{z}_e$ is necessarily an isolated minimum of the Hamiltonian, and the Lagrange-Dirichlet Theorem can be applied.

As is well known, the main limitation of the above approach is that in general there is no converse conclusion available, and critical points of the

Hamiltonian that are neither maxima nor minima may either be stable or unstable (in the full nonlinear sense). It is easy to show that when (1.5) has only positive eigenvalues, (1.4) has only pure imaginary eigenvalues, which is certainly compatible with our conclusion of nonlinear stability via the Lagrange-Dirichlet Theorem. Nevertheless it is possible that (1.4) can have only pure imaginary eigenvalues while (1.5) has negative eigenvalues, which allows the possibility that critical points other than minima can be stable. However consideration of the eigenvalue problems (1.4) and (1.5) can provide some partial converses to the Lagrange-Dirichlet Theorem. The first converse applies in the case that the Hamiltonian decouples into a sum of a potential that depends only on the state variables $\mathbf{q}$, and a kinetic energy that is a positive definite quadratic form of the momenta $\mathbf{p}$, i.e.

$$(1.6) \qquad H(\mathbf{q}, \mathbf{p}) = \tfrac{1}{2}\mathbf{p} \cdot K(\mathbf{q})\mathbf{p} + V(\mathbf{q}),$$

where K is a symmetric positive definite matrix whose entries are functions of the state variables $\mathbf{q}$. All critical points of Hamiltonians of this type are of the form $\mathbf{z}_e = (\mathbf{q}_e, 0)$ where $\mathbf{q}_e$ is a critical point of the potential $V(\mathbf{q})$. The Hamiltonian can have no maximum, and the isolated minima arise in correspondence with points $\mathbf{q}_e$ that are isolated minima of the potential. In fact the index of a nondegenerate critical point $\mathbf{z}_e$, i.e. the number of negative eigenvalues of (1.5), is determined by the number of negative eigenvalues of the the matrix $\nabla^2 V(\mathbf{q}_e)$ (or equivalently by the index of $\mathbf{q}_e$ viewed as a critical point of $V(\mathbf{q})$). It is also simple to calculate that at any nondegenerate critical point $\mathbf{z}_e$ for which $\mathbf{q}_e$ is not a minimum of the potential, the eigenvalue problem (1.4) has one or more positive real eigenvalues (in fact the number of positive real eigenvalues coincides with the index of $\mathbf{q}_e$). Consequently the linearized dynamics has a growing mode and $\mathbf{z}_e$ is therefore a (nonlinearly) unstable solution of the dynamics (1.1). This conclusion fails for Hamiltonians only slightly more general than (1.6). For example, the addition of terms linear in $\mathbf{p}$ (often called gyroscopic terms) can lead to stable critical points that are not minima of the Hamiltonian. Moreover even for the decoupled Hamiltonian (1.6), instability of degenerate critical points that are not minima has only been proven in particular cases (cf. for example, the discussion in Arnol'd et al, 1988, p. 271).

The second converse to the Lagrange-Dirichlet Theorem is less well-known, but Kelvin was already aware of one version (see for example the discussion in Chetayev, 1961, Chpt. 5). It states that for an arbitrary Hamiltonian, any nondegenerate critical point of odd index is an unstable solution. The proof is immediate from the observation $|\mathcal{J}\nabla^2 H(\mathbf{z}_e)| = |\mathcal{J}||\nabla^2 H(\mathbf{z}_e)| = |\nabla^2 H(\mathbf{z}_e)|$ which is negative by assumption. But the spectrum of $\mathcal{J}\nabla^2 H(\mathbf{z}_e)$ is symmetric both in the real and imaginary axes, so $|\mathcal{J}\nabla^2 H(\mathbf{z}_e)|$ can only be negative if there is a real positive eigenvalue, which in turn implies instability of $\mathbf{z}_e$.

In the cases of nondegenerate critical points with an even index, and any degenerate critical point, there is no simple converse to the Lagrange-Dirichlet Theorem available. Some progress can be made in concluding instability in the presence of dissipative perturbations of the conservative system (1.1), which is an idea that again traces back at least to Kelvin, who introduced the notions of secular and ordinary stability based on whether or not the time scale was such that dissipation was significant (cf. Chetayev, *op. cit.*). In fact the effect of dissipative perturbations is still a topic of contemporary interest (see, for example, MacKay, 1991, or Bloch *et al*, 1992). We shall not discuss such approaches here, and instead defer to Maddocks & Overton (1994). In that work it is shown that in the presence of an appropriate class of dissipations, the number of unstable modes of the dynamics linearized at a nondegenerate equilibrium exactly coincides with the index of the solution regarded as a critical point of the Hamiltonian. The limiting behavior of eigenvalues in the case of vanishing dissipation is also fully analyzed.

The situation that is of primary concern in this work arises when the Hamiltonian system has one or more integrals in addition to the Hamiltonian itself. It will be useful to differentiate between two ways in which additional integrals can arise. The first is for classic Hamiltonian systems that fit in the framework (1.2). Then by an appropriate version of Noether's Theorem (cf. Olver, 1986, p. 399) there is a correspondence between infinitesimal (Hamiltonian) symmetries of the Hamiltonian and integrals. The second way that integrals can arise is when the matrix $\mathcal{J}$ appearing in (1.1) may itself be dependent upon the phase variable $\mathbf{z}$, and $\mathcal{J}(\mathbf{z})$ may be singular. Then there can be integrals, called Casimir or distinguished functions (cf. Olver, *op. cit.*, p. 380), that are integrals of (1.1) that are not associated with any symmetry of the Hamiltonian.

It is well known that in the presence of several integrals of motion very special trajectories of (1.1) arise when the initial data in phase space represent a constrained critical point of one integral subject to prescribed values of the others. The purpose of this note is to demonstrate ways in which such constrained variational principles can be used to examine stability properties of these special trajectories. In Sections 2 and 3, respectively, we shall consider the analogues of the variational characterization (1.3), and the symmetric and nonsymmetric eigenvalue problems (1.5) and (1.4) that arise in the presence of Casimir and non-Casimir integrals. The arguments that we describe provide the natural generalization of the results described above, which conclude stability of equilibria that are actually minima of the Hamiltonian. However in order to obtain useful conclusions the role of integrals as constraints on the motion must usually be exploited to the full extent possible. This is the case because the requirement of a critical point being a *constrained* minimum is less stringent than being an *unconstrained* minimum. Consequently, both in principle and in practice, the stability properties of a larger class of solutions of the dynamics can be analyzed if constrained variational principles are exploited. However there is a price to be paid because in many examples the analysis required to characterize a critical point as being a constrained minimum is considerably more delicate than that required to characterize a critical point as being an unconstrained minimum. This fact can substantially complicate the arguments required to conclude stability or instability.

One more or less trivial, but nevertheless instructive, example is presented by the Hamiltonian

$$(1.7) \qquad H(q_1, q_2, p_1, p_2) = V_1(q_1) - V_2(q_2) + \tfrac{1}{2}p_1^2 - \tfrac{1}{2}p_2^2$$

with the standard symplectic structure (1.2). If the two potentials V_1 and V_2 have nondegenerate minima at zero then the origin (in phase space) is a nondegenerate critical point of the Hamiltonian with index two. However, because of the decoupling in the system, the origin is clearly nonlinearly stable. One way to see this to observe that in addition to the Hamiltonian there is another independent integral, namely

$$(1.8) \qquad\qquad I(q_2, p_2) = V_2(q_2) + \tfrac{1}{2}p_2^2.$$

It is then apparent that the origin is an isolated constrained minimum of H subject to the constraint $I = 0$, which fact allows us to conclude stability of the equilibrium amongst the restricted class of nearby initial data satisfying $I = 0$. More pertinently the origin is also an unconstrained, nondegenerate minimum of the function

$$(1.9) \qquad H + \gamma I = V_1(q_1) + \tfrac{1}{2}p_1^2 + (\gamma - 1)\left(V_2(q_2) + \tfrac{1}{2}p_2^2\right), \qquad \gamma > 1.$$

Because (1.9) is a linear combination of integrals it is also an integral, and so can be used to prove nonlinear stability of the origin amongst *all* nearby initial conditions. The rather pedantic treatment of this trivial example captures the two essential features of the stability analysis that we will present below. First we will present an effective strategy to classify those critical points that are *constrained* minima of the Hamiltonian subject to prescribed values of the other integrals. A stability result amongst a *restricted* class of perturbations is then implied, but is often not of any physical interest. Then we show that standard variational arguments, essentially due to Hestenes, can be used to characterize any *constrained* minimum as an *unconstrained* minimum of an appropriate integral of the motion. Consequently the Lagrange-Dirichlet Theorem can be applied to conclude stability amongst *arbitrary* small perturbations of the initial conditions.

We also remark that the Hamiltonian (1.7) is closely related to a standard example in KAM theory, and it is well known that small perturbations in the Hamiltonian that introduce coupling can destroy the conservation of (1.8) and destabilize the origin. However such questions of *structural, strong* or *parametric* stability will not be of concern here.

The constrained variational arguments pertaining to stability properties are actually somewhat simpler in the (perhaps less familiar) case of a
noncanonical Hamiltonian structure for which the additional integrals arise
as Casimir functions, so we first describe that situation in §2. The example
that we discuss there is the motion of an asymmetric, heavy, rigid body
tumbling about a fixed point. The more familiar case of canonical systems
with the symplectic matrix (1.2) and multiple conserved quantities is described in §3. The complication in the variational analysis of the canonical
case is that due to the continuous symmetries associated with the integrals
of motion, the critical points cannot be isolated, and so must be degenerate
(in the classic sense). As a consequence, the Lagrange-Dirichlet Theorem
cannot immediately be applied in its simplest form. However the difficulty
of nonisolation can be overcome provided that we weaken the notion of stability in an appropriate sense that will be made precise below. Of course it
is possible to construct examples in which both Casimir and non-Casimir
integrals arise (for example the *symmetric*, heavy rigid body or Lagrange
top), but the required analysis is then a straightforward combination of the
two cases considered separately in Sections 2 and 3, so such examples are
not explicitly considered. We do describe some elementary finite dimensional examples that are typical of problems where the integrals of motion
are all in involution. In particular the stability analysis of the relative equilibria in the finite dimensional examples we consider directly parallels that
of the 1- and 2-soliton solutions to the KdV equation detailed in Maddocks
& Sachs (1993).

The general framework that we construct in Section 3 is restricted to
the case where all of the additional integrals are in involution. Much can
be said about variational principles for relative equilibria in Hamiltonian
systems where all of the integrals are not in involution, and such examples
are certainly of interest, e.g. the Lagrange equilateral solutions of the three
body problem, but we defer to a later presentation for a general analysis
from the viewpoint adopted here. One place to find a general discussion of
constrained variational characterizations of relative equilibria in rotating
systems is Zombro (1993).

Before closing the Introduction we make two observations. First, we
wish to stress that much of what we describe is either classic and well
known, or classic and not well known. The use of variational principles
involving integrals of the motion to examine dynamic stability properties
can be viewed from several different standpoints, all apparently having at
least some origins in the work of Kelvin, Routh and Lagrange. We make
no attempt at a comprehensive survey of the literature. However particularly closely related works include Rubanovskii & Stepanov (1969) and
Rumyantsev (1966, 1968) from the Russian school, and from a more geometrical viewpoint, Holm *et al* (1985), Simo *et al* (1990) and Simo *et al*
(1991). The novel feature of the treatment described here is the manner
in which we exploit the natural embedding of critical points into families

to classify those critical points that are actually constrained minima. The same embedding also provides considerable information about the eigenvalue problem arising from the associated linearized dynamics. In this article our goal is to combine a description of this new technique with an essentially self-contained, straightforward and consistent account of how constrained variational principles arise in the characterization and associated stability analysis of certain special solutions in Hamiltonian systems.

Second, we here confine ourselves to the case of dynamics governed by systems of ordinary differential equations for which phase space is finite dimensional. However we do this for simplicity only. The general structure of the arguments that are exploited carries over to a wide class of Hamiltonian partial differential equations where the phase space is infinite dimensional with the only difference being the (considerably higher) level of technical argument necessary to rigorously justify each step. Indeed much of the particular viewpoint of general Hamiltonian systems that we espouse was crystallized during our analysis of the stability properties of multi-soliton solutions of the Korteweg-de Vries equation (Maddocks & Sachs, 1993). That analysis pivots on the new approach to constrained second variations that is described below, as well as on the general viewpoint of Hamiltonian systems that we advocate.

2. Noncanonical systems. In our terminology a noncanonical Hamiltonian system of ordinary differential equations means a set of equations of the form

$$(2.1) \qquad \qquad \mathbf{z}' = \mathcal{J}(\mathbf{z}) \nabla H(\mathbf{z}),$$

in other words a system of the form (1.1) in which the skew-symmetric matrix $\mathcal{J}$ can depend upon the phase space variable variable $\mathbf{z}(t) \in \Re^n$ (and n need not be even). We further suppose that the Poisson bracket of any two functions $F(\mathbf{z})$ and $G(\mathbf{z})$ induced by $\mathcal{J}$, namely

$$(2.2) \qquad \qquad \{F, G\} \equiv \nabla F \cdot \mathcal{J} \nabla G,$$

satisfies the Jacobi identity

$$(2.3) \qquad \{\{F, G\}, K\} + \{\{G, K\}, F\} + \{\{K, F\}, G\} = 0.$$

Of course when $\mathcal{J}$ is any skew-symmetric constant matrix, i.e. $\mathcal{J}$ is independent of $\mathbf{z}$, the Jacobi identity is automatically satisfied.

When $\mathcal{J}$ satisfies (2.3) and is nonsingular, the classic form of Darboux's Theorem states that there is a local change of variable under which the system (2.1) transforms into a Hamiltonian system with the canonical symplectic matrix (1.2) (which is not to say that is either easy or desirable to actually carry out the change of variable). However in this section we shall primarily be interested in the case in which the matrix $\mathcal{J}$ has a nontrivial nullspace. Then a version of Darboux's Theorem still applies (cf. Olver, 1986, p. 394), and in particular it may be concluded that if the matrix $\mathcal{J}(\mathbf{z})$ is locally of constant rank, then the nullspace varies in such a way that there are functions $C_i(\mathbf{z})$, $i = 1, \ldots, p$ with the property that at each $\mathbf{z}$, $\nabla C_i(\mathbf{z})$ forms a (linearly independent) basis of the nullspace. As each $\nabla C_i(\mathbf{z})$ lies in the nullspace of the matrix $\mathcal{J}$, the quantities $C_i(\mathbf{z})$ are easily seen to be integrals, or conserved quantities, of the dynamics (2.1). Because conservation of the C_i arises from degeneracies of $\mathcal{J}$, they are conserved independent of the particular Hamiltonian, and are consequently called *Casimir* or *distinguished* functions (see, for example, Olver, 1986).

The equilibrium solutions $\mathbf{z}_e$ of (2.1) arise when $\nabla H(\mathbf{z}_e)$ is an element of the nullspace of $\mathcal{J}(\mathbf{z}_e)$. However in contrast to the canonical case, $\nabla H(\mathbf{z}_e)$ need not vanish. Rather $\nabla H(\mathbf{z}_e)$ must lie in the span of the $\nabla C_i(\mathbf{z}_e)$, i.e.

$$(2.4) \qquad \nabla H(\mathbf{z}_e) = \sum_{i=1}^{p} \mu_i \, \nabla C_i(\mathbf{z}_e),$$

for some constants μ_i. Equivalently the equilibria of the dynamics (2.1) are the critical points of the function

$$(2.5) \qquad F(\mathbf{z}, \mu_i) \equiv H(\mathbf{z}) - \sum_{i=1}^{p} \mu_i \, C_i(\mathbf{z}).$$

Clearly F, being a linear combination of first integrals, is itself a first integral of the dynamics, so by the Lagrange-Dirichlet Theorem any critical point $\mathbf{z}_e$ of F that is actually a nondegenerate, local minimum, is necessarily stable. For example, this was essentially the argument followed by Arnol'd (1966) in his analysis of the stability of certain ideal fluid flows. However, even when $\mathbf{z}_e$ fails to be a minimum of (2.5), it is often possible to deduce a stability result when $\mathbf{z}_e$ is a nondegenerate, *constrained* local minimum of the variational problem

$$(2.6) \qquad \text{minimize } H(\mathbf{z}) \quad \text{subject to} \quad C_i(\mathbf{z}) = k_i, \quad i = 1, \ldots, p.$$

Here the *data* k_i are chosen to be $C_i(\mathbf{z}_e)$, the μ_i are to be interpreted as Lagrange multipliers associated with the constraints, and $F(\mathbf{z}, \mu_i)$ is the "Lagrangian" associated with the constrained variational principle (2.6). (Given the present context of Hamiltonian mechanics it is perhaps unfortunate that within the optimization literature the function (2.5) is referred to as the Lagrangian, but we shall retain this usage.) We remark that being a constrained minimum of (2.6) is a strictly weaker requirement than being an unconstrained minimum of (2.5), and our interest is focussed on the sharper constrained characterization.

The question we shall now discuss is the following: What can be concluded about the stability properties of $\mathbf{z}_e$ as an equilibrium solution of (2.1) by dint of it being a nondegenerate, local, constrained minimum of

(2.6)? The answer is that the arguments leading to the Lagrange-Dirichlet Theorem can be refined to imply that $\mathbf{z}_e$ is indeed Lyapunov stable. There are at least two distinct routes to this conclusion. The customary approach recognizes that the Lagrangian (2.5) cannot be used as a Lyapunov function controlling a neighborhood of initial data in the entire phase space, but (assuming that the constraint gradients $\nabla C_i(\mathbf{z}) = 0$ are linearly independent) attention may be restricted to the dynamics on the invariant manifold defined by

$$(2.7) \qquad \mathcal{M}_k = \{\mathbf{z} : C_j(\mathbf{z}) = k_j, j = 1, \ldots, p\,\}.$$

On this manifold, $\mathbf{z}_e$ is, by construction, a nondegenerate, local minimum of the restriction of H to $\mathcal{M}_k$. Then, by the Lagrange-Dirichlet Theorem applied to the Hamiltonian H, it can be seen that $\mathbf{z}_e$ is stable relative to nearby trajectories that begin on $\mathcal{M}_k$ (and that therefore remain on $\mathcal{M}_k$). Note that as yet this argument implies nothing about the ultimate behavior of trajectories that begin near $\mathbf{z}_e$, but which realize different values of the Casimirs. Such trajectories do not lie on the manifold $\mathcal{M}_k$, but rather on another, nearby, invariant manifold corresponding to perturbed values of the data. Some authors refer to this restricted notion of stability as "conditional stability." To strengthen such a result to a statement of full Lyapunov stability, one needs to consider initial data with slightly perturbed values of the constraints. When there are known to be other, nearby, conditionally stable equilibria $\tilde{\mathbf{z}}_e$ satisfying any perturbed values of the constraints, then full Lyapunov stability of the point $\mathbf{z}_e$ is implied by application of the Lagrange-Dirichlet argument to the equilibria with the perturbed values $\tilde{k}_j$ of the constraints arising from the given initial data. This circumstance commonly arises when $\mathbf{z}_e$ is one member of an entire family of constrained minimizers parametrized by the data k_j.

The second, and in some ways more elegant, approach to conclude stability of $\mathbf{z}_e$ is to explicitly construct a Lyapunov function for the full dynamics. The existence of such a functional follows immediately from the theory of the *augmented Lagrangian* (see, for example, Hestenes, 1975), which we summarize in:

Lemma 1. *Suppose that $\mathbf{z}_e$ is a nondegenerate constrained minimum of (2.6) at which the following constrained, second-order conditions hold,*
(2.8)
$$\mathbf{h} \cdot \nabla^2 F(\mathbf{z}_e)\,\mathbf{h} > 0, \qquad \forall\,\mathbf{h} \neq 0 \quad s.t. \ \nabla C_i(\mathbf{z}_e) \cdot \mathbf{h} = 0, \qquad i = 1, \ldots, p.$$

Then there exists a (large, positive) constant M such that $\mathbf{z}_e$ is a nondegenerate, unconstrained minimum of the augmented Lagrangian

$$(2.9) \qquad H(\mathbf{z}) - \sum_{i=1}^{p} \mu_i C_i(\mathbf{z}) + \frac{M}{2} \sum_{i=1}^{p} (C_i(\mathbf{z}) - k_i)^2.$$

Application of the Lagrange-Dirichlet Theorem with the augmented Lagrangian (2.9) playing the role of Lyapunov function immediately implies the desired stability properties of the equilibrium z_e. It should be remarked that the above idea of augmentability yields a sharp characterization of applicability of the approach of Chetayev (see e.g. Rumyantsev, 1966, 1968) and the energy-Casimir method (see Holm et al, 1985), which both provide general prescriptions for proving stability in various classes of dynamical systems. We also remark that in the finite-dimensional context of ordinary differential equations the augmented Lagrangian (2.9) could just be regarded as an un-augmented Lagrangian (2.5), but with new Casimir integrals chosen to be certain quadratic functions of the original ones. However in the examples known to us, a stability analysis based directly on this observation turns out to be cumbersome. Moreover, in the infinite dimensional context the Lagrangian and augmented Lagrangian are distinct objects. For then each Casimir function is an integral of a density (depending on the phase variables) over an appropriate spatial domain, and because the Lagrangian (2.5) is a linear combination of the Casimir functions, it also has an associated density. In contrast the augmented Lagrangian (2.9) is a quadratic function of definite integrals and so cannot be written as an integral of a single density.

With Lemma 1 in hand, attention is focussed on the classification of those critical points z_e at which the constrained second order conditions (2.8) are satisfied. It is important to realize that (2.8) is not immediately related to the number of negative eigenvalues of the second variation, or Hessian, $\nabla^2 F(z_e)$ of the Lagrangian. Rather condition (2.8) is equivalent to

$$(2.10) \qquad \mathbf{h} \cdot P \nabla^2 F(z_e) P \mathbf{h} > 0, \quad \forall \mathbf{h} \neq 0,$$

where the matrix P is the orthogonal projection onto $R(\mathcal{J}(z_e))$, namely the range of $\mathcal{J}(z_e)$. Of course some simple conclusions can be drawn immediately. For example if $\nabla^2 F(z_e) > 0$ then condition (2.10) certainly holds. Alternatively, if the number of negative eigenvalues of $\nabla^2 F(z_e)$ exceeds the number of constraints, then $P \nabla^2 F(z_e) P \not> 0$. However, in general, and with the exception of special problems where $R(\mathcal{J}(z_e))$ happens to be simply related to the eigenspaces of $\nabla^2 F(z_e)$, a direct analysis of (2.8) or (2.10) is rather complicated. An indirect approach that finesses these difficulties, and that has proven successful in several different Hamiltonian systems, is described below. Before describing that technique it is useful to recall that whenever

$$(2.11) \qquad N(\nabla^2 F(z_e)) = \{0\},$$

i.e. the nullspace of $\nabla^2 F(z_e)$ is trivial, then the number of negative eigenvalues of $\nabla^2 F(z_e)$ provides a Morse index of z_e when regarded as an (unconstrained) critical point of the Lagrangian (2.5). Similarly when $P \nabla^2 F(z_e) P$

is nondegenerate in the sense

$$(2.12) \qquad N(P\nabla^2 F(\mathbf{z}_e)P) = N(\mathcal{J}(\mathbf{z}_e)),$$

i.e. the nullspace of $P\nabla^2 F(\mathbf{z}_e)P$ exactly coincides with the nullspace of $\mathcal{J}(\mathbf{z}_e)$, then the number of negative eigenvalues of $P\nabla^2 F(\mathbf{z}_e)P$ provides a Morse index of $\mathbf{z}_e$ regarded as a (constrained) critical point of the Hamiltonian $H(\mathbf{z})$ on the constraint manifold $\mathcal{M}_k$ defined in (2.7). The second order conditions (2.8) can then be paraphrased as the requirement that $\mathbf{z}_e$ be a nondegenerate critical point of constrained Morse index zero.

At this juncture we pause to consider the analogue of the eigenvalue problem (1.4) that is obtained after separation of variables in the linearization of (2.1). Because of the defining property of Casimir integrals, namely $\mathcal{J}(\mathbf{z})\nabla C_i(\mathbf{z}) = 0$, we may linearize at $\mathbf{z}_e$ to obtain the identity

$$(2.13) \qquad \nabla\mathcal{J}(\mathbf{z}_e)\nabla C_i(\mathbf{z}_e) = -\mathcal{J}(\mathbf{z}_e)\nabla^2 C_i(\mathbf{z}_e).$$

Consequently the nonsymmetric eigenvalue problem (analogous to (1.4)) obtained from linearization of (2.1) at $\mathbf{z}_e$ can be written in the form

$$(2.14) \qquad \rho\mathbf{g} = \mathcal{J}(\mathbf{z}_e)\nabla^2 F(\mathbf{z}_e)\mathbf{g}.$$

It turns out that the two nondegeneracy conditions (2.11) and (2.12) on $\mathbf{z}_e$ regarded as an unconstrained and constrained critical point, imply that (2.14) has zero as an eigenvalue of both algebraic and geometric multiplicity p. In other words there are p eigenvectors in the nullspace and no generalized nullvectors. The vectors $\nabla C_i(\mathbf{z}_e)$ form a basis for the left nullspace, and, by the assumed nonsingularity of $\nabla^2 F(\mathbf{z}_e)$, a basis for the right nullspace is provided by vectors η_i that are (uniquely) defined as solutions to the inhomogeneous linear equations

$$(2.15) \qquad \nabla^2 F(\mathbf{z}_e)\eta_i = \nabla C_i(\mathbf{z}_e), \quad i = 1, \ldots, p.$$

It is now possible to remark that, in direct analogy with the case of equilibria of canonical systems that was discussed in the Introduction, there is a partial instability result available, namely that nondegenerate constrained critical points of odd index are necessarily unstable solutions under the dynamics (2.1). This conclusion follows from a straightforward argument that demonstrates that at such $\mathbf{z}_e$ there is necessarily a positive real eigenvalue of (2.14) (cf. Maddocks, 1991, or the end of Section 3 below).

We now turn to a description of an effective and general approach to allow the computation of the constrained index. The basic building block in this technique is an index theory that traces back to Hestenes (1951) and which was later developed by Maddocks (1985). This theory provides a relation between three quantities, namely the index of an unconstrained quadratic form (here the second variation $\nabla^2 F(\mathbf{z}_e)$), and two constrained indices that are obtained when the quadratic form is restricted to two complementary subspaces that are orthogonal in the (indefinite) inner product induced by the quadratic form. Consequently any one index is determined by the other two. In the particular context of Hamiltonian systems that is of direct concern here, the unconstrained index coincides with the Morse index of z_e regarded as an unconstrained critical point of the Lagrangian (2.5), one subspace is the range of $J(z_e)$ (which is also the tangent space at z_e to the manifold (2.7)) with the associated constrained index coinciding with the (constrained) Morse index of z_e regarded as a critical point of (2.6), and the second subspace is the right nullspace of $J(z_e)\nabla^2 F(z_e)$ (i.e. the subspace spanned by the vectors η_i defined in (2.15)) with the corresponding constrained index being the number of negative eigenvalues of the $p \times p$ symmetric matrix

$$(2.16) \quad \eta_i \cdot \nabla^2 F(z_e)\eta_j = \eta_i \cdot \nabla C_j(z_e) = \nabla C_i(z_e) \cdot \left\{\nabla^2 F(z_e)\right\}^{-1} \nabla C_j(z_e).$$

So far we seem no closer to verification of conditions (2.8). However the above arguments demonstrate that the constrained Morse index at a particular critical point $\hat{z}_e$ (equivalently the number of negative eigenvalues of $P\nabla^2 F(\hat{z}_e)P$) can be determined in terms of two other indices. The final observation is that these two other indices can often be explicitly calculated if, as is typically the case, the given critical point $\hat{z}_e$ is embedded in an entire *family* of critical points. The embedding arises because as the multipliers μ_i in (2.4) (and consequently the data k_i in (2.6)) are varied, a family of critical points is generated. Thus even if we are primarily interested in a particular extremal $\hat{z}_e$ that is a critical point for given $\hat{\mu}_i$ and data $\hat{k}_i$, we shall embed $\hat{z}_e$ in a family of extremals $z_e(\mu)$ to constrained variational principles with data $k_i(\mu) = C_i(z_e(\mu))$. When $\nabla^2 F(\hat{z}_e)$ is nonsingular, i.e. $\hat{z}_e$ is a nondegenerate critical point of (2.5), the implicit function theorem applied to (2.4) guarantees the existence of such an embedding. Degenerate critical points can usually be embedded in at least one solution family, but we shall not address such issues here. We also remark that given a smooth embedding, conditions (2.4) can be differentiated with respect to μ_i, which allows us to conclude that the vectors η_i defined in (2.15) have the explicit representation

$$(2.17) \qquad\qquad \eta_i(\mu) = \frac{\partial z_e(\mu)}{\partial \mu_i}.$$

Given such an embedding of a particular critical point $\hat{z}_e$ in a family of critical points $z_e(\mu)$, the Hamiltonian, Lagrangian, and Casimir functions can be evaluated along the family of critical points to yield (implicitly defined, possibly multi-valued) functions $H(z(\mu_i))$, $F(z(\mu_i), \mu_i)$ and $C_j(z(\mu_i))$ each of which maps $\Re^p \to \Re$. There are then two *distinguished bifurcation diagrams* consisting of the surfaces in $\Re^{p+1}$ that are obtained by plotting the Lagrangian F as a function of the multipliers μ_i, and by plotting the Hamiltonian H as a (parametrically defined) function of the integrals C_j. An explicit example of the two plots for a problem with $p = 2$ is illustrated in Figure 1. These surfaces have (at points of smoothness) $p \times p$ Hessian matrices

(2.18)
$$\left\{ \frac{\partial^2 F}{\partial \mu_i \partial \mu_j} \right\},$$

and

(2.19)
$$\left\{ \frac{\partial^2 H}{\partial C_i \partial C_j} \right\}.$$

3. Figure caption. The figure depicts the two distinguished bifurcation diagrams arising in the variational analysis of the steady motions of a heavy rigid body that is described in Section 2. The two multi-sheeted surfaces are each a projection into R^3 of the set of critical points of the appropriate constrained variational principle. In the top picture the Hamiltonian (H) is evaluated along the family of critical points, and plotted against the values of the two constraints (K1 and K2). In the bottom picture the (variational) Lagrangian (F), namely the sum of the Hamiltonian and the constraints weighted with two Lagrange multipliers, is plotted against the multipliers themselves (MU1 and MU2). In each projection there are four distinct sheets. In the top view it should be recognized that there are two smooth sheets (one entirely green and one entirely blue), and two bi-color, folded sheets that are cusped. The color coding of the surface indicates the constrained index of each critical point, which is also related to the stability properties of the steady motions when regarded as solutions of the associated dynamics. Points in the dark green segments correspond to constrained minima, have constrained index 0, and are dynamically stable. Points in the pale yellow segments, have constrained index 1 and are dynamically unstable, and points in the blue segments have constrained index 2, and their dynamic stability properties are undetermined from the purely variational arguments exploited here. Notice that the exchange of stability, i.e. a change in constrained index, as visualized with a change in color, arises at solutions on a cusp line in the top projection, but which lie on a line where a principal curvature vanishes in the bottom projection. The graphics were generated as still frames of the output of an interactive computation and visualization package called MC^2. Some of the features of the bifurcation diagrams are comparatively hard to recognize from still views, but are easily visualized when the surfaces can be seen while rotating on the computer screen or on video.

In fact it can be shown that (2.19) is invertible at nondegenerate critical points, and that

$$(3.1) \qquad \left\{ \frac{\partial^2 F}{\partial \mu_i \partial \mu_j} \right\} = - \left\{ \frac{\partial C_i}{\partial \mu_j} \right\} = - \left\{ \frac{\partial^2 H}{\partial C_i \partial C_j} \right\}^{-1},$$

so that there is a form of Legendre duality between the two bifurcation diagrams. Consequently the inertias (i.e. the numbers of positive and negative eigenvalues) of the matrices (2.18) and (2.19) are related. Moreover the signs of the eigenvalues of the Hessian matrices (2.18) and (2.19) determine the signs of the principal curvatures of the two distinguished bifurcation surfaces. It is also a simple calculation to verify that the (i, j)th entry of the matrix (2.18) is (2.16).

The test for the validity of the second order conditions (2.8) that can be explicitly exploited in several applications is provided by the following:

Lemma 2. *Suppose that a nondegenerate extremal $(\hat{z}_e, \hat{\mu})$ is (locally and smoothly) embedded in a family of extremals $z_e(\mu)$. Then condition (2.8) is satisfied at $\hat{z}_e(\hat{\mu}_i)$ if and only if the number of positive principal curvatures of the solution surface $F(\mu)$ at $\hat{\mu}$ equals the number of negative eigenvalues of $\nabla^2 F(\hat{z}_e)$.*

In several applications the last two quantities can be computed directly. The theory underpinning Lemma 2 was developed by Maddocks (1987,

1991, 1993). In the case $p = 1$ of a single constraint many authors have obtained similar results (a selection of such references is made in Maddocks, 1987, to which can be added Lewis, 1989, and Zombro & Holmes, 1993). For the case of multiple constraints, i.e. $p > 1$, the only related work known to us is a version of Lemma 2 found independently by Grillakis, Shatah, and Strauss (1990) in the specific context of solitary wave problems with symmetry, and a rather special case that was described by Oh (1987).

Lemma 2 (or closely related versions thereof) has been applied in Maddocks (1991) and Maddocks & Sachs (1993) to obtain stability results in certain particular Hamiltonian systems. We remark that at first sight it is by no means clear that the test for the second order conditions (2.8) provided by Lemma 2 is any more tractable than a direct attack. This issue can only be resolved in the context of concrete applications, and the answer is that, at least in some examples, and in particular for the KdV multi-soliton problem treated in Maddocks & Sachs (*op. cit.*), the indirect approach provided by Lemma 2 has proven to be tractable, while a direct assault appears to be hopeless. We also remark that Lemma 2 is of a form that is particularly suitable for combination with numerical continuation techniques for generating solution families of equations of the form (2.4), namely the first order conditions for stationarity of a parameter dependent Lagrangian.

Lemma 2 is also a natural result from the view point of bifurcation theory, because it represents the appropriate manifestation of the "principle of exchange of stability" (cf. Poincaré, 1885, Crandall and Rabinowitz, 1973, Maddocks, 1987, Sattinger, 1972, and Weinberger, 1978) in the context of *constrained* variational principles. Poincaré originally observed that the type (or Morse index) of a critical point in a family of *unconstrained* variational principles can only change at singularities (i.e. fold or bifurcation points) in what would now be called a bifurcation diagram. A consequence of Lemma 2 is that a change in the type (or constrained Morse index) of a *constrained* critical point can be associated with either a fold or a switch in sign of a principal curvature of one of the distinguished bifurcation surfaces described above. It can also be shown that a switch in sign of a curvature in the plot of the Lagrangian is (generically) associated with a fold in the plot of the Hamiltonian, while a switch in the sign of a curvature in the plot of the Hamiltonian is (generically) associated with a fold in the plot of the Lagrangian.

Points of zero curvature in the Lagrangian bifurcation diagram arise when the matrix (2.18) has zero as an eigenvalue. It is apparent from Lemma 2, that such critical points are transitional from the point of view of the variational analysis and the determination of the constrained index. For example, by direct construction it can be shown that the constrained nondegeneracy condition (2.12) must fail at critical points with a vanishing curvature. Consequently it should also be the case that the corresponding special solutions of (2.1) are transitional from the perspective of the dynamics. It has already been remarked that at nondegenerate critical points, the eigenvalue problem (2.14) arising from the linearized dynamics has the vectors η_i defined in (2.15) as a basis for the right nullspace with no associated generalized nullspace. We shall now show that at critical points with a zero curvature there is (at least one) additional chain of two generalized nullvectors, so that the algebraic multiplicity of zero as an eigenvalue of (2.14) increases by (an even number that is at least) two. Suppose therefore that (2.18) (equivalently (2.16)) has a nullvector $\alpha \in \Re^p$, and consider the following systems of equations

$$(3.2) \qquad \mathcal{J}\nabla^2 F(\mathbf{z}_e)\xi = \sum_i \alpha_i \eta_i,$$

and

$$(3.3) \qquad \mathcal{J}\nabla^2 F(\mathbf{z}_e)\zeta = \xi + \sum_i \gamma_i \eta_i,$$

which characterize additional generalized eigenvectors. The ∇C_i are a basis for the left nullspace of $\mathcal{J}\nabla^2 F(\mathbf{z}_e)$, so the solvability conditions for the variable ξ in (3.2) are equivalent to the requirement that α is a nullvector of (2.18). Accordingly a solution ξ exists, but is only uniquely defined up to an element of the right nullspace of $\mathcal{J}\nabla^2 F(\mathbf{z}_e)$, i.e. up to a linear combination of the η_i. This linear combination must then be picked so that the solvability conditions for the variable ζ in equation (3.3) are satisfied. These solvability conditions in turn require that the coefficients γ_i in (3.3) satisfy the $(p \times p)$ symmetric system

$$(3.4) \qquad \sum_i \gamma_i \eta_i \cdot \nabla C_j = \nabla C_j \cdot \xi.$$

Now the matrix appearing on the left hand side of (3.4) is nothing but (2.18), which, by assumption, is singular with nullvector α. Consequently the solvability condition for the system (3.4) is

$$(3.5) \qquad \sum_i \alpha_i \nabla C_i \cdot \xi = 0.$$

Finally we note that (3.5) is automatically satisfied for any solution ξ of (3.2), as can be seen by taking the scalar product of (3.2) with the vector $\nabla^2 F(\mathbf{z}_e)\xi$, and exploiting the symmetry of $\nabla^2 F(\mathbf{z}_e)$ and the definition (2.15) of the η_i.

The above arguments demonstrate that a purely variational analysis can provide sharp control of a stability exchange (in the sense of the linearized dynamics) that involves the passage through zero of an eigenvalue of (2.14). The reason that a variational analysis does not give sharp control of stability exchange in the case of critical points with a (positive) even, constrained index, is that a transition can arise in which two non-vanishing pair of imaginary eigenvalues of the nonsymmetric eigenvalue problem (2.14) coalesce and move directly from the imaginary axis to form a complex quadruplet, with no associated change in any variational index. The motion of a heavy, rigid body tumbling about a fixed point is one classic, physical system that exhibits such behavior in certain parameter regimes. This example is further described below, but only from the variational viewpoint. One place to find a more complete discussion of the associated linearized dynamics of the example is Maddocks (1991).

We also remark that the eigenvalues of the Hessian (2.18) (or equivalently (2.19)) can be used to characterize the size of the constant M appearing in Lemma 1. In particular it is shown in Maddocks & Sachs (1993) that any M satisfying $M > \max(\frac{1}{\lambda_{\max}}, 0)$ suffices. Here $\lambda_{\max}$ denotes the greatest eigenvalue of (2.18). However in the general approach to a stability analysis that we advocate, it suffices to know that the critical point is an unconstrained minimum of the augmented Lagrangian (2.9) for some M and knowledge of the explicit value is not essential to the argument.

As a final note, it is perhaps of interest that the bifurcation diagram obtained by evaluation of the Hamiltonian and other integrals along families of critical points satisfying (2.4) has been exploited in another context by Fomenko (1991). He uses such bifurcation diagrams to classify the topological structure of phase space for low-dimensional completely integrable systems. Other examples of this approach can be found in Arnol'd *et al* (1988, e.g. p. 105) which considers the Hamiltonian system describing the motion of a heavy rigid body moving about a fixed point. We shall now illustrate the use of Lemma 2 within the same example. That is we describe the variational characterization of the steady motions of a heavy asymmetric rigid body tumbling about a fixed point, with a particular emphasis on the associated stability properties. The heavy rigid body system is one of the classic problems of mechanics, but in other than the case of a symmetric, or Lagrange, top, the characterization of steady spins and associated stability properties is quite intricate. The pertinent classic literature is reviewed in Leimanis (1965). The treatment described here is a much abbreviated version of that given in Maddocks (1991), which can be consulted for more details. Subsequently another contemporary treatment of the same problem was made by Lewis *et al* (1992).

After nondimensionalization, the equations of motion are

$$(3.6) \qquad m' = m \times \mathbf{I}^{-1} m + n \times \chi, \qquad n' = n \times \mathbf{I}^{-1} m.$$

Equations (3.6) are respectively the equations of Euler and of Poisson. They comprise a system of six first-order differential equations for the six unknowns $m(t) \in \Re^3$ and $n(t) \in \Re^3$. The variables $m(t)$ and $n(t)$ are the components with respect to the principal inertia axes of, respectively, the angular momentum and the unit vertical. There are also five independent parameters implicit in (3.6): three are the nondimensional coordinates $\chi \geq$

0 of the center of mass, and the remaining two characterize the positive, diagonal inertia tensor $\mathbf{I}$ with trace one. It is now widely realized that equations (3.6) are of the noncanonical Hamiltonian form (2.1) with $\mathbf{z} \equiv [m, n] \in \Re^6$,

$$(3.7) \qquad H(\mathbf{z}) = H(m, n) = \tfrac{1}{2} m \cdot \mathbf{I}^{-1} m + n \cdot \chi ,$$

and

$$(3.8) \qquad J(\mathbf{z}) = J(m, n) = \begin{pmatrix} m\times & n\times \\ n\times & 0 \end{pmatrix} ,$$

where, in turn, $m\times$ and $n\times$ denote the (3×3) skew-symmetric matrices associated with the vector product in $\Re^3$, namely
(3.9)

$$m\times = \begin{pmatrix} 0 & -m_3 & m_2 \\ m_3 & 0 & -m_1 \\ -m_2 & m_1 & 0 \end{pmatrix} \quad \text{and} \quad n\times = \begin{pmatrix} 0 & -n_3 & n_2 \\ n_3 & 0 & -n_1 \\ -n_2 & n_1 & 0 \end{pmatrix} .$$

Whenever $n^2 \neq 0$ the nullspace of J is two dimensional, and is spanned by the vectors

$$(3.10) \qquad \nabla C_1(\mathbf{z}) \equiv [n, m] \quad \text{and} \quad \nabla C_2(\mathbf{z}) \equiv [0, n] .$$

Naturally, there are two associated Casimir integrals of the dynamics (3.6), namely

$$(3.11) \qquad C_1 = m \cdot n \quad \text{and} \quad C_2 = \tfrac{1}{2} n^2 .$$

In the absence of further symmetries of the Hamiltonian (3.7), that is in the absence of special relations between the parameters $\mathbf{I}$ and χ, there are no further conserved quantities.

Accordingly we are in the case $p = 2$ of the general framework described in this section. The first order conditions (2.4) take the explicit form

$$(3.12) \quad \begin{pmatrix} 1/I_1 & -\mu_1 & 0 & 0 & 0 & 0 \\ -\mu_1 & -\mu_2 & 0 & 0 & 0 & 0 \\ 0 & 0 & 1/I_2 & -\mu_1 & 0 & 0 \\ 0 & 0 & -\mu_1 & -\mu_2 & 0 & 0 \\ 0 & 0 & 0 & 0 & 1/I_3 & -\mu_1 \\ 0 & 0 & 0 & 0 & -\mu_1 & -\mu_2 \end{pmatrix} \begin{pmatrix} m_1 \\ n_1 \\ m_2 \\ n_2 \\ m_3 \\ n_3 \end{pmatrix} = \begin{pmatrix} 0 \\ -\chi_1 \\ 0 \\ -\chi_2 \\ 0 \\ -\chi_3 \end{pmatrix} .$$

Because in this example the Hamiltonian and Casimir functions are quadratic functions of the dependent variables, conditions (3.12) are linear and so it can be calculated that there is a unique solution for each pair (μ_1, μ_2) satisfying

$$(3.13) \qquad \mu_2 + I_i \mu_1^2 \neq 0, \qquad i = 1, 2, 3.$$

The solutions are

$$(3.14) \qquad n_i = \frac{\chi_i}{(\mu_2 + I_i\,\mu_1^2)}, \qquad m_i = \mu_1\,I_i\,n_i, \qquad i = 1, 2, 3.$$

There are additional solutions when inequalities (3.13) are violated provided that the appropriate parameter χ_i vanishes, but we will not consider
those cases here. Equations (3.12) determine the possible body coordinates n_i of the vertical vector $\mathbf{n}$, and further state that the angular velocity
$\mathbf{I}^{-1}\mathbf{m} = \omega$ is parallel to $\mathbf{n}$. Thus the relative equilibria are all steady, or permanent, rotations about the vertical. The special directions (in the body
frame) that can be the axis of a steady, or *permanent*, rotation are called
the *axes of Staude*. Each axis has (up to time reversal) a unique angular
velocity associated with it. This example is typical in that the noncanonical structure (2.1) is obtained from a canonical Hamiltonian system by
changing to a moving coordinate system and reducing by a symmetry action (in this case rotation about the vertical $\mathbf{n}$). Thus the equilibria of the
set of ordinary differential equations (3.6) are really steady motions in an
inertial frame and are accordingly often termed relative equilibria.

We remark that the familiar case of a symmetric or Lagrange top is a
degenerate limit of the problem considered here. In the Lagrange top the
set of relative equilibria are spins about the symmetry axis of the body.
For an asymmetric body this single spin axis with arbitrary rotation rate
is unfolded to become a one parameter family of allowable steady-spin
directions each with a specified rotation rate.

Our main focus of attention is on the stability analysis. In this example
we have the explicit representations

$$(3.15) \qquad \nabla^2 F(\mathbf{z}_e) = \begin{pmatrix} 1/I_1 & -\hat{\mu}_1 & 0 & 0 & 0 & 0 \\ -\hat{\mu}_1 & -\hat{\mu}_2 & 0 & 0 & 0 & 0 \\ 0 & 0 & 1/I_2 & -\hat{\mu}_1 & 0 & 0 \\ 0 & 0 & -\hat{\mu}_1 & -\hat{\mu}_2 & 0 & 0 \\ 0 & 0 & 0 & 0 & 1/I_3 & -\hat{\mu}_1 \\ 0 & 0 & 0 & 0 & -\hat{\mu}_1 & -\hat{\mu}_2 \end{pmatrix}$$

and

$$(3.16) \qquad \nabla C_1(\hat{\mathbf{z}}_e) = \begin{pmatrix} \hat{n}_1 \\ \hat{\mu}_1\,I_1\hat{n}_1 \\ \hat{n}_2 \\ \hat{\mu}_1\,I_2\hat{n}_2 \\ \hat{n}_3 \\ \hat{\mu}_1\,I_3\hat{n}_3 \end{pmatrix}, \qquad \nabla C_2(\hat{\mathbf{z}}_e) = \begin{pmatrix} 0 \\ \hat{n}_1 \\ 0 \\ \hat{n}_2 \\ 0 \\ \hat{n}_3 \end{pmatrix},$$

where $\hat{n}$ is expressed in terms of $\hat{\mu}_1$ and $\hat{\mu}_2$ in (3.14). The expression (3.15)
for $\nabla^2 F(\mathbf{z}_e)$ is particularly simple because the Hamiltonian and Casimir
functions are quadratic functions. Consequently $\nabla^2 F(\mathbf{z}_e)$ is independent of

$\mathbf{z}_e$ and is just the matrix appearing on the left-hand side of equations (3.12). The block diagonal form of $\nabla^2 F(\mathbf{z}_e)$ means that it is trivial to determine the signs of its eigenvalues. In particular, because of (3.13), $\nabla^2 F(\mathbf{z}_e)(\mu)$ is always nonsingular, and there are precisely zero, one, two or three negative eigenvalues depending upon the location of the point $(\hat{\mu}_1, \hat{\mu}_2)$ compared to the three parabolas $\mu_2 = -I_i \mu_i^2$.

While the unconstrained second variation has a block diagonal structure, the constraints (3.16) are not partitioned in a compatible fashion so there is no simple structure in the projected Hessian $P\nabla^2 F(\mathbf{z}_e)P$. Accordingly we aim to invoke Lemma 2. When expressed as functions of the multipliers, the Lagrangian, Casimir integrals, and Hamiltonian become

$$(3.17) \qquad F(\mu_1, \mu_2) = \tfrac{1}{2} \sum_{i=1}^{3} \frac{\chi_i^2}{(I_i \mu_1^2 + \mu_2)},$$

(3.18)

$$C_1(\mu_1, \mu_2) = \sum_{i=1}^{3} \frac{\mu_1 I_i \chi_i^2}{(\mu_2 + I_i \mu_1^2)^2}, \qquad C_2(\mu_1, \mu_2) = \tfrac{1}{2} \sum_{i=1}^{3} \frac{\chi_i^2}{(\mu_2 + I_i \mu_1^2)^2},$$

and

$$(3.19) \qquad H(\mu_1, \mu_2) = \tfrac{1}{2} \sum_{i=1}^{3} \frac{(3\mu_1^2 I_i + 2\mu_2)\chi_i^2}{2(I_i \mu_1^2 + \mu_2)^2}.$$

The surface (3.17), for a particular choice of χ and $\mathbf{I}$, is plotted in the bottom of Figure 1, which, in accord with common practice, will be referred to as a bifurcation diagram (although no bifurcations actually arise in this example). Similarly the Hamiltonian plotted against the two Casimir integrals is shown in the top part of Figure 1. These two piecewise smooth, multi-component surfaces are each a projection of the manifold of all critical points. The actual pictures shown here were generated by an interactive numerical continuation and 3D graphics code called MC^2 (an acronym for Multiplier and Constraint Continuation) that has been developed at the University of Maryland by C.K. Mesztenyi and the first author. MC^2 is designed to numerically solve equations of the general form (2.4) and to simultaneously display the two distinguished solution surfaces with associated information about the constrained index as determined by application of Lemma 2. The use of MC^2 in this example is in some ways unnecessary because the linearity of (3.12) allows the explicit parametrization of the solution detailed in (3.14). Nevertheless it is useful to exploit MC^2 just to obtain the graphical output displayed in Figure 1.

The issue at hand is to determine the constrained index of each critical point. We aim to apply Lemma 2. It is simplest to consider the representation of the solution set provided in the plot of the Lagrangian F versus the two multipliers μ_1 and μ_2. According to our previous deliberations the four smooth sheets appearing in the bottom part of Figure 1 correspond (from right foreground to left background) to patches of four disconnected components of the solution manifold at which $\nabla^2 F(\mathbf{z}_e)$ has respectively zero, one, two, and three negative eigenvalues. Accordingly in order to apply Lemma 2 it only remains to determine the effect of the constraints as encoded in the shape of the solution surface. The signs of the principal curvatures are determined by the signs of the eigenvalues of the (2×2) Hessian of the solution surface $F(\mu_1, \mu_2)$, which from (3.1) and (3.18) can be calculated to be

$$(3.20) \qquad F_{\mu\mu} = \begin{pmatrix} -\sum_{i=1}^{3} \frac{(3I_i \mu_1^2 - \mu_2)\mu_2 \chi_i^2}{(I_i \mu_1^2 + \mu_2)^3} & \sum_{i=1}^{3} \frac{\mu_1 \chi_i^2 (I_i \mu_1^2 - \mu_2)}{(I_i \mu_1^2 + \mu_2)^3} \\ \sum_{i=1}^{3} \frac{\mu_1 \chi_i^2 (I_i \mu_1^2 - \mu_2)}{(I_i \mu_1^2 + \mu_2)^3} & \sum_{i=1}^{3} \frac{\mu_1^2 \chi_i^2}{(I_i \mu_1^2 + \mu_2)^3} \end{pmatrix}$$

For any given (μ_1, μ_2) the signs of the eigenvalues of (3.20) can straightforwardly (if tediously) be determined. Moreover a little work demonstrates that (3.20) has, respectively, 0, 1 or 0, and 1 positive eigenvalue on the rightmost, middle two and leftmost components. Consequently on the rightmost segment both the unconstrained and constrained index is 0, and accordingly the associated solutions $\mathbf{z}_e$ are dynamically stable, which is encoded in Figure 1 by green shading. On the leftmost segment the unconstrained index is three , $F_{\mu\mu}$ has one positive eigenvalue and so the constrained index is $3 - 1 = 2$, which is depicted by blue shading. As the constrained index is even, the dynamic stability properties of the associated solutions is undetermined by this analysis. On the rightmost portion of each of the middle two sheets $F_{\mu\mu}$ has one positive eigenvalue, and there is exactly one switch to a region of no positive eigenvalue on the leftmost portion of each patch. The actual changeover point is hard to determine analytically and was calculated numerically in Figure 1 (the parameter mesh size is visible upon close inspection of the regions of color transition). On the second sheet from the right the unconstrained index is everywhere 1, and the constrained index is either $1 - 1 = 0$ (the green portion), or $1 - 0 = 1$ (the pale yellow segment). Similarly on the remaining segment the unconstrained index is everywhere 2, while the constrained index is either 1 or 0. Solutions with constrained index 1 are necessarily unstable solutions of the dynamics.

Notice that the color switch on the two bi-color segments occurs where the sign of a principle curvature is switching. In the top portion of Figure 1 the same color coding of constrained index is retained in the plot of the Hamiltonian against the values of the Casimir functions. The two smooth single color segments are mapped to smooth single color segments. However the two bi-color segments in the Lagrangian projection are mapped to cusped segments in the Hamiltonian projection. It can be observed in this example, and can be proven in general, that the change in constrained index, as coded in the color change, that arises at points of zero curvature in the Lagrangian projection occurs precisely at the cusped fold points in the Hamiltonian projection.

4. Canonical systems. We now return to the canonical case of autonomous Hamiltonian systems of the form (1.1) with the matrix $\mathcal{J}$ being the standard symplectic matrix (1.2). Because we are restricting attention to autonomous systems, the Hamiltonian is automatically a first integral. We shall assume that there are an additional r integrals $I_i(\mathbf{z})$, $i = 1, \ldots, r$, which, as a matter of convenience, we distinguish from the Hamiltonian. Each integral $I_i(\mathbf{z})$ is associated with a symmetry of the Hamiltonian $H(\mathbf{z})$. In contrast to the case of Casimir integrals, $\mathcal{J}\nabla I_i \neq 0$, and in fact the $\mathcal{J}\nabla I_i$ are the infinitesimal generators of the associated symmetry of the Hamiltonian. By definition the integrals I_i all commute (or are in involution) with the Hamiltonian, and we will consider the case in which they also all commute with each other, i.e.

$$(4.1) \qquad \{I_i, I_j\} = 0, \quad \forall \, i, j = 1, \ldots, r,$$

where as before $\{I_i, I_j\}$ denotes the Poisson bracket induced by $\mathcal{J}$. In general the number of degrees of freedom will exceed $r + 1$ so we are not restricting attention to the case of completely integrable systems. We re-iterate that our focus is on topics directly related to the characterization of constrained critical points and the stability analysis of special solutions, and, in particular, on issues related to continuation methods. For a wider viewpoint of Hamiltonian systems in the presence of symmetry we recommend Arnol'd *et al*, (1988).

The analogue of equations (2.4) is

$$(4.2) \qquad \nabla H(\mathbf{z}_s) = \sum_{i=1}^{r} \mu_i \, \nabla I_i(\mathbf{z}_s),$$

which identifies points $\mathbf{z}_s$ in phase space where the gradients of the integrals become dependent. Of course (4.2) are the first-order conditions associated with the variational principle

$$(4.3) \qquad \text{minimize } H(\mathbf{z}) \quad \text{subject to} \quad I_i(\mathbf{z}) = k_i, \quad i = 1, \ldots, r,$$

with associated Lagrangian

$$(4.4) \qquad F(\mathbf{z}, \mu_i) \equiv H(\mathbf{z}) - \sum_{i=1}^{r} \mu_i \, I_i(\mathbf{z}).$$

The case of critical points of the Hamiltonian has already been discussed in the Introduction, so we will here consider only solutions of (4.2) for which $\sum_{i=1}^{r} \mu_i \, \nabla I_i(\mathbf{z}_s) \neq 0$. Such points are not equilibria of the dynamics, but the orbit $\mathbf{z}(t)$ of (1.1) with initial data $\mathbf{z}(0) = \mathbf{z}_s$ is special because it is trapped on the (critical) level set of (4.4). In particular such critical points cannot be isolated. In point of fact, with the commutativity assumption (4.1), the symmetries $\mathcal{J}\nabla I_i$ of the Hamiltonian associated with the integrals $I_i(\mathbf{z})$ generate (for fixed μ_i) an r dimensional family of critical points $\mathbf{z}_s(\delta_i)$, $i = 1, \ldots, r$, where the δ_i will be referred to as phases. Accordingly $|\nabla^2 F(\mathbf{z}_s)| = 0$, and the analogue of the nondegeneracy condition (2.11) assumed in Section 2 must fail. Explicitly, we find by linearizing (4.1) that

$$(4.5) \qquad \{\nabla I_i \mathbf{h}, I_j\} + \{I_i, \nabla I_j \mathbf{h}\} = 0, \qquad \forall i, j, \text{ and } \mathbf{h}.$$

Identity (4.5) is also valid when one of the integrals is replaced by the Hamiltonian H, so by taking appropriate linear combinations, and exploiting (4.2), it may be concluded that $\mathcal{J}\nabla I_i \in N(\nabla^2 F(\mathbf{z}_s))$. This calculation is just constructing that part of the nullspace of the second variation that is generated by the symmetries associated with the integrals. Consequently it can be seen that there is still an appropriate notion of nondegeneracy, namely that

$$(4.6) \qquad N(\nabla^2 F(\mathbf{z}_s)) = \text{span} \{\mathcal{J}\nabla I_i\}, \quad i = 1, \ldots, r,$$

i.e. the $\mathcal{J}\nabla I_i$ form a basis for the *entire* nullspace of $\nabla^2 F(\mathbf{z}_s)$. In the context of variational principles with symmetry, condition (4.6) is often described as $\nabla^2 F(\mathbf{z}_s)$ being transversely, or Bott, nondegenerate. Equivalently $\mathbf{z}_s$ is nondegenerate after the symmetry action is factored out. For such extremals the number of negative eigenvalues of the Hessian $\nabla^2 F(\mathbf{z}_s)$ yields an index of $\mathbf{z}_s$ regarded as an unconstrained critical point of the Lagrangian (4.4). The analogue of the constrained nondegeneracy condition (2.12) is then

$$(4.7) \qquad N(P\nabla^2 F(\mathbf{z}_s)P) = \text{span} \{\mathcal{J}\nabla I_i, \nabla I_i\}, \quad i = 1, \ldots, r,$$

i.e. the entire nullspace of the projection $P\nabla^2 F(\mathbf{z}_s)P$ of $\nabla^2 F(\mathbf{z}_s)$ onto the tangent plane to the constraints, is generated by the projection and the nullspace (4.6) of $\nabla^2 F(\mathbf{z}_s)$. With the constrained nondegeneracy condition (4.7) the number of negative eigenvalues of $P\nabla^2 F(\mathbf{z}_s)P$ provides a constrained index of $\mathbf{z}_s$ regarded as a critical point of (4.3).

Notice that the involution conditions (4.1) taken with (4.6) imply that nondegenerate extremals can be embedded in families of extremals in *two* ways. First, there is an r parameter family of critical points of (4.4) described by the phases δ_i along which the multipliers μ_i and integrals $I_i(\mathbf{z}_s)$ are constant. Second, there is another r parameter family of critical points as the multipliers μ_i (and consequently the integrals I_i) are varied. Accordingly the critical points $\mathbf{z}_s$ arise as a $2r$ parameter family $\mathbf{z}_s(\mu_i, \delta_i)$, $i = 1, \ldots, r$. Once a particular set of integrals $I_i(\mathbf{z})$ have been selected, there is no remaining freedom in the choice of the multipliers μ_i. In contrast there is considerable freedom in the parametrization of the set of critical points that arise for fixed μ_i. If the phases δ_i are picked in such a way that we have the identities

(4.8)
$$\frac{\partial \mathbf{z}_s}{\partial \delta_i} = \mathcal{J} \nabla I_i(\mathbf{z}_s),$$

the associated special trajectories of (1.1) are

(4.9)
$$\hat{\mathbf{z}}(t; \mu_i, \delta_i) = \mathbf{z}_s\left(\mu_i, (\mu_i t + \delta_i)\right).$$

As time evolves the trajectory (4.9) traverses a one dimensional family in the r dimensional manifold of critical points corresponding to prescribed values of the multipliers μ_i (and therefore of the integrals $I_i = k_i$). (Notice that in a slight abuse of notation the symbols δ_i are playing the dual roles of parametrizing the family of critical points $\mathbf{z}_s(\mu_i, \delta_i)$, and also of determining the initial data $\hat{\mathbf{z}}(0)$ of the special solution.) We also remark that the choice of parametrization (4.8) leading to (4.9) clearly depends upon the initial choice of the integrals I_i, and in any case might not be the most natural representation in any given problem. For example, with the standard hierarchy of integrals for the KdV equation, the analogue of (4.9) does not yield the standard Hirota representation of KdV multi-solitons (cf. e.g. Maddocks & Sachs, 1993).

Just as in the analogous analysis described in §2 there are two questions of primary concern. First, are solutions (4.9) that evolve along a set of critical points that are actually constrained minima in any sense stable? Second, how can those critical points that are constrained minima be classified? The resolution of both of these questions is straightforward, and directly parallels the analysis presented in §2. The essential differences are perhaps best explained in the context of two simple examples. These two simple model problems have also been designed so that the phenomena they exhibit exactly mimics the properties of the constrained variational characterization of 1- and 2-soliton solutions of the KdV partial differential equation.

The first example is the two degree of freedom system with Hamiltonian

(4.10)
$$H(x_1, x_2, y_1, y_2) = \tfrac{1}{2}(x_2^2 - y_1^2 + y_2^2).$$

Here (x_1, x_2) describe the location and (y_1, y_2) the momenta of a particle sliding along a one-dimensional "rain-gutter." It would seem more natural to take the Hamiltonian

(4.11)
$$H(x_1, x_2, y_1, y_2) = \tfrac{1}{2}(x_2^2 + y_1^2 + y_2^2),$$

but, just as in the example with Hamiltonian (1.7) that was treated in the Introduction, the dynamics generated by (4.10) and (4.11) are equivalent, and it turns out that the choice (4.10) more closely mimics the more complicated, but analogous, case of a KdV 1-soliton. Since H is independent of

x_1, the momentum $I_1 = y_1$ is an integral of the motion, and the solutions of the variational principle (4.2) (with $r = 1$) are

(4.12) $z_s = (x_1, x_2, y_1, y_2) = (\delta_1, 0, -\mu_1, 0)$.

In this example the value k_1 of the integral $I_1(z_s)$, and the multiplier μ_1 are explicitly related through the simple relation $k_1 = -\mu_1$. Notice that each critical point is embedded in a two parameter family, first as the phase δ_1 is varied, and second as the multiplier μ_1 (or equivalently k_1) is varied.

The analogue of the augmented Lagrangian (2.9) is

(4.13)

$$H(z) - \mu_1 I_1(z) + \frac{M}{2}(I_1(z) - k_1)^2 = \tfrac{1}{2}(x_2^2 - y_1^2 + y_2^2) - \mu_1 y_1 + \frac{M}{2}(y_1 + \mu_1)^2.$$

This example has been specifically designed so that the Hessians that arise are diagonal matrices. Consequently it is easy to verify that for any $M > 1$, and for a fixed value of the multiplier μ_1, the line of critical points (4.12) represent nonisolated, unconstrained minima of the augmented Lagrangian (4.13). The fact that the second variation has a zero eigenvalue is of no consequence because, due to the associated symmetry, the Lagrangian is exactly constant along the nullspace direction, and there is no need to control the behavior of (4.13) in that direction.

The special solutions of the dynamics (1.1) associated with the critical points of (4.3) (or equivalently of (4.13)) are the steady translations

(4.14) $\hat{z}(t; \mu_1, \delta_1) = ((-\mu_1 t + \delta_1), 0, -\mu_1, 0)$.

The time evolution of these solutions simply traverses the one-dimensional line of critical points that arise for the prescribed values of the multiplier μ_1. Of course the solution (4.14) is an equilibrium in the appropriate steadily translating frame, and is consequently (perhaps the simplest example of) a *relative equilibrium*.

The solutions (4.14) can be seen not to be Lyapunov stable by consideration of the exact solution $\hat{z}(t; \tilde{\mu}_1, \tilde{\delta}_1)$, where $\tilde{\mu}_1$ and $\tilde{\delta}_1$ are small perturbations of μ_1 and δ_1. Accordingly the initial values of the original and perturbed solutions can be chosen to be arbitrarily close, but, provided that $\mu_1 \neq \tilde{\mu}_1$, the two trajectories will move arbitrarily far apart in their x_1-components for $|t| \gg 1$. On the other hand it is apparent from the characterization of the critical points (4.12) as unconstrained minimizers of the time invariant functional (4.13), that any trajectory with initial conditions close to (4.12) will remain close to *the set of all minimizers*. Moreover the solution (4.14) is *orbitally* stable, because we have already concluded that any perturbed motion stays close to the set of minimizers, and in this example the set of minimizers is one dimensional and coincides with the orbit of the unperturbed motion.

As a second example we consider an analogous problem with three degrees of freedom, and two additional integrals, namely

(4.15) $H(x_1, x_2, x_3, y_1, y_2, y_3) = \tfrac{1}{2}(x_3^2 - y_1^2 + y_2^2 + y_3^2)$.

The two additional (commuting) integrals are $I_1 = y_1$ and $I_2 = y_2$, and the associated four dimensional set of nondegenerate critical points of (4.3) (in the case $r = 2$) are

(4.16) $z_s = (x_1, x_2, x_3, y_1, y_2, y_3) = (\delta_1, \delta_2, 0, -\mu_1, \mu_2, 0).$

The associated special solutions of the dynamics are

(4.17) $\hat{z}(t; \mu_1, \mu_2, \delta_1, \delta_2) = ((-\mu_1 t + \delta_1), (\mu_2 t + \delta_2), 0, \mu_1, \mu_2, 0).$

As in the previous example it is straightforward to verify that the critical points (4.16) can be characterized as nondegenerate, unconstrained minima of the augmented Lagrangian
(4.18)
$$H(z) - \mu_1 I_1(z) - \mu_2 I_2(z) + \tfrac{M}{2}\left\{(I_1(z) - k_1)^2 + (I_2(z) - k_2)^2\right\} =$$

$$\tfrac{1}{2}(x_3^2 - y_1^2 + y_2^2 + y_3^2) - \mu_1 y_1 - \mu_2 y_2 + \tfrac{M}{2}\left\{(y_1 + \mu_1)^2 + (y_2 - \mu_2)^2\right\}, \quad M > 1.$$

(In fact the additional penalization terms $(y_2 - \mu_2)^2$ are unnecessary in this example because the coefficient of y_2^2 in the Hamiltonian (4.15) is positive, but the expression (4.18) is the appropriate special case of (4.13).) Then, just as before, (4.18) can be used as a Lyapunov function which allows us to conclude that the entire two-dimensional set of critical points determined by (4.16) with the multipliers μ_i prescribed is stable under the dynamics (1.1) in the sense that any trajectory starting close to the manifold of critical points must remain close. Now, however, the special solutions (4.17) are neither Lyapunov stable nor orbitally stable, as can be seen by consideration of the exact solution $\hat{z}(t; \tilde{\mu}_1, \tilde{\mu}_2, \delta_1, \delta_2)$ with $\tilde{\mu}_1 \tilde{\mu}_2 \neq \mu_1 \tilde{\mu}_2$. The absence of orbital stability is explained by the fact that now the set of constrained critical points is two dimensional, while the orbit of any one trajectory under the dynamics is still only one dimensional. The orbit of the perturbed solution is a straight line that remains close to the original plane of critical points, but which diverges from the line of the orbit of the original trajectory.

The information to be gleaned from the above examples is that the manifold of critical points that is parametrized by the phases δ_i (for fixed values of the multipliers μ_i) is, in an appropriate sense, stable under the dynamics (1.1) provided that it can be characterized as representing unconstrained minima of an augmented Lagrangian. Unfortunately, whenever the manifold of critical points is of dimension two or more, the sense of stability is comparatively weak because it does not imply stability (orbital or otherwise) of any one trajectory under the time dynamics with Hamiltonian H. Nevertheless stability of the manifold of critical points is the best that can be concluded from purely variational arguments of a Lyapunov (or Lagrange-Dirichlet) type. However in recent work Bona & Soyeur (1992) have combined a variational approach with some additional analysis to obtain control over the phase of perturbed solutions in certain solitary wave problems where the manifold of relative equilibria is one dimensional. This approach can also be generalized to obtain some control on the phases of solutions that are close to multi-dimensional manifolds of relative equilibria, see Bona *et al* (1993).

Stability of the manifold of critical points can also be regarded from another viewpoint. Because the integrals I_i are all in involution, the entire manifold of critical points is the "orbit" of given initial data under the r different dynamical systems in which each of the integrals $I_i(\mathbf{z})$ is taken as Hamiltonian. (Because of the first order conditions (4.2) the original Hamiltonian flow is a linear combination of these r flows.) In this sense the r dimensional manifold of critical points is the "orbit" of the dynamics (or equivalently of the symmetry group), and consequently stability of the manifold is sometimes also described as "orbital" stability. We shall not adopt this usage.

With the appropriate notion of stability firmly in hand, it is now of interest to return to the question of classifying those constrained critical points, i.e. solutions $\mathbf{z}_s$ of (4.2), that can actually be classified as unconstrained minima of an augmented Lagrangian of the form (4.13). The two examples above are sufficiently simple that it is easy to verify from direct consideration of the appropriate Hessian matrices that the family of critical points are in fact unconstrained minimizers of the augmented Lagrangian. However in general it is of interest to seek alternative characterizations. With the commutativity assumption (4.1), it turns out that there is a direct parallel between available results for the case of (Bott or transversely) nondegenerate (nonisolated) critical points of (4.3), and the case of nondegenerate (isolated) critical points of (2.6) that was described in the previous section. The first observation in this direction is that a direct analogue of Lemma 1 is valid. In particular when the adjective nondegenerate is interpreted in the sense of (4.6) and (4.7), and the second order conditions (2.8) are replaced by

(4.19)
$$\mathbf{h}\cdot\nabla^2 F(\mathbf{z}_s)\,\mathbf{h} > 0, \quad \forall\,\mathbf{h} \neq 0 \text{ s.t. } \nabla I_i(\mathbf{z}_e)\cdot\mathbf{h} = \mathcal{J}\nabla I_i(\mathbf{z}_e)\cdot\mathbf{h} = 0, \quad i = 1,\ldots,r.$$

then the conclusion of Lemma 1 is valid as stated.

With the analogue of Lemma 1 in hand, attention shifts to an analysis of the second order conditions (4.19). In the two examples given above it is again simple to analyze these conditions directly. For instance, in the second problem the Hessian $\nabla^2 F(\mathbf{z}_s)$ is the diagonal matrix with entries $(0, 0, 1, -1, 1, 1)$, while

(4.20)
$$\nabla I_1(\mathbf{z}_s) = \mathbf{e}_4, \quad \nabla I_1(\mathbf{z}_s) = \mathbf{e}_5, \quad \mathcal{J}\nabla I_1(\mathbf{z}_s) = \mathbf{e}_1, \quad \text{and} \quad \mathcal{J}\nabla I_1(\mathbf{z}_s) = \mathbf{e}_2.$$

Here $\mathbf{e}_i$ denotes the usual elementary 6-tuple with 1 in the ith row being the only nonzero entry. Because of the simple form (4.20) of the gradients, the second order conditions (4.19) immediately reduce to consideration of a (2×2) submatrix of $\nabla^2 F(\mathbf{z}_s)$, which in this case is the (2×2) identity matrix. Thus the second order conditions are clearly satisfied. However in many examples there is no analogous simplification, e.g. in the KdV equation. Consequently a direct attack on the second order conditions (4.19) is not feasible. Fortunately, given the commutativity conditions (4.1) and nondegeneracy condition (4.6), the direct analogue of Lemma 2 (along with the associated preamble and subsequent remarks) remains valid, and provides an alternative method of classifying constrained minima. While it is not necessary to apply Lemma 2 in the two above examples, its conclusions must certainly still be valid, and it is perhaps of some interest to consider the associated explicit calculations. For instance, in the second problem the diagonal form of the Hessian $\nabla^2 F(\mathbf{z}_s)$ shows that there is one negative eigenvalue and so the manifold of relative equilibria are critical points of the Lagrangian (4.4) with unconstrained index 1. The explicit representation (4.17) of the critical points can then be used to evaluate the Lagrangian (4.4) purely as a function of the multipliers, to yield

$$(4.21) \qquad\qquad F(\mu_1, \mu_2) = \tfrac{1}{2}(\mu_1^2 - \mu_2^2),$$

which reveals that there is precisely one positive curvature of the distinguished Lagrangian bifurcation surface. Thus, according to the conclusion of Lemma 2, the constrained index is $1 - 1 = 0$, and the critical points are constrained minima. We also remark that the largest eigenvalue of the second variation of (4.21) is one, which, according to a remark made after the statement of Lemma 2, confirms the lower bound $M > 1$ on the constant appearing in (4.18).

As has been previously mentioned, the two examples described above are directly analogous to the variational characterization of the multi-soliton solutions of the KdV equation. In the first example there is a two dimensional set of critical points parametrized by one multiplier and one phase. Each of these critical points has unconstrained index 1 and constrained index 0. Consequently the corresponding dynamic solution is orbitally stable. These are precisely the conclusions reached by Benjamin (1972) in his seminal analysis of the stability of solitary wave (i.e. soliton) solutions of the KdV equation. For the KdV solitary wave the multiplier determines the wave speed, and the phase, appropriately enough, determines the phase. However we remark that Benjamin certainly couches his conclusions in markedly different terms from those adopted here. The case of general multi-solitons is considered in Maddocks & Sachs (1993) (which can be consulted for more details). In the 2-soliton case there is a four dimensional family of critical points arising from two multipliers (equivalently a two parameter family of wave speeds) and two phases. Each of the critical points has unconstrained index 1 and constrained index 0. This situation directly parallels the second example described above. Consequently the corresponding manifold of critical points with prescribed speeds is dynamically stable while no individual trajectory is itself orbitally stable. The situation for n-soliton solutions is that they arise in $2n$-parameter families with n speeds and n phases. Each critical point can be shown to have unconstrained index $[n/2]$ (where $[n/2]$ denotes the integral part of $n/2$), but it can also be calculated that the associated Lagrangian surface has $[n/2]$ positive curvatures. Consequently, according to Lemma 2, the constrained index is always 0, and the manifold of n-solitons with prescribed speeds is stable by the arguments detailed earlier in this Section.

To complete the analogy between the analysis of this Section, and that of the Introduction and Section 2, it remains to relate the constrained indices and associated symmetric eigenvalue problems that arise in the variational analysis of critical points, to an appropriate nonsymmetric eigenvalue problem arising from separation of time in the linearized dynamics. The eigenvalue problem, analogous to (1.4), and (2.14), that arises from linearization of (2.1) about the special solution $\hat{z}(t)$ described in (4.9) is

$$(4.22) \qquad \rho \mathbf{g} = \mathcal{J} \nabla^2 H(\hat{z}) \mathbf{g}.$$

However the variational analysis described in this Section naturally provides information about the nonsymmetric eigenvalue problem

$$(4.23) \qquad \rho \mathbf{g} = \mathcal{J} \nabla^2 F(\mathbf{z}_s) \mathbf{g},$$

which arises after linearization about the stationary solution $\mathbf{z}_s$ of the dynamical system in which the Lagrangian (4.4) is taken as Hamiltonian. There are a variety of ways to argue that the two eigenvalue problems (4.22) and (4.23) are related. For example the flows generated by H and F certainly commute. Consequently if (4.23) has a positive eigenvalue so that the F flow is unstable, we anticipate that the H flow will also be unstable. Alternatively, coordinates can be picked such that the integrals $I_i(\mathbf{z})$ are all linear in the phase variables (cf. the two examples earlier in this Section) in which case $\nabla^2 H(\hat{z}) = \nabla^2 F(\mathbf{z}_s)$ and the two eigenvalue problems coincide. In any case we shall consider results relating the spectrum of (4.23) to the constrained index of the associated manifold of critical points.

The first step in this direction is to identify the left and right nullspaces of (4.23). The invertibility of $\mathcal{J}$ shows that the right nullspace of (4.23) coincides with the nullspace of $\nabla^2 F(\mathbf{z}_s)$, which, by (4.6), is precisely the span of the vectors $\mathcal{J} \nabla I_i$. The associated left nullspace is spanned by the vectors ∇I_i. There are also generalized nullspaces. Because of the nondegeneracy condition (4.6), and the involution assumption (4.1), vectors η_i can be defined as solutions of

$$(4.24) \qquad \nabla^2 F(\mathbf{z}_s) \eta_i = \nabla I_i(\mathbf{z}_s), \quad i = 1, \ldots, r,$$

(cf. (2.15)). Of course these η_i are only defined up to an element of the nullspace of $\nabla^2 F(\mathbf{z}_s)$, i.e. up to a linear combination of the $\mathcal{J} \nabla I_i(\mathbf{z}_s)$. Differentiation of the first order conditions (4.2) demonstrates, in analogy with (2.17), that one choice of the η_i is

$$(4.25) \qquad \eta_i = \frac{\partial \mathbf{z}_s}{\partial \mu_i}.$$

Multiplication of equations (4.24) by the matrix $\mathcal{J}$ reveals that the vectors η_i are a basis for the right generalized nullspace. Furthermore $\mathcal{J}^{-1} = -\mathcal{J}$, so the vectors $\mathcal{J}\eta_i$ can be seen to form a basis for the left generalized nullspace. After comparing the analogous results for the noncanonical problem considered in Section 2, we see that the vectors η_i change roles from being right nullvectors to being right generalized nullvectors. The role of right nullvectors is assumed by the vectors $\mathcal{J}\nabla I_i(\mathbf{z}_s)$, which vanish in the noncanonical case. The vectors $\nabla I_i(\mathbf{z}_s)$ appear as left nullvectors in both circumstances. The vectors $\mathcal{J}^{-1}\eta_i$ that span the left generalized nullspace of (4.23), are undefined in the noncanonical case.

With the definition (4.25) of the η_i, the analogues of formulæ (3.1) and (2.16) remain valid (provided that the symbol $\nabla^2 F(\mathbf{z}_s)^{-1}$ is interpreted as specifying the pre-image). Then, just as before, it can be verified that the constrained nondegeneracy condition (4.7), implies that the matrix

$$(4.26) \qquad F_{\mu_i \mu_j} \equiv \{\eta_j \cdot \nabla I_i(\mathbf{z}_s)\},$$

that determines the signs of the curvatures is nonsingular. Consequently there is no generalized nullvector other than the η_i. This fact follows because any additional generalized eigenvector must satisfy the system of equations

$$(4.27) \qquad \mathcal{J}\nabla^2 F(\mathbf{z}_s)\xi = \sum_i \alpha_i \eta_i + \sum_i \gamma_i \mathcal{J}\nabla I_i(\mathbf{z}_s),$$

that are analogous to (3.2), and the associated solvability conditions cannot be satisfied. In other words, for critical points satisfying the two nondegeneracy conditions (4.6) and (4.7), the generalized nullspace of (4.23) comprises r chains made up of a nullvector and precisely one generalized nullvector. On the other hand when (4.26) has a zero eigenvalue $\alpha \in \Re^r$, and the constrained nondegeneracy condition (4.7) fails, the system of equations (4.27) and

$$(4.28) \qquad \mathcal{J}\nabla^2 F(\mathbf{z}_s)\zeta = \xi + \sum_i \gamma_i \eta_i.$$

can be solved to construct a chain of a nullvector and three generalized eigenvectors. The analysis of the solvability conditions for (4.27) and (4.28) directly parallels that for (3.2) and (3.3), and so will not be repeated.

Finally, we prove that whenever the constrained index of $\mathbf{z}_s$ is odd, the eigenvalue problem (4.23) has a positive real eigenvalue. One way to see this is the following homotopy argument, which is based on an idea introduced by Maddocks & Overton (1994) to treat the case of dissipative perturbations of Hamiltonian systems. With the projection onto the orthogonal complement of the $\nabla I_i(\mathbf{z}_s)$ denoted (as before) by P, it can be verified that at nondegenerate critical points the family of eigenvalue problems

$$(4.29) \qquad \rho \mathbf{g} = (\tau \mathcal{J} - (1 - \tau)P) \nabla^2 F(\mathbf{z}_s) \mathbf{g}, \qquad 0 \leq \tau \leq 1,$$

has the same generalized nullspaces as (4.23), independent of the value of the homotopy parameter τ. In particular the right generalized nullspace is spanned by the vectors $\{\nabla I_i(\mathbf{z}_s), \mathcal{J} \nabla I_i(\mathbf{z}_s)\}$. Consequently, because $P \geq 0$ with nullspace spanned by the $\nabla I_i(\mathbf{z}_s)$, $\nabla^2 F(\mathbf{z}_s)\bar{\mathbf{g}} \cdot P \nabla^2 F(\mathbf{z}_s)\mathbf{g} > 0$ for any eigenvector of (4.29) corresponding to a non-zero eigenvalue. Then for $0 \leq \tau < 1$, the scalar product of (4.29) with $\nabla^2 F(\mathbf{z}_s)\bar{\mathbf{g}}$ yields a contradiction on the assumption that there is a pure imaginary eigenvalue, for the only real term in the ensuing expression is $(1 - \tau)\nabla^2 F(\mathbf{z}_s)\bar{\mathbf{g}} \cdot P \nabla^2 F(\mathbf{z}_s)\mathbf{g}$ which is nonzero. Consequently, as τ varies the number of eigenvalues of (4.29) (strictly) in the right half-plane is invariant. But for $\tau = 0$ the eigenvalues of (4.29) are real, and the nonzero eigenvalues coincide with the nonzero eigenvalues of the matrix $-P \nabla^2 F(\mathbf{z}_s)P$. By hypothesis the number of negative eigenvalues of $P \nabla^2 F(\mathbf{z}_s)P$, namely the constrained index of $\mathbf{z}_s$, is odd, say σ^-, and so for every τ, $0 \leq \tau < 1$, there are σ^- eigenvalues of (4.29) in the right half-plane. However we are interested in the case $\tau = 1$, and as $\tau \to 1$ eigenvalues may well approach the imaginary axis. The saving observation is that the matrices in (4.29) are all real, so that complex eigenvalues of (4.29) must arise in complex conjugate pairs. Accordingly if the constrained index σ^- is odd, so that there is an odd number of eigenvalues of (4.29) in the right half plane, there must be at least one positive real eigenvalue for all $0 \leq \tau < 1$. The desired conclusion is reached once it is observed that while complex conjugate pairs of eigenvalues can approach the imaginary axis in the limit $\tau \to 1$, the real eigenvalue cannot approach zero because the dimension of the generalized nullspace is fixed independent of τ. We remark that the above argument relies on little more than continuity of the point spectrum with respect to a parameter and is therefore not an intrinsically finite dimensional technique. A similar proof yields the odd index instability result for noncanonical systems that was described in Section 2.

5. Acknowledgements. It is a pleasure to thank Professors J.K. Moser and P.J. Olver for illuminating conversations concerning the ramifications of Darboux's Theorem, and Dr. B.W. Zombro for a thoughtful critique of an early version of the manuscript. The two figures in the paper were generated by a numerical continuation and interactive graphics software package called MC^2 (an acronym for Multiplier and Constraint Continuation) that was largely programmed by C.K. Mesztenyi of the Advanced Visualization Laboratory at the University of Maryland. S. James, R. Paffenroth and K. Rogers also assisted in the production of the graphics. The development of MC^2 is partially funded by a grant from the Scientific Innovators Program of Digital Equipment Corporation.

REFERENCES

V.I. Arnol'd (1966). Sur un principe variationnel pour les écoulements stationaires des liquides parfaits et ses applications aux problèmes des stabilité non linéaires, *J. Mécanique* **5**, pp. 29–43.

V.I. Arnol'd (1978). *Mathematical Methods of Classical Mechanics*, Springer-Verlag, New York.

V.I. Arnol'd, V.V. Kozlov, and A.I. Neishtadt (1988). Mathematical aspects of classical and celestial mechanics, in *Dynamical Systems III*, *Encyclopaedia of Mathematical Sciences* V. 3, Editor, V.I. Arnol'd.

T.B. Benjamin (1972). The stability of solitary waves. *Proc. Roy. Soc. Lond. A* **328**, pp. 153–183.

A. Bloch, P.S. Krishnaprasad, J.E. Marsden, and T.S. Ratiu (1992). Dissipation induced instabilities, *Univ. of Maryland SRC Tech. Report # TR 92–76*. To appear in *Annales de L'Institut Henri Poincare: Analyse Non Lineare* **11**, pp. 1-54.

J.L. Bona, and A. Soyeur (1994). On the stability of solitary-wave solutions of model equations for long waves, *J. Nonlinear Science*, in press.

J.L. Bona, J.H. Maddocks, and R.L. Sachs (1993). On the control of phases in the stability analysis of relative equilibria, *forthcoming*.

N.G. Chetayev (1961). *The Stability of Motion*, Pergamon.

M.C. Crandall and P.H. Rabinowitz (1973). Bifurcation, perturbation of simple eigenvalues and linearized stability, *Arch. Rat. Mech. Anal.* **52**, pp. 161–192.

A.T. Fomenko (1991). Topological classification of all integrable Hamiltonian differential equations of general type with two degrees of freedom, in *The Geometry of Hamiltonian Systems*, Ed. T. Ratiu, MSRI Publications **22**, Springer-Verlag.

M. Grillakis, J. Shatah, and W.A. Strauss (1990). Stability theory of solitary waves in the presence of symmetry, II, *J. Func. Anal.* **94**, pp. 308–348.

M.R. Hestenes (1951). Applications of the theory of quadratic forms in Hilbert space to the calculus of variations, *Pacific J. Math.* **1**, pp. 525–581.

M.R. Hestenes (1975). *Optimization Theory, The Finite Dimensional Case*, John Wiley, New York. Reprint, Krieger, 1982.

D.D. Holm, J.E. Marsden, T. Ratiu, and A. Weinstein. (1985). Nonlinear stability of fluid and plasma equilibria, *Phys. Rep.* **123**, pp. 1–116.

E. Leimanis (1965). *The General Problem of the Motion of Coupled Rigid Bodies About a Fixed Point*, Springer–Verlag.

D.K. Lewis (1989). Nonlinear stability of a rotating planar liquid drop, *Arch. Rat. Mech. Anal.* **106**, pp. 287–333.

D.K. Lewis, T.S. Ratiu, J.C. Simo, and J.E. Marsden (1992). The heavy top: a geometric treatment, *Nonlinearity* **5**, pp. 1–48.

R.S. MacKay (1991). Movement of eigenvalues of Hamiltonian equilibria under non-Hamiltonian perturbations, *Phys. Lett.* **A155**, pp. 266–268.

J.H. Maddocks (1985). Restricted quadratic forms and their application to bifurcation and stability in constrained variational principles, *SIAM J. Math. Anal.* **16**, pp. 47–68; Errata **19**, pp. 1256–1257 (1988).

J.H. Maddocks (1987). Stability and folds, *Arch. Rat. Mech. Anal.* **99**, pp. 301–328.

J.H. Maddocks (1991). On the stability of relative equilibria, *IMA J. Appl. Math.* **46**, pp. 71–99.

J.H. Maddocks (1993). On second-order conditions in constrained variational principles, *JOTA*, forthcoming.

J.H. Maddocks and M.L. Overton (1994). Stability theory for dissipatively perturbed Hamiltonian systems, *Comm. Pure Appl. Math.*, submitted.

J.H. Maddocks and R.L. Sachs (1993). On the stability of KdV multi-solitons, *Comm. Pure Applied Math.* **46**, pp. 867–901.

Y.-G. Oh (1987). A stability theory for Hamiltonian systems with symmetry, *J. Geometry and Physics* **4**, pp. 163–182.

P.J. Olver (1986). *Applications of Lie Groups to Differential Equations*, Springer–Verlag, New York.

H. Poincaré (1885). Sur le equilibre d'une masse fluide animée d'un mouvement de rotation, *Acta Math.* **7**, pp. 259–380.

V.N. Rubanovskii and S.Ia. Stepanov (1969). On the Routh Theorem and the Chetaev method for constructing the Liapunov function from the integrals of the equations of motion, *J. of Applied Math. Mech.* (PMM) **33**, pp. 882–890.

V.V. Rumyantsev (1966). On the stability of steady-state motions, *J. of Applied Math. Mech.* (PMM) **30**, pp. 1090–1103.

V.V. Rumyantsev (1968). On the stability of steady-state motions, *J. of Applied Math. Mech.* (PMM) **32**, pp. 517–521.

D.H. Sattinger (1972). Stability of solutions of nonlinear equations, *J. Math. Anal. and Appl.* **39**, pp. 1–12.

C.L. Siegel and J.K. Moser (1971). *Lectures on Celestial Mechanics*, Springer-Verlag, New York.

J.C. Simo, T.A. Posbergh, and J.E. Marsden (1990). Stability of coupled rigid bodies and geometrically exact rods: block diagonalization and the energy momentum method, *Phys. Rep.* **193**, pp. 280–360.

J.C. Simo, D. Lewis, and J.E. Marsden (1991). Stability of relative equilibria: Part 1. The reduced energy-momentum method, *Arch. Rat. Mech. and Anal.* **115**, pp. 15–59.

H.F. Weinberger (1978). The stability of bifurcating solutions, in *Nonlinear Analysis, dedicated to Eric Rothe*, Academic Press.

B. Zombro (1993). Relative Equilibria in Rotationally Symmetric Hamiltonian Systems, Ph.D. Dissertation, Cornell University.

B. Zombro and P. Holmes (1993). Reduction, stability, instability and bifurcation in rotationally symmetric Hamiltonian systems, *Dynamics and Stability of Systems*, **8** # 1, pp. 41–71.

THE GLOBAL PHASE STRUCTURE OF THE THREE DIMENSIONAL ISOSCELES THREE BODY PROBLEM WITH ZERO ENERGY

KENNETH R. MEYER* AND QUIDONG WANG†

Abstract. We study the global flow defined by the three-dimensional isosceles three-body problem with zero energy. A new set of coordinates and a scaled time are introduced which alow the phase space to be compactified by adding boundary manifolds. Geometric argument gives an almost complete sketch of the global phase portrait of this gravitational system.

1. Introduction. In this paper, we apply the method developed in [1] to study the flow on the collision manifold of the three-dimensional isosceles three-body problem. The isosceles three-body problem has played an important role in the study of the Newtonian gravitational system for the last twenty years [2–5]. Our goal here is a clear picture of the phase structure of this flow.

We assume that three particles with masses $m_1 = m_2 = 1/2$, $m_3 = \beta$ move in Euclidian 3-space with coordinates (x, y, z_2), $(-x, -y, z_2)$, $(0, 0, z_1)$ and initial velocities $(dx/dt, dy/dt, dz_2/dt)$, $(-dx/dt, -dy/dt, dz_2/dt)$, $(0, 0, dz_1/dt)$ respectively. The symmetries of the motion will be maintained forever.

Use x, y and $z = z_1 - z_2$ as the position coordinates. Fix the center of mass at the origin, $\beta z_1 + z_2 = 0$, and therefore linear momentum is zero, $\beta dz_1/dt + dz_2/dt = 0$. The kinetic energy of this system is

$$T = \frac{1}{2}\left\{ \left(\frac{dx}{dt}\right)^2 + \left(\frac{dy}{dt}\right)^2 + \left(\frac{dz_2}{dt}\right)^2 + \beta\left(\frac{dz_1}{dt}\right)^2 \right\}$$

$$= \frac{1}{2}\left\{ \left(\frac{dx}{dt}\right)^2 + \left(\frac{dy}{dt}\right)^2 + \left(\frac{\beta}{1+\beta}\right)\left(\frac{dz}{dt}\right)^2 \right\}$$

and the potential function expressed in x, y, z is

$$U = \frac{1}{8}\frac{1}{(x^2 + y^2)^{1/2}} + \beta\frac{1}{(x^2 + y^2 + z^2)^{1/2}}.$$

Let $q = (x, y, z)^T$ and $p = M(dx/dt, dy/dt, dx/dt)^T$ where M is the 3×3 matrix $M = \mathrm{diag}(1, 1, \epsilon)$ and $\epsilon = \beta/(1 + \beta)$. The equations of motion of

* Institute for Dynamics, Department of Mathematics, University of Cincinnati, Cincinnati, Ohio 45221-0025. This research partially supported by grants from the National Science Foundation.

† Institute for Dynamics, Department of Mathematics, University of Cincinnati, Cincinnati, Ohio 45221-0025.

the system are

$$\frac{dq}{dt} = M^{-1}p, \qquad \frac{dp}{dt} = \nabla U(q).$$

Total energy $T - U$ is an integral and so equal to a constant h. Also the z-component of angular monentum is an integral and thus $k \cdot (q \times p) = c$ is constant.

In the next section we will introduce a sequence of changes of variables which are used in section 3 to define the collision manifold. This sequence of transformations will make the phase space a compact cube in three dimensional space with a non-trivial fictitious flow on the boundaries. In section 4 we will discuss the final evolution of the solutions. In the last section a sketch of the phase portrait will be given.

2. Changing variables. In order to attach a boundary manifold to the phase space we shall perform a series of changes of variables. First, we scale the position vector p, the momentum vector q, and time t by the potential U to introduce new variables u, F, G and a new time τ by

(2.1)

$$u = (2U(q))^{-1}, \ d\tau = u^{-3/2}dt,$$

$$F = u^{-1}q, \qquad G = u^{1/2}p$$

Note that F has only the dimension of mass and measures the geometry of the system, whereas u has the dimension of length/mass and so measures the size of the system. The variable u plays the role of I, the moment of inertia, in McGehee's scaling. The equations for u, F, G, are

(2.2)

$$\frac{du}{d\tau} = -2(M^{-1}G, \nabla U(F))u,$$

$$\frac{dF}{d\tau} = M^{-1}G + 2(M^{-1}G, \nabla U(F))F,$$

$$\frac{dG}{d\tau} = \nabla U(F) - (M^{-1}G, \nabla U(F))G,$$

where

(2.3)

$$G^T M^{-1} G = 1, \qquad U(F) = 1/2 - uh$$

and

$$U(F) = \frac{1}{8} \frac{1}{(F_1^2 + F_2^2)^{1/2}} + \beta \frac{1}{(F_1^2 + F_2^2 + F_3^2)^{1/2}}.$$

In these coordinates the conservation of angular mometum yields: $F_1 G_2 - F_2 G_1 = cu^{-1/2}$.

Next let

$$F_1 = r \sin \phi \qquad G_1 = R \sin \vartheta$$

(2.4)

$$F_2 = r \cos \phi \qquad G_2 = R \cos \vartheta.$$

The equations for (r, R, ϕ, ϑ) are

$$\frac{dr}{d\tau} = R \cos(\vartheta - \phi) + 2(M^{-1}G, \nabla U(F))r$$

$$\frac{dR}{d\tau} = -(1/8r^2 + \beta r/(r^2 + F_3^2)^{3/2}) \cos(\vartheta - \phi) - (M^{-1}G, \nabla U(F))R$$

(2.5)

$$\frac{d\phi}{d\tau} = \frac{R \sin(\vartheta - \phi)}{r}$$

$$\frac{d\vartheta}{d\tau} = \frac{[(1/8r^2 + \beta r/(r^2 + F_3^2)^{3/2}) \sin(\vartheta - \phi)]}{R}$$

Note that, for the complete set of equations of motion, we also need the equations for F_3, G_3 and u given in 2.2.

We again have the following restrictions

(2.6) $$R^2 + \epsilon^{-1}G_3^2 = 1, \qquad 1/8r + \beta/(r^2 + F_3^2)^{3/2} = 1/2 - uh$$

and the angular momentum integral

$$u^{1/2}(F_1G_2 - F_2G_1) = c.$$

The manifold defined by 2.6 with $u = 0$ is the celebrated collision manifold. The flow defined by 2.5 extends to this boundary and the boundary is invariant. The flow defined by 2.5 on the boundary is called the fictitious flow. It reflects the near-collision behavior of the gravitational system.

It is easy to see that setting $h = 0$ in 2.6 will give us the same set of equations as obtained by setting $u = 0$. So what has been found on the collision manifold is not that fictitious after all. It is exactly the flow of the original problem with zero energy. From now on, we will only discuss this flow. We see that with $u = 0$ (or $h = 0$), 2.6 gives two constrains on the variables $(r, R, \phi, \vartheta, F_3, G_3)$. Also notice that the equations in 2.5 are independent of u, and furthermore ϕ and ϑ appear only as $\vartheta - \phi$. These observations will reduce the dimension of our problem to three.

Again, change variables by

$$r = w \sin \xi, \quad R = \sin \eta,$$

$$F_3 = w \cos \xi, \; G_3 = \epsilon^{1/2} \cos \eta,$$

$$\Phi = \vartheta - \phi.$$

We see that $w = (1 + 8\beta \sin \xi)/4 \sin \xi$. The equations for (ξ, η, Φ) are

$$(2.7) \quad \begin{aligned} \frac{d\xi}{d\tau} &= \frac{4\sin\xi(\cos\xi\sin\eta\cos\Phi - \epsilon^{-1/2}\sin\xi\cos\eta)}{(1 + 8\beta\sin\xi)}, \\[2mm] \frac{d\eta}{d\tau} &= \frac{16[\beta\epsilon^{-1/2}\sin\eta\cos\xi\sin^2\xi - \cos\eta(1/8 + \beta\sin^3\xi)\cos\Phi]}{(1 + 8\beta\sin\xi)^2}, \\[2mm] \frac{d\Phi}{d\tau} &= \frac{16\sin\Phi[(1/8 + \beta\sin^3\xi) - (1/4 + 2\beta\sin\xi)\sin^2\eta]}{(1 + 8\beta\sin\xi)^2\sin\eta}. \end{aligned}$$

Let $\sin\eta\, d\tau' = d\tau$ and rewrite τ' as τ.

The equations of motion become

$$(2.8) \quad \begin{aligned} \frac{d\xi}{d\tau} &= \frac{4\sin\xi\sin\eta(\cos\xi\sin\eta\cos\Phi - \epsilon^{-1/2}\sin\xi\cos\eta)}{(1 + 8\beta\sin\xi)}, \\[2mm] \frac{d\eta}{d\tau} &= \frac{16\sin\eta[\beta\epsilon^{-1/2}\sin\eta\cos\xi\sin^2\xi - \cos\eta(1/8 + \beta\sin^3\xi)\cos\Phi]}{(1 + 8\beta\sin\xi)^2}, \\[2mm] \frac{d\Phi}{d\tau} &= \frac{16\sin\Phi[(1/8 + \beta\sin^3\xi) - (1/4 + 2\beta\sin\xi)\sin^2\eta]}{(1 + 8\beta\sin\xi)^2}. \end{aligned}$$

The three quantities ξ, η, Φ are the coordinates we shall use to study this problem. The angle ξ is half the vertex angle at m_3 of the configuration triangle. The angle Φ is the angle between the radius vector and the velocity vector of particle m_1. The dimensionless quanity η measures the ratio of the velocity of particle m_3 and the radial velocity of particle m_1. Refer to Figure 2.1.

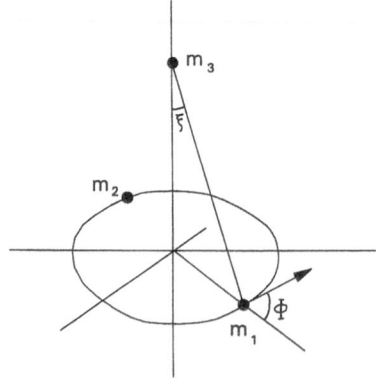

FIG. 2.1. *Coordinates for the isosceles problem in three-space.*

For any point of the original phase space D_0

$$D_0 = (0, \pi) \times (0, \pi) \times (0, \pi)$$

there is a solution of $h = 0$. However, 2.8 now has been defined on D:

$$D = [0, \pi] \times [0, \pi] \times [0, \pi],$$

which is the closure of D_0. This is our compactified phase space.

3. The boundary manifolds. In this section we shall look at the flow on the various boundaries. First set $\xi = 0$ in 2.8 so that the equations become

$$\frac{d\eta}{d\tau} = -2 \sin \eta \cos \eta \cos \Phi = -\sin 2\eta \cos \Phi,$$

$$\frac{d\Phi}{d\tau} = (2 - 4 \sin^2 \eta) \sin \Phi = 2 \cos 2\eta \sin \Phi.$$

These equations admit the integral $\sin 2\eta \sin \Phi = C$, so the phase portrait is as given in Figure 3.1. This is the flow in the limit when the particle m_3 goes upward to infinity. The flow on the boundary $\xi = \pi$ is exactly the same.

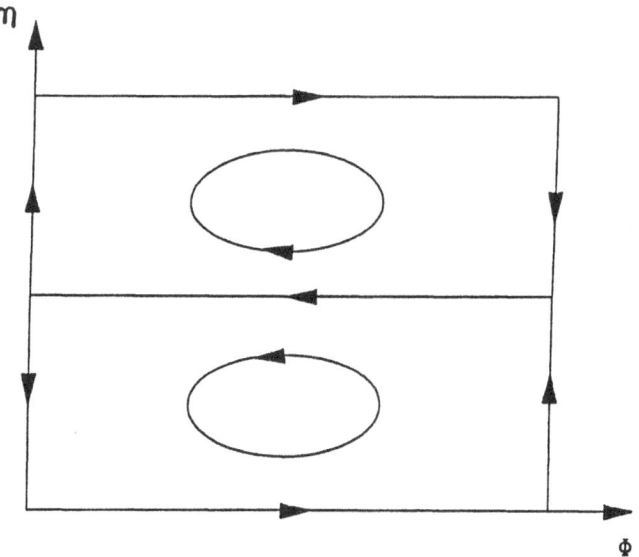

FIG. 3.1. *Escape boundary,* $\xi = 0$ *or* π.

Consider the boundary defined by setting $\Phi = 0$ in 2.8. The equations on this boundary are

$$\frac{d\xi}{d\tau} = \frac{4 \sin \xi \sin \eta (\cos \xi \sin \eta - \epsilon^{-1/2} \sin \xi \cos \eta)}{(1 + 8\beta \sin \xi)}$$

$$\frac{d\eta}{d\tau} = \frac{16 \sin \eta [\beta \epsilon^{-1/2} \sin \eta \cos \xi \sin^2 \xi - \cos \eta (1/8 + \beta \sin^3 \xi)]}{(1 + 8\beta \sin \xi)^2}.$$

This is half of the triple collision manifold for the planar isosceles three-body problem. It corresponds to the collision part—see Figure 3.2.

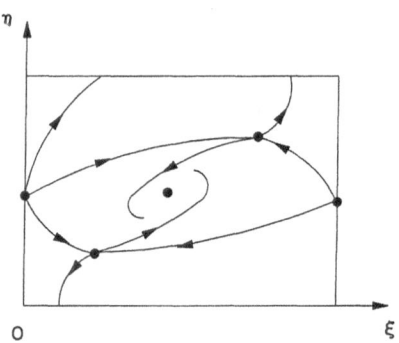

FIG. 3.2. *Collision boundary,* $\Phi = 0$.

The other half, the boundary $\Phi = \pi$, corresponds to the ejection part—see Figure 3.3.

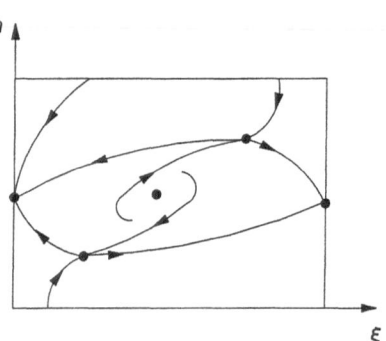

FIG. 3.3. *Ejection boundary,* $\Phi = \pi$.

Consider the boundary defined by $\eta = 0$ or π. When $\eta = 0$ or π the equations are

$$\frac{d\xi}{d\tau} = 0$$

$$\frac{d\Phi}{d\tau} = \frac{16 \sin \Phi (1/8 + \beta \sin^3 \xi)}{(1 + 8\beta \sin \xi)^2}.$$

Figure 3.4 gives the phase portrait for these boundaries.

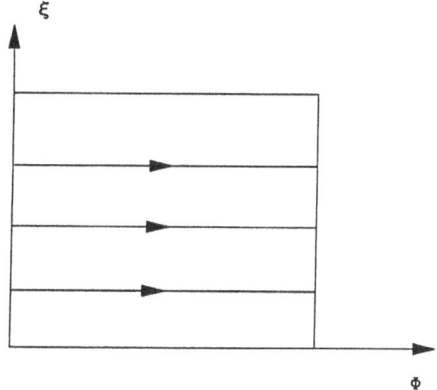

FIG. 3.4. *Boundary $\eta = 0$ or π.*

Putting all these pictures together, we have Figure 3.5.

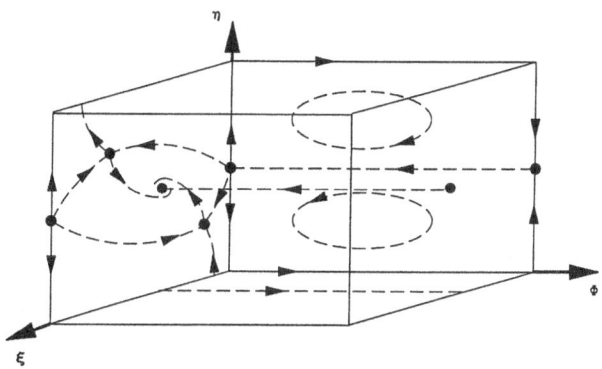

FIG. 3.5. *All boundaries.*

4. The final evolution. First we claim that there is no rest point inside the phase space D_0. By 2.8 an internal rest point would satisfy

$$(4.1) \qquad \cos \xi \sin \eta \cos \Phi - \epsilon^{-1/2} \sin \xi \cos \eta = 0,$$

$$(4.2) \qquad \beta \epsilon^{-1/2} \sin \eta \cos \xi \sin^2 \xi - \cos \eta (1/8 + \beta \sin^3 \xi) \cos \Phi = 0,$$

$$(4.3) \qquad (1/8 + \beta \sin^3 \xi) - (1/4 + 2\beta \sin \xi) \sin^2 \eta = 0.$$

From 4.1 and 4.2

$$(4.4) \qquad \beta \cos^2 \xi \sin^2 \eta \sin \xi = (1/8 + \beta \sin^3 \xi) \cos^2 \eta.$$

Combining 4.3 and 4.4 gives

$$\beta \cos^2 \xi \sin^2 \eta \sin \xi = \sin^2 \eta (1/4 + 2\beta \sin \xi) \cos^2 \eta.$$

Therefore,

$$(4.5) \qquad \beta \cos^2 \xi \sin \xi = (1/4 + 2\beta \sin \xi) \cos^2 \eta$$

Together 4.5 and 4.3 yield $1/8 + \beta \sin \xi = 1/4 + 2\beta \sin \xi$, so $1/8 + \beta \sin \xi = 0$. Since $\xi \in (0, \pi)$ and $\beta > 0$ this is a contradiction.

Next we claim that the flow is gradient-like. Define

$$P = \frac{(F, G)}{(F^T M F)^{-1/4}}.$$

From 2.5 we have

$$\frac{dP}{d\tau} = \frac{\sin \eta [1 - (F, G)^2/(F^T M F)]}{(F^T M F)^{-1/4}}.$$

Let $W = M^{1/2} F$ and $Z = M^{-1/2} G$. Since $G^T M G = 1$ we have $\| Z \| = 1$, $| (F, G)^2/(F^T M F) | = | (W/ \| W \|, Z) | \le 1$. So $dP/d\tau \ge 0$. Thus the flow is gradient-like.

Note that

$$P = \chi(\xi)(\sin \xi \sin \eta \cos \Phi + \epsilon^{1/2} \cos \xi \cos \eta) \sin^{-1/2} \xi$$

where

$$\chi(\xi) = \frac{(1 + 8\beta \sin \xi)^{1/2}}{2(\sin^2 \xi + \epsilon \cos^2 \xi)^{1/4}}.$$

Now what are the α- and ω-limit set of the solutions? Since the flow is gradient-like and there is no rest point inside of the phase space, the α- and ω-limit sets of the solutions will be in the boundaries of the cube.

All the points on the lines $\eta = 0, \pi$; $\Phi = 0, \pi$; $\xi \in (0, \pi)$ are rest points. Denote the collection of these lines by L.

PROPOSITION 4.1. *No solution in D_0 has its α- or ω-limit set in the set L.*

Proof. We will prove the statement for the line $\eta = 0, \Phi = 0, \xi \in (0, \pi)$. The proof is similar for the other cases. Take any surface of the form

$$S_{\eta_0} = \{(\xi, \eta, \Phi) : \eta = \eta_0, \Phi < \Phi_0, \xi \in (0, \pi)\}$$

where $\eta_0 < \epsilon, \Phi_0 < \epsilon$ for sufficiently small $\epsilon > 0$. See Figure 4.1. According to 2.8, we can take ϵ so small that $d\eta/d\tau < 0$ for any point on S_{η_0}.

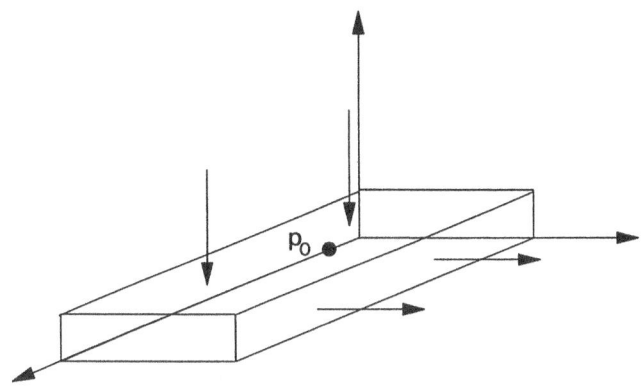

FIG. 4.1. *Flow near P_0.*

Similarly, take

$$S_{\Phi_0} = \{(\xi, \eta, \Phi) : \Phi = \Phi_0, \eta < \eta_0, \xi \in (0, \pi)\}$$

where $\eta_0 < \epsilon, \Phi_0 < \epsilon$. We can again make ϵ so small that for any point on $S_{\Phi_0}, d\Phi/d\tau > 0$. So η is decreasing and Φ is increasing inside the interior of the box with boundaries S_{η_0} and S_{Φ_0} and a solution entering this box of on the side S_{η_0} must leave it on the side S_{Φ_0}.

Now assume that a solution takes point p_0 on the line L as an ω-limit point. This solution will intersect S_{η_0} infinitely many times, therefore it will have an ω-limit point on the surface $\Phi = 0$ with $\eta = \eta_0$. Because P increases along any non-trivial orbit on $\Phi = 0$ and the flow is gradient-like, this is impossible. □

PROPOSITION 4.2. *The α- or ω-limit set of a solution inside of D_0 is one of the following:*

    i) One of the rest point inside of the boundaries $\Phi = 0$ or $\Phi = \pi$.
    ii) On the boundary $\xi = 0$.
    iii) On the boundary $\xi = \pi$.

The proof of this proposition follows from the discussion above.

Look at the hyperbolic final evolution of the system. Assume that the ω-limit set of a solution is on the boundary $\xi = 0$. We see that as $t \to \infty$, the particle m_3 goes upward forever. So $z(t)$ increases and $dz(t)/dt > 0$ decreases monotonically. Also $P(t) \to \infty$ as $t \to \infty$ for this solution.

On the boundary $\xi = 0$, we have a special collection of solutions. It is the lines $\Phi = 0, \eta \in (0, \pi/2)$; $\Phi = \pi$, $\eta \in (0, \pi/2)$; $\eta = 0$ and $\pi/2$, $\Phi \in (0, \pi)$ together with four rest points $(0, \pi/2, 0)$; $(0, 0, 0)$; $(0, \pi/2, \pi)$; $(0, 0, \pi)$. Denote it as ℓ. See Figure 3.1 where ℓ is the lower loop made up of the four rest points and the four trajectories connecting them. We claim that if one point of ℓ is a limit point of a solution, so will be all the other points of ℓ .

Now assume that such a solution exists. $P \to \infty$ and $\xi \to 0$ as $\tau \to \infty$. By the definition of P we see that $\cos \eta/(\tan \xi)^{1/2} \to \infty$. Go back to the original definition of η and ξ in section 2. We will have $z^{1/2} dz/dt \to \infty$. So for any $C > 0, z(t) > Ct^{2/3}$ for $t > T$ sufficiently large. This can happen only when the motion of m_3 is hyperbolic. See [6]. Therefore $dz(\infty)/dt = lim_{t\to\infty}(dz/dt) > 0$ for this solution.

But we will have $\tau_n \to \infty$, such that $\eta(\tau_n) \to \pi/2$. This simply means $(dx/dt)^2 + (dy/dt)^2 \mid_{\tau_n} \to \infty$. But this is impossible and so the points of ℓ are not limit points. Let $Q = \sin 2\eta \sin \Phi$. From 2.8

$$dQ/d\tau = 16\beta \sin^2 \eta \sin \Phi \sin$$
$$2\xi(\epsilon^{-1/2} \cos 2\eta \sin \xi - \cos \xi \sin 2\eta \sin \Phi)/(1 + 8\beta \sin \xi)^2.$$

PROPOSITION 4.3. *If the ω-limit of a solution is on the boundary $\xi = 0$, then $lim_{\tau\to\infty} Q(\tau) = Q_0$ exists and $0 < Q_0 < 1$.*

Proof. This solution can not take ℓ as its limit. So there is an $\epsilon' > 0$, such that $\xi(\tau) \to 0$, $\Phi(\tau) > \epsilon'$ and $\eta(\tau) > \epsilon'$ for $\tau \in [0, \infty)$. $P(\tau) \to \infty$ as $\tau \to \infty$. Set $U = 1/P$ so

$$\frac{dU}{dt} = -P^{-2}\frac{dP}{dt} = \frac{-[1 - (F, G)^2/(F^T M F)]}{P^2 (F^T M F)^{1/4}}.$$

Since $1 - (F, G)^2/(F^T M F) = 1 - (\sin \xi \sin \eta \cos \Phi + \epsilon^{-1/2} \cos \xi \cos \eta)^2/ (\sin^2 \xi + \epsilon \cos^2 \xi)$ there is $aM(\epsilon') > 0$, such that $\mid 1 - (F, G)^2/(F^T M F) \mid > M(\epsilon')$. It will turn out that $\mid dQ/dU \mid < M'(\epsilon')$. Where $M'(\epsilon')$ is another constant related to $M(\epsilon')$. According to this inequality, the limit of Q exists as $\tau \to \infty$ ($U \to 0$). Q_0 can not be 1 since nearby $Q = 1, dQ/d\tau < 0$.

By this proposition, a solution takes either a rest point on the boundary $\Phi = 0$ (the parabolic case) or a periodic solution on the boundary $\xi = 0$ (or π) (the hyperbolic case) as its α- or ω-limit set.

The boundary manifolds $\eta = 0$ and π are rather artificial. They appear when we break the collision manifold of the planar problem into two pieces and paste them on the boundaries $\Phi = 0$ and $\Phi = \pi$. Taking the surface, say $\eta = 0$, as the singular element, we can use the conception of block regularization to simplify the picture. This means we can identify the lines $\eta = 0, \Phi = 0$ and $\eta = 0, \Phi = \pi$ and ignore the surface $\eta = 0$. The phase space is topologically $D^2 \times I$ and the flow on its boundary $S^1 \times I$ will be exactly the flow on the collision manifold of the planar problem. Here $I = [0, 1]$, and D^2 is the two dimensional disk. $\square$

5. The McGehee manifold. Set $\Phi = 0$ in 2.7. The unbroken collision manifold is $S^1 \times I$, and the fictitious flow on it is

$$\frac{d\xi}{d\tau} = \frac{4 \sin \xi (\cos \xi \sin \eta - \epsilon^{-1/2} \sin \xi \cos \eta)}{(1 + 8\beta \sin \xi)}$$

$$\frac{d\eta}{d\tau} = \frac{16[\beta \epsilon^{-1/2} \sin \eta \cos \xi \sin^2 \xi - \cos \eta (1/8 + \beta \sin^3 \xi)]}{(1 + 8\beta \sin \xi)^2}$$

The phase picture of this flow is given in Figure 5.1.

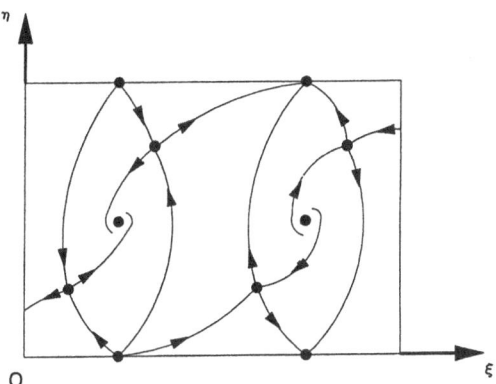

FIG. 5.1. *Flow on total collision manifold.*

Figure 5.1 looks different from the familiar picture of McGehee's collision manifold for the planar problem. The difference is caused by the different treatments of binary collision. McGehee regularizes binary collision. While we have a sub-boundary of binary collisions.

We will need more information about the flow in which binary collision is regularized. So we are going to study the picture on McGehee's manifold first, then come back to that of ours later.

The flow depends on β. Refer to [5]. Its picture is depicted in Figure 5.2 for small β. We will concentrate only on this picture from now on. For the case of larger β, the picture is different but the analysis is similar.

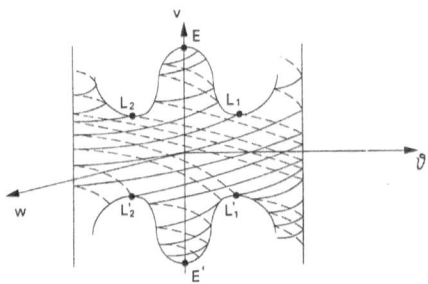

FIG. 5.2. *Flow on McGehee manifold.*

The 2-manifold in Figure 5.2 is defined in (ϑ, w, v)-space by equation

$$(v^2 \cos^2 \vartheta + w^2)/2 + U(\vartheta) \cos^2 \vartheta = 0.$$

See [2] for the details of this equation.

There are six rest points in Figure 5.2. Those labeled, E, E' are the Eulerian points, and L_1, L_1', L_2, L_2' are the Langrangian points. E is a sink, E' is a source and L_1, L_1', L_2, L_2' are all saddles. Two curves $\gamma^+(L_1)$ and $\gamma^-(L_1)$ in Figure 5.3 form the stable manifold of L_1.

Denote by K as the curve $\vartheta = 0$ on this manifold and

K^+ = all the points of K with $w > 0$,

K^- = all the points of K with $w < 0$,

a = the last intersection of $\gamma^+(L_1)$ with K^+,

b = the last intersection of $\gamma^-(L_1)$ with K^+.

Similarly, let $\gamma^+(L_2)$ be the stable curve of L_2 depicted in Figure 5.3 and

c = the last intersection of $\gamma^+(L_2)$ with K^+.

We still need two more crucial points:

d = the last intersection of $\gamma^+(L_1')$ with K^+,

b_1 = the last intersection of $\gamma^+(L_2')$ with K^+.

$\gamma(L_1')$ is the stable curve of L_1' and $\gamma(L_2')$ is that of L_2'—see Figure 5.3.

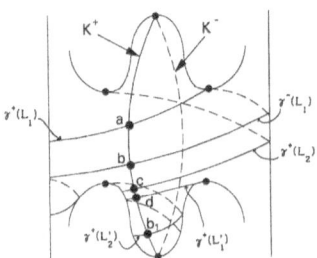

FIG. 5.3. *Stable and unstable manifolds with key points.*

Let $F : K^+ \to K^+$ be the flow-defined Poincare map, and

$$\Gamma = \{a, b, c, d, b_1\} \cup \{F^{-n}(a), F^{-n}(c), F^{-n}(d), F^{-n}(b_1) : n \in Z^+\}.$$

In their natural order from E to E', the points of Γ can be listed as

$$E, a, b, c, d; f^{-1}(a), b_1, f^{-1}(c), f^{-1}(d); \ldots;$$

$$f^{-n}(a), f^{-n+1}(b_1), f^{-n}(c), f^{-n}(d); \ldots E'.$$

We also have a similar sequence on K^- See Figure 5.4.

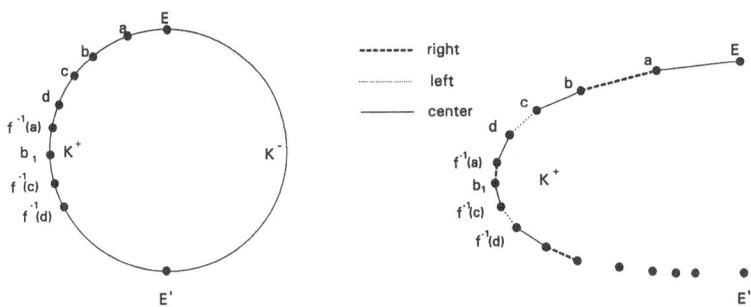

FIG. 5.4.   *Boundary of*
Poincare Section.

FIG. 5.5.   *Orbits on*
Poincare Section.

The points on $K^+ \backslash \Gamma$ have three possible destinations as $\tau \to \infty$ namely the right arm, the left arm or the center. The set Γ divides K^+ into countable many intervals (see Figure 5.5. The destiny of these intervals are in the order of

$$center, right, center, left, center, right, center, left \ldots$$

points on the intervals $(f^{-n+1}(b_1), f^{-n}(c))$ and $(f^{-n}(d), f^{-n-1}(a))$ will go to the center, $(f^{-n}(a), f^{-n+1}(b_1))$ to the right and $(f^{-n}(c), f^{-n}(d))$ to the left.

Any two points on the same interval have the same intersection number (see its definition at the beginning of the next segment), and the intersection numbers of these intervals are in the order of

$$\infty, 0, \infty, 0; \infty, 1, \infty, 1; \ldots; \infty, n, \infty, n; \infty, n+1, \infty, n+1; \ldots$$

6. The global picture. We now go back to the three-dimensional picture of the phase space (ξ, η, Φ) in Figure 3.5 and define the intersection number. For a point p in D_0, let

$$p = (\xi(p), \eta(p), \Phi(p)),$$

$O(\tau, p) = $ the solution at time τ which starts from p at time $\tau = 0$,

$$O(p) = \{O(\tau, p) : 0 < \tau < \infty\},$$

$$S = \{p : p \in D_0 \text{ and } \xi(p) = \pi/2\}.$$

S will serve as a global Poincare section in our discussion. From now on, by a S-neighborhood U of p, we will mean that $p \in S$ and U is a neighborhood of p in S. Divide S into three parts by

$$S^+ = \{p : p \in S, \eta(p) > \pi/2\},$$

$$S^- = \{p : p \in S, \eta(p) < \pi/2\},$$

$$T = \{p : p \in S, \eta(p) = \pi/2\}.$$

We see that $S = S^+ \cup S^- \cup T$, and T itself is a solution of 2.8. See Figure 6.1.

DEFINITION 6.1. *For a given point $p \in S$, let $N(p) = $ cardinality of $\{O(p) \cap S\}$. $N(p)$ is the intersection number of p.*

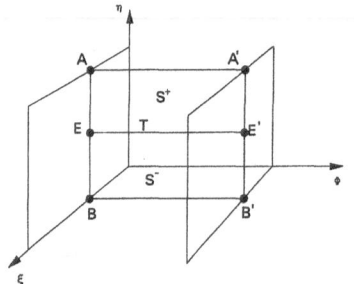

FIG. 6.1.   *Poincare Section.*

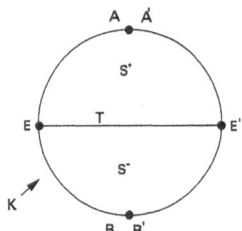

FIG. 6.2.   *Boundary of Poincare Section.*

For the six rest points, an easy calculation shows that E is a sink, E' is a source, L_1, L_2 are hyperbolic points with 2-dimensional stable manifold and

1-dimensional unstable manifold and L'_1, L'_2 are that with 1-dimensional stable manifold and 2-dimensional unstable manifold.

Denote the stable manifolds of L_1 and L_2 as $W^s(L_1)$ and $W^s(L_2)$ and $\Lambda = W^s(L_1) \cup W^s(L_2)$. In D, $W^s(L_1)$ $(W^s(L_2))$ is a manifold with $\gamma^+(L_1)$ and $\gamma^-(L_1)$ $(\gamma^+(L_2)$ and $\gamma^-(L_2))$ as its boundary. Clearly we have:

PROPOSITION 6.2. *For any* $p \in S, p \notin \Lambda$, *there is a S-neighborhood U of p, such that for any* $p' \in U, N(p') = N(p)$.

The solutions which tend to the boundaries $\xi = 0$ form an open set in the phase space. So do the solutions which tend to the boundary $\xi = \pi$ and to the Eulerian rest points. So the intersection number changes only on $\Lambda \cap S$.

Since the vector field is transversal to S^+ and S^-, and Λ is an embedded two dimensional manifold inside of the phase space, we have:

PROPOSITION 6.3. $\Lambda \cap S^+$ *and* $\Lambda \cap S^-$ *are one-dimensional embedded manifold in* S^+ *and* S^- *respectively.*

As we noted at the end of section 2, the surfaces $\eta = 0$ and $\eta = \pi$ can be ignored and the phase space could be regarded, topologically, as $D^2 \times I$. Then S becomes the disk in Figure 6.2. Its boundary is exactly the circle K on McGehee's manifold which we discussed in section 5. We see that EAA'E' $= K^+$, EBB'E' $= K^-$ as well.

Now let us look to the global structure of the flow. For a given C^∞-curve c in S, $c(t) : (-\infty, \infty) \rightarrow S$, the α- and ω- limit set of $c(t)$ are defined by

$$\omega(c) = \{p : p \in \bar{S}; \text{ there is } t_n \rightarrow \infty \text{ such that } c(t_n) \rightarrow p\},$$

$$\alpha(c) = \{p : p \in \bar{S}; \text{ there is } t_n \rightarrow -\infty \text{ such that } c(t_n) \rightarrow p\},$$

where $\bar{S}$ is the closure of S. We also say that $c(t)$ is a curve connecting $\alpha(c)$ and $\omega(c)$.

Since $\Lambda \cap S = (\Lambda \cap S^+) \cup (\Lambda \cap S^-) \cup (\Lambda \cap T)$ and $\Lambda \cap T = \emptyset$, we see that $\Lambda \cap S = (\Lambda \cap S^+) \cup (\Lambda \cap S^-)$. Take B as a connected branch of $\Lambda \cap S$, either $B \subset \Lambda \cap S^+$ or $B \subset \Lambda \cap S^-$. Without loss of generality, we always assume $B \subset \Lambda \cap S^+$. B is a one-dimensional embedded manifold of S^+ by Proposition 6.3.

PROPOSITION 6.4. B *has no limit point on* T.

Remember that T is a solution connecting E and E'. For any $q \in T$, there is a S-neighborhood U_q of q, such that all the points $q' \in U_q$ approch E. Therefore q can not be a limit point of B. Also, we have:

PROPOSITION 6.5. *For two given points* $p_1, p_2 \in B; N(p_1) = N(p_2)$.

For a given point $p \in B$, take a small S-neighborhood U_p of p. For any $p \in U_p \cap B, O(p')$ will stay close to $O(p)$ therefore end up with L_1(or L_2) by the same intersection number. This means $N(p)$ can not be changed at any point of B. So we must have $N(p_1) = N(p_2)$.

PROPOSITION 6.6. B *does not take E or E' as its limit point.*

Proof. Since the flow is gradient like and $P(E) > P(L_{1,2})$, E can not be a limit point of B.

E' is a spiral-source. For any $n > 0$, there is a S-neighborhood U_n of E' such that $N(q) > n$ if $q \in U_n$. E' can not be a limit point of B because the intersection number of B is finite.

No limit points are located in S^+ since B is an embedded manifold in S^+. Also, B can not take any of the points of $K^+ \backslash \Gamma$ as its limit since the sets of hyperbolic solutions and going-to-center solutions are open.

But Γ is a countable set of discrete point. So w have the following for $\Lambda \cap S^+$ □

PROPOSITION 6.7. *B is a curve connecting two points of* Γ.

From this proposition, B can only be one of the following:

(1) A curve which takes the same point of Γ as its α- and ω-limit set. We will call it a loop.

(2) A curve connecting two different points of Γ. We will call it a regular branch.

PROPOSITION 6.8. *For any given point p of* Γ, *there exists at least one branch B which takes p as its limit point.*

In any given neighborhood of $q \in \Gamma$, there are at least two points with different destinations. Take a curve connecting these two points. There must be a point q' on this curve, such that $q' \in \Lambda$. So the branch we claimed exists.

PROPOSITION 6.9. *The limit points of B must have the same intersection number as that of the points on B.*

Now let us see the possible branches of Λ for the point a in Figure 5.3:

(1) No loops since a loop destroys the hyperbolic structure of L_1.

(2) B can not connect a and c since they have different destinations.

(3) B can not connect a and d.

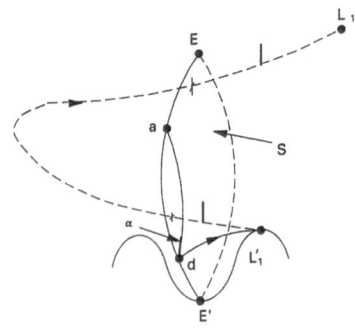

FIG. 6.3. *Connection between points a and d.*

If it does. Take a segment α of B that ends with d (refer to Figure 6.3). Under the action of our flow, α will go forward with one of its ends on $\gamma^+(L_1')$. After reaching L_1', it will go along with an orbit connecting L_1' and

L_1. But there is no such connection with intersection number zero. So the points of α will have intersection number at least 2. But the intersection number of a is zero.

(4) Also, B can not connect b_1 by the reason of different intersection numbers.

(5) Similarly other points of Γ, except b, can not be connected to a.

So we have:

PROPOSITION 6.10. *There is only one regular branch connecting a and b, and it is the only branch for a and b.*

Similarly, there is only one branch of Λ which connects c and d. Now the picture looks like Figure 6.4. Notice we have the same thing on K^-.

What happens for d might be complicated. However, we see that

(1) The possible connections are between d and $f^{-1}(a)$, and b_1.

(2) The connection among $d, f^{-1}(a), b_1$ on K^+ forms the fundamental diagrams. Transfering this digrams by the flow defined Poincare map backward will give the whole picture of the $\Lambda \cap S$.

What about the connection among $d, f^{-1}(a), b_1$?

(1) The ideal picture is in Figure 6.5. There would be one connection between $f^{-1}(a)$ and b_1, no connection between d and $f^{-1}(a)$. It would be true if there were only two connecting orbits for L_1, L_2, L'_1, L'_2. One joins L'_1 and L_2 and another joins L'_2 and L_1.

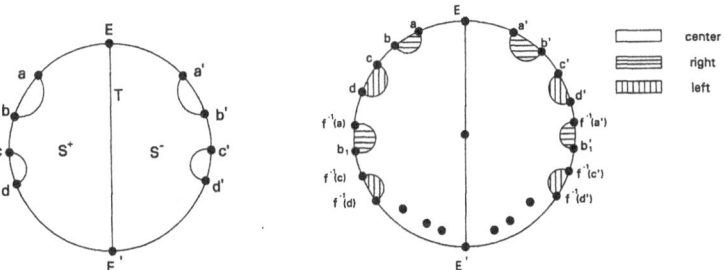

FIG. 6.4.   *Connection of*   FIG. 6.5. *Ideal phase struc-*
points.   ture.

(2) Unfortunately, what really happens might not be so nice. We might have loops around d and b_1 if there are non-transversal intersections among $W^s(L_{1,2})$ and $W^u(L'_{1,2})$. The number of the regular curves connecting d and b_1 is determined by the number of transversal intersections between $W^s(L_{1,2})$ and $W^u(L'_{1,2})$.

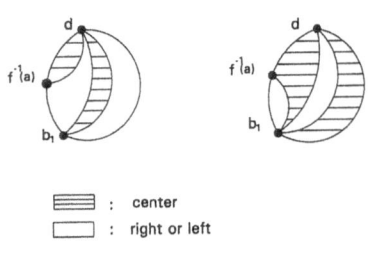

$\begin{array}{ll}\underline{\underline{}} & : \quad \text{center}\\ \boxed{} & : \quad \text{right or left}\end{array}$

FIG. 6.6. *Other possibilities.*

However, we know that there will be only one regular curve terminating in $f^{-1}(a)$. The destinations of the points on different sides of any branch of $\Lambda \cap S$ will be different, and one should be the center. Refer to Figure 6.6. Since there is virtually no method available to analyze the intersections of $W^s(L_{1,2})$ and $W^u(L'_{1,2})$, these are all the cases we can do.

REFERENCES

[1] K.R. MEYER, Q.D. WANG, *The global phase structure of the restricted isosceles three-body problem with positive energy*, to appear in Trans. Amer. Math. Soc. 338 (1) (1993), 311–336.

[2] R. MOECKEL, *Heteroclinic phenomena in the isosceles three-body problem*, SIAM J. Math. 15 (5) (1984), 857–876.

[3] R. DEVANEY, *Triple collision in the plannar isosceles three-body problem*, Inv. Math. 60 (1980), 249–267.

[4] Z.H. XIA, *The existence of non-collision singularity in newtonian system*, Annals of Mathematics 135 (1992), 411–468.

[5] C. SIMO, *Analysis of triple collision in the isosceles three-body problem*, Classical Mechanics and Dynamical Systems, Marcel Dekker, New York 1980, 203–224.

[6] C. MARCHEL, D. SAARI, *On the final evolution of the n-body problem*, J. Diff. Eq. 20 (1976), 150–186.

NON-CANONICAL TRANSFORMATIONS OF NONLINEAR HAMILTONIANS

BRUCE R. MILLER* AND VINCENT T. COPPOLA[‡][†]

Abstract. We describe a new method based on Lie transformations for simplifying perturbed Hamiltonians in one degree of freedom. A non-canonical Lie transformation is used to eliminate terms from the perturbation that are not of the same form as those in the main part, or even to eliminate the perturbation entirely. The system is thus transformed into a modified version of the principal part. In conjunction with a time transformation which recovers the dynamical equations, the procedure synchronizes the motions of the perturbed system onto those of the unperturbed part. The method is most useful when the unperturbed part has solutions in non-elementary functions. Applications of the method are described.

Key words. Duffing equation – Hamiltonian systems – Lie transformation – non-canonical transformations – perturbation theory – synchronization.

1. Introduction. Dynamical systems can often be expanded as perturbation problems to great advantage. But in many cases the expansions must be made to accommodate the method rather than the problem. That a perturbed harmonic oscillator is relatively easy to treat partly explains the popularity of expanding around harmonic oscillators.

But what if the system is poorly represented as a harmonic oscillator? What if the frequency of the response depends strongly, rather than weakly, on the amplitude? What if a quartic term is equally as important as the quadratic? Then, in principle, that quartic term should be included in the principal part. However, this approach leads to great difficulties as a reflection of the fact that the principal part is now a Duffing system, rather than an harmonic oscillator. That system has solutions in elliptic functions; the action-angle variables are quite awkward. A traditional normalization[1] by Lie transformations will be hobbled by the repeated integrations, or 'averagings', — an operation over which the elliptic functions are not closed.

A recent exchange in Celestial Mechanics[2,3,4] has led to the evolution of an alternative method for finding approximate analytic solutions to perturbed systems where the principal part has solutions in terms of non-elementary functions. The key idea is to forego canonical transformations

---

* National Institute of Standards and Technology, Gaithersburg, MD 20899 USA. email: miller@cam.nist.gov (Internet).

‡The authors would like to thank André Deprit for many worthwhile discussions on the subject. This work was supported in part by the Applied Mathematics Program at the Defense Advanced Research Program Agency (Washington, DC) (B. R. M.) and by the Office of Naval Technology's Postdoctoral Fellowship Program (Naval Research Laboratory, Washington, DC) (V. T. C.).

† Department of Aerospace Engineering, University of Michigan, Ann Arbor, MI 48109-2109 USA. email: coppola@krylov.engin.umich.edu (Internet).

to obtain greater liberty in simplification. Along with this idea is the recognition that, for any one degree of freedom transformation, the Hamiltonian dynamical equations can be recovered by a time transformation.

We will see that in many interesting cases, the liberty afforded by a non-canonical simplification will allow the perturbation to be either eliminated entirely, or at least reduced to a form identical to the zero-th order. This has the effect of giving to the perturbed system the same form as the unperturbed part. In reference to the time transformation, we say that the motions of the system have been *synchronized* to the motions of the principal part.

We will first describe how the dynamical equations are recovered by the time transformation. Then we will outline the advantages of non-canonical transformations. Finally, we will illustrate some applications of the method.

2. Time transformation. Consider a Hamiltonian system $\mathcal{H}(\mathbf{x})$, where the vector $\mathbf{x}$ includes both the coordinates and momenta. The differential equations for the system are

$$\frac{d\mathbf{x}}{dt} = \mathbf{J}\frac{\partial \mathcal{H}}{\partial \mathbf{x}},$$

where $\mathbf{J}$ is the antisymmetric matrix.

Now consider an arbitrary, not necessarily canonical, transformation $\mathbf{x} \leftarrow \mathbf{x}'$ of the coordinates and momentum. The time derivatives of these new coordinates are

$$\frac{d\mathbf{x}'}{dt} = \{\mathbf{x}'; \mathcal{H}\} = \{\mathbf{x}'; \mathbf{x}'\}\frac{\partial \mathcal{H}}{\partial \mathbf{x}'}.$$

where $\{\mathbf{x}'; \mathbf{x}'\}$ stands for the matrix of Poisson brackets. Since the matrix of Lagrange brackets $[\mathbf{x}'; \mathbf{x}']$ is the negative of the inverse of the matrix of Poisson brackets, we may write

$$[\mathbf{x}'; \mathbf{x}']\frac{d\mathbf{x}'}{dt} = -\frac{\partial \mathcal{H}}{\partial \mathbf{x}'}.$$

It turns out that for one degree of freedom, this matrix of Lagrange brackets is proportional to $\mathbf{J}$.

(2.1) $[\mathbf{x}'; \mathbf{x}'] = \mathcal{K}\mathbf{J}$, with $\mathcal{K} = [x'; X'] = \dfrac{\partial x}{\partial x'}\dfrac{\partial X}{\partial X'} - \dfrac{\partial x}{\partial X'}\dfrac{\partial X}{\partial x'}$,

where x and X are the coordinate and momentum, respectively. Given the transformation $\mathbf{x}(\mathbf{x}')$, $\mathcal{K}$ is easily computed.

Noting that $\mathbf{J}^{-1} = -\mathbf{J}$, the canonical character of the system can thus be recovered by way of a time transformation

(2.2) $dt = \mathcal{K}\ dt'$,

so that

$$\frac{d\mathbf{x}'}{dt'} = \mathbf{J}\frac{\partial \mathcal{H}}{\partial \mathbf{x}'}.$$

Thus, $\mathcal{H}'(\mathbf{x}') = \mathcal{H}(\mathbf{x}(\mathbf{x}'))$ is viewed as a Hamiltonian with t' as independent parameter.

The transformation $\mathbf{x} \leftarrow \mathbf{x}'$ is presumably chosen to simplify $\mathcal{H}'$, so that one will be able to solve the system for $\mathbf{x}'(t')$ in closed form. It is interesting to note that, without doing anything further, one can determine the shapes of the orbits of $\mathcal{H}$. That is to say, with the transformation $\mathbf{x}(\mathbf{x}')$, one has the orbits $\mathbf{x}(t')$ of the original system as parameterized by the pseudo-time t'. This may be sufficient to determine certain properties of the system under study, such as critical points, bifurcations and so forth.

If explicit solutions in t are required, however, the solutions $\mathbf{x}'(t')$ are inserted into $\mathcal{K}$ and integrated — if possible — to give

$$t - t_0 = \int \mathcal{K}(\mathbf{x}'(t'))\, dt'.$$

Viewed as implicitly defining $t'(t)$, this may be inverted by Lie transformation [1,5], and used to give explicit solutions $\mathbf{x}(t)$.

This recovery of the canonical equations results because the $[\mathbf{x}';\mathbf{x}']$ is proportional to $\mathbf{J}$. This is true for any transformation of one degree of freedom, providing the time transformation is not singular. In the case of a Lie transformation, the transformation is near-identity and so $\mathbf{x} = \mathbf{x}' + \mathcal{O}(\epsilon)$, and thus $\mathcal{K} = 1 + \mathcal{O}(\epsilon)$. Thus, within the region of validity, the transformation is not singular.

For higher degrees of freedom, the recovery of canonical equations implies a constraint on the transformation. In principle, it should be possible to incorporate this constraint in such systems to some advantage; in practice, however, sorting out the implications upon the transformation are forbidding.

3. **Why non-canonical?** But, what advantage might one expect with a non-canonical Lie transformation of a Hamiltonian? In this case, the general form of the Lie transformation[5] is employed, where the generator is a vector $\mathbf{W}$ with components (U, V) corresponding to (x, X). The Lie derivative of a scalar is given by

(3.1) $$\mathcal{L}_{\mathbf{W}} F = \nabla F \cdot \mathbf{W} = U\frac{\partial F}{\partial x} + V\frac{\partial F}{\partial X}.$$

In the Lie simplification of $\mathcal{H}$, at each order, we have computed the 'tentative' n-th order term of the transformed $\mathcal{H}'$, based on $\mathcal{H}_n$ and Lie derivatives of $\mathcal{H}_m$ for $m < n$; we'll call it $\tilde{\mathcal{H}}_n$. The generator is chosen at each order n so as to simplify

(3.2) $$\mathcal{H}'_n = \tilde{\mathcal{H}}_n + \mathcal{L}_{\mathbf{W}}\mathcal{H}_0 = \tilde{\mathcal{H}}_n + U_n\frac{\partial \mathcal{H}_0}{\partial x} + V_n\frac{\partial \mathcal{H}_0}{\partial X}.$$

We will consider two typical forms for $\mathcal{H}_0$ to see what benefits arise.

Many systems are in the form of a kinetic energy plus a potential

$$(3.3) \qquad \mathcal{H}_0 = \frac{1}{2}X^2 + \mathcal{V}(x).$$

Then (3.2) becomes

$$\mathcal{H}'_n = \tilde{\mathcal{H}}_n + U_n \mathcal{V}' + X V_n.$$

By partitioning $\tilde{\mathcal{H}}_n$ into the form $Xf(x,X) + g(x)$, the momentum can be eliminated from the perturbation by choosing $V_n = -f$. How U_n is chosen depends on the form of $\mathcal{V}$ and the algebra of g; in the next section we will describe two classes of potential.

Another typical form is commonly encountered when action-angle-like variables are used. In this case, the principal part depends only on the momentum.

$$(3.4) \qquad \mathcal{H}_0 = h(r).$$

And so, (3.2) takes the form

$$\mathcal{H}'_n = \tilde{\mathcal{H}}_n + \frac{dh}{dr} V_n.$$

By choosing $V_n = -\tilde{\mathcal{H}}/(dh/dr)$, the perturbation can be eliminated entirely. In the next section, we will describe a more refined approach which respects the periodicity of orbits.

4. Applications

4.1. Ideal resonance problem and critical inclination. The ideal resonance problem concerns the system

$$\mathcal{H} = B(y) - \epsilon^2 A(y) \cos x,$$

where $dB/dy = 0$ at some y^*. By expanding in y around some y^*, Henrard and Wauthier [4] obtain a system in the form of (3.3) with the potential of the form $\mathcal{V}(x) = \cos x$. The perturbation is polynomial in X and a trigonometric sum in x. In that case, the Lie derivative to be used in (3.2) is $\mathcal{L}_{\mathbf{W}}\mathcal{H}_0 = XV_n - U_n \sin x$. Using trigonometric identities, the tentative term of $\mathcal{H}$ can be put into the form

$$\tilde{\mathcal{H}} = Xa(x,X) + b \cos x + c(x) \sin x,$$

with b constant. By choosing $V_n = -a$ and $U_n = c$, the perturbation can be reduced simply to a term proportional to $\cos x$ — that is of the same form as in $\mathcal{H}_0$. [This is not exactly the algorithm used by Henrard and Wauthier, but the results are equivalent.]

Preliminary research by the first author indicates that the method can be adapted to the problem of the artificial satellite near the critical inclination and related problems[6]. After the satellite Hamiltonian is averaged over the short period motions, reduced to one degree of freedom and transformed to a slow time scale, it has a similar form to the ideal resonance problem

$$\mathcal{H} = B(G) - J_2 A(G) \cos 2g,$$

where G is the angular momentum, g is the argument of perigee and J_2 is the planet's oblateness. At the critical inclination, $G = \sqrt{5}H$ (H is the polar projection of the angular momentum), $dB/dG = 0$ which gives rise to singularities if the system is treated by a simple averaging over g.

The strategy is to expand near the critical inclination: transform from $G = \sqrt{5}H + \sqrt{J_2}\gamma$ to γ and then expand in powers of $\sqrt{J_2}$, rather than J_2. The system then takes a form with the principal part being

$$\mathcal{H}_0 = -\frac{\sigma}{2}\gamma^2 + \frac{\rho}{2}\cos 2g,$$

and with the perturbation being polynomial in γ and with trigonometric terms of even multiples of g. The method described above can be adapted to treat this system as well. Not only are explicit solutions $g(t)$ and $G(t)$ obtained, but it is possible to lift these solutions to the original six dimensional phase space, and integrate to obtain solutions for $\ell(t)$ and $h(t)$ as well.

4.2. Polynomial potentials $\mathcal{V}(x)$. In a recent study[7], we developed an algorithm for simplifying systems where $\mathcal{V}(x)$ is a polynomial in x.

$$\mathcal{V}(x) = \sum_{i=\ell}^{m} \frac{\alpha_i}{i} x^i,$$

We showed that for polynomial potentials $\mathcal{V}$, any polynomial perturbation in x and X with no monomials of lower degree than the ℓ (or when $\ell = 2$), can be reduced to a polynomial in x with degree less than that of $\mathcal{V}$. That is, the perturbation is reduced to the same form as the potential with different coefficients.

After eliminating the momentum as described in Section 2, (3.2) becomes

$$\mathcal{H}'_n = g(x) + U_n \sum_{i=\ell}^{m} \alpha_i x^{i-1}.$$

We may let U_n be a polynomial in x and choose the coefficients as follows. Assume the highest degree term in g is $g_k x^k$. Choosing the highest degree term in U_n to be $-(g_k/\alpha_m)x^{k-m+1}$ eliminates the term in g. The remaining terms in the sum can be added back into g and the next highest term eliminated the same way. If the iteration is terminated when the highest degree in g is reduced to m, the result is of the same form as $\mathcal{V}(x)$. Again, we have reduced the entire perturbation to the same form as the zero-th order. We note that the entire simplification of $\mathcal{H}$ is carried out in polynomial arithmetic.

In [7], we applied this algorithm to a perturbed Duffing system. In that case, the solutions were given by Jacobian elliptic functions. The resulting time transformation equation was integrated with relative ease, and inverted to obtain $t'(t)$. We were able to compute the solutions to high order for all regions of phase space, both those containing periodic orbits and those containing hyperbolic trajectories.

4.3. Periodic orbits. In problems with periodic solutions, one often uses action-angle variables, or at least action-angle-like variables say (r, ϕ), to the extent that $\mathcal{H}_0$ can be written in terms of the momentum r alone. For a normalization, the equation corresponding to (3.2) becomes

$$\mathcal{H}'_n = \tilde{\mathcal{H}}_n + \frac{\partial \mathcal{H}_0}{\partial r} \frac{\partial W}{\partial \phi},$$

and the normalization becomes, in effect, an averaging. The new term $\mathcal{H}'_n$ is the average of the tentative term $\tilde{\mathcal{H}}_n$ and the generator W is obtained by integrating the difference.

This approach generally brings the functions of the solution into play — not so bad with a harmonic oscillator, where the solutions are trigonometric, but the Duffing system has solutions in elliptic functions. The real problem comes because in normalization the result of the integration to get the generator at one order becomes folded into the system at the next order to be integrated again, and so on, order by order — and elliptic functions are not closed over integration. The method developed in [8] accepts the manipulations of those functions but avoids the repeated integration by simplifying the system via a non-canonical transformation.

To avoid introducing secular terms into the transformation, it is necessary to arrange for the Lie transformation of a function, say f, periodic in ϕ to be also periodic in ϕ. This is not automatically the case because, in general non-linear problems, the period of the solutions depend on the momentum.

Taking the components of $\mathbf{W}$ to be (U, V), corresponding to (ϕ, r), the Lie derivative of a scalar function is

$$\mathcal{L}_{\mathbf{W}} f = U \frac{\partial f}{\partial \phi} + V \frac{\partial f}{\partial r}.$$

Although $\partial f/\partial \phi$ is periodic, since the period T depends on r, $\partial f/\partial r$ is not periodic. We show in [8] that $\phi = T\psi$, where ψ is the angle corresponding to the action J. We may define a derivative $\delta f/\delta r$, standing for the derivative of f with respect to r holding ψ, rather than ϕ, fixed. This derivative is periodic in ψ, and ϕ. Of course, in practice this derivative may be computed simply as

$$\frac{\delta f}{\delta r} = \frac{\partial f}{\partial r} - \Gamma\phi\frac{\partial f}{\partial \phi}, \text{ with } \Gamma = -\frac{1}{T}\frac{\partial T}{\partial r}.$$

In terms of this derivative, it turns out that the Lie derivative can be written as

$$\mathcal{L}_\mathbf{W} f = \frac{\partial f}{\partial \phi}[U + \Gamma\phi V] + V\frac{\delta f}{\delta r}.$$

Now, choosing V periodic in ϕ, and $U = -\Gamma\phi V$ guarantees that the Lie transformation of any periodic function will be periodic.

Having established a transformation which respects the periodicity of the functions, we have reduced (3.2) to

$$\mathcal{H}'_n = \tilde{\mathcal{H}}_n + V\frac{\delta \mathcal{H}_0}{\delta r}.$$

Choosing $V = -\tilde{\mathcal{H}}/\Omega$ where $\Omega = \delta\mathcal{H}_0/\delta r$, eliminates the perturbation entirely. The completion of the time transformation and its inversion then proceeds as described in Section 2.

5. Conclusion. Synchronization opens the door to a number of applications that have been off limits in the past. Granted, the method is limited to one degree of freedom, and granted also that such systems can be handled by numerical quadratures, nonetheless it is often important to obtain solutions in analytic form. This is especially true for systems which one obtains by normalization from a multi-dimensional problem.

Synchronization enables one to treat systems when the principal part has solutions in non-elementary functions, and this to an arbitrarily high order. One avoids the repetitive integration associated with traditional averaging by Lie transformations. Thus, one evades the risk of creating quadratures outside a given domain of functions. Perhaps more significantly, the method often transforms a perturbed system into a modified copy of the unperturbed system. The dynamical equations are recovered from the non-canonical transformation by a time transformation. The perturbed system is thus synchronized to the principal part.

REFERENCES

[1] A. DEPRIT, *Canonical transformations depending on a small parameter*, Celestial Mechanics 1 (1969), 12–30.
[2] R.A. HOWLAND, *A new approach to the librational solution in the ideal resonance problem*, Celestial Mechanics 44 (1988), 209–226.

[3] R.A. HOWLAND, *A note on the application of a generalized canonical approach to non-linear Hamiltonian systems*, Celestial Mechanics 45 (1988), 407–412.

[4] J. HENRARD, P. WAUTHIER, *A geometric approach to the ideal resonance problem*, Celestial Mechanics 44 (1988), 227–238.

[5] K.R. MEYER, *Lie Transform Tutorial – II*, Computer Aided Proofs in Analysis (1991), 190–210, Springer-Verlag.

[6] B.R. MILLER, *Explicit solutions near the Critical Inclination*, in preparation.

[7] B.R. MILLER, V.T. COPPOLA, *Synchronization of perturbed nonlinear hamiltonians*, Celestial Mechanics 55 (1993), 331–350.

[8] V.T. COPPOLA, B.R. MILLER, *Synchronization: an alternative to normalization for perturbed nonlinear Hamiltonians*, in preparation.

LINEAR STABILITY ANALYSIS OF SOME SYMMETRICAL CLASSES OF RELATIVE EQUILIBRIA

RICHARD MOECKEL*

Abstract. The linear stability of several classes of symmetrical relative equilibria of the Newtonian n-body problem are studied. Most turn out to be unstable; however, a ring of at least seven small equal masses around a sufficiently large central mass is stable.

1. Introduction. A *relative equilibrium* of the Newtonian n-body problem is a planar configuration of the n point masses such that the acceleration vector of each mass produced by the gravitational attraction of the other $n - 1$ masses is directed toward the center of mass with magnitude proportional to the distance from the center of mass. This acceleration can be exactly balanced by an outward centrifugal acceleration if the configuration is uniformly rotated at the appropriate angular velocity. This balance of forces is delicate and relative equilibria are very special configurations. The study of relative equilibria began with work of Euler [6] and Lagrange [8] in the eighteenth century. Euler found the collinear relative equilibria in the three-body problem; Lagrange discovered the equilateral triangular relative equilibria.

Each relative equilibrium determines a periodic orbit of the n-body problem in which the configuration rotates rigidly. This paper is concerned with the stability of these periodic solutions. In uniformly rotating coordinate system, the periodic solution becomes a restpoint. The first step in the stability analysis for such a restpoint is to study the eigenvalues of the linearization of the differential equation, that is, to study the question of *linear stability*. This step will be carried out here for several well-known classes of relative equilibria.

After the work of Euler and Lagrange, many special examples of relative equilibria were found. The collinear configurations, discovered in the case $n = 3$ by Euler [6], were generalized to arbitrary n by Lehmann-Filhes [9] and analyzed more fully by Moulton [15]. The regular n-gon discussed in section 3.3 and regular n-gon with a central mass studied in section 3.4 appear in [7]. However, the latter was studied earlier by Maxwell in his work on the rings of Saturn [10,11].

Maxwell's paper also contains an analysis of the linear stability of the regular n-gon with a large central mass. He concluded that the motion was linearly stable provided the central mass is sufficiently large. In section 3.4 this is proved under the additional hypothesis that $n \geq 7$. The reason that Maxwell missed this necessary condition will be explained below.

---

* School of Mathematics, University of Minnesota, 127 Vincent Hall, Minneapolis, MN 55455.
Research supported by the NSF and the Sloan Foundation.

The linear stability of the equilateral configurations was studied by Routh [19]. These results are rederived in section 3.2. The question of linear stability was taken up systematically by Andoyer [1] who formulated the eigenvalue problem for a general relative equilibrium. He also analyzed the collinear relative equilibria, obtaining the factorization of the characteristic polynomial into quadratics discussed in section 3.1. M. G. Meyer completed the analysis of the collinear case by determining the nature of the roots of Andoyer's quadratic factors[12]. Brumberg studied the stability of several symmetrical configurations of four masses obtaining the $n = 4$ cases of the many the results in this paper[3]. Simó studied linear stability of more general relative equilibria in the four-body problem [21].

If a relative equilibrium is linearly stable, it is natural to investigate the higher order terms in the Birkhoff normal form [2]. This step was carried out for the equilateral triangular relative equilibrium in the restricted three-body problem by Deprit and Deprit [5]. If the higher order terms are nondegenerate one can use the KAM theory to prove the existence of quasiperiodic motions near the relative equilibrium, but the actual stability would still be in doubt. Indeed, the question of the generic stability or, more likely, instability of linearly stable restpoints in conservative dynamical systems is a major open problem.

Most of the examples, however, turn out to be linearly unstable. The only exception is Maxwell's regular n-gon with a large central mass. For further examples of linearly stable relative equilibria see [14].

2. Relative equilibria and their stability. In this section, relative equilibria of the Newtonian n-body problem are defined and the conditions for their linear stability are formulated.

2.1. Relative equilibria. For $j = 1 \ldots n$, let $q_j \in \mathbf{R}^3$ be the position of a point particle of mass $m_j \in \mathbf{R}^+$. Let $q = (q_1, \ldots, q_n) \in \mathbf{R}^{3n}$. Newton's equations for the n-body problem are:

$$M\ddot{q} = \nabla U(q)$$

where M is the $3n \times 3n$ mass matrix, $\mathrm{diag}(m_1, m_1, m_1, \ldots, m_n, m_n, m_n)$ and $U(q)$ is the Newtonian potential function:

$$U(q) = \sum_{\substack{(i,j) \\ i<j}} \frac{m_i m_j}{|q_i - q_j|}.$$

Here ∇ denotes the Euclidean gradient operator of $\mathbf{R}^{3n}$.

A *relative equilibrium* is a configuration q which becomes an equilibrium of Newton's equations in a uniformly rotating coordinate system. Without loss of generality, one can consider a rotation about one of the axes in $\mathbf{R}^3$. Let $R(\theta)$ denote the $3n \times 3n$ block diagonal matrix with n equal 3×3 blocks of the form:

$$
\begin{bmatrix}
\cos\theta & -\sin\theta & 0 \\
\sin\theta & \cos\theta & 0 \\
0 & 0 & 1
\end{bmatrix}.
$$

Then define a new coordinate vector $x \in \mathbf{R}^{3n}$ by $q(t) = R(\omega t)x(t)$, where ω is a constant. Substitution of this definition into Newton's equation yields:

$$
(2.1) \qquad M\ddot{x} + 2\omega \hat{J} M\dot{x} = \nabla U(x) + \omega^2 M \hat{I} x
$$

where $\hat{J} = R^{-1}(\theta)R'(\theta)$ and $\hat{I} = R^{-1}(\theta)R''(\theta)$ are $3n \times 3n$ block diagonal matrices with blocks:

$$
\begin{bmatrix}
0 & -1 & 0 \\
1 & 0 & 0 \\
0 & 0 & 0
\end{bmatrix}
\qquad
\begin{bmatrix}
1 & 0 & 0 \\
0 & 1 & 0 \\
0 & 0 & 0
\end{bmatrix}
$$

respectively. In the derivation of this equation, use has been made of the fact that the matrices R, $\hat{I}$, and $\hat{J}$ commute with M.

A configuration x is a relative equilibrium provided that for some ω the constant function with value x solves 2.1. Since $\dot{x} = \ddot{x} = 0$ one finds:

$$
(2.2) \qquad \nabla U(x) = -\omega^2 M \hat{I} x.
$$

If x is a relative equilibrium according to this definition then the center of mass must lie at the origin. To see this note that 2.2 gives:

$$
\sum m_k x_k = -\tfrac{1}{\omega^2} \sum \frac{\partial U}{\partial x_k} = 0
$$

where the last equality is easily checked.

It also follows from 2.2 that the configuration is planar, that is, $x_k \in \mathbf{R}^2 \times 0$ for all k. If this were not so then consider the mass m_k which is farthest above the plane. Let the components of x_j be denoted by (x_{j1}, x_{j2}, x_{j3}). Then the $k3$ component of 2.2 is:

$$
\sum_{j \neq k} \frac{m_j m_k (x_{k3} - x_{j3})}{d_{jk}^3} = 0
$$

where $d_{jk} = |x_j - x_k|$. By choice of m_k, every term of this sum is nonnegative. It follows that all of the masses have the same third component and since the center of mass is at the origin, this component must vanish.

The constant ω is determined by:

$$
(2.3) \qquad \omega^2 = \frac{U(x)}{x^T M x}
$$

which follows from 2.2 by multiplying on the left by x^T and using the homogeneity of the Newtonian potential together with the planarity of x.

Relative equilibria can be characterized as critical points of the restriction of the Newtonian potential to a certain ellipsoid in configuration space. Since the configuration is planar there is no harm in ignoring the vanishing third coordinates and supposing $x_k \in \mathbf{R}^2$ and $x \in \mathbf{R}^{2n}$. Then equation 2.2 simplifies to:

$$(2.4) \qquad\qquad \nabla U(x) = -\omega^2 M x.$$

This can be viewed as the condition for a critical point of the restriction of the potential to $\{x \in \mathbf{R}^{2n} : x^T M x = c\}$ for any constant c; $-\omega^2$ plays the role of a Lagrange multiplier. This point of view suggests a Morse-theoretical approach to the question of existence of relative equilibria [13,16,17]. For example, for each choice of n and m the restricted potential attains its minimum value at some relative equilibrium.

2.2. Linear and spectral stability. A relative equilibrium x is *linearly stable* if the origin is a stable solution of the linearization at x of the differential equation 2.1:

$$(2.5) \qquad\qquad M\ddot{\xi} + 2\omega \hat{J} M \dot{\xi} = (D\nabla U(x) + \omega^2 M \hat{I})\xi.$$

Call a complex number λ an *eigenvalue of the relative equilibrium x* if:

$$\det[M^{-1}D\nabla U(x) + \omega^2 \hat{I} - \lambda^2 I - 2\omega\lambda\hat{J}] = 0.$$

Here I is the identity matrix. Then a necesary condition for x to be linearly stable is that all of its eigenvalues are either zero or purely imaginary. Only this weaker condition, which could be called *spectral stability*, will be considered here. It is somewhat simpler to consider what might be called the normalized eigenvalues $\mu = \lambda/|\omega|$. These satisfy the equation:

$$(2.6) \qquad\qquad \det[\omega^{-2}M^{-1}D\nabla U(x) + \hat{I} - \mu^2 I - 2\mu\hat{J}] = 0$$

Since the normalization factor is real and positive, the stability condition is unchanged.

Note that $J^T M = MJ^T = -MJ$ and $(M^{-1}D\nabla U(x))^T M = D\nabla U(x)^T$ $= D\nabla U(x) = M(M^{-1}D\nabla U(x))$. These equations show that J is antisymmetric and $M^{-1}D\nabla U(x)$ is symmetric with respect to the inner product determined by M. Thus in any M-orthonormal basis, the matrix of J is antisymmetric and that of $M^{-1}D\nabla U(x)$ is symmetric. Using such a basis and taking the transposes in 2.6 one finds that the determinant is an even function of μ. Let $z = \mu^2$ and let $F(z)$ be the unique polynomial of degree $3n$ such that:

$$F(\mu^2) = \det[\omega^{-2}M^{-1}D\nabla U(x) + \hat{I} - \mu^2 I - 2\mu\hat{J}].$$

Then x is spectrally stable if and only if all of the roots of $F(z)$ are either zero or real and negative. $F(x)$ will be called the *stability polynomial of x*.

The matrix $D\nabla U(x)$ can be viewed as an $n \times n$ array of 3×3 blocks:

$$D\nabla U(x) = \begin{bmatrix} A_{11} & \cdots & A_{1n} \\ \vdots & & \ddots \\ A_{n1} & \cdots & A_{nn} \end{bmatrix}$$

The off-diagonal blocks are given by:

$$A_{jk} = \frac{m_j m_k}{d_{jk}^3} [I - 3u_{jk} u_{jk}^T] \quad j \neq k$$

where $u_{jk} = \frac{x_k - x_j}{d_{jk}}$ is the unit vector in the direction from m_j to m_k. The diagonal blocks are given by:

$$A_{kk} = - \sum_{j \neq k} A_{jk}.$$

The fact that x is planar implies that each of these 3×3 blocks takes the form:

$$A_{jk} = \begin{bmatrix} & B_{jk} & 0 \\ & & 0 \\ 0 & 0 & \frac{m_j m_k}{d_{jk}^3} \end{bmatrix}$$

where B_{jk} is a 2×2 block. Multiplication by M^{-1} as in 2.6 preserves this block structure. Since the matrices $\hat{I}$ and $\hat{J}$ share this structure, the problem of determining spectral stability splits into a planar part and a "normal" part.

To formulate the problem of planar spectral stability, adopt the point of view that $x_k \in \mathbf{R}^2$ and $x \in \mathbf{R}^{2n}$ and introduce $2n \times 2n$ matrices M, I, J, and B as follows. M is the diagonal matrix $\text{diag}(m_1, m_1, \ldots, m_n, m_n)$ and I is the identity matrix. J and B are $n \times n$ arrays of 2×2 blocks. J is block diagonal with blocks:

(2.7)
$$\begin{bmatrix} 0 & -1 \\ 1 & 0 \end{bmatrix}.$$

The blocks B_{jk} of B are the submatrices of the A_{jk} mentioned above. Explicitly,

(2.8)
$$B_{jk} = \frac{m_j m_k}{d_{jk}^3} [I - 3u_{jk} u_{jk}^T] \quad j \neq k$$

where now $u_{jk} \in \mathbf{R}^2$ and

(2.9)
$$B_{kk} = - \sum_{j \neq k} B_{jk}.$$

Define a *planar stability polynomial* $G(z)$ of degree $2n$ by:

(2.10) $G(\mu^2) = \det[\omega^{-2}M^{-1}B + (1 - \mu^2)I - 2\mu J]$.

Then x will be called *spectrally stable in the plane* if all of the roots of $G(z)$ are either zero or real and negative.

The problem of normal spectral stability can be formulated in a similar way. Define an $n \times n$ matrix C with components

(2.11)
$$C_{jk} = \frac{m_k}{d_{jk}^3} \qquad j \neq k$$

$$C_{kk} = -\sum_{j \neq k} \frac{m_k}{d_{jk}^3}$$

and let

$$H(z) = \det[\omega^{-2}C - zI].$$

Then x will be called *normally spectrally stable* if all of the roots of $H(z)$ are either zero or real and negative. Clearly, this is so if and only if the eigenvalues of C are all of this form. Note that zero is an eigenvalue of C with eigenvector $(1, \ldots, 1)$. If all other eigenvalues are strictly negative, x will be called *normally nondegenerate*.

PROPOSITION 2.1. *Every relative equilibrium is normally spectrally stable and normally nondegenerate.*

Proof. This follows from the fact that C is dominated by its negative diagonal:

$$|C_{kk}| \geq \sum_{j \neq k} |C_{jk}|.$$

In fact equality holds in this formula. Since C is symmetric with respect to M, all of its eigenvalues are real. Suppose ζ is an eigenvalue and $Cv = \zeta v$ where $v \in \mathbf{R}^n$. Let v_k be the component of v which is largest in absolute value. One may assume $v_k > 0$ without loss of generality. Then:

(2.12) $\zeta v_k = \sum_j C_{kj} v_j \leq C_{kk} v_k + \sum_{j \neq k} |C_{kj} v_j| \leq \left(-|C_{kk}| + \sum_{j \neq k} |C_{jk}|\right) v_k \leq 0$

which shows that $\zeta \leq 0$. Hence x is normally spectrally stable.

To see that x is also normally nondegenerate note that if v is in the kernel of C and v_k is chosen as above then one must have equalities in 2.12. But this happens if and only if $v_j = v_k$ for all j and then v is proportional to $(1, \ldots, 1)$.   □

The block structure of the matrices in 2.6 implies that the three stability polynomials satisfy $F(z) = G(z)H(z)$ and so the relative equilibrium x is spectrally stable if and only if it is both normally spectrally stable and spectrally stable in the plane. In light of the proposition, one can concentrate on spectral stability in the plane.

2.3. Spectral stability in the plane. In order to study spectral stability in the plane for the examples in section 3 it will be necessary to use special structural properties of the matrix B to obtain factorizations of the planar stability polynomial, $G(z)$. The proposition below shows that such a factorization can be obtained by finding subspaces of $\mathbf{R}^{2n}$ which are simultaneously invariant for the linear transformations J and $M^{-1}B$.

PROPOSITION 2.2.   *Let $S \subset \mathbf{R}^{2n}$ be a subspace such that $JS = M^{-1}BS = S$ and let $S^{\perp} = \{v \in \mathbf{R}^{2n} : v^T M w = 0 \text{ for all } w \in S\}$ be the orthogonal complement of S with respect to M. Then $JS^{\perp} = M^{-1}BS^{\perp} = S^{\perp}$ and the splitting $\mathbf{R}^{2n} = S \oplus S^{\perp}$ induces a factorization $G(z) = G_1(z)G_2(z)$ where $G_1(z)$ and $G_2(z)$ are given by 2.10 with the operators restricted to S and $S^{\perp}$ respectively.*

Proof. As noted above, using an M-orthonormal basis makes the matrix of J antisymmetric and that of $M^{-1}B$ symmetric. The invariance of $S^{\perp}$ is a standard fact about such matrices.

Since the linear transformation in 2.10 involves only $M^{-1}B$, J and the identity, it splits into a direct sum of its restrictions to the two invariant subspaces S and $S^{\perp}$. This lead to a factorization of $G(\mu^2)$ into two polynomial in μ. It remains to show that both of these are even functions. But the proof of the evenness depended only on the antisymmetry of J and the symmetry of $M^{-1}B$ and these properties apply equally to their restrictions.   □

The evenness of the factors as functions of μ shows that any such invariant subspace, S, must have even dimension. The simplest case would be dimension two. Since $M^{-1}B$ is symmetric with respect to M, all of its eigenvalues are real and the eigenvectors can be chosen M-orthogonal. Any $M^{-1}B$-invariant subspace S will contain an eigenvector. Let $v \in S$ be any eigenvector. If S is also J-invariant, it will contain $w = Jv$ as well and moreover, w is M-orthogonal to v. Thus w is also an eigenvector of $M^{-1}B$.

PROPOSITION 2.3.   *Suppose that $S \subset \mathbf{R}^{2n}$ is a two-dimensional subspace which is simultaneously invariant for $M^{-1}B$ and J. Let the eigenvalues of the restriction of $\omega^{-2}M^{-1}B$ be a and b. Then $G(z)$ has a quadratic factor:*

$$Q(z) = z^2 + \alpha z + \beta$$

where $\alpha = 2 - a - b$ and $\beta = (1 + a)(1 + b)$. The corresponding roots are real and negative if and only if:

$$\alpha > 0 \qquad \beta > 0 \qquad \alpha^2 - 4\beta \geq 0.$$

Proof. By proposition 2.2, there will be a quadratic factor corresponding to the subspace S. To find it, introduce a basis of eigenvectors $\{v, w\}$ for $M^{-1}B$ with $w = Jv$. The matrix of the restriction of $\omega^{-2}M^{-1}B$ is simply $\mathrm{diag}(a, b)$ and the matrix of J is 2.7. From 2.10:

$$Q(\mu^2) = \det \begin{bmatrix} a + 1 - \mu^2 & 2\mu \\ -2\mu & b + 1 - \mu^2 \end{bmatrix}.$$

The proposition follows easily.   □

This proposition can be applied at least twice to every relative equilibrium. For the first application let $v = (1, 0, \ldots, 1, 0)$ and $w = (0, 1, \ldots, 0, 1)$. It follows from 2.9 that these are in the kernel of B. Substituting $a = b = 0$ into $Q(z)$ yield the repeated root $z = -1$. The corresponding eigenvalues of the relative equilibrium are $\lambda = \pm i$.

Next consider $v = x$ and $w = Jx$. Equation 2.4 together with the homogeneity of the Newtonian potential give:

$$M^{-1} D\nabla U(x)v = -2M^{-1}\nabla U(x) = 2M^{-1}\omega^2 Mx = 2\omega^2 v.$$

So v is an eigenvector of $M^{-1}B$ with eigenvalue $2\omega^2$. On the other hand $w = R'(0)x$ where $R(\theta)$ is the $2n$-dimensional version of the rotation operator introduced above. Now

$$\nabla U(R(\theta)x) = R(\theta)\nabla U(x)$$

because of the rotation invariance of the potential. Differentiation of this formula at $\theta = 0$ gives:

$$M^{-1} D\nabla U(x)w = M^{-1} J\nabla U(x) = -M^{-1} J\omega^2 Mx = -\omega^2 w$$

which means that w is also an eigenvector, with eigenvalue $-\omega^2$. Substituting $a = 2$ and $b = -1$ into $Q(z)$ gives the roots $z = 0$ and $z = -1$.

The root $z = 0$ found above was present because of the rotational symmetry of the potential. If the other $2n - 1$ roots of $G(z)$ are nonzero then x will be called *nondegenerate in the plane.*

Although proposition 2.3 will find further application below, it usually happens that when v is an eigenvector of $M^{-1}B$, $w = Jv$ is not. Then the two-dimensional subspace which they span will not be $M^{-1}B$-invariant.

Another simple situation, and one which will occur in the examples, is that $M^{-1}B$ may admit two invariant planes which are transformed into one another by J. Suppose C and D are two-dimensional subspaces of $\mathbf{R}^{2n}$ which are $M^{-1}B$-invariant and also satisfy $D = JC$. Then $S = C \oplus D$ is simultaneously invariant for J and $M^{-1}B$. In the following proposition, the assumption about the form of the restrictions of $M^{-1}B$ to the subspaces are motivated by the examples in section 3; it is not generally possible to make both restrictions diagonal without unduly complicating the matrix of J.

PROPOSITION 2.4. *Suppose C and D are two-dimensional subspaces of $\mathbf{R}^{2n}$ which satisfy $C \cap D = \{0\}$, $(M^{-1}B)C = C$, $(M^{-1}B)D = D$ and*

$D = JC$. Then $S = C \oplus D$ is a four-dimensional subspace which is simultaneously invariant for J and $M^{-1}B$. Choose bases $\{v, v'\}$ for C and $\{Jv, Jv'\}$ for D and suppose the matrices of the restrictions of $\omega^{-2}M^{-1}B$ to C and D are:

$$\begin{bmatrix} a & c \\ d & b \end{bmatrix} \qquad \begin{bmatrix} b & d \\ c & a \end{bmatrix}$$

respectively. Then $G(z)$ has a quartic factor:

$$\begin{aligned} Q(z) &= z^4 + 2\alpha z^3 + (\alpha^2 + 2\beta)z^2 + (2\alpha\beta + 4(c+d)^2)z + \beta^2 \\ &= (z^2 + \alpha z + \beta)^2 + 4(c+d)^2 z \end{aligned}$$

where

$$\begin{aligned} \alpha &= 2 - a - b \\ \beta &= (a+1)(b+1) - cd. \end{aligned}$$

Proof. From the assumption $D = JC$ and the choice of basis, one finds that the matrix of J on S is:

$$J = \begin{bmatrix} 0 & -I \\ I & 0 \end{bmatrix}$$

where I is the 2×2 identity matrix.

Proposition 2.2 shows that $G(z)$ has a quartic factor:

$$Q(\mu^2) = \det \begin{bmatrix} a+1-\mu^2 & c & 2\mu & 0 \\ d & b+1-\mu^2 & 0 & 2\mu \\ -2\mu & 0 & b+1-\mu^2 & d \\ 0 & -2\mu & c & a+1-\mu^2 \end{bmatrix}.$$

Evaluating this 4×4 determinant leads to the formulas in the proposition. □

Because of its special form, it is possible to state relatively simple necessary and sufficient conditions for this quartic to have real and negative roots.

PROPOSITION 2.5. *Suppose $c + d \neq 0$. Then the roots of the quartic:*

$$Q(z) = (z^2 + \alpha z + \beta)^2 + 4(c+d)^2 z$$

are all real and negative if and only if

$$\alpha > 0 \qquad \beta \neq 0 \qquad \alpha^2 - 4\beta \geq 0 \qquad \Delta \geq 0$$

where Δ is the discriminant of $Q(z)$:

$$\Delta = (c+d)^4 \left(\beta(\alpha^2 - 4\beta)^2 + \alpha(\alpha^2 - 36\beta)(c+d)^2 - 27(c+d)^4 \right).$$

Proof. The fourth condition involves the discriminant about which the following facts are needed [4]: $\Delta = 0$ if and only if $Q(z)$ has repeated roots, $\Delta < 0$ if and only if $Q(z)$ has two distinct real roots and a pair of nonreal roots, and $\Delta > 0$ if and only if $Q(z)$ has four distinct roots either all real or all nonreal.

First it will be shown that the four conditions are necessary. So suppose all of the roots are real and negative. Note that 2α is the coefficient of z^3 in $Q(z)$, so it is minus the sum of the roots; hence $\alpha > 0$. If β were equal to zero then $z = 0$ would be a root of $Q(z)$, contrary to the assumption. Since all roots are real, $\Delta \geq 0$. It remains to show that $\alpha^2 - 4\beta \geq 0$. From Rolle's theorem one sees that all of the roots of the derivatives of $Q(z)$ must be real. But $Q''(z) = 12z^2 + 12\alpha z + 2\alpha^2 + 4\beta$ which has real roots if and only if $\alpha^2 - 4\beta \geq 0$.

For the converse, suppose that all four conditions hold. The first three conditions show that $z^2 + \alpha z + \beta$ has at least one real, negative root, $r < 0$. Since $c + d \neq 0$, $Q(r) = 4(c+d)^2 r < 0$. Since $\beta \neq 0$, $Q(0) > 0$, and it follows that $Q(z)$ has at least two distinct negative real roots. If the other roots were not real one would have $\Delta < 0$. Hence all four roots are real. Since $Q(z) > 0$ for $z \geq 0$, all of them must be negative as required.   □

The observation that the real roots of this kind of quartic are always negative provided $\beta \neq 0$ and $c + d \neq 0$ will be used repeatedly below.

3. Examples. In this section, several simple kinds of relative equilibrium will be analyzed for linear stabilty in the plane. These examples are chosen primarily because the results of section 2.3 apply. The first two sections describe known results.

3.1. Collinear relative equilibria. Suppose $x_j \in \mathbf{R} \times 0$ for all j and $x_j < x_k$ for $j < k$. In 1767, Euler solved equation 2.4 in the case $n = 3$ finding a unique solution up to scaling [6]. Lehmann-Filhes and Moulton generalized this result to arbitrary n [9,15]. Unfortunately, the x_j are rather complicated algebraic functions of the masses. The first stability analysis of these relative equilibria can be attributed to Andoyer [1] and M. G. Meyer [12].

Since the difference vectors $x_j - x_k \in \mathbf{R} \times 0$ the unit vectors u_{jk} which appear in equation 2.8 all reduce to $(\pm 1, 0)$. Hence:

$$B_{jk} = \frac{m_j m_k}{d_{jk}^3} \begin{bmatrix} -2 & 0 \\ 0 & 1 \end{bmatrix}.$$

It is clearly convenient to permute the coordinates of $\mathbf{R}^{2n}$ so that the x_{k1} are together and the x_{k2} are together (recall $x_k = (x_{k1}, x_{k2})$). Then $M^{-1}B$ becomes block diagonal with two $n \times n$ blocks:

$$M^{-1}B = \begin{bmatrix} -2C & 0 \\ 0 & C \end{bmatrix}$$

where C is the matrix 2.11 which appeared in the analysis of normal spectral stability. It is clear that if $(v_1, \ldots, v_n)$ is an eigenvector of $w^{-2}C$ with eigenvalue a then $v = (0, \ldots, 0, v_1, \ldots, v_n)$ and $w = -(v_1, \ldots, v_n, 0, \ldots, 0)$ are eigenvectors of $w^{-2}M^{-1}B$ with eigenvalues a and b respectively, where $b = -2a$. Moreover, $w = Jv$.

Applying proposition 2.3 one finds that the planar stability polynomial has a quadratic factor:

$$Q(z) = z^2 + (2+a)z + (1+a)(1-2a).$$

The eigenvalues $a = 0$ and $a = -1$ of $w^{-2}C$ give the root $z = 0$ and the triply repeated root $z = -1$ already known from section 2.3. M. G. Meyer claims that the other eigenvalues satisfy $a < -1$ but the author was not able to fill in the details in the proof [12]. A more recent proof due to Conley can be found in [16]. Using this fact, one finds that the quadratic factor has real roots of opposite sign. In particular, x is not stable in the plane. The following table classifies the roots of the planar stability polynomial other than the root $z = 0$ and the triply repeated $z = -1$.

Roots of the stability polynomial for a collinear relative equilibrium			
n	Negative	Positive	Nonreal
$n \geq 3$	n-2	n-2	0

3.2. Equilateral triangle. In this section, the famous equilateral relative equilibrium discovered by Lagrange will be analyzed. The results are well-known [20] but the derivation below is considerably simpler than previously published ones.

Let $n = 3$. Then no matter how the masses m_1, m_2, and m_3 are chosen, the equilateral triangular configurations with the center of mass at the origin are relative equilibria.

The planar stability polynomial has $2n = 6$ roots. The root $z = 0$ and a triply repeated root $z = -1$ were found in section 2.3. It remains to determine the other two roots. Recall that the four known roots were found by constructing two planes in $\mathbf{R}^6$ which were simultaneously invariant under J and $M^{-1}B$. The M-orthogonal complement of the direct sum of these two planes is another plane, S, which is simultaneously invariant under J and $M^{-1}B$. If the eigenvalues a and b of the restriction of $w^{-2}M^{-1}B$ to S were known, proposition 2.3 could be applied to determine the last two roots of the stability polynomial. It is possible to deduce the eigenvalues a and b with a minimal amount of computation.

First one needs to find the matrix $w^{-2}M^{-1}B$ itself. Assume that the particles are at positions $x_j = x'_j - c$ where c is the center of mass and

$$x'_1 = (1,0) \qquad x'_2 = (-1/2, \sqrt{3}/2) \qquad x'_3 = (-1/2, -\sqrt{3}/2).$$

The three mutual distances are $d_{jk} = \sqrt{3}$, the potential energy is $U(x) = (m_1m_2 + m_1m_3 + m_2m_3)/\sqrt{3}$ and the moment of inertia is $x^T M x = 3(m_1m_2 + m_1m_3 + m_2m_3)/\mu$ where $\mu = m_1 + m_2 + m_3$. From 2.3, $\omega^2 = \mu/3\sqrt{3}$. The formulas 2.8 for the second derivative matrix, B, are translation invariant so one can use the x'_{jk} instead of the x_{jk}. One finds that $\omega^{-2}M^{-1}B$ is:

$$\frac{1}{4\mu}\begin{bmatrix} 5(m_2+m_3) & 3\sqrt{3}(m_3-m_2) & -5m_2 & 3\sqrt{3}m_2 & -5m_3 & -3\sqrt{3}m_3 \\ 3\sqrt{3}(m_3-m_2) & -(m_2+m_3) & 3\sqrt{3}m_2 & m_2 & -3\sqrt{3}m_3 & m_3 \\ -5m_1 & 3\sqrt{3}m_1 & 5m_1-4m_3 & -3\sqrt{3}m_1 & 4m_3 & 0 \\ 3\sqrt{3}m_1 & m_1 & -3\sqrt{3}m_1 & -m_1+8m_3 & 0 & -8m_3 \\ -5m_1 & -3\sqrt{3}m_1 & 4m_2 & 0 & 5m_1-4m_2 & 3\sqrt{3}m_1 \\ -3\sqrt{3}m_1 & m_1 & 0 & -8m_2 & 3\sqrt{3}m_1 & -m_1+8m_2 \end{bmatrix}.$$

(3.1)

The four eigenvalues of this matrix found in section 2.3 are $0, 0, 2, -1$. Taking the trace of 3.1 gives $2 + (-1) + a + b = 2$ or $a + b = 1$. Thus the first parameter in proposition 2.3 is:

$$\alpha = 2 - a - b = 1.$$

To find $\beta = (a + 1)(b + 1) = ab + 2$ one needs to compute the product ab. The sum of all possible products of pairs of eigenvalues of 3.1 is $-2 + 2a + 2b - a - b + ab = ab - 1$. On the other hand one can find this sum as one-half of the difference of the square of the trace of 3.1 and the trace of its square. Carrying out this computation yields:

$$ab - 1 = (-12(m_1^2 + m_2^2 + m_3^2) + 3(m_1m_2 + m_1m_3 + m_2m_3))/4\mu^2$$

which leads to:

$$\beta = 27(m_1m_2 + m_1m_3 + m_2m_3)/4\mu^2.$$

Since $\alpha > 0$ and $\beta > 0$, the equilateral triangle is spectrally stable if and only if:

$$27(m_1m_2 + m_1m_3 + m_2m_3) < \mu^2.$$

For the majority of mass triples, this inequality fails to hold and the stability polynomial has two nonreal roots.

3.3. Regular polygons. Suppose $m_1 = \ldots = m_n = 1$. Then the regular n-gon, $x_k = (\cos\theta_k, \sin\theta_k)$, $\theta_k = \frac{2\pi k}{n}$ is a relative equilibrium since the symmetry implies that the acceleration vector of each mass is directed towards the origin and that the magnitudes of these vectors are all equal. Since the case $n = 2$ gives a collinear configuration, assume for the rest of this section that $n \geq 3$.

The mutual distances are given by:

$$d_{j(j+k)} = d_{nk} = 2\sin\frac{\pi k}{n}$$

where indices are interpretted modulo n. The formula $d_{nk}^2 = 2(1 - \cos \theta_k)$ will also be useful. The potential is:

$$U(x) = \frac{n}{2} \sum_{k=1}^{n-1} \frac{1}{d_{nk}} = \frac{n}{4} \sum_{k=1}^{n-1} \csc \frac{\pi k}{n}$$

and by 2.3 the rotation frequency is given by:

$$\omega^2 = \frac{1}{2} \sum_{k=1}^{n-1} \frac{1}{d_{nk}} = \frac{1}{4} \sum_{j=1}^{n-1} \csc \frac{\pi k}{n}.$$

Palmore was the first to notice that the matrix B has an especially simple structure in this case [17]. To elucidate this structure it is convenient to express all of the unit vectors u_{jk} of 2.8 in terms of the $u_{nk} = d_{nk}^{-1}(\cos \theta_k - 1, \sin \theta_k)$. Let

$$R_k = \begin{bmatrix} \cos \theta_k & -\sin \theta_k \\ \sin \theta_k & \cos \theta_k \end{bmatrix}.$$

Then $u_{j(j+k)} = R_j u_{nk}$. Thus all of the blocks B_{jk} of B can be expressed in terms of the B_{nk}:

$$(3.2) \qquad B_{j(j+k)} = R_j B_{nk} R_j^T.$$

and a short computation shows:

$$(3.3) \qquad B_{nk} = \frac{1}{2d_{nk}^3} \begin{bmatrix} -1 + 3 \cos \theta_k & 3 \sin \theta_k \\ 3 \sin \theta_k & -1 - 3 \cos \theta_k \end{bmatrix}$$

LEMMA 3.1. *Let* $\rho_l = \cos \theta_l + i \sin \theta_l$ *be any n-th root of unity and define a 2×2 matrix*

$$B_{\rho_l} = \sum_{k=1}^{n} \rho_l^k B_{nk} R_k = \begin{bmatrix} P_l - 3Q_l & iR_l \\ -iR_l & P_l + 3Q_l \end{bmatrix}$$

where

$$(3.4) \qquad \begin{aligned} P_l &= \sum_{k=1}^{n-1} \frac{1 - \cos \theta_k \cos \theta_{kl}}{2d_{nk}^3} \\ Q_l &= \sum_{k=1}^{n-1} \frac{\cos \theta_k - \cos \theta_{kl}}{2d_{nk}^3} \\ R_l &= \sum_{k=1}^{n-1} \frac{\sin \theta_k \sin \theta_{kl}}{2d_{nk}^3}. \end{aligned}$$

If $u_0 \in \mathbf{C}^2$ and $u = (\ldots, \rho_l^k R_k u_0, \ldots) \in \mathbf{C}^{2n}$, then

$$Bu = (\ldots, \rho_l^k R_k B_{\rho_l} u_0, \ldots).$$

In particular, if u_0 is an eigenvector of B_{ρ_l} then u is an eigenvector of B with the same eigenvalue.

Proof. It is not difficult to find the explicit formula for B_{ρ_l} by using equations 3.3 and 2.9. Using equation 3.2 one finds that the j-th pair of components of Bu is:

$$\sum_{k=1}^{n} R_j B_{nk} R_j^T u_{j+k} \;=\; \rho_l^j R_j \sum_{k=1}^{n} \rho_l^k B_{nk} R_k u_0 = \rho_l^j R_j B_{\rho_l} u_0$$

as claimed.   □

This lemma will be used to find a splitting of $\mathbf{R}^{2n}$ into two-dimensional B-invariant subspaces. Fixing an integer l and using the initial vectors $u_0 = (1, 0)$ and $u_0' = (0, 1)$ leads to the construction of two complex vectors $u = v + iw$ and $u' = w' - iv'$ where:

$$
\begin{aligned}
v_k &= \cos\theta_{kl}(\cos\theta_k, \sin\theta_k) \\
w_k &= \sin\theta_{kl}(\cos\theta_k, \sin\theta_k) \\
w_k' &= \cos\theta_{kl}(-\sin\theta_k, \cos\theta_k) \\
v_k' &= \sin\theta_{kl}(\sin\theta_k, -\cos\theta_k).
\end{aligned}
$$

It is not difficult to show that these four vectors are linearly independent except when $l = 0$ and when n is even and $l = \frac{n}{2}$. In both cases $w = v' = 0$ and the other two vectors are independent and satisfy the hypotheses of proposition 2.3. By symmetry, one need only consider l in the range $0 \leq l \leq \frac{n}{2}$.

Setting aside these two exceptional cases and taking real and imaginary parts of the formula in the lemma one finds that the subspaces $C_l = <v, v'>$ and $D_l = <w', w>$ are B-invariant. Moreover the restrictions of B to these subspaces have matrices:

$$
\begin{bmatrix} P_l - 3Q_l & -R_l \\ -R_l & P_l + 3Q_l \end{bmatrix}
\qquad
\begin{bmatrix} P_l + 3Q_l & -R_l \\ -R_l & P_l - 3Q_l \end{bmatrix}.
$$

Note also that $Jv = w'$ and $Jv' = w$ so $JC = D$. Thus the hypotheses of proposition 2.4 hold. Dividing the matrices above by ω^2 shows that the parameters a, b, c, and d of the proposition are $a = p_l - 3q_l$, $b = p_l + 3q_l$ and $c = d = -r_l$ where $P_l = \omega^2 p_l$, $Q_l = \omega^2 q_l$ and $R_l = \omega^2 r_l$. So the planar stability polynomial has quartic factors as in proposition 2.4 with:

$$(3.5) \quad \alpha_l = 2(1 - p_l) \qquad \beta_l = (1 + p_l)^2 - 9q_l^2 - r_l^2 \qquad (c_l + d_l)^2 = 4r_l^2.$$

The exceptional case $l = 0$ leads to a quadratic factor as in proposition 2.3. Equations 3.4 show that $R_0 = 0$ and $Q_0 = -P_0$ where:

$$(3.6) \qquad P_0 = \sum_{k=1}^{n-1} \frac{1 - \cos \theta_k}{2d_{nk}^3} = \frac{1}{4} \sum_{k=1}^{n-1} \frac{1}{d_{nk}} = \frac{1}{2}\omega^2.$$

Proposition 2.3 gives the quadratic factor:

$$G_0(z) = z^2 + z.$$

The other exceptional case, when n is even and $l = \frac{n}{2}$, can be handled similarly. One finds $R_{\frac{n}{2}} = 0$. Applying proposition 2.3 leads to a quadratic factor with:

$$(3.7) \qquad \alpha_{\frac{n}{2}} = 2(1 - p_{\frac{n}{2}}) \qquad \beta_{\frac{n}{2}} = (1 + p_{\frac{n}{2}})^2 - 9q_{\frac{n}{2}}^2.$$

Now the roots of these factors will be studied. The quadratic factor with $l = 0$ gives the roots $z = 0$ and $z = -1$ found already in section 2.3. Next consider the quartic factor which arise from the case $l = 1$. One has $Q_1 = 0$ and $R_1 = P_1$ with:

$$P_1 = \sum_{k=1}^{n-1} \frac{1 - \cos^2 \theta_k}{2d_{nk}^3}$$

From this it follows that the quartic factor can be factorized as:

$$(z + 1)^2 (z^2 + 2(1 - 2p_1)z + (1 + 2p_1)^2).$$

The repeated root $z = -1$ was known already from section 2.3 but the other quadratic factor leads to nontrivial roots. Since p_1 is clearly positive, these roots are nonreal and so the regular n-gon is always linearly unstable.

To understand the rest of the roots, the following lemmas are required. The first lemma concerns a sum which arises frequently in the study of the regular n-gon.

LEMMA 3.2. *Let*

$$(3.8) \qquad \begin{aligned} A_n &= \frac{1}{n} \sum_{k=1}^{n-1} \frac{1}{d_{nk}} = \frac{1}{2n} \sum_{k=1}^{n-1} \csc \frac{\pi k}{n} \\ A_n^- &= \frac{1}{2\pi} \ln \frac{1 + \cos \frac{\pi}{n}}{1 - \cos \frac{\pi}{n}} + \frac{1}{2n \sin \frac{\pi}{n}} \end{aligned}$$

Then for all $n \geq 3$:

$$A_n > A_n^-.$$

Moreover A_n^- *is monotonically increasing with* n.

Proof. The sum:

$$\sum_{k=1}^{n-1} \csc \frac{\pi k}{n}$$

can be estimated as follows. Since the function $f(u) = \csc \frac{\pi u}{n}$ is convex on $[1, n-1]$ the trapezoidal rule gives an upper bound for its integral over this interval, that is:

$$\int_1^{n-1} f(x)\, dx < \tfrac{1}{2}f(1) + f(2) + \ldots + f(n-2) + \tfrac{1}{2}f(n-1).$$

This leads directly to the formula for A_n^-.

To show that A_n^- is monotonically increasing one can replace $\frac{\pi}{n}$ by a continuous variable u and verify that the derivative of the resulting function is negative. This computation will be omitted. $\square$

The next lemma compares the values of the quantities P_l and Q_l, for fixed n and varying l.

LEMMA 3.3. *All of the following sequences are monotonically increasing for $1 \le l \le \frac{n}{2}$:*

$$P_l \qquad Q_l \qquad Q_l - P_l.$$

Furthermore $P_l > 0$ for $1 \le l \le \frac{n}{2}$ and $Q_l > 0$ for $2 \le l \le \frac{n}{2}$.

Proof. The method of proof, which involves the differences of the given sequences, is similar to one used in [18]. Introduce the notation $\Delta P_l = P_{l+1} - P_l$, $\Delta^2 P_l = \Delta P_{l+1} - \Delta P_l$, $\Delta^3 P_l = \Delta^2 P_{l+1} - \Delta^2 P_l$, and similarly for the differences of Q_l.

First it will be shown that $\Delta P_l \ge 0$ for $1 \le l \le \frac{n}{2} - 1$. A computation using 3.4 and the trigonometric identities for the sine and cosine of a sum and difference gives:

$$\Delta^3 P_l = -\tfrac{1}{4} \sum_{k=1}^{n-1} \sin \tfrac{\pi k(2l+1)}{n} + \sin \tfrac{\pi k(2l+5)}{n} = -\tfrac{1}{4}(\cot \tfrac{\pi(2l+1)}{2n} + \cot \tfrac{\pi(2l+5)}{2n}).$$

This is negative for $1 \le l \le \frac{n}{2} - 3$ and so the sequence ΔP_l is concave for $1 \le l \le \frac{n}{2} - 1$. Thus to show $\Delta P_l \ge 0$ for $1 \le l \le \frac{n}{2} - 1$ it suffices to check the endpoints of the range.

First consider $l = 1$. From 3.4 and more trigonometry:

$$\Delta P_1 \;=\; P_2 - P_1 \;=\; \sum_{k=1}^{n-1} \frac{\cos \theta_k (\cos \theta_k - \cos \theta_{2k})}{2 d_{nk}^3}$$

$$=\; \tfrac{3}{4} \sum_{k=1}^{n-1} \frac{1}{d_{nk}} - \tfrac{3}{8} \sum_{k=1}^{n-1} d_{nk} - \tfrac{1}{4} \sum_{k=1}^{n-1} \cos \theta_k d_{nk}.$$

The first sum can be estimated using 3.8 while the last two can be summed explicitly. In fact:

(3.9)
$$\sum_{k=1}^{n-1} d_{nk} = 2\cot\tfrac{\pi}{2n}$$

$$\sum_{k=1}^{n-1} \cos\theta_k d_{nk} = \cot\tfrac{3\pi}{2n} - \cot\tfrac{\pi}{2n}.$$

Using $\cot u \le 1/u$ one finds $\Delta P_1 \ge \tfrac{3n}{4}A_n^- - \tfrac{7n}{6\pi}$. For $\Delta P_1 \ge 0$ it is sufficient that $A_n^- \ge \tfrac{14}{9\pi} = 0.4951\ldots$. But A_n^- is an increasing sequence and $A_5^- = 0.5279\ldots$. Hence $\Delta P_1 \ge 0$ for $n \ge 5$. It can be verified numerically for $n = 4$.

Next consider $l = [\tfrac{n}{2} - 1]$ where the square brackets denote the greatest integer function. When n is odd,

$$\Delta P_{[\frac{n}{2}-1]} = P_{\frac{n+1}{2}} - P_{\frac{n-1}{2}} = 0$$

by symmetry. When n is even another trigonometric computation reveals that:

$$\Delta P_{[\frac{n}{2}-1]} = P_{\frac{n}{2}} - P_{\frac{n}{2}-1} = \sum_{k=1}^{n-1} \frac{(-1)^{k+1}\cos\theta_k}{4d_{nk}}$$

$$= \tfrac{1}{4}\sum_{k=1}^{n-1}\frac{(-1)^{k+1}}{d_{nk}} - \tfrac{1}{8}\sum_{k=1}^{n-1}(-1)^{k+1}d_{nk}$$

By symmetry, the first sum can be estimated using:

$$\sum_{k=1}^{n-1}\frac{(-1)^{k+1}}{d_{nk}} \ge 2\sum_{k=1}^{\frac{n}{2}}\frac{(-1)^{k+1}}{d_{nk}}$$

if $\tfrac{n}{2}$ is even and as:

$$\sum_{k=1}^{n-1}\frac{(-1)^{k+1}}{d_{nk}} \ge 2\sum_{k=1}^{\frac{n}{2}-1}\frac{(-1)^{k+1}}{d_{nk}}$$

if $\tfrac{n}{2}$ is odd. In both cases the estimates are alternating with decreasing terms and so are positive. The second sum is found to be:

$$\sum_{k=1}^{n-1}(-1)^{k+1}d_{nk} = -2\cot\tfrac{\pi}{2n} + 4\cot\tfrac{\pi}{n} \le -\frac{\pi}{n}$$

where the last inequality is obtained by using $\tfrac{1}{u} - \tfrac{u}{2} \le \cot u \le \tfrac{1}{u}$. In particular, this sum is negative and it follows that $\Delta P_{[\frac{n}{2}-1]} > 0$. This completes the proof that P_l is increasing.

Next it will be shown that $\Delta Q_l \geq \Delta P_l$ for $1 \leq l \leq \frac{n}{2} - 1$. From this it will follow that Q_l and $Q_l - P_l$ are increasing. Yet another trigonometric computation gives:

$$\Delta Q_l - \Delta P_l = \sum_{k=1}^{n-1} \frac{(1 - \cos \theta_k)(\cos \theta_{k(l+1)} - \cos \theta_{kl})}{2d_{nk}^3}$$

$$= \frac{1}{4} \sum_{k=1}^{n-1} \sin \frac{\pi k (2l+1)}{n} = \frac{1}{4} \cot \frac{\pi(2l+1)}{2n}$$

which is positive for $1 \leq l \leq \frac{n}{2} - 1$ as claimed.

It is clear from the definitions that $P_l > 0$ and $Q_1 = 0$. Since Q_l is monotonically increasing, $Q_l > 0$ for $2 \leq n \leq \frac{n}{2}$. □

The last lemma concerns the quantities α_l and β_l of equations 3.5 and 3.7 and the discriminant as in proposition 2.5.

LEMMA 3.4. *The coefficients α_l and β_l, $2 \leq l \leq \frac{n}{2}$ and the discriminant Δ_l, $2 \leq l < \frac{n}{2}$ are negative except in the following cases:.*

$$\begin{array}{ll}
\alpha_2 > 0 & 4 \leq n \leq 7 \\
\alpha_3 > 0 & n = 6 \\
\beta_2 > 0 & 4 \leq n \leq 9 \\
\Delta_2 > 0 & n = 5, 6.
\end{array}$$

Proof. The proof is a combination of numerical evaluation of the quantities involved for small n and estimates for large n. For $n \leq 13$, one can verify the signs by evaluating the finite sums involved with a computer. It will now be shown that $\alpha_l < 0$ and $\beta_l < 0$ for $n \geq 14$. The fact that the discriminant is negative in these cases follows from the formula for Δ_l in proposition 2.5.

Since $\alpha_l = 2(1 - p_l)$, it follows from lemma 3.3 that α_l is monotonically decreasing in l. Therefore it suffices to show that $\alpha_2 < 0$. This is equivalent to showing $P_2 - \omega^2 = P_2 - 2P_0 > 0$. A computation like those in the proof of the last lemma gives:

$$P_2 - 2P_0 = \frac{3}{4} \sum_{k=1}^{n-1} \frac{1}{d_{nk}} - \frac{1}{2} \sum_{k=1}^{n-1} d_{nk} - \frac{1}{4} \sum_{k=1}^{n-1} \cos \theta_k d_{nk}$$

$$= \frac{3n}{4} A_n - \frac{3}{4} \cot \frac{\pi}{2n} - \frac{1}{4} \cot \frac{3\pi}{2n}$$

It is therefore sufficient to show that $A_n^- \geq \frac{20}{9\pi} = 0.7074\ldots$. By lemma 3.2, A_n^- is increasing and $A_9^- = 0.7148\ldots$ so the sufficient condition holds for $n \geq 9$.

The quantity $\beta_l = (1 + p_l)^2 - 9q_l^2 - r_l^2$ can be estimated by a somewhat crude upper bound for β_l obtained by ignoring the last term:

$$\beta_l \leq (1 + p_l + 3q_l)(1 + p_l - 3q_l).$$

The first factor is always positive, so to prove that $\beta_l < 0$ it suffices to show that the second factor is negative. This factor can be written: $1 - (q_l - p_l) - 2q_l$. Using lemma 3.3 one sees that this is monotonically decreasing in l. Hence it suffices to consider the case $l = 2$. One finds, by means of more trigonometric computation that:

$$
\begin{aligned}
\omega^2 + P_2 - 3Q_2 &= -\frac{1}{2}\sum_{k=1}^{n-1}\frac{1}{d_{nk}} + \frac{1}{4}\sum_{k=1}^{n-1} d_{nk} - \frac{1}{4}\sum_{k=1}^{n-1}\cos\theta_k d_{nk} \\
&= -\frac{n}{2}A_n + \frac{3}{4}\cot\frac{\pi}{2n} - \frac{1}{4}\cot\frac{3\pi}{2n}.
\end{aligned}
$$

For this to be negative it suffices to have $A_n^- + \frac{1}{2n}\cot\frac{3\pi}{2n} > \frac{3}{\pi}$. the left hand side is increasing and for $n = 14$ the inequality is $0.9575\ldots > 0.9549\ldots$. Hence $\beta_l < 0$ for $n \geq 14$ as claimed. $\square$

Now one can determine the nature of the roots of the quadratic and quartic factors of the stability polynomial. The quadratic factor arising from the case $l = \frac{n}{2}$ gives two complex roots when $n = 4$ and gives one positive and one negative real root for $n = 2k$, $k \geq 3$. Recall that any quartic factor with a negative discriminant has two real roots which are distinct and two nonreal ones. Also recall that when $\beta \neq 0$ any real root of these quartic factors must be negative. Hence, by the lemma, all of the quartic factors with $n \geq 7$ give two distinct negative real roots and two nonreal roots. It remains to analyze the quartic factors with $l = 2$ and $n = 5, 6$. Since the discriminant is positive, they either have four real (and hence negative) roots or four nonreal ones. But in both cases $\alpha_2^2 - 4\beta_2 < 0$ so by proposition 2.5 the roots cannot all be real and negative. Hence these two quartics have four nonreal roots. These results are summarized in the following table which classifies the roots of the planar stability polynomial. The roots $z = 0$ and the triply repeated $z = -1$ found in section 2.3 have been omitted.

Roots of the stability polynomial for the regular n-gon			
n	Negative	Positive	Nonreal
3	0	0	2
4	0	0	4
5	0	0	6
6	1	1	6
odd$\geq$ 7	n-3	0	n-1
even$>$ 7	n-3	1	n-2

For $n \geq 7$ and counting the triple repeated $z = -1$, one finds that exactly half of the roots are real and negative. This is the same as for the collinear relative equilibria. Note that for $n = 3, 4, 5$ the n-gon exhibits the maximum possible instability.

3.4. Regular polygons with a central mass. Suppose $n \geq 3$ and $m_1 = \ldots = m_n = 1$ as in the previous section but add another mass $m_0 = m$. Then, whatever the value of m, the configuration with $x_0 = (0,0)$ and the other masses forming a regular n-gon is a relative equilibrium.

The mutual distances are given by the formulas of the last section together with $d_{0k} = 1$. The potential is

$$U(x) = nm + \tfrac{n}{4} \sum_{k=1}^{n-1} \csc \frac{\pi k}{n}$$

and the rotation frequency is given by:

$$\omega'^2 = m + \omega^2$$

where ω is the frequency of the regular n-gon without the central mass. For later use define $\epsilon = \omega^2/\omega'^2$. Then $m/\omega'^2 = 1 - \epsilon$.

Let B' denote the second derivative matrix of the N-gon with the central mass. The matrix B will have the same meaning as in the last section. Then the 2×2 blocks of B' are given by:

$$\begin{aligned}
B'_{jk} &= B_{jk} & 1 \leq j \neq k \leq n \\
B'_{kk} &= B_{kk} - B'_{0k} & k \neq 0 \\
B'_{00} &= -\sum_{k=1}^{n} B'_{0k}
\end{aligned}$$

where

$$B'_{0k} = m \begin{bmatrix} 1 - 3\cos^2\theta_k & -3\sin\theta_k\cos\theta_k \\ -3\sin\theta_k\cos\theta_k & 1 - 3\sin^2\theta_k \end{bmatrix}.$$

The invariant subspaces C_l and D_l found in the last section have analogues here. If $l \neq 1$ one simply embeds the vectors v, w, v' and w' into $\mathbf{R}^{2n+2}$ by making their first two components 0. A short computation then shows that the corresponding subspaces are $M^{-1}B'$-invariant. In the nonexceptional cases, $1 < l < \frac{n}{2}$, the matrix is the restriction of $M^{-1}B'$ to C_l becomes:

$$\begin{bmatrix} P_l - 3Q_l + 2m & -R_l \\ -R_l & P_l + 3Q_l - m \end{bmatrix}$$

while the matrix of its restriction to D_l becomes:

$$\begin{bmatrix} P_l + 3Q_l - m & -R_l \\ -R_l & P_l - 3Q_l + 2m \end{bmatrix}.$$

The additional diagonal terms arise from the difference between B_{kk} and B'_{kk}. Dividing by ω'^2 and recalling the definition of ϵ shows that the parameters in proposition 2.4 are:

$$a = 2 + \epsilon(p_l - 3q_l - 2) \qquad b = -1 + \epsilon(p_l + 3q_l + 1) \qquad c = d = -\epsilon r_l$$

and so the planar stability polynomial now has a quartic factor as in proposition 2.4 with:

$$
\begin{aligned}
\alpha_l &= 1 + \epsilon(1 - 2p_l) \\
\beta_l &= \epsilon(1 + p_l + 3q_l)(3 + \epsilon(p_l - 3q_l - 2)) - \epsilon^2 r_l^2 \\
(c_l + d_l)^2 &= 4\epsilon^2 r_l^2.
\end{aligned}
$$

(3.10)

Next consider the exceptional case $l = 0$. The matrix B on the two-dimensional space $C_0 \oplus D_0$ was $\operatorname{diag}(2\omega^2, -\omega^2)$ and the matrix of $M^{-1}B'$ is $\operatorname{diag}(2\omega^2 + 2m, -\omega^2 - m) = (2\omega'^2, -\omega'^2)$. Once again this leads, via proposition 2.3 to the quadratic factor:

$$
G_0(z) = z^2 + z
$$

and the roots $z = 0$ and $z = -1$.

The other exceptional case is when n is even and $l = \frac{n}{2}$. Once again $C \oplus D$ reduces to a two-dimensional subspace and the matrix of $M^{-1}B'$ on this space is:

$$
\operatorname{diag}(P_{\frac{n}{2}} - 3Q_{\frac{n}{2}} + 2m, P_{\frac{n}{2}} + 3Q_{\frac{n}{2}} - m)
$$

which leads to the quadratic factor as in proposition 2.3 with:

$$
\begin{aligned}
\alpha_{\frac{n}{2}} &= 1 + \epsilon(1 - 2p_{\frac{n}{2}}) \\
\beta_{\frac{n}{2}} &= \epsilon(1 + p_{\frac{n}{2}} + 3q_{\frac{n}{2}})(3 + \epsilon(p_{\frac{n}{2}} - 3q_{\frac{n}{2}} - 2))
\end{aligned}
$$

(3.11)

Finally, the case $l = 1$ requires a different embedding of the subspaces into $\mathbf{R}^{2n+2}$. Make the first two components of the vectors v and v' be $(-n/2m, 0)$ and those of w and w' be $(0, -n/2m)$. Then the subspaces $C_1 = \langle v, v' \rangle$ and $D_1 = \langle w', w \rangle$ are $M^{-1}B'$-invariant. Computation shows that the matrix is the restriction of $M^{-1}B'$ to C_1 is:

$$
\begin{bmatrix}
P_1 + 2m + n & -P_1 + n \\
-P_1 - n/2 & P_1 - m - n/2
\end{bmatrix}
$$

while the matrix of its restriction to D_1 is:

$$
\begin{bmatrix}
P_1 - m - n/2 & -P_1 - n/2 \\
-P_1 + n & P_1 + 2m + n
\end{bmatrix}.
$$

Applying proposition 2.4 with

$$
\begin{aligned}
a &= 2 + \epsilon(p_1 + n/\omega^2 - 2) & b &= -1 + \epsilon(p_1 - n/2\omega^2 + 1) \\
c &= \epsilon(-p_1 + n/\omega^2) & d &= \epsilon(-p_1 - n/2\omega^2)
\end{aligned}
$$

gives a quartic factor with:

$$
\begin{aligned}
\alpha_1 &= 1 + \epsilon(1 - 2p_1 - n/2\omega^2) \\
\beta_1 &= \epsilon(1 + p_1 - n/2\omega^2)(3 + \epsilon(p_1 + n/\omega^2 - 2)) \\
&\quad - \epsilon^2(p_1 - n/\omega^2)(p_1 + n/2\omega^2) \\
(c_1 + d_1)^2 &= \epsilon^2(2p_1 - n/2\omega^2)^2.
\end{aligned}
$$

(3.12)

Finally, the stability polynomial of the n-gon with a central mass has two more roots which do not arise from the invariant subspaces C_l and D_l. These are the repeated $z = -1$ associated with the plane spanned by $(1, 0, \ldots, 1, 0)$ and $(0, 1, \ldots, 0, 1)$ as in section 2.3.

With these formulas for all of the the factors of the stability polynomial in hand, one can analyze the spectral stability in the plane of the n-gon with a central mass. The analysis will be confined to the following cases: a sufficiently small central mass ($\epsilon \approx 1$), all masses equal ($\epsilon = \omega^2/(1 + \omega^2)$), and a sufficiently large central mass ($\epsilon \approx 0$).

When the central mass is small, the quadratic and quartic factors with $l \neq 1$ approach those found for the n-gon without the central mass. The roots of these factors were classified in section 3.3. On the other hand, the quartic factor with $l = 1$ has four complex roots as will now be shown. In the limiting case, $\epsilon = 1$, this quartic can be factored as:

$$
\left(z^2 + 2(1 - n/2\omega^2)z + (1 + n/2\omega^2)^2\right) \left(z^2 + 2(1 - 2p_1)z + (1 + 2p_1)^2\right).
$$

The discriminant $\Delta_1 > 0$ and this property persists for ϵ sufficiently close to 1. Hence there are either four complex roots or four real ones. As noted above, the real roots would have to be negative. The possibility of real, negative roots will be eliminated by showing that $\alpha_1 < 0$. Now when $\epsilon = 1$, $\alpha_1 = 2(1 - p_1) - n/2\omega^2$. Note that

$$
\begin{aligned}
P_1 &= \sum_{k=1}^{n-1} \frac{1 - \cos^2 \theta_k}{2d_{nk}^3} = \sum_{k=1}^{n-1} \frac{1 + \cos \theta_k}{4d_{nk}} \\
&= \tfrac{1}{2}\sum_{k=1}^{n-1} \frac{1}{d_{nk}} - \tfrac{1}{8}\sum_{k=1}^{n-1} d_{nk} = \omega^2 - \tfrac{1}{4}\cot\tfrac{\pi}{2n}
\end{aligned}
$$

from which it follows that:

(3.13) $p_1 = 1 - \frac{1}{4\omega^2}\cot\frac{\pi}{2n} > 1 - \frac{n}{\pi\omega^2}.$

Using this one finds $\alpha_1 < (\frac{2}{\pi} - 1)\frac{n}{2\omega^2} < 0$. This will continue to hold for ϵ sufficiently close to 1.

Combining this information about the roots associated to $l = 1$ with the results of section 3.3 leads to the following table of roots. Once again, the root $z = 0$ and the triply repeated $z = -1$ have been omitted.

Roots of the stability polynomial for the regular n-gon with a small central mass			
n	Negative	Positive	Nonreal
3	0	0	4
4	0	0	6
5	0	0	8
6	1	1	8
odd$\geq$ 7	n-3	0	n+1
even$>$ 7	n-3	1	n

This differs from the results for the n-gon only in the addition of 2 extra nonreal roots. Once again one finds that for $n = 3, 4, 5$ the n-gon exhibits the maximum possible instability.

Next consider the case $m = 1$, that is, $n + 1$ equal masses. A lemma similar to 3.4 is needed.

LEMMA 3.5. *When* $\epsilon = \omega^2/(1+\omega^2)$, *the coefficients* α_l *and* β_l, $1 \leq l \leq \frac{n}{2}$ *and the discriminant* Δ_l, $1 \leq l < \frac{n}{2}$ *are negative except in the following cases:.*

$$\alpha_1 > 0 \qquad 3 \leq n \leq 5$$
$$\alpha_2 > 0 \qquad 4 \leq n \leq 9$$
$$\alpha_3 > 0 \qquad n = 6, 7$$
$$\beta_1 > 0 \qquad n \geq 4$$
$$\beta_2 > 0 \qquad 4 \leq n \leq 19$$
$$\beta_3 > 0 \qquad 4 \leq n \leq 8$$
$$\beta_4 > 0 \qquad n = 8$$
$$\Delta_1 > 0 \qquad n \geq 4$$
$$\Delta_2 > 0 \qquad 5 \leq n \leq 12$$
$$\Delta_3 > 0 \qquad n = 7, 8.$$

Proof. This is similar to the proof of lemma 3.4 in that the results for small n are left to the computer. First it will be shown that $\alpha_l < 0$ for $2 \leq l \leq \frac{n}{2}$, when $n \geq 12$. Since α_l is decreasing with l in this range, it suffices to show $\alpha_2 < 0$. This is equivalent to showing $P_2 - \omega^2 > \frac{1}{2}$. In the proof of lemma 3.4, it was shown that $P_2 - \omega^2 > 0$. Modifying the estimates there shows that a sufficient condition for $\alpha_2 < 0$ is $A_n^- > \frac{20}{9\pi} + \frac{2}{3n}$. Since the left side of this inequaltity is increasing and the right decreasing it is enough to verify it for $n = 12$: $0.7786\ldots > 0.7680\ldots$.

Next it will be shown that $\beta_l < 0$ for $2 \leq l \leq \frac{n}{2}$ and $n \geq 28$. As in the proof of lemma 3.4, one finds a simple upper bound by ignoring r_l:

$$\beta_l \leq \epsilon(1 + p_l + 3q_l)(3 + \epsilon(p_l - 3q_l - 1)).$$

Hence it suffices to show that the second factor is negative and as it is decreasing with l one can consider $l = 2$. Recalling the value of ϵ reduces the required condition to $3 + \omega^2 + P_2 - 3Q_2 < 0$. Except for the 3 this is what was shown in the proof of lemma 3.4. Modifying the formulas for that proof leads to the sufficient condition $A_n^- + \frac{1}{2n} \cot \frac{3\pi}{2n} > \frac{3}{\pi} + \frac{6}{n}$. Since the right side is increasing and the left decreasing with n one need only check $n = 28$ for which one has $1.1811\ldots > 1.1692\ldots$. The few remaining statements about $2 \leq l \leq \frac{n}{2}$ can be verified numerically.

It remains to consider the claims about $l = 1$. First it will be shown that $\alpha_1 < 0$ for $n \geq 6$. From 3.13 one finds

$$
\begin{aligned}
\frac{\alpha_1}{\epsilon} &= \frac{1}{\epsilon} + 1 - 2p_1 - \frac{n}{2\omega^2} \\
&= \frac{1}{\omega^2} + 2(1 - p_1) - \frac{n}{2\omega^2} \\
&= \frac{1}{\omega^2} + \frac{1}{2\omega^2} \cot \frac{\pi}{2n} - \frac{n}{2\omega^2}
\end{aligned}
$$

Estimating the cotangent gives:

$$
(3.14) \qquad \left(\tfrac{2}{\pi} - 1\right)\tfrac{n}{2\omega^2} < \tfrac{\alpha_1}{\epsilon} < \left(\tfrac{2}{\pi} - 1 + \tfrac{1}{n}\right)\tfrac{n}{2\omega^2}.
$$

The upper bound will be negative provided $n > \frac{2\pi}{\pi - 2} = 5.5038\ldots$ hence certainly $\alpha_l < 0$ for $n \geq 6$.

From 3.12 and the value of ϵ:

$$
\tfrac{\beta_1}{\epsilon} = 1 + 2p_1 + \tfrac{n}{2\omega^2} + \tfrac{np}{\omega^2} + \tfrac{3}{\omega^2}\left(1 + p_1 - \tfrac{n}{2\omega^2}\right).
$$

From 3.13 and the fact that $n/2\omega^2 = 1/A_n$ it follows that the last term is positive provided $A_n^- > 1 + \frac{1}{\pi}$ which holds for $n \geq 13$. Dropping it and using 3.13 gives the lower bound:

$$
\tfrac{\beta_1}{\epsilon} > 3 + \left(3 - \tfrac{2}{\pi} - \tfrac{n}{\pi\omega^2}\right)\tfrac{n}{2\omega^2}.
$$

Using $n/(\pi\omega^2) < 2/(\pi A_n^-)$ leads to the simpler bound:

$$
(3.15) \qquad \tfrac{\beta_1}{\epsilon} > 3 + \tfrac{n}{2\omega^2}
$$

for $n \geq 13$. In particular $\beta_1 > 3 > 0$ as claimed. It remains to consider the discriminant Δ_1.

Each of the terms of the second factor in the formula for Δ_1 in propositon 2.5 will have to be estimated. Assume $n \geq 13$. Then combining the bounds 3.14 and 3.15 shows that $4\beta_1^2 - \alpha_1^2 > 12\epsilon^2$. It follows from this and the fact that $\alpha_1 < 0$ and $\beta_1 > 0$ that the term $\alpha_1(\alpha_1^2 - 36\beta_1) > 0$. Dropping it gives the lower bound:

$$
(c_1 + d_1)^{-4}\Delta_1 > 144\epsilon^2\beta_1 - 27(c_1 + d_1)^4.
$$

Now 3.12 and 3.13 show that $(c_1 + d_1)^4 < 16\epsilon^2$. Using $\beta_1 > 3$ finally yields

$$
(c_1 + d_1)^{-4}\Delta_1 > 432\epsilon^2 - 432\epsilon^2 = 0
$$

for $n \geq 13$. The results for smaller n can be checked numerically.   □

Using this lemma one can determine the nature of the roots of the planar stability polynomial for the equal mass case. When a quartic has a negative discriminant, which comprises most of the cases with $l \neq 1$, there are two real, negative roots and two nonreal roots. The case $l = 1$ gives four nonreal roots when $n \geq 4$. Finally the quadratic associated to $l = \frac{n}{2}$ gives one negative and one positive root when $\beta_{\frac{n}{2}} < 0$. Combining these observations with a special study of the exceptional cases gives the following table:

Roots of the stability polynomial for the regular n-gon with an equal central mass			
n	Negative	Positive	Nonreal
3	2	0	2
$4 \leq n \leq 7$	0	0	2n-2
8	0	2	12
9	4	0	12
10	5	1	12
11	6	0	14
12	7	1	14
odd$\geq$ 13	n-3	0	n+1
even$>$ 13	n-3	1	n

This time all of the configurations with $4 \leq n \leq 8$ are maximally unstable.

Finally, there is the case of a large central mass. This is the case studied by Maxwell [10]. In this classic study of the rings of Saturn, he introduces special solutions of the linearized differential equations 2.5 which can be viewed as arising from the invariant subspaces $C_l \oplus D_l$ employed here. He obtains quartic equations for the eigenvalues of the equilibrium and finds imaginary eigenvalues when the central mass is sufficiently large. Maxwell's analysis seems to be correct except in the case $l = 1$. The quartic factor 3.12 is associated to the subspace $C_1 \oplus D_1$ which is spanned by vectors with nonvanishing 0-th components. In other words, one must consider perturbations which change the position of the central mass. When Maxwell derives the linearized equations of motion for the perturbed ring, he assumes that the central mass is always fixed at the origin. This is valid if $l \neq 1$ because in these cases, the perturbation of the ring produces no net effect on the central mass. But when $l = 1$ the perturbation shifts the center of mass of the ring away from the origin producing an acceleration on the central mass.

For the quartic factors arising from the cases $1 < l < \frac{n}{2}$, the coefficients 3.10 satisfy the estimates: $\alpha_l = 1 + O(\epsilon)$, $\beta_l = 3\epsilon(1 + p_l + 3q_l) + O(\epsilon^2)$, $\alpha_l^2 - 4\beta_l = 1 + O(\epsilon)$, and $\Delta_l = (c_l + d_l)^4(\beta + O(\epsilon^2))$. Since the coefficient of

ϵ in β is positive, the hypotheses of proposition 2.5 hold and all four roots are real and negative for $\epsilon > 0$ sufficiently small.

If n is even then the case $l = \frac{n}{2}$ gives a quadratic factor whose coefficients 3.11 satisfy the estimates: $\alpha_{\frac{n}{2}} = 1 + O(\epsilon)$, $\beta_{\frac{n}{2}} = 3\epsilon(1 + p_{\frac{n}{2}} + 3q_{\frac{n}{2}}) + O(\epsilon^2)$, and $\alpha_{\frac{n}{2}}^2 - 4\beta_{\frac{n}{2}} = 1 + O(\epsilon)$. Hence both roots are real and negative for $\epsilon > 0$ sufficiently small.

Finally, the quartic factor arising from the cases $l = 1$ has the coefficients 3.12 satisfying: $\alpha_1 = 1 + O(\epsilon)$, $\beta_1 = 3\epsilon(1 + p_1 - n/2\omega^2) + O(\epsilon^2)$, $\alpha_1^2 - 4\beta_1 = 1 + O(\epsilon)$, and $\Delta_1 = (c_1 + d_1)^4(\beta + O(\epsilon^2))$. If $1 + p_1 - n/2\omega^2 > 0$ then all of the roots will be real and negative for $\epsilon > 0$ sufficiently small; otherwise, there will be two real, negative roots and two nonreal roots.

One can check numerically that the inequality fails to hold when $n \leq 6$ and does hold for $n = 7$. To see that it holds for $n \geq 8$, one computes from equation 3.13 that $A_n^- > \frac{1}{4n} \cot \frac{\pi}{2n} + \frac{1}{2} > \frac{1}{2\pi} + \frac{1}{2}$ is sufficient. For $n = 8$ one finds that this holds: $0.6773\ldots > 0.6592\ldots$ and so it holds for $n \geq 8$ as well.

Thus the regular n-gon with a large central mass is spectrally stable if $n \geq 7$ and otherwise has two nonreal roots. Note that it is the factor with $l = 1$, the one analyzed incorrectly by Maxwell, which leads to the necessary condition $n \geq 7$ for stability.

Roots of the stability polynomial for the regular n-gon with a large central mass			
n	Negative	Positive	Nonreal
$3 \leq n \leq 6$	2n-4	0	2
$n \geq 7$	2n-2	0	0

REFERENCES

[1] M. H. Andoyer. Sur les solutiones périodiques voisines des position d'équilibre relatif dans le probleme des n corps. *Bull. Astron.*, 23:129–146, 1906.

[2] G. D. Birkhoff. *Dynamical Systems.* AMS, Providence, 1927.

[3] V. A. Brumberg. Permanent configurations in the problem of four bodies and their stability. *Soviet Astron.*, 1,1:57–79, 1957.

[4] W. S. Burnside and A. R. Panton. *Theory of Equations.* Dublin Univ. Press, Dublin, 1899.

[5] A. Deprit and A. Deprit-Bartolomé. Stability of the triangular Lagrange points. *Astron. J.*, 72:173–179, 1967.

[6] L. Euler. De motu rectilineo trium corporum se mutuo attahentium. *Novi Comm. Acad. Sci. Imp. Petrop.*, 11:144–151, 1767.

[7] R. Hoppe. Erweiterung der bekannten Speziallösungen des Dreikörpersproblem. *Archiv. der Math. und Phys.*, 64:218–223, 1879.

[8] J. L. Lagrange. *Ouvres, vol.6.* Gauthier-Villars, Paris, 1873.

[9] R. Lehmann-Filhes. Über zwei Fälle des Vielkörpersproblem. *Astron. Nach.*, 127, 1891.

[10] J. C. Maxwell. Stability of the motion of Saturn's rings. In W. D. Niven, editor, *The Scientific Papers of James Clerk Maxwell*. Cambridge University Press, Cambridge, 1890.

[11] J. C. Maxwell. Stability of the motion of Saturn's rings. In S. Brush, C. W. F. Everitt, and E. Garber, editors, *Maxwell on Saturn's Rings*. MIT Press, Cambridge, 1983.

[12] M. G. Meyer. Solutiones voisines des solutiones de lagrange dans le problème des n corps. *Ann. Obs. Bordeaux*, 17:77–252, 1933.

[13] R. Moeckel. On central configurations. *Math. Annalen*, 205:499–517, 1990.

[14] R. Moeckel. Linear stability of relative equilibria with a dominant mass. submitted to *Jour.Dyn.Diff.Eq.*, 1992.

[15] F. R. Moulton. The straight line solutions of the n-body problem. *Annals of Math.*, II. Ser. 12:1–17, 1910.

[16] F. Pacella. Central configurations of the n-body problem via the equivariant Morse theory. *Arch. Rat. Mech.*, 97:59–74, 1987.

[17] J. Palmore. Measure of degenerate relative equilibria, I. *Annals of Math.*, 104:421–429, 1976.

[18] L. M. Perko and E. L. Walter. Regular polygon solutions of the n-body problem. *Proc. AMS*, 94,2:301–309, 1985.

[19] E. J. Routh. On Laplace's three particles with a supplement on the stability of their motion. *Proc. Lond. Math. Soc.*, 6:86–97, 1875.

[20] C. Siegel and J. Moser. *Lectures on Celestial Mechanics*. Springer-Verlag, New York, 1971.

[21] C. Simó. Relative equilibria of the four body problem. *Cel. Mech.*, 18:165–184, 1978.

[22] E. T. Whittaker and G. N. Watson. *A Course in Modern Analysis*. Cambridge University Press, Cambridge, 1950.

IDENTICAL MASLOV INDICES FROM DIFFERENT SYMPLECTIC STRUCTURES

W. SCHERER*

Abstract. For the symplectic manifold $\mathbf{R}^{2n}$ it is shown how to obtain the same Maslov index from different symplectic structures.

1. Introduction. Let (Γ, ω_0) be a $2n$-dimensional symplectic manifold, i.e. Γ is a differentiable manifold of dimension $2n$ and ω_0 is a differential 2-form which is antisymmetric:

$$(1.1) \qquad \omega_0(X, Y) = -\omega_0(Y, X) \quad \forall X, Y \in T\Gamma$$

(here and in what follows $T\Gamma$ denotes the tangent bundle of Γ and $T^*\Gamma$ denotes its dual), closed:

$$(1.2) \qquad d\omega_0 = 0$$

and non-degenerate:

$$(1.3) \qquad \omega_0(X, Y) = 0 \, \forall Y \in T\Gamma \Rightarrow X = 0.$$

Let $H : \Gamma \to \mathbf{R}$ be a differentiable function which is a Hamiltonian for some dynamical system. The dynamical system is given by a vector field $X_H \in T\Gamma$ which is obtained via Hamilton's equations in geometric form:

$$(1.4) \qquad \imath(X_H)\omega_0 = dH_0.$$

X_H is then a Hamiltonian system w.r.t ω_0 with Hamiltonian (generating) function H_0. Due to the non-degeneracy (1.3) of ω_0 a given H_0 determines X_H uniquely. Changing H_0 and leaving ω_0 unchanged yields a new dynamics. The same happens in general if we change ω_0 but retain H_0. If, however, ω_0 and H_0 are changed simultaneously to a symplectic form ω_1 and a generating function H_1 such that their changes compensate each other we still obtain the same dynamics X_H. In this case X_H is called a bi-Hamiltonian system. In other words X_H is bi-Hamiltonian if there exists a symplectic form $\omega_1 \neq c\,\omega_0$ ($c = const$) and a function $H_1 : \Gamma \to \mathbf{R}$ such that

$$(1.5) \qquad \imath(X_H)\omega_1 = dH_1.$$

That such a situation is not uncommon can be seen from the various ways to obtain bi-Hamiltonian formulations [5,8,9]. Since every symplectic structure ω on Γ yields an associated Poisson bracket

$$(1.6) \qquad \{f, g\}_\omega := \omega(X_f, X_g)$$

---
* Institut für Theoretische Physik A, TU Clausthal, W-38678 Clausthal-Zellerfeld, Germany.

for $f, g \in C^\infty(\Gamma)$ (where X_f is defined via $\imath(X_f)\omega = df$ and X_g likewise) this means that the equations of motion for a bi-Hamiltonian system can be obtained with either $\{\cdot, \cdot\}_0$ and H_0 or $\{\cdot, \cdot\}_1$ and H_1. The Hamiltonian description of the given dynamics X_H is thus not unique and since canonical quantization utilizes Poisson brackets (or in its geometric version symplectic structures) this leads to the question how much the quantum system depends on the chosen structure. As a first step towards answering this question one can look at the semiclassical approximation [1] obtained with the help of different Hamiltonian descriptions. In this approximation there appears the Maslov index which depends on the symplectic structure [1,2,7]. In this paper we give sufficient conditions such that the Maslov index is independent of the symplectic structure (proposition 3.1) in the case the manifold is $\mathbf{R}^{2n}$. The following Section contains the material necessary to prove this result in Section 3. Section 4 contains a discussion of the result.

   2. Almost Kähler structures. In this Section we discuss some general results concerning two symplectic structures, their associated almost Kähler structures (see below) and their Lagrangian distributions. These will be statements about relations between various structures (symplectic, metric, almost complex) and they are completely independent of any dynamics and hold for any symplectic manifold Γ.

   Given two symplectic structures ω_s, $s = 0, 1$ we can define

$$(2.1) \qquad \omega_s^\flat : T\Gamma \to T^*\Gamma$$

via

$$(2.2) \qquad T\Gamma \ni X \mapsto \omega_s^\flat(X) := \omega_s(X, \cdot) : T\Gamma \to \mathbf{R}.$$

Because of the non-degeneracy (1.3) we also have the existence of:

$$(2.3) \qquad \omega_s^\sharp := (\omega_s^\flat)^{-1}.$$

With the help of these mappings we define the $(1, 1)$-tensor

$$(2.4) \qquad R := \omega_0^\sharp \circ \omega_1^\flat : T\Gamma \to T\Gamma.$$

An equivalent way to define R is

$$(2.5) \qquad \omega_1(X, Y) = \omega_0(RX, Y) \quad \forall X, Y \in T\Gamma.$$

Provided our manifold Γ is paracompact (which we assume) we can for any symplectic structure ω on Γ always find an almost complex structure [4,7]

$$(2.6) \qquad J : T\Gamma \to T\Gamma, \quad J^2 = -id_{T\Gamma}$$

such that

$$(2.7) \qquad \omega(JX, JY) = \omega(X, Y) \quad \forall X, Y \in T\Gamma$$

and g defined as

$$(2.8) \qquad g(X,Y) := \omega(X,JY) \quad \forall X,Y \in T\Gamma$$

is a positive definite Riemannian metric (which is necessarily also invariant or "hermitian" with respect to J). We shall call such a triple (ω, J, g) an almost Kähler structure (aKs). Note that a given ω does not uniquely determine an aKs whereas the converse is evident. Let (ω_0, J_0, g_0) and (ω_1, J_1, g_1) be two aKs containing the given symplectic structures ω_0 and ω_1. Then we have the following

LEMMA 2.1. Let $X,Y \in T\Gamma$ then for $s = 0$ or 1:

$$(2.9) \qquad \omega_s(RX,Y) = \omega_s(X,RY)$$
$$(2.10) \qquad g_s(RX,Y) = g_s(X,RY) \quad \Leftrightarrow \quad J_s R = R J_s$$
$$(2.11) \qquad g_1(X,J_1 Y) = g_0(RX,J_0 Y).$$

Proof. (2.9) follows for $s = 0$ from

$$\omega_0(RX,Y) = \omega_1(X,Y) = -\omega_1(Y,X) = -\omega_0(RY,X) = \omega_0(X,RY).$$

For $s = 1$ it follows from

$$\omega_1(RX,Y) = \omega_0(R^2 X,Y) = \omega_0(RX,RY) = \omega_1(X,RY).$$

To prove (2.10) we observe that

$$g_s(RX,Y) = \omega_s(RX,J_s Y) = \omega_s(X,RJ_s Y)$$

and

$$g_s(X,RY) = \omega_s(X,J_s RY).$$

Hence, $\forall X,Y \in T\Gamma$

$$\begin{aligned} g_s(X,RY) &= g_s(RX,Y) \\ \Leftrightarrow \omega_s(X,RJ_s Y) &= \omega_s(X,J_s RY) \\ \Leftrightarrow RJ_s &= J_s R. \end{aligned}$$

Finally we have $\forall X,Y \in T\Gamma$

$$\begin{aligned} g_1(X,J_1 Y) &= \omega_1(X,J_1^2 Y) = -\omega_0(RX,Y) \\ &= \omega_0(RX,J_0^2 Y) \\ &= g_0(RX,J_0 Y). \end{aligned}$$

$\square$

Next, we introduce the concept of a Lagrangian distribution. A ω_s-Lagrangian distribution V is a map

$$(2.12) \qquad V : \Gamma \ni x \mapsto V_x \in \{n - dim.\ subspaces\ of\ T_x \Gamma\}$$

such that V_x is a ω_s-Lagrangian subspace of $T_x\Gamma$:

(2.13) $\omega_s \big|_{V_x} = 0$

i.e. $\forall X, Y \in V_x \ \ \omega_s(X,Y) = 0$.

LEMMA 2.2. *Let V_x be a ω_s-Lagrangian subspace of $T_x\Gamma$ and $\{e_1,\ldots,e_n\}$ be a basis of V_x. Then*

1. *J_sV_x is the g_s-orthogonal complement to V_x and*

$$T_x\Gamma = V_x \oplus J_sV_x$$

2. *$\{e_1,\ldots,e_n, J_se_1,\ldots, J_se_n\}$ is a basis of $T_x\Gamma$.*

Proof. Let $V_x = span\{e_1,\ldots,e_n\}$ be a ω_s-Lagrangian subspace. Then

$$g_s(e_k, J_se_j) = \omega_s(e_k, e_j) = 0$$

since V_x is Lagrangian. Moreover, $\{J_se_1,\ldots, J_se_n\}$ are linearly independent, since

$$\sum_{j=1}^{n} \lambda_j J_s e_j = 0 \ \Rightarrow \ J_s\left(\sum_{j=1}^{n} \lambda_j J_s e_j\right) = 0$$

$$\Rightarrow -\sum_{j=1}^{n} \lambda_j e_j = 0 \ \Rightarrow \ \lambda_j = 0 \ \ \forall j = 1,\ldots, n.$$

□

So far we have only considered subspaces which are Lagrangian with respect to one symplectic structure (either ω_0 or ω_1). The fact that a subspace is Lagrangian with respect to ω_0 and ω_1 implies that R leaves this space invariant as is shown below. If we require that R commutes with J_0 then R takes a particularly simple form.

PROPOSITION 2.1. *Let $RJ_0 = J_0R$ and let V_x be an ω_0 and ω_1-Lagrangian subspace of $T_x\Gamma$. Then there is a g_0-orthonormal basis $\{e_1,\ldots,e_n\}$ of V_x such that in the basis $\{e_1,\ldots,e_n, J_0e_1,\ldots,J_0e_n\}$ of $T_x\Gamma$ R at $x \in \Gamma$ has the form*

(2.14) $R = \begin{pmatrix} a(x) & 0 \\ 0 & a(x) \end{pmatrix}$

where $a(x) \in GL(n, \mathbf{R})$ is symmetric.

Proof. Let $RJ_0 = J_0R$ and let $\{e_1,\ldots,e_n\}$ be a g_0 orthonormal basis of the ω_0-Lagrangian subspace V_x. From lemma 2.2 we know that $\{e_1,\ldots,e_n, J_0e_1,\ldots, J_0e_n\}$ is a g_0-orthonormal basis of $T_x\Gamma$. Hence we can write:

(2.15) $Re_k = \sum_{j=1}^{n} (a_{jk}e_j + b_{jk}J_0e_j)$

where a, b are real $n \times n$ Matrices (which depend on the point $x \in \Gamma$). Moreover,

$$(2.16) \qquad R J_0 e_k = J_0 R e_k = \sum_{j=1}^{n} (-b_{jk} e_j + a_{jk} J_0 e_j)$$

such that in the basis $\{e_1, \ldots, e_n, J_0 e_1, \ldots, J_0 e_n\}$ R has the form

$$(2.17) \qquad R = \begin{pmatrix} a & -b \\ b & a \end{pmatrix}.$$

Since V_x is by assumption also ω_1-Lagrangian we must have

$$\begin{aligned} 0 &= \omega_1(e_k, e_j) = \omega_0(R e_k, e_j) = \sum_{l=1}^{n} b_{lk} \omega_0(J_0 e_l, e_j) \\ &= -b_{jk} \quad \forall j, k = 1, \ldots, n \end{aligned}$$

where we have used that $\omega_0(e_k, e_j) = 0$ (V_x is ω_0-Lagrangian) and $\omega_0(J_0 e_l, e_j) = -g_0(e_l, e_j) = -\delta_{lj}$ ($\{e_1, \ldots, e_n\}$ is g_0-orthonormal). Hence, R at the point $x \in \Gamma$ is of the form asserted in (2.14). Since by its definition R is invertible a has to be a regular $n \times n$ matrix. Moreover, (2.10) tells us that R has to be symmetric with respect to g_0 and since $\{e_1, \ldots, e_n\}$ is g_0-orthonormal it follows that a must be symmetric. $\square$

3. **Identical Maslov indices from ω_0 and ω_1.** From now on we restrict ourselves to the case $\Gamma = \mathbf{R}^{2n}$. Let (ω_0, J_0, g_0) be the canonical aKs, i.e. in the canonical basis $\{e_1^0, \ldots, e_{2n}^0\}$ of $\mathbf{R}^{2n}$ we have for $k, j = 1, \ldots, 2n$: $g_0(e_k^0, e_j^0) = \delta_{kj}$ and for *for* $k = 1, \ldots, n$: $\omega_0(e_k^0, e_j^0) = 0$, $\omega_0(e_k^0, e_{j+n}^0) = \delta_{kj}$ and $J_0 e_k^0 = e_{k+n}^0$, $J_0 e_{k+n}^0 = -e_k^0$. Given a completely integrable dynamical system X_H

$$(3.1) \qquad \imath(X_H)\omega_0 = H_0$$

with independent constants of motion $f_1, \ldots, f_n$ in involution there exist action angle variables $(I_1^0, \ldots, I_n^0, \theta_1^0, \ldots, \theta_n^0)$ such that the I_k^0 are functionally independent constants of motion as well. If the level sets

$$(3.2) \qquad M_c := \{x \in \Gamma \mid f_j(x) = c_j\}$$

are compact they are isomorphic to n-tori T^n. The semiclassical approximation of the quantum theory associated with the dynamical system X_H uses the corrected Bohr-Sommerfeld quantization condition [1,3] which states that the quantum mechanical observable $\hat{I}_j^0$ asssociated with the classical observable actions I_j^0 can take only the values

$$(3.3) \qquad \hat{I}_j^0 = \hbar(n_j + \frac{\mu_j^0}{4})$$

where $n_j \in \mathbf{Z}$ and μ_j^0 is the Maslov index of the closed curve γ_j winding around the j-th factor of S^1 in $TM_c \simeq T^n$ (the $\gamma_1, \ldots, \gamma_n$ form a homology basis of M_c). It can be shown that this index can be calculated as follows [6]. Let

$$(3.4) \qquad\qquad Y_k^0 = \omega_0^\sharp(df_k)$$

i.e. $\imath(Y_k^0)\omega_0 = df_k$ and thus (see (1.6))

$$\omega_0(Y_k^0, Y_j^0) = \imath(Y_k^0)df_j = \{f_k, f_j\}_0 = 0.$$

Hence,

$$(3.5) \qquad\qquad T_x M_c = span\{Y_1^0, \ldots, Y_n^0\}$$

(which is the tangent space of M_c at the point $x \in M_c$) is a Lagrangian subspace of $T_x\Gamma$ and the assignement (3.5) gives a Lagrangian distribution (at least on the submanifold M_c). Let $\{t_1^0, \ldots, t_n^0\}$ be a g_0-orthonormal basis of $T_x M_c$ then it can be shown that the $n \times n$ matrix [6]

$$(3.6) \quad D_{kj}^0(x) := g_0(e_k^0, t_j^0) - i\omega_0(e_k^0, t_j^0) = \omega_0(e_k^0, J_0 t_j^0) - i\omega_0(e_k^0, t_j^0)$$

is unitary, i.e. $D^0(x) \in U(n)\ \forall x \in M_c$. Since $\gamma_j : S^1 \to M_c$ is a closed curve $(\det D^0(\gamma_j))^2$ traces out a closed curve on $S^1 = \{z \in \mathbf{C} \mid |z| = 1\}$. As α winds around S^1 $(\det D^0(\gamma_j(\alpha)))^2$ winds μ_j^0 times around $S^1 \subset \mathbf{C}\backslash\{0\}$, i.e. μ_j^0 is the winding number $\mathcal{W}_{\gamma_j}(D^0)$

$$(3.7) \qquad \mu_j^0 = \mathcal{W}_{\gamma_j}(D^0) := \frac{1}{2\pi i} \oint_{\gamma_j} \frac{d}{d\alpha} \left(\ln(\det D^0(\gamma_j(\alpha)))^2\right) d\alpha.$$

μ_j^0 depends on the structure ω_0 and seems to depend on either J_0 or g_0 in addition. However, using a homotopy argument one can show [7] that μ_j^0 is independent of (J_0, g_0) as long as the triple (ω_0, J_0, g_0) is aKs. Hence, μ_j^0 depends only on the structure ω_0. Suppose now X_H were a bi-Hamiltonian system, i.e. in addition to (3.1) we have

$$(3.8) \qquad\qquad \imath(X_H)\omega_1 = dH_1$$

and let (ω_1, J_1, g_1) be the corresponding aKs. The following lemma states the condition that the f_j are also in involution with respect to the Poisson bracket coming from ω_1.

LEMMA 3.1.

$$(3.9) \qquad RT_x M_c = T_x M_c \Leftrightarrow \{f_k, f_j\}_1 = 0 \quad \forall k, j = 1, \ldots, n.$$

Proof. Let $Y_k^0 = \omega_0^\sharp(df_k)$ and $Y_k^1 = \omega_1^\sharp(df_k)$. Hence, $\omega_0^\flat(Y_j^0) = df_j = \omega_1^\flat(Y_j^1)$ which implies

$$Y_j^0 = \omega_0^\sharp \circ \omega_1^\flat(Y_j^1) = RY_j^1.$$

Note that $T_x M_c = span\{Y_1^0, \ldots, Y_n^0\}$. Let $RT_x M_c = T_x M_c$, consequently since $\forall j \ Y_j^0 \in T_x M_c$ we have $Y_j^1 = R^{-1} Y_j^0 \in T_x M_c \Leftrightarrow \imath(Y_j^1) df_k = 0 \Leftrightarrow \{f_k, f_j\}_1 = 0$. Conversely, let $\{f_k, f_j\}_1 = 0$ as above it follows that $Y_j^1 \in T_x M_c$. Since the df_k are linearly independent so are the Y_k^1 (and the Y_k^0) hence $T_x M_c = span\{Y_1^1, \ldots, Y_n^1\} \Rightarrow RT_x M_c = span\{RY_1^1, \ldots, RY_n^1\} = span\{Y_1^0, \ldots, Y_n^0\} = T_x M_c$. □

Let us now assume that the f_k are in involution with respect to $\{\cdot, \cdot\}_1$ as well and that $V = span\{e_1^0, \ldots, e_n^0\}$ is also a ω_1-Lagrangian distribution. Then we pick a g_1-orthonormal basis $\{e_1^1, \ldots, e_n^1\}$ of V and a g_1-orthonormal basis $\{t_1^1, \ldots, t_n^1\}$ of $T_x M_c$. It follows that $\forall x \in M_c$

$$(3.10) \qquad D_{kj}^1(x) = g_1(e_k^1, t_j^1) - i\omega_1(e_k^1, t_j^1) \in U(n)$$

and we can define the index μ_j^1 as

$$(3.11) \qquad \mu_j^1 = \mathcal{W}_{\gamma_j}(D^1)$$

which depends only on the structure ω_1.

PROPOSITION 3.1. *Let ω_0 be the canonical and ω_1 any symplectic structure on $\mathbf{R}^{2n}$ such that*

1. *$V = span\{e_1^0, \ldots e_n^0\}$ is ω_1 and ω_0-Lagrangian*
2. *the f_j are independent integrals in involution with respect to $\{\cdot, \cdot\}_0$ and $\{\cdot, \cdot\}_1$*
3. *there exists an almost complex structure $J_1 = J_0$ making both (ω_0, J_0, g_0) and (ω_1, J_1, g_1) to almost Kähler structures*

then

$$(3.12) \qquad \mu_j^1 = \mu_j^0.$$

Proof. Note that if $A, B \in GL(n, \mathbf{R})$ then

$$(3.13) \qquad \mathcal{W}_{\gamma_j}(AD^1 B) = \mathcal{W}_{\gamma_j}(D^1)$$

and since $span\{e_1^0, \ldots, e_n^0\} = V = span\{e_1^1, \ldots, e_n^1\}$ and $span\{t_1^1, \ldots, t_n^1\} = T_x M_c = span\{t_1^1, \ldots, t_n^1\}$ we have

$$(3.14) \qquad e_j^1 = \sum_{k=1}^n A_{kj}(x) e_k^0$$
$$(3.15) \qquad t_j^1 = \sum_{k=1}^n B_{kj}(x) t_k^0$$

wher $A(x), B(x) \in GL(n, \mathbf{R})$. Consequently, with (3.11), (3.13)-(3.15)

$$(3.16) \qquad \mu_j^1 = \mathcal{W}_{\gamma_j}(C^1)$$

where

$$(3.17) \qquad C_{kj}^1(x) := g_1(e_k^0, t_j^0) - i\omega_1(e_k^0, t_j^0).$$

Now, $J_1 = J_0$ implies $\forall X, Y \in T_x M_c : \omega_1(J_0 X, J_0 Y) = \omega_1(X, Y)$. On the other hand $\omega_1(X, Y) = \omega_0(RX, Y)$ and $\omega_1(J_0 X, J_0 Y) = \omega_0(RJ_0 X, J_0 Y) = -\omega_0(J_0 RJ_0 X, Y)$ and thus

$$(3.18) \qquad\qquad RJ_0 = J_0 R.$$

Moreover, from (2.11) we have with $J_1 = J_0$

$$(3.19) \qquad\qquad g_1(X, Y) = g_0(RX, Y) \quad \forall X, Y \in T\Gamma$$

which gives

$$(3.20) \qquad\qquad C^1_{kj} = g_0(Re^0_k, t^0_j) - i\omega_0(Re^0_k, t^0_j).$$

Using the results of proposition 2.1 this gives the matrix equation

$$(3.21) \qquad\qquad C^1 = a^T D^0 = a D^0$$

since $a^T = a \in GL(n, \mathbf{R})$ (see (2.14)). Since $\mathcal{W}_{\gamma_j}(a^T D^0) = \mathcal{W}_{\gamma_j}(D^0)$ we have

$$
\begin{aligned}
\mu^1_j &= \mathcal{W}_{\gamma_j}(C^1) = \mathcal{W}_{\gamma_j}(a^T D^0) = \mathcal{W}_{\gamma_j}(D^0) \\
(3.22) \qquad\qquad &= \mu^0_j
\end{aligned}
$$

$$\square$$

Note that conditions 1 and 2 of proposition 3.1 are necessary to define μ^0 and μ^1 because it is necessary for the definition of the Maslov index that the distributions be Lagrangian. Since we would like to compare μ^1 and μ^0 for the same pair of distributions V and $T_x M_c$ (and, of course, the same curves γ_j) we need 1 and 2. Once μ^1 and μ^0 can be defined it is only condition 3 which implies (3.12).

4. Discussion. That the Maslov indices do not always coincide can be seen from the following simple counterexample. Given ω_0 take $\omega_1 = -\omega_0$ and $J_1 = -J_0$ then one has $g_1 = g_0$ and $D^1 = \overline{D^0}$ (⁻ denotes complex conjugation) and thus $\mu^1_j = -\mu^0_j$.

Proposition 3.1 gives a sufficient criterium to have the same Maslov index from two different symplectic structures. However, that does not yet imply that the semiclassical approximations coincide since the action variables to be quantized also depend on the symplectic structure. Under which assumptions identical semiclassical approximations and ultimately equivalent quantizations result is still investigated.

The definition of the Maslov index (which is equivalent to (3.7) for $\Gamma = \mathbf{R}^{2n}$) for arbitray symplectic manifolds Γ is more involved than (3.7) [7]. In the definition (3.7) given here (which is due to Arnold [2]) the trivial topology of the phase space $\mathbf{R}^{2n}$ is essential. Consequently the proof of proposition 3.1 cannot be generalized in a straightforward manner to arbitrary manifolds.

Acknowledgements. It is a pleasure to thank the organizers for such a fruitful meeting. In particular the author would like to thank for their financial support which made his attendance possible.

REFERENCES

[1] V. P. Maslov and M. V. Fedoriuk, Semiclassical Approximations in Quantum Mechanics, Reidel, Boston (1989).

[2] V. I. Arnold, Functional Anal. Appl. 1(1967) 1.

[3] R. G. Littlejohn and J. M. Robbins, Phys. Rev. A **36** (1987) 2953.

[4] A. Lichnerowicz, Global Theory of Connections and Holonomy Groups, Nordhoff International Publishing (1976).

[5] S. De Filippo, G. Vilasi, G. Marmo, M. Salerno, Nuovo Cim. **83 B** (1984) 97.

[6] W. Scherer, J. Math. Phys. **33** (1992) 3746.

[7] I. Vaisman, Symplectic Geometry and Secondary Characteristic Classes, Birkhäuser, Boston/Basel (1987).

[8] J. F. Cariñena, M. F. Rañada, J. Math. Phys. **31** (1990) 801.

[9] F. Gandolfi, G. Marmo, C. Rubano, Nuovo Cim. **66 B** (1981) 34.

DISCRETIZATION OF AUTONOMOUS SYSTEMS AND RAPID FORCING

JÜRGEN SCHEURLE*

Abstract. Given any finite dimensional autonomous first order system of ordinary differential equations and any one step discretization method consistent of order $p \geq 1$, the diffeomorphism associated to step size ε is the period map of the autonomous system with a certain periodic forcing. The forcing amplitude is of order $O(\varepsilon^p)$ as ε tends to zero, and the forcing frequency is $1/\varepsilon$. In this sense, the forcing is rapid. We give a proof of this lemma (from Fiedler and Scheurle [1991]). Also, based on recent work on splitting of separatrices for rapidly forced systems, we discuss consequences concerning the numerical simulation of the long-time behaviour of autonomous systems near homoclinic orbits. Especially, we want to make the point that discretizing a homoclinic orbit, generically breaks the homoclinic orbit and can lead to chaos. This will be discussed for both Hamiltonian and general systems. Also, upper bounds in terms of ε for the splitting effects are presented. Under appropriate smoothness assumptions, those are of arbitrarily high algebraic order in ε, as ε tends to zero. In particular, in the analytic case, they are exponentially small. A proof of these results is outlined at the end of the paper.

1. Introduction and lemma. Recently the investigation of rapidly forced systems has become quite popular in connection with the phenomenon of exponentially small splitting of separatrices. There are various contexts where such systems come up naturally, e.g. averaging or singular perturbations (cf. Holmes, Marsden and Scheurle [1988]; Hunter and Scheurle [1988]). In this paper, we consider another one, namely the discretization of autonomous differential equations by finite difference schemes. In particular, we discuss the phenomenon of exponentially small splitting of separatrices in this context.

We consider a general autonomous equation of the form

$$(1.1) \qquad \dot{x} = f(x) \qquad \left(".\," = \frac{d}{dt} \right),$$

where either $x \in \mathbb{R}^n$ or $x \in \mathbb{C}^n$, and $t \in \mathbb{R}$.

To describe the kind of discretization schemes which we are going to deal with, let us introduce the flow map $F(t, x)$, i.e.

$$x(t) = F(t, x(0)) \qquad (t \in \mathbb{R})$$

denotes the solution of (1.1) with initial value $x(0)$ at $t = 0$. We discretize this flow by a diffeomorphism $x \mapsto \phi_f(\varepsilon, x)$ associated to step size ε, such that for some $p \geq 1$

$$(1.2) \qquad \phi_f(\varepsilon, x) - F(\varepsilon, x) = O(\varepsilon^{p+1}), \qquad as \;\; \varepsilon \to 0,$$

---

* Institut für Angewandte Mathematik, Universität Hamburg, Bundesstrasse 55, D–20146 Hamburg, Germany.

locally uniformly with respect to x.

This implies in particular,

(1.3)
$$\phi_f(0, x) = x$$
$$D_\varepsilon \phi_f(0, x) = f(x) ,$$

i.e. the scheme

(1.4) $$x_{n+1} = \phi_f(\varepsilon, x_n) \quad (n = 0, 1, 2, \ldots)$$

describes a general numerical one-step discretization method with fixed step-size ε for (1.1), which is consistent of order p. An example is the explicit Euler method

(1.5) $$x_{n+1} = x_n + \varepsilon f(x_n) \quad (n = 0, 1, 2, \ldots) ,$$

where $\phi_f(\varepsilon, x) = x + \varepsilon f(x)$ and $p = 1$. The scheme (1.4) is **symplectic**, if the x-space is symplectic and $x \mapsto \phi_f(\varepsilon, x)$ preserves the corresponding symplectic two-form when f is a Hamiltonian vector field. We refer to this situation as the **symplectic case**. An example is the implicit midpoint rule

(1.6) $$x_{n+1} = x_n + \varepsilon f(\frac{1}{2}(x_n + x_{n+1})) \quad (n = 0, 1, 2, \ldots)$$

in $\mathbb{R}^{2n}$ (or $\mathbb{C}^{2n}$), equipped with the canonical symplectic structure. Here, ϕ_f is defined implicitly by $\phi_f(\varepsilon, x) = x + \varepsilon f(\frac{1}{2}(x + \phi_f(x)))$ (cf. Marsden [1992]).

Lemma   *Assume that f and ϕ_f are given as above and sufficiently smooth. Let be $0 < \varepsilon \leq \varepsilon^0$, ε^0 sufficiently small depending on compact subsets for x. Then there exists a smooth vector field $g(\varepsilon, \tau, x)$ with the following properties: If $G(t, s, \varepsilon, x(s))$ denotes the (local) principal fundamental solution of the equation*

(1.7) $$\dot{x} = f(x) + \varepsilon^p g(\varepsilon, \frac{t}{\varepsilon}, x),$$

then $\phi_f(\varepsilon, x) = G(\varepsilon, 0, \varepsilon, x)$ for all x. Furthermore, $g(\varepsilon, \tau, x)$ is 1-periodic with respect to $\tau \in \mathbb{R}$.

*In particular, in the **analytic case**, i.e. when both $f(x)$ and $\phi_f(\varepsilon, x)$ are complex analytic in x and $\phi_f(\varepsilon, x)$ is real analytic in ε, $g(\varepsilon, \tau, x)$ is of class C^∞ and together with all its τ-derivatives analytic in x and ε.*

Thus, any diffeomorphism ϕ_f as above can be viewed as the period map (with $s = 0$) of a non-autonomous periodic system of type (1.7), i.e. the corresponding numerical scheme reproduces the dynamics of such a system rather than that of (1.1). Note that it is not possible to embed any diffeomorphism into the flow of an autonomous system (cf. Kopell [1968]). In the sense of the definition given in the abstract, (1.7) is a rapidly forced system. Also, it is obvious from the construction in the next section that $x \mapsto G(t, s, \varepsilon, x)$ is volume preserving in the symplectic case for all t, s and ε, i.e. the vector field in (1.7) is divergence-free in this case.

2. Proof of the lemma. The basic idea of the proof is very simple. Namely, between each pair of points y and $\phi_f(\varepsilon, y)$ in $\mathbb{R}^n$ we interpolate by a curve parametrized by some function $t \mapsto \tilde{G}(t, \varepsilon, y)$, $0 \leq t \leq \varepsilon$, where

$$\frac{\partial^n}{\partial t^n}[\tilde{G}(t, \varepsilon, y) - F(t, y)]|_{t=0+} = 0$$

$$(2.1) \qquad \frac{\partial^n}{\partial t^n}[\tilde{G}(t, \varepsilon, y) - F(t - \varepsilon, \phi_f(\varepsilon, y))]|_{t=\varepsilon-} = 0$$

for all $n \in \mathbb{N}$ for which the n-th order t-derivative of $F(t, y)$ exists. Moreover, we require the map $y \mapsto \tilde{G}(t, \varepsilon, y)$ to be a diffeomorphism for all $t \in [0, \varepsilon]$. Then, for $0 \leq t \leq \varepsilon$, the vector field in (1.7) is given as the tangent vector field corresponding to these interpolation curves. Its specific form is due to the p-th order consistency assumption for the discretization scheme. By (2.1), the vector field can be extended to a smooth, ε-periodic vector field for all $t \in \mathbb{R}$, and

$$G(\varepsilon, 0, \varepsilon, x) = \tilde{G}(\varepsilon, \varepsilon, x) = \phi_f(\varepsilon, x)$$

holds for the corresponding principal fundamental solution $G(t, s, \varepsilon, x)$.

We now define $\tilde{G}(t, \varepsilon, x)$ more explicitly. To this end, we introduce a C^∞ cut-off function $\chi_0(\tau)$, $\tau \in \mathbb{R}$, such that $\chi_0(\tau) \equiv 1$ for $\tau \leq 0$, $\chi_0(\tau) \equiv 0$ for $\tau \geq 1$. In particular, we can assume that χ_0 is analytic outside $\tau = 0$ and $\tau = 1$. Also, we introduce $\chi_1(\tau) = 1 - \chi_0(\tau)$. Note that all derivatives of χ_0 and χ_1 vanish at $\tau = 0$ and $\tau = 1$. For appropriate $y \in \mathbb{R}^n$, we set

$$(2.2) \quad \tilde{G}(t, \varepsilon, y) = \chi_0(\frac{t}{\varepsilon})F(t, y) + \chi_1(\frac{t}{\varepsilon})F(t - \varepsilon, \phi_f(\varepsilon, y)) \quad (0 \leq t \leq \varepsilon).$$

Note that for each x, the equation

$$(2.3) \qquad\qquad x = \tilde{G}(t, \varepsilon, y)$$

has a unique solution $y = y(\varepsilon, \frac{t}{\varepsilon}, x)$ provided that ε is sufficiently small. To see this, we rescale time by $t = \varepsilon\tau$, $0 \leq \tau \leq 1$, and look at the equation $x = \tilde{G}(\varepsilon\tau, \varepsilon, y) = \tilde{\tilde{G}}(\tau, \varepsilon, y)$. For $\varepsilon = 0$, this equation has the trivial solution $y = y(0, \tau, x) = x$. Furthermore, $D_y\tilde{\tilde{G}}(\tau, 0, y) = id$. Hence, the implicit function theorem applies to yield a unique solution $y = y(\varepsilon, \tau, x)$ for sufficiently small ε. There is a uniform upper bound for ε in each compact subset of x. This solution is as smooth as $\tilde{\tilde{G}}$, i.e. in particular, analytic in the analytic case. This proves that the map $y \mapsto \tilde{G}(t, \varepsilon, y)$ is a diffeomorphism. Therefore, there is a well defined vector field

$$(2.4) \qquad\qquad h(\varepsilon, \frac{t}{\varepsilon}, x) = \frac{\partial}{\partial t}\tilde{G}(t, \varepsilon, y)$$

given by the tangent vectors to the curves in (2.2), where $y = y(\varepsilon, \frac{t}{\varepsilon}, x)$ is the unique solution of (2.3). This vector field is as smooth as f in all variables. Moreover, one easily checks that $\tilde{G}$ satisfies (2.1). So, h can be smoothly extended to a 1-periodic vector field with respect to $\tau = \frac{t}{\varepsilon} \in \mathbb{R}$. Note, however, that in the analytic case, this extended vector field is only C^∞ with respect to τ at $\tau = 0$ and $\tau = 1$.

It remains to prove that h can be written in the form of the right-hand side of equation (1.7). This amounts to say that the vector field defined by

$$(2.5) \qquad g(\varepsilon, \tau, x) = \varepsilon^{-p}(-f(x) + h(\varepsilon, \tau, x))$$

has a well-defined limit as $\varepsilon \to 0$.

To prove this, we use the assumption that $\phi_f(\varepsilon, x)$ represents a p-th order discretization scheme, i.e.

$$\phi_f(\varepsilon, x) = F(\varepsilon, x) + \varepsilon^{p+1} r_1 \;,$$

where $r_1 = O(1)$ as $\varepsilon \to 0$, uniformly in compact subsets of x. Therefore, by (2.2), it follows that

$$(2.6) \qquad \begin{aligned} \tilde{\tilde{G}}(\tau, \varepsilon, y) &= \chi_0(\tau) F(\varepsilon\tau, y) + \chi_1(\tau) F(\varepsilon\tau - \varepsilon, \phi_f(\varepsilon, y)) \\ &= F(\varepsilon\tau, y) + \varepsilon^{p+1} r_2 \end{aligned}$$

with $r_2 = O(1)$. But this implies

$$F(-\varepsilon\tau, \tilde{\tilde{G}}(\tau, \varepsilon, y)) = y - \varepsilon^{p+1} r_3$$

with $r_3 = O(1)$, i.e. for $x = \tilde{\tilde{G}}(\tau, \varepsilon, y)$, we have

$$(2.7) \qquad F(-\varepsilon\tau, x) = y(\varepsilon, \tau, x) - \varepsilon^{p+1} r_3,$$

where $y(\varepsilon, \tau, x)$ is the unique solution of (2.3). Using (2.6) and (2.7), we obtain

$$\begin{aligned} h(\varepsilon, \tau, x) &= \frac{1}{\varepsilon} \frac{\partial}{\partial \tau} \tilde{\tilde{G}}(\tau, \varepsilon, y(\varepsilon, \tau, x)) \\ &= f(F(\varepsilon\tau, y(\varepsilon, \tau, x))) + \varepsilon^p \frac{\partial}{\partial \tau} r_2 \\ &= f(F(\varepsilon\tau, F(-\varepsilon\tau, x))) + \varepsilon^p r_4 \\ &= f(x) + \varepsilon^p r_4 \;, \end{aligned}$$

where $r_4 = O(1)$ as $\varepsilon \to 0$. Thus, the limit of g in (2.5) exists, and the proof of the lemma is finished.

3. Splitting of separatrices. In this section, we discuss some consequences of the Lemma. Namely, we want to make the point that generically, there is a considerable difference between the long-time dynamics described by an autonomous equation of type (1.1) and the one produced by a corresponding discretization as considered above. In a sense, this is bad news for straight forward numerical simulations.

For simplicity of the exposition, we restrict ourselves to the case $n = 2$ here. Assume, for instance, that the equation in (1.1) has a hyperbolic equilibrium point (saddle point) at $x = 0$ and a corresponding homoclinic orbit Γ, along which the stable and unstable manifolds of $x = 0$ coincide: $\Gamma = \Gamma(t) \to 0$ as $t \to \pm\infty$. Then, for any small $\varepsilon > 0$, the period map $\tilde{G}(\varepsilon, \varepsilon, x)$ of the perturbed equation (1.7) has a hyperbolic fixed point near $x = 0$, which corresponds to an ε-periodic solution of (1.7). This fixed point is also of saddle type. So, it has a local stable and a local unstable invariant manifold W_ε^s and W_ε^u, respectively. In general, these do not form a homoclinic orbit, though, rather they totally miss each other. So, it is not to be expected that the homoclinic orbit of the continuous problem will be reproduced by a discretization scheme as above. This is even true in the symplectic case. But the invariant manifolds cannot totally miss each other in this case, since $\tilde{G}(\varepsilon, \varepsilon, \cdot) = \phi_f(\varepsilon, \cdot)$ preserves area. Instead, generically they intersect each other transversally at various points (Zehnder [1973] and Genecand [1993]). This then leads to a stochastic layer in a region close to the homoclinic orbit Γ, where the discretized system has horseshoes and behaves chaotically.

On the other hand, if ϕ_f is symplectic and very smooth in x, then, for small ε, the splitting of the separatrix Γ is a tiny and also delicate phenomenon, and the corresponding chaos occurs over rather long time scales only. So, there is also good news for numerical simulations. In fact, one has the following upper estimates for the relevant splitting quantities. Let us introduce the **splitting distance** $d(\varepsilon)$ which measures the maximal absolute distance between W_ε^s and W_ε^u along one-dimensional transversal sections, say normal to Γ, in a fixed compact region near Γ, say B, which does not include $x = 0$. Given any $k \in \mathbb{N}, d(\varepsilon) \le C\varepsilon^k$ holds in the symplectic case provided that ϕ_f is sufficiently smooth and ε is sufficiently small. Here, C is some constant. Moreover, if ϕ_f is symplectic and analytic, then the splitting distance is exponentially small with respect to ε, i.e. there exist constants C and $\eta > 0$, such that $d(\varepsilon) \le Ce^{-2\pi\eta/\varepsilon}$ holds for sufficiently small ε. In these cases, similar upper estimates are valid for the **splitting angle** $\omega(\varepsilon)$, i.e. the maximal (absolute) angle between the tangent lines of W_ε^s and W_ε^u at points of intersection in B. (See Fontich and Simó [1990 a, b]; cf. also Neisthadt [1984]; Holmes, Marsden and Scheurle [1988]; Fontich [1993]; Remarks ii) below.)

It is to be noted that nontrivial lower estimates for $d(\varepsilon)$ and $\omega(\varepsilon)$ are very hard to obtain here (cf. Holmes, Marsden and Scheurle [1988]; Scheurle [1989]; Lazutkin, Schachmannski and Tabanov [1989]; Scheurle,

Marsden and Holmes [1991]; Ellison, Kummer and Sáenz [1993]; Delshams and Seara [1992]).

To see that the occurence of chaos through discretization of a homoclinic orbit is also an issue for nonconservative systems, let us now consider a general smooth one-paramerter family of equations

$$(3.1) \qquad \dot{x} = f(\lambda, x) \quad (\lambda \in \mathbb{R}, x \in \mathbb{R}^2).$$

Assume that for $\lambda = 0$, this equation has a hyperbolic saddle point and a corresponding homoclinic orbit $\Gamma = \Gamma(t)$ as in the previous case. Here we assume in addition, that Γ is nondegenerate in the sense that the local stable and unstable invariant manifolds of the perturbed saddle points $x = x(\lambda)$ cross each other with non-vanishing velocity (in normal direction with respect to Γ) as λ increases through zero. To formulate this **nondegeneracy condition** on Γ analytically rather than geometrically, we consider the operator

$$
\begin{aligned}
L \; : \; C^1(\mathbb{R}, \mathbb{R}^2) \; &\rightarrow \; C^0(\mathbb{R}, \mathbb{R}^2) \\
x(\cdot) \; &\mapsto \; \dot{x}(\cdot) - A(\cdot)x(\cdot)\,,
\end{aligned}
$$

where $A(t) = D_x f(0, \Gamma(t))$, and C^k denotes the space of functions with uniformly bounded derivatives up to order k, endowed with the usual supnorm. It follows that L is a Fredholm operator with index zero (see Palmer [1984]).

We assume

$$C^0(\mathbb{R}, \mathbb{R}^2) = range\,(L) \oplus span\,(D_\lambda f(0, \Gamma(\cdot)))\,.$$

This implies that the homoclinic bifurcation at $\lambda = 0$ is robust under small autonomous perturbations of $f(\lambda, x)$. However, it turns out that generically this homoclinic bifurcation is not reproduced by a discretization scheme as considered here. Rather, it gives rise to a chaotic regime in the (λ, ε)-plane, where ε measures the step size of the discretization scheme again.

To state theorems from Fiedler and Scheurle [1991] concerning this claim, we need to introduce some notation first. For $f = f(\lambda, x)$ we set

$$\phi_f(\varepsilon, \lambda, x) = \phi_{f(\lambda, \cdot)}(\varepsilon, x)$$

for all λ. If f has the properties as in (3.1), then $W^s_{\varepsilon, \lambda}, W^u_{\varepsilon, \lambda}$ and $d(\varepsilon, \lambda)$ are well defined as before for small $\varepsilon > 0$ and $|\lambda|$. In addition, we introduce the **splitting interval**

$$I(\varepsilon) = \{\, \lambda \in [-\lambda^0, \lambda^0] \mid W^s_{\varepsilon, \lambda} \cap W^u_{\varepsilon, \lambda} \cap B \neq \emptyset \,\}\,,$$

where λ^0 is some positive constant. So, for $\lambda \in I(\varepsilon)$ the map $\phi_f(\varepsilon, \lambda, \cdot)$ has homoclinic points, and we can also define $\omega(\varepsilon, \lambda)$ as before.

THEOREM 3.1. *Assume that f is given as in (3.1). Then for small $\varepsilon > 0$, $I(\varepsilon)$ either contains just one point or is an interval. There exists a continuous curve $\lambda = \lambda_0(\varepsilon)$, $0 \leq \varepsilon < \varepsilon^0$, smooth for positive ε, such that $\lambda_0(\varepsilon) \in I(\varepsilon)$ for all ε, and $|\lambda_0(\varepsilon)| \leq C\varepsilon^p$, where ε^0 and C are constants and p is the order of consistency of the discretization scheme.*

Moreover, given any $k \in \mathbb{N}$, length $I(\varepsilon) \leq C\varepsilon^k$ holds provided that ϕ_f is sufficiently smooth. In particular, if ϕ_f is jointly real analytic in ε and complex analytic in λ and x, then there are constants $\varepsilon^0, \eta > 0$ and C such that for $0 < \varepsilon < \varepsilon^0$,

$$\text{length } I(\varepsilon) \leq C\, e^{-2\pi\eta/\varepsilon}$$

holds. For all $\lambda \in I(\varepsilon)$, the same upper estimates are valid for $d(\varepsilon, \lambda)$ and $\omega(\varepsilon, \lambda)$.

We shall outline a proof of this theorem in the next section. The next theorem which we state without proof, says that for generic, analytic f and any small $\varepsilon > 0$, $I(\varepsilon)$ really is an interval, inside which $d(\varepsilon, \lambda)$ and $\omega(\varepsilon, \lambda)$ are positive. Consequently, the homoclinic bifurcation in the continuous problem (3.1) gives rise to a tongue-like region in the (λ, ε)-plane emenating at the origin, inside which the discretized system behaves chaotically, even though the width of this region shrinks to zero exponentially fast as $\varepsilon \to 0$. As a matter of fact, it is not easy to hit this region at all numerically (cf. Beyn [1987]).

Let C^ω be the space of vector fields $f(\lambda, x)$ which are complex analytic for $|\lambda| \leq \lambda^0$ and $x = (x_1, x_2) \in \mathbb{C}^2$. In particular, we may write $f \in C^\omega$ as a power series

$$f(\lambda, x) = \sum_{k_0, k_1, k_2 \in \mathbb{N}_0} a_k \lambda^{k_0} x_1^{k_1} x_2^{k_2} \qquad (a_k \in \mathbb{C}^2).$$

We define a strong Whitney type topology on C^ω by introducing open neighborhoods $U_\delta(f)$, $\delta = (\delta_k)$ with $\delta_k > 0$ for all $k \in \mathbb{N}_0$. An element $\tilde{f} \in C^\omega$, the power series of which has coefficients $\tilde{a}_k$, lies in $U_\delta(f)$ if and only if $|a_k - \tilde{a}_k| < \delta_k$ holds for all k. Equipped with the topology generated by these open sets, C^ω becomes a Baire space, i.e. countable intersections of open and dense sets are still dense. Such intersections are called **residual subsets**.

Note that for each $f_0 \in C^\omega$ with the properties of f in (3.1), there exists a neighborhood V in C^ω, such that any $f \in V$ also has these properties after a possible translation of the variables λ and x.

THEOREM 3.2. *Assume that $\phi_f(\varepsilon, \cdot, \cdot)$ is contained in C^ω and real analytic in ε for any $f \in C^\omega$. In addition, assume that the map*

$$\phi \,:\, C^\omega \to C^\omega \,,\, f \mapsto \phi_f(\varepsilon, \cdot, \cdot)$$

maps arbitrarily small neighborhoods of f onto neighborhoods of $\phi_f(\varepsilon, \cdot, \cdot)$ for any f and any fixed positive $\varepsilon \leq \varepsilon^0(f)$. (E.g., note that this additional assumption is clearly satisfied in case of Euler's scheme.) Also, suppose that $f_0 \in C^\omega$ has the properties of f in (3.1). Then there is a neighborhood V of f_0 in C^ω and an ε^0 such that for f in some residual subset of V, i.e. for generic f in V, we have

$$length\ I(\varepsilon) > 0 \quad for\ all\ \varepsilon \in (0, \varepsilon^0).$$

Moreover, for any $\varepsilon \in (0, \varepsilon^0)$, $d(\varepsilon, \lambda) > 0$ and $\omega(\varepsilon, \lambda) > 0$ holds for some $\lambda \in I(\varepsilon)$.

Remarks

i) The above theorems remain true if $\mathbb{R}^2$ is replaced by $\mathbb{R}^n$ with arbitrary n. See Fiedler and Scheurle [1991]. However, there the analytic case is considered only. Also, cf. Gruendler [1991] who studies multi-parameter homoclinic bifurcations of autonomous systems in arbitrary dimensions.

ii) The before mentioned estimates for the symplectic case can be deduced from Theorem 3.1 as follows. Given a Hamiltonian system (1.1) with Hamiltonian function $H = H(x)$, $x \in \mathbb{R}^2$, and a homoclinic orbit $\Gamma = \Gamma(t)$ as above, we introduce an artifical parameter $\lambda \in \mathbb{R}$ as follows:

$$(3.2) \qquad \dot{x} = f(x) + \lambda\ grad\ H(x)$$

It is easy to see that Γ is a nondegenerate homoclinic orbit for this one-parameter family of equations. Hence, Theorem 3.1 yields upper estimates for $d(\varepsilon, \lambda)$ and $\omega(\varepsilon, \lambda)$ for all $\lambda \in I(\varepsilon)$. But we know a-priori that $\lambda = 0$ is contained in $I(\varepsilon)$ for all ε, since $\phi_f(\varepsilon, \cdot)$ is symplectic.

iii) Our discussion strongly suggests to use direct methods to compute homoclinic orbits numerically, i.e. one sets up a defining functional equation and solves this numerically (see e.g. Beyn [1990]).

4. Proof of Theorem 3.1. Following Fiedler and Scheurle [1991], we prove Theorem 3.1 by a Liapunov-Schmidt reduction. We just outline the basic ideas. For details, we refer to that paper.

In Section 2, we have proved that we can view ϕ_f as the period map of an equation of type (1.7). Using this equation, we derive a representation $\lambda = \lambda(\varepsilon, \alpha)$ for $\lambda \in I(\varepsilon)$. Here, α parametrizes the homoclinic orbit Γ. From that representation, the desired estimates will follow. Note that those λ-values are characterized by the fact that there exists a solution of (1.7) close to Γ, which connects the perturbed saddle point near $x = 0$ with itself. To construct such solutions, we introduce the variable $z = x - \Gamma(t + \alpha)$ as well as a shift $\beta \in \mathbb{R}$ of the fast-time variable $\tau = \frac{t}{\varepsilon}$ and consider the equation

$$(L(\alpha)z)(t) - \mathcal{F}(\varepsilon, \lambda, \alpha, \beta, z(\cdot))(t) = 0$$

$$(4.1) \qquad \langle z, \dot{\Gamma}(\cdot + \alpha) \rangle = \int_{-\infty}^{\infty} (z(t), \dot{\Gamma}(t + \alpha))dt = 0,$$

where

$$
\begin{aligned}
(L(\alpha)z)(t) &= \dot{z}(t) - A(t+\alpha)\,z(t)\,, \\
\mathcal{F}(\varepsilon,\lambda,\alpha,\beta,z(\cdot))(t) &= f(\lambda,\Gamma(t+\alpha)+z(t)) - f(0,\Gamma(t+\alpha)) \\
&\quad - A(t+\alpha)z(t) + \varepsilon^p\,g(\varepsilon,\lambda,\frac{t}{\varepsilon}+\beta,\Gamma(t+\alpha)+z(t))\,, \\
A(t) &= D_x f(0,\Gamma(t))\,,
\end{aligned}
$$

and $(\cdot,\cdot)$ denotes the standard Hermitean inner product. We look for solutions $z = z(\varepsilon,\lambda,\alpha,\beta,\cdot) \in C^1(\mathbb{R},\mathbb{R}^2)$ ($\in C^2(\mathbb{R},\mathbb{C}^2)$ in the analytic case) of this equation, which are small in the C^1-norm. It is easy to see that for $\beta = 0$, any such solution represents a solution of (1.7) of the desired type and vize versa.

The operator $L(\alpha) : C^1(\mathbb{R},\mathbb{R}^2) \to C^0(\mathbb{R},\mathbb{R}^2)$ defined above is a bounded linear Fredholm operator of index zero, which depends smoothly on α. Its kernel is spanned by $\dot{\Gamma}(t+\alpha)$, its co-kernel with respect to the pairing $\langle\cdot,\cdot\rangle$ by $\Psi(t+\alpha)$, where $\Psi(t)$ ($t \in \mathbb{R}$) is the unique (up to a scalar multiple) nontrivial bounded solution of the adjoint equation

$$
\dot{\Psi} = -\overline{A(t)}^T\,\Psi\ .
$$

Define the projection operator P_α onto the co-kernel of $L(\alpha)$ by

$$
P_\alpha z = \langle z,\Psi(\cdot+\alpha)\rangle\,\Psi(\cdot+\alpha)\,.
$$

Note that the second equation in (4.1) is just a normalizing condition for z.

We now follow the usual Liapunov-Schmidt procedure and decompose (4.1) into an equivalent system of equations

$$
\begin{aligned}
(Id - P_\alpha)\left(L(\alpha)z - \mathcal{F}(\varepsilon,\lambda,\alpha,\beta,z)\right) &= 0\,, & \langle z,\dot{\Gamma}(\cdot+\alpha)\rangle &= 0 \\
P_\alpha\left(L(\alpha)z - \mathcal{F}(\varepsilon,\lambda,\alpha,\beta,z)\right) &= -P_\alpha\,\mathcal{F}(\varepsilon,\lambda,\alpha,\beta,z) &= 0
\end{aligned}
$$

(4.2)

For (ε,λ,z) near $(0,0,0)$ and arbitrary α and β, the first of these equations can be solved uniquely for z by the implicit function theorem. In particular, the solution z has period 1 in β and satisfies

$$
(4.3) \qquad z(\varepsilon,\lambda,\alpha,\beta,t+\tilde{t}) = z(\varepsilon,\lambda,\alpha+\tilde{t},\beta+\frac{\tilde{t}}{\varepsilon},t)
$$

for all $\varepsilon,\lambda,\alpha,\beta$ as above and real $t,\tilde{t}$. This identity is a consequence of the particular normalization condition for z in (4.2). Thus, it remains to solve the second equation in (4.2). This is equivalent to the scalar **reduced equation**

$$
(4.4) \qquad\qquad S(\varepsilon,\lambda,\alpha,\beta) = 0
$$

with

$$S(\varepsilon, \lambda, \alpha, \beta) = \int_{-\infty}^{\infty} \overline{\Psi(t + \alpha)}^T \, \mathcal{F}(\varepsilon, \lambda, \alpha, \beta, z(\varepsilon, \lambda, \alpha, \beta, \cdot))(t) \, dt.$$

Clearly, S has period 1 in β. Furthermore, using (4.3) (with $\tilde{t} = -\beta\,\varepsilon$), we obtain

$$S(\varepsilon, \lambda, \alpha, \beta) = S(\varepsilon, \lambda, \alpha - \varepsilon\,\beta, 0).$$

Hence, S has period ε in α. The fast-time shift β has been introduced solely in order to prove this. From now on, we set $\beta = 0$. Since S is ε-periodic and in α as smooth as f, it possesses a Fourier expansion

$$S(\varepsilon, \lambda, \alpha, 0) = S(\varepsilon, \lambda, \alpha) = \sum_{m \in \mathbb{Z}} S_m(\varepsilon, \lambda) \, e^{2\pi i m \, \alpha/\varepsilon},$$

where

$$(4.5) \qquad S_m(\varepsilon, \lambda) = \frac{1}{\varepsilon} \int_0^\varepsilon S(\varepsilon, \lambda, \alpha, 0) \, e^{-2\pi i m \, \alpha/\varepsilon} \, d\alpha.$$

By partial integration it follows that, given any $k \in \mathbb{N}$, there is a constant c_k such that

$$(4.6) \qquad |S_m(\varepsilon, \lambda)| \le c_k \left(\frac{\varepsilon}{|m|} \right)^k$$

holds for all $m \ne 0$. Moreover, in the analytic case the homoclinic orbit $\Gamma = \Gamma(t)$ has an analytic extension to a complex strip $|Im\,t| \le \tilde{\eta}$ for some $\tilde{\eta} > 0$. Therefore, both the solution z of the first equation in (4.2) and S extend analytically to the complex strip $|Im\,\alpha| \le \tilde{\eta}$. Since S has real period ε in α, we can use Cauchy's theorem to shift the path of integration in (4.5) from $[0, \varepsilon]$ to the straight line from $0 \pm i\tilde{\eta}$ to $\varepsilon \pm i\tilde{\eta}$, respectively. This then yields the uniform exponential estimate

$$(4.7) \qquad |S_m(\varepsilon, \lambda)| \le C \, e^{-2\pi|m|\tilde{\eta}/\varepsilon}$$

for all $m \ne 0$, with some constant $C = C(\tilde{\eta})$. The estimates of the Fourier coefficients of S, which are based on both smoothness and ε-periodicity of S in α, are crucial for what follows. Namely, they immediately imply that S is of the form

$$(4.8) \qquad S(\varepsilon, \lambda, \alpha) = S_0(\varepsilon, \lambda) + R(\varepsilon, \lambda, \alpha),$$

where $R = O(\varepsilon^k)$ for $k \ge 2$, respectively, $R = O(e^{-2\pi\eta/\varepsilon})$ with $0 < \eta < \tilde{\eta}$, uniformly in λ and α as $\varepsilon \to 0$. In particular, $S(0, \lambda, \alpha) = S_0(0, \lambda)$ is independent of α. By definition of S, it follows that

$$(4.9) \qquad S_0(0, \lambda) = S(0, \lambda, 0) = 0$$

for $\lambda = 0$. Furthermore, by nondegeneracy of Γ,

$$(4.10) \quad D_\lambda S_0(0,0) = D_\lambda S(0,0,0) = \int_{-\infty}^{\infty} \overline{\Psi(t)}^T D_\lambda f(0, \Gamma(t)) \, dt \neq 0 \, .$$

Also,

$$(4.11) \qquad\qquad |S_0(\varepsilon, \lambda) - S_0(0, \lambda)| = O(\varepsilon^p) \, .$$

Now, we first consider the equation

$$(4.12) \qquad\qquad S_0(\varepsilon, \lambda) = 0 \, .$$

By (4.9), (4.10) and (4.11), the implicit function theorem applies to yield a solution curve $\lambda = \lambda_0(\varepsilon) = O(\varepsilon^p)$. Next we prove $\lambda_0(\varepsilon) \in I(\varepsilon)$. Indeed, by definition,

$$(4.13) \qquad 0 = S_0(\varepsilon, \lambda_0(\varepsilon)) = \frac{1}{\varepsilon} \int_0^\varepsilon S(\varepsilon, \lambda_0(\varepsilon), \alpha) \, d\alpha \, .$$

Hence, the integrand cannot be strictly of one sign. Since it is continuous in α, there exists an $\alpha = \alpha_0(\varepsilon)$ for which the integrand vanishes. Thus, we have solved the reduced equation for $\lambda = \lambda_0(\varepsilon)$, $\alpha = \alpha_0(\varepsilon)$ and $\beta = 0$, which implies $\lambda_0(\varepsilon) \in I(\varepsilon)$. In fact, because of (4.9) and (4.10), the implicit function theorem yields a unique solution $\lambda = \lambda(\varepsilon, \alpha)$ of the reduced equation for $\beta = 0$, any $\alpha \in \mathbb{R}$, and sufficiently small ε. Therefore, $I(\varepsilon) = \{\lambda = \lambda(\varepsilon, \alpha) \mid \alpha \in \mathbb{R}\}$. If $\lambda(\varepsilon, \alpha) = \lambda_0(\varepsilon)$ for all α, then $I(\varepsilon)$ contains just one point. Otherwise it is an interval, since $\lambda(\varepsilon, \alpha)$ is smooth in α. Because of (4.13), the length of $I(\varepsilon)$ is of the same order with respect to ε as $R(\varepsilon, \lambda, \alpha)$ in (4.8), as $\varepsilon \to 0$. Also, for $\lambda \in I(\varepsilon)$, $d(\varepsilon, \lambda)$ is of this same order. This follows from the fact that in B, $W_{\varepsilon,\lambda}^s$ and $W_{\varepsilon,\lambda}^u$ cross each other with finite speed as λ increases. This speed is of order $O(1)$ uniformly in λ, as $\varepsilon \to 0$. Therefore, up to a constant factor, *length* $I(\varepsilon)$ is an upper bound for $d(\varepsilon, \lambda)$. But this is also an upper bound for $\omega^2(\varepsilon, \lambda)$, since $W_{\varepsilon,\lambda}^s$ and $W_{\varepsilon,\lambda}^u$ have uniformly bounded curvature inside B. Thus, Theorem 3.1 follows with appropriate constants ε^0, C and η. For a sharper estimate of $\omega(\varepsilon, \lambda)$ we refer to Fiedler and Scheurle [1991].

REFERENCES

Beyn W.J. [1987] The effect of discretizations of homoclinic orbits, *in* "Bifurcation: Analysis, Algorithms, Applications", T. Küpper et al. (eds.), Birkhäuser-Verlag, Basel, 1–8.

Beyn W.J. [1990] The numerical computation of connecting orbits, *IMA J. Numer. Analysis* **9**, 379–405.

Delshams A. and Seara T.M. [1992] An asymptotic expansion for the splitting of separatrices of the rapidly forced pendulum, *Comm. Math. Phys.* **150**, 433–463.

Ellison J.A., Kummer M. and Sáenz A.W. [1993] Transcendentally small transversality in the rapidly forced pendulum, *J. Dyn. Diff. Equat.* **5(2)**, 241.

Fiedler B. and Scheurle J. [1991] Discretization of homoclinic orbits, rapid forcing and "invisible" chaos, to appear in *Memoirs of the AMS*.

Fontich E. [1993] Exponentially small upper bounds for the splitting of separatrices for high frequency periodic perturbation, *J. Nonlinear Analysis* **20(6)**, 733–744.

Fontich E. and Simó C. [1990 a] Splitting of separatrices for analytic diffeomorphisms, *Ergod. Theor. Dyn. Sys.* **10**, 295–318.

Fontich E. and Simó C. [1990 b] Invariant manifolds for near identity differentiable maps and splitting of separatrices, *Ergod. Theor. Dyn. Sys.* **10**, 319–346.

Genecand Ch. [1993] Transversal homoclinic orbits near elliptic fixed points of area preserving diffeomorphisms of the plane, *Dynamics Reported* **2** (New Series), 1–30.

Holmes P.J., Marsden J.E. and Scheurle J. [1988] Exponentially small splittings of separatrices with applications to KAM theory and degenerate bifurcations, *Cont. Math.* **81**, 213–244.

Hunter J. and Scheurle J. [1988] Existence of perturbed solitary wave solutions to a model equation for water waves, *Physica D* **32**, 253–268.

Kopell N. [1968] PhD thesis, *University of California, Berkeley*.

Lazutkin V.F., Schachmannski I.G. and Tabanov M.B. [1989] Splitting of separatrices for standard and semistandard mappings, *Physica D* **40**, 235–248.

Marsden J.E. [1992] Lectures on Mechanics, *London Math. Soc. Lecture Notes Series* **174**, *University Press, Cambridge*.

Neisthadt A.I. [1984] The separation of motion in systems with rapidly rotating phase, *J. Appl. Math. Mech.* **48 (2)**, 133–139.

Palmer K.J. [1984] Exponential dichotomies and transversal homoclinic points, *J. Diff. Equat.* **55**, 225–256.

Scheurle J. [1989] Chaos in a rapidly forced pendulum equation, *Cont. Math.* **97**, 411–419.

Scheurle J., Marsden J.E. and Holmes P.J. [1991] Exponentially small estimates for separatrix splittings, *in "Asymptotics beyond all orders"*, *H. Segur et al. (eds.), Plenum Press, New York*, 187–195.

Zehnder E. [1973] Homoclinic points near elliptic fixed points, *Comm. Pure Appl. Math.* **26**, 131–182.

COMPUTING THE MOTION OF THE MOON ACCURATELY*

DIETER S. SCHMIDT†

Abstract. The difficulties which one encounters in calculating the motion of the moon are typical for an entire class of problems where the solution is found approximately with the help of Poisson series. On the one side these difficulties are of a practical nature and have to do with the large number of terms which one has to manipulate. On the other side these difficulties also have to do with theoretical issues concerning the convergence of these series.

1. Introduction. Three hundred years after Newton published his *'Philosophiae Naturalis Principia Mathematica,'* calculating the motion of the moon accurately remains a challenging problem.

Newton's theory of gravitation explains the laws of Kepler, and Keplerian ellipses provide a good first order approximation to the motion of the planets around the sun. They serve well as the starting point for an analytic treatment of the planetary motion. On the other hand, when one starts with the moon on an elliptic orbit around the earth and then tries to incorporate the effects of the sun and other bodies with the help of a perturbation analysis, the results are disappointing. Newton himself knew of these difficulties. In his correspondence he mentioned several times that just thinking about the lunar problem gave him headaches.

The reasons for the disappointing results are not hard to find. The gravitational force of the sun on the moon is twice as large as the gravitational force of the earth on the moon. This means that the sun should not be treated as a perturbing force on the motion of the moon around the earth. Instead one has to consider a three body problem where the following two simplifying assumptions can be made:

1. The earth and the moon stay close together for all time.
2. The center of mass of the earth-moon system moves on an elliptic orbit around the sun.

This problem is known as the main problem of lunar theory. Finding an accurate solution for it is the first step in obtaining a complete solution for the motion of the moon.

Let the largest term in the solution to the main problem be normalized to one. The following table then shows the magnitude of the largest terms as they are attributed to the different perturbations of the moon:

---

* This research was supported by a grant from the Applied and Computational Mathematics Program of DARPA, administered by NIST.

† Department of Computer Science, University of Cincinnati, Cincinnati, Ohio 45221.

Effect	Magnitude
Main problem	1
Direct planetary perturbations	10^{-5}
Indirect planetary perturbations	10^{-6}
Oblateness of earth (J_2 term)	10^{-6}
Shape of earth (J_3 and J_4 terms)	10^{-8}
Mass concentrations on moon	10^{-8}
General relativity	10^{-9}
Tidal forces	$5 \cdot 10^{-9}$

We will not discuss the methods which might be used to calculate these effects analytically. Let us only say that J. Chapront and M. Chapront-Touzé at the Bureau des Longitudes de Paris [2] have carried out such a calculation. Currently their work is checked against numerical integration and has already been compared to actual observations of the moon's motion.

Although numerical integration of the equations of the motion of the moon suffices for most practical needs today, interest in an analytic solution remains high. An analytic solution gives a better understanding of the dynamical behavior of the moon. It allows one to detect effects that are sensitive to changes in the physical parameters. Many of these parameters could be determined more accurately if a better analytic solution were available. The solution of Chapront may satisfy this need some day, but at the moment it requires an independent verification of its own. This can be accomplished with another analytic solution which is developed independently, preferably by different methods and on computers with different hardware.

As far as the observations of the moon are concerned the distance between the earth and the moon can be measured today with an accuracy of about 10 cm. This is accomplished by laser ranging to reflectors which have been placed on the surface of the moon. The moon moves roughly 1 second of arc in the sky for every 2 seconds of time. At a distance of approximately 400,000 km this gives the moon a relative velocity of 2 km/sec. Occultations of fixed stars can be measured with an accuracy of about 10^{-4} seconds. This means that the position of the moon in its orbit can be fixed to within 20 cm. In a geocentric coordinate system the position of the moon can therefore be fixed to within 10 to 20 cm. An improvement of the observational accuracy by another order of magnitude is likely. If one wants an analytic solution to achieve this accuracy of 1 to 2 cm then all effects have to be included which produce normalized terms greater than $5 \cdot 10^{-11}$.

An analytic solution to the motion of the moon is usually computed as an approximation by a Poisson series of the form

$$(1.1) \qquad \sum A_{k,l} r_1^{k_1} \cdots r_m^{k_m} \begin{matrix} \cos \\ \sin \end{matrix} (l_1 \theta_1 + \cdots + l_n \theta_n)$$

where summation is over non-negative integers k_j and any integers l_j.

If the desired accuracy is to be obtained it is obvious that we have to account for the accumulation of roundoff errors when the series are evaluated. Therefore the accuracy of $5 \cdot 10^{-11}$ has to be increased by several orders of magnitude in order to achieve our goal. The number of terms in (1) tells us by how much to increase the accuracy. Ideally the worst case in evaluating (1) should be considered but practically one is satisfied if the mean square error is below the given threshold.

2. Historical notes. The problem which is discussed here is how to solve a nonintegrable Hamiltonian system of differential equations to a very high degree of accuracy. In working on such a problem one encounters two difficulties. The first one is of a practical nature and has to do with the large number of terms which one has to manipulate. The other difficulty is of a more theoretical nature and concerns the mathematical convergence (or divergence) of these series.

We believe that our experience with the lunar problem is representative of this entire class of problems where one tries to find solutions in the form of (1.1). Furthermore, the lunar problem should at least be of historical interest to the field of computer algebra. André Deprit was one of the first to write a computer program for symbolic computations. He used his program to check Delaunay's theory of the moon. Since Delaunay's work influenced those who attempted the lunar problem after him, and since Deprit suggested to us how to approach the main problem of lunar theory, we will mention briefly the findings of Deprit [4].

Delaunay [3] spent 20 years of his life computing the motion of the moon in a new set of variables. He computed all terms through order 7 and even a few to higher order. When he substituted the numerical values for the parameters he had to realize that his solution did not approach the observational accuracy which could be achieved then. His solution was only good to about 100 km. When Deprit repeated these calculations he found that Delaunay had made only one mistake in computing these huge series expansions by hand. The term in error is at order 7. It should have been $33/16 m^3 \gamma^2 e'^2$, whereas Delaunay wrote down the factor as $23/16$. Unfortunately, it is not just a misprint, as the wrong coefficient is used in later calculations. Fortunately the error has an insignificant effect on the accuracy of the final result.

The reason for the disappointing accuracy of Delaunay's work was that he had not gone far enough in the expansion. Deprit of course could go further by machine. He calculated terms through order 20. Nevertheless, even this extension did not provide the accuracy which is needed today.

The failure of Delaunay's work provided several valuable lessons for those who followed him. The first lesson was recognized by Hill [6] who saw that lunar theory had to be done differently from planetary theory. Hill saw a new way to account for the combined effects of the sun and earth on the moon when he constructed an 'intermediate orbit' which is used as the starting point for the perturbation analysis.

The second lesson was learned by Brown [1], who carried out the calculations following the ideas of Hill. Due to the poor convergence of the series with respect to one of the parameters, he substituted its numerical value into the differential equations a priori. This meant that he calculated the coefficients of the series not as rational numbers, but as numbers in fixed decimal arithmetic. Brown's solution achieves an accuracy of about 4 km.

The third lesson was that the tedious calculations should be left to computers or machines. We cite from Brown (1905): '*All computations which could with advantage be turned over to a computer have again be done by Mr. Ira Sterner. His speed and accuracy have been fully maintained, and have contributed in no small degree to an earlier conclusion of the work than I had hoped.*' Not only was it a heroic effort by Delaunay and Brown to get these huge series expansions, but it was also a substantial effort to use these series in calculating the ephemerides of the moon for each year. In effect it stifled celestial mechanics because too many people (i.e. computers) were busy calculating and checking the evaluation of these series. Already in 1932 Comrie proposed the use of Hollerith Tabulating Machines in evaluating Brown's tables of the moon. The application of electronic computers to the lunar problem dates back to Eckert around 1954. Eckert had planned to redo and improve Brown's work, but when he died the project was not finished. M. Gutzwiller had directed an assistant of Eckert to complete it. This work was not published because Gutzwiller had heard of our own effort and how it had surpassed the goal of Eckert. It was at this point that Gutzwiller and Schmidt [5] got together and wrote a joint paper, in part as a tribute to Eckert's pioneering efforts.

3. The equations of motion. Although the equations of motion have been derived elsewhere, we will give a brief presentation, as the choice of parameters has some influence on how well the solution will converge. We believe that there exists a natural choice of the parameters. It is a set which has physical significance, but it differs from that used by Brown and others. Hill had already started to experiment by working with different parameters but it deserves more exploration.

The general three body problem in barycentric coordinates is given by

$$H = \frac{S+E+M}{2S(E+M)}|X_2|^2 + \frac{E+M}{2EM}|X_1|^2$$
$$- \frac{GEM}{|x_1|} - \frac{GSE}{|x_2 + \frac{M}{E+M}x_1|} - \frac{GSM}{|x_2 - \frac{M}{E+M}x_1|}$$

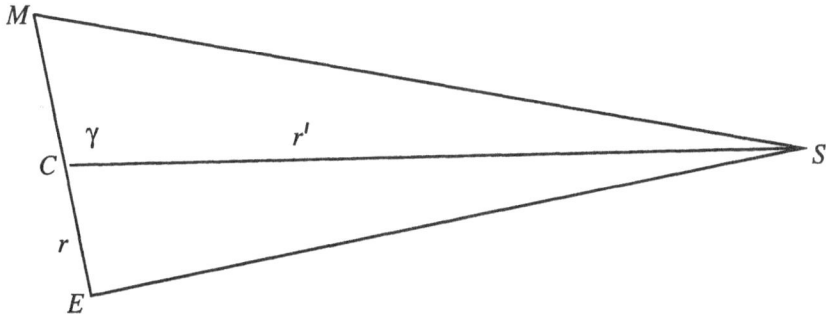

FIG. 3.1. *The main problem of lunar theory: Sun, Earth and Moon. C is the center of gravitation for the Earth–Moon system*

where S, E, and M are the masses of the three bodies, G the gravitational constant; see Figure 3.1. In a fixed but otherwise unspecified coordinate system for R^3, the vector x_1 points from the earth to the moon and x_2 points from the center of mass of the earth-moon system to the sun. The vectors X_1 and X_2 are the corresponding momenta. The Hamiltonian function gives rise to the following twelve first order differential equations

$$\frac{dx_j}{dt} = \frac{\partial H}{\partial X_j}, \quad \frac{dX_j}{dt} = -\frac{\partial H}{\partial x_j} \qquad j = 1, 2.$$

Since $|x_1|/|x_2| \approx 1/400$, one expands the last two terms in H with the help of the Legendre functions P_j and arrives at

$$H = K' + K$$

(3.1) $$K' = \frac{S + E + M}{2S(E + M)} \,|\, X_2 \,|^2 - \frac{GS(E + M)}{|x_2|}$$

and

(3.2) $$K = \frac{E + M}{2EM}|X_1|^2 - \frac{GEM}{|x_1|} - \frac{GSEM}{E + M} \sum_{j=2} c_j \frac{|x_1|^j}{|x_2|^{j+1}} P_j(\cos \gamma)$$

where γ is the angle between the two position vectors and the coefficients

$$c_j = \frac{E^{j-1} - (-M)^{j-1}}{(E + M)^{j-1}}$$

are functions of the mass ratio $\mu = M/E$.

The next step is to justify the decoupling of the corresponding systems of differential equations. This means that the Hamiltonian for H can be written as two separate Hamiltonian functions as is done in (3.1) and (3.2). The Hamiltonian K' describes a two body problem consisting of the sun and the center of mass of the earth-moon system. The solution to this problem is a Keplerian ellipse. The plane in which this orbit lies is known as the ecliptic. We now choose our coordinate frame such that its x-y coordinate plane coincides with the ecliptic.

The shape of the ellipse is determined by the semi major axis a' and the eccentricity e'. The orientation is given by g', which is the angle between the x-axis and the perigee. The initial position of the sun on the ellipse at the time $t = t_0$ is given by the mean anomaly l'_0.

In lunar theory it is a common convention to denote everything that refers to the sun with a "$'$". The radius vector r' and the true longitude ϕ' are given by

$$r' = |x_2| = a'(1 - e' \cos l' + \cdots)$$

$$\phi' = g' + l' + 2e' \sin l' - \cdots$$

where l' is the mean anomaly given by

$$l' = n'(t - t_0) + l'_0$$

with n' the mean motion of the sun around the earth-moon system. It is related to a' by Kepler's law

(3.3) $$n'^2 a'^3 = G(M + E + S).$$

We therefore have to consider the Hamiltonian function K which is given in (3.2). It gives rise to only six first order differential equations

$$\frac{dx_1}{dt} = \frac{\partial H}{\partial X_1}, \qquad \frac{dX_1}{dt} = -\frac{\partial H}{\partial x_1}$$

but they are now dependent on time and they contain several parameters.

It is natural to introduce a rotating coordinate system in the ecliptic such that its positive x-axis always points to the mean position of the sun. With $\phi = g' + l'$, which is known as the mean longitude of the sun, this transformation is given by

$$x_1 = b \begin{pmatrix} x \cos \phi - y \sin \phi \\ x \sin \phi + y \cos \phi \\ z \end{pmatrix}, \quad X_1 = b(n-n') \frac{EM}{E + M} \begin{pmatrix} X \cos \phi - Y \sin \phi \\ X \sin \phi + Y \cos \phi \\ Z \end{pmatrix}.$$

The scale factor b will be determined later. The additional factors in front of the momenta have been chosen so that the new Hamiltonian depends only on essential parameters.

The transformation is canonical provided that the Hamiltonian function is multiplied by $(E+M)/(EM(n-n')b^2)$. The constant n denotes the mean motion of the moon around the earth. An additional factor $1/(n-n')$ occurs because we have also changed to a new time by $\tau = (n-n')(t-t_0)$.

With the notation

$$m = \frac{n'}{n-n'}, \qquad r^2 = x^2 + y^2 + z^2, \qquad \alpha = \frac{b}{a'}$$

the Hamiltonian derived from (3.2) is

$$K = \frac{1}{2}(X^2 + Y^2 + Z^2) - m(xY - yX) - \frac{G(E+M)}{b^3(n-n')^2}\frac{1}{r}$$
$$- \frac{GS}{a'^3(n-n')^2} \sum_{j=2} c_j \alpha^{j-2} \left(\frac{a'}{r'}\right)^{j+1} r^j P_j(\cos\gamma).$$

In analogy to Kepler's law a quantity a is defined by

$$n^2 a^3 = G(E+M).$$

The terms in K which will be affected by the choice of b are the coefficient of $1/r$

$$A = \frac{G(E+M)}{b^3(n-n')^2} = \frac{a^3}{b^3}(1+m)^2$$

the coefficient in front of the summation sign

$$B = \frac{GS}{a'^3(n-n')^2} = m^2\left(1 - \frac{E+M}{S+E+M}\right)$$

and

$$\alpha = \frac{b}{a'}.$$

We will discuss three choices for b:

Case 1) The most common choice is $b = a$. It leads to

$$A = (1+m)^2, \qquad B = m^2 - (1+m)^2\alpha^3, \qquad \text{and} \quad \alpha = a/a'.$$

As will be seen in the next section, it is important for the development of the solution that the perturbing function, i.e. B, has the factor m^2. The usual trick is then to introduce a new independent parameter for the mass ratio, say $q = (E + M)/(S + E + M)$ so that $B = m^2(1 - q)$. But q and α obviously depend on each other and we question this approach for a high accuracy solution.

Case 2) To overcome the above shortcoming we introduce instead of α the parameter β which is defined by

$$\beta^3 = \frac{E + M}{S + E + M}.$$

We choose b so that $A = 1$. We find that $B = m^2(1 - \beta^3)$ and $\alpha = m^{2/3}\beta$. This choice has the advantage that β has numerically the same order of magnitude as the other parameters. On the other hand, the form of α shows that if the solution is developed in terms of m it has to be done in powers of $m^{1/3}$. Such a solution has not yet been attempted. Due to the poor convergence of the series with respect to m, it is common practice to substitute the numerical value of m into the differential equations a priori.

Case 3) We could choose b so that $\alpha = \beta$. This eliminates the fractional powers of m and leads to $A = m^2$ and $B = m^2(1 - \beta^3)$. Unfortunately m^2/r is of the wrong order of magnitude for the contribution of the earth's potential.

Therefore, we adopt the second selection and set out to solve the differential equations defined by

$$K = \frac{1}{2}(X^2 + Y^2 + Z^2) - m(xY - yX) - \frac{1}{r} - \Omega$$

where

$$\Omega = m^2(1 - \beta^3)\sum_{j=2} c_j\beta^{j-2}m^{2(j-2)/3}\left(\frac{a'}{r'}\right)^{j+1} r^j P_j(\cos\gamma).$$

The function Ω exhibits the following dependence $\Omega = \Omega(\tau, x, y, z; m, \beta, e', \mu)$. The lowest order terms in Ω are

$$\Omega_2 = m^2 r^2 P_2(\frac{x}{r}) = \frac{m^2}{2}(3x^2 - r^2).$$

Hill discovered that the moon moves close to a 2π-periodic orbit of K with Ω reduced to Ω_2. This so-called intermediate orbit is symmetric with respect to the x-axis and plays a role which is equivalent to that of a Keplerian ellipse in the planetary problem. If we write $\Omega = \Omega_2 + \Omega'$, the equations of motion are

(3.4)
$$
\begin{array}{llllll}
\ddot{x} & -2m\dot{y} & -m^2 x & -2m^2 x & +x/r^3 & = \partial\Omega'/\partial x \\
\ddot{y} & +2m\dot{x} & -m^2 y & +m^2 y & +y/r^3 & = \partial\Omega'/\partial y \\
\ddot{z} & & & +m^2 z & +z/r^3 & = \partial\Omega'/\partial z
\end{array}
$$

The equations are written in such a way that the different physical forces can be identified in each column: Inertial, Coriolis, centrifugal, sun's gravitational (reduced to a quadrupole), earth's gravitational force and perturbations.

The behavior of the trajectories in the six dimensional phase space is more easily visualized for an autonomous system. This happens when $e' = 0$, since then $r' = a'$ and $\cos\gamma = x/r$, and thus Ω does not depend on τ. In this case the trajectories of (3.4) lie on a five dimensional submanifold given by $K = $ const. Since no additional integrals exist, it is to be expected that the solutions behave ergodically on these submanifolds.

This is not the case for the intermediate orbit, nor for the trajectories in its neighborhood. Each of these orbits lies on a three dimensional submanifold. It is a 3-torus which is described by Delaunay's angles l, F and τ. Two radial variables, e and k, are needed to fix the 3-torus as it lies in the five dimensional manifold which is given by $K = $ const (or by the value of m).

For small $e' \neq 0$, these 3-tori are not destroyed, but they change their size synchronously with the mean motion l' of the sun. These explanations should justify the form in which one tries to find the solution:

$$
\begin{pmatrix} x \\ y \\ z \end{pmatrix} = \sum A_j \begin{pmatrix} \cos(j_1 l + j_2 F + j_3 l' + (j_4 + 1)\tau) \\ \sin(j_1 l + j_2 F + j_3 l' + (j_4 + 1)\tau) \\ \sin(j_1 l + j_2 F + j_3 l' + j_4 \tau) \end{pmatrix}
$$

with

$$
A_j = \sum C_j^i e^{i_1} k^{i_2} e'^{i_3} \beta^{i_4}.
$$

Summation is over all integer tuples $i = (i_1, i_2, i_3, i_4)$ and $j = (j_1, j_2, j_3, j_4)$ with the restriction that

(3.5) $i_1 \geq |j_1|, \quad i_2 \geq |j_2|, \quad$ and $\quad i_3 \geq |j_3|.$

The angles are

$$
l = c\tau + l_0
$$

$$
F = g\tau + g_0
$$

$$
l' = m\tau + l_0'
$$

$$\tau = (n - n')(t - t_0).$$

The frequencies c and g in the angles l and F have to be determined also and they have the form

$$c = c_0 + \sum c_i e^{i_1} k^{i_2} e^{i_3} \beta^{i_4}$$

$$g = g_0 + \sum g_i e^{i_1} k^{i_2} e^{i_3} \beta^{i_4}$$

For later it is important to point out that all coefficients C_j^i and those in c and g are functions of m and also of μ, although the latter dependency is not too important for what follows.

4. Method of solution. We will keep the discussion brief and concentrate on what is needed for the error analysis in the next section.

It is more convenient to work with complex power series than with Fourier series, in particular when the calculations have to be done efficiently by machine. For this reason we introduce the complex-valued variable

$$\zeta = e^{i\tau}$$

and adopt the following conventions

$$\zeta^c \quad \text{will mean} \quad e^{i(c\tau + l_0)}$$
$$\zeta^g \quad \text{will mean} \quad e^{i(g\tau + g_0)}$$
$$\zeta^m \quad \text{will mean} \quad e^{i(m\tau + l'_0)}$$

The basic operations like adding or multiplying series are preserved by this convention. Also, differentiating is preserved if one uses the differential operator D with the property

$$D\zeta^n = n\zeta^n$$

and which is related to the derivative with respect to τ by

$$\frac{d}{d\tau} = \zeta \frac{d}{d\zeta} = iD.$$

With the complex position coordinates

$$\zeta u = x + iy \qquad \text{and} \qquad \zeta^{-1} v = x - iy$$

the differential equations to solve are then

$$(D + m + 1)^2 u + \frac{1}{2}m^2 u + \frac{3}{2}m^2 \zeta^{-2} v - \frac{u}{r^3} \quad = \quad -2\frac{\partial \Omega'}{\partial v}$$

(4.1) $\qquad (D - m - 1)^2 v + \frac{3}{2}m^2 \zeta^2 u + \frac{1}{2}m^2 v - \frac{v}{r^3} \quad = \quad -2\frac{\partial \Omega'}{\partial u}$

$$D^2 z - m^2 z - \frac{z}{r^3} \quad = \quad -\frac{\partial \Omega'}{\partial z}$$

The intermediate orbit begins with

$$u_0 = 1 - \frac{2}{3}m + (\frac{3}{16}\zeta^2 + \frac{7}{18} - \frac{19}{16}\zeta^{-2})m^2 + \cdots$$
$$z_0 = 0$$

The general solution has terms of the form

$$C_j^i e^{i_1} k^{i_2} e'^{i_3} \beta^{i_4} \zeta^{j_1 c + j_2 g + j_3 m + j_4},$$

where $e^{i_1} k^{i_2} e'^{i_3} \beta^{i_4}$ is called the characteristic and $j_1 l + j_2 F + j_3 l' + j_4 \tau$ is the argument of a term.

From terms with characteristic e in u one finds (i.e. see [8])

(4.2)
$$c_0 = 1 + m - \tfrac{3}{4}m^2 - \tfrac{201}{32}m^3 - \tfrac{2367}{128}m^4 - \tfrac{111749}{2048}m^5$$
$$- \tfrac{4095991}{24576}m^6 - \cdots .$$

Similarly terms of characteristic k in z give

(4.3)
$$g_0 = 1 + m + \tfrac{3}{4}m^2 - \tfrac{33}{32}m^3 - \tfrac{105}{128}m^4 + \tfrac{43}{2048}m^5$$
$$+ \tfrac{2567}{24576}m^6 + \cdots .$$

In a general context integrating the differential equations leads to a quadrature of the form

$$(x, y, z) = \int \sum A_j {\cos \atop \sin} (j_1 l + j_2 F + j_3 l' + j_4 \tau) d\tau$$

or in terms of powers of ζ

$$(u, v, z) = D^{-1} \sum A_j \zeta^{j_1 c + j_2 g + j_3 m + j_4}$$
$$= \sum \frac{A_j}{j_1 c + j_2 g + j_3 m + j_4} \zeta^{j_1 c + j_2 g + j_3 m + j_4}.$$

Since it is impractical to keep all the different denominators during the calculations and most of c and g are not known anyway, we write

$$c = c_0 + c_1,$$

$$g = g_0 + g_1$$

and develop these denominators with the help of the binomial formula, that is

(4.4) $(j_1 c + j_2 g + j_3 m + j_4)^{-1} =$

$$(j_1 c_0 + j_2 g_0 + j_3 m + j_4)^{-1}(1 + \frac{j_1 c_1 + j_2 g_1}{j_1 c_0 + j_2 g_0 + j_3 m + j_4})^{-1}$$

It is apparent that the binomial expansion of the last term may not be valid. Its coefficient can be greater than 1 in absolute value for certain quadruples (j_1, j_2, j_3, j_4). Thus we have to conclude that there is a limiting order where the formal solution can not be constructed any further because the binomial expansion becomes formally divergent. This point has not yet been reached for the lunar problem!

Consider next the first term on the right hand side of (4.4) as a function of m

$$j_1 c_0 + j_2 g_0 + j_3 m + j_4 = j_1 + j_2 + j_4 +$$
$$m(j_1 + j_2 + j_3) - \frac{3}{4}m^2(j_1 - j_2) - \frac{3}{32}m^3(67j_1 + 11j_2)$$
$$-\frac{3}{128}m^4(789j_1 + 35j_2) + \cdots$$

Depending on the quadruple (j_1, j_2, j_3, j_4), some of the low order terms in the denominators can be zero. This means that the denominators can have a factor m of order 0,1,2 or 3, but not of higher order. Since Ω' has the factor m^2, the first three cases are acceptable but not the divisor m^3. In this case it appears possible that our solution has a singularity at $m = 0$. These are the considerations presented in Kovalevski [7]. We will show through a more careful analysis which terms lead to a serious loss of accuracy in the solution and we will list their arguments.

5. Investigations into the accuracy of the solution. In order to discuss the accuracy with which individual terms of the solution can be computed we have to look more carefully at the determining equations.

Let λ denote a certain characteristic, say $\lambda = e^{i_1} k^{i_2} e'^{i_3} \beta^{i_4}$. We denote all terms in u with this characteristic by u_λ and those in z by z_λ. The form of equation (4.1) determines that u_λ has only terms with even i_2, and z_λ has only terms with odd i_2.

With $v_\lambda = \bar{u}_\lambda$ denoting the conjugate complex value, we obtain from (4.1) the following linear set of differential equations

$$(5.1) \quad \begin{aligned} (D + m + 1)^2 u_\lambda + M u_\lambda + N v_\lambda &= A_\lambda \\ (D - m - 1)^2 v_\lambda + N u_\lambda + M v_\lambda &= \overline{A}_\lambda \\ D^2 z_\lambda - 2M z_\lambda &= C_\lambda. \end{aligned}$$

The order of a term is $i_1 + i_2 + i_3 + i_4$. We assume that all terms of order lower than λ have been computed. Collectively they contribute terms with characteristic λ to the right hand sides in (5.1) via the perturbing function Ω' and the nonlinearities in (4.1). The functions M and N depend only on the intermediate orbit and are given by

$$\begin{aligned} M &= \tfrac{1}{2}(m^2 + (u_0 v_0)^{-3/2}) & = \sum M_{2i} \zeta^{2i} \\ N &= \tfrac{3}{2}(m^2 \zeta^{-2} + u_0^{-1/2} v_0^{-3/2}) & = \sum N_{2i} \zeta^{2i}. \end{aligned}$$

For later we list all terms through m^4

$$M_0 = \tfrac{1}{2} + m + \tfrac{5}{4}m^2 - \tfrac{9}{64}m^4 + \cdots$$

$$M_2 = M_{-2} = \tfrac{3}{4}m^2 + \tfrac{19}{8}m^3 + \tfrac{10}{3}m^4 + \cdots$$

$$M_4 = M_{-4} = \tfrac{33}{32}m^4 + \cdots$$

$$N_0 = \tfrac{3}{2} + 3m + \tfrac{9}{4}m^2 - \tfrac{417}{128}m^4 + \cdots$$

$$N_2 = \tfrac{69}{16}m^2 + \tfrac{29}{2}m^3 + \tfrac{2137}{96}m^4 + \cdots \qquad N_{-2} = \tfrac{27}{16}m^2 - \tfrac{1}{4}m^3 - \tfrac{217}{96}m^4 + \cdots$$

$$N_4 = \tfrac{4497}{512}m^4 + \cdots \qquad\qquad\qquad N_{-4} = \tfrac{123}{512}m^4 + \cdots$$

Many arguments $cj_1 + gj_2 + mj_3 + j_4$ can go with a particular characteristic provided that the inequalities (3.5) are met. They restrict the triple (j_1, j_2, j_3) to finitely many values, whereas j_4 can range through all even or all odd integers. For such a triple we write

$$\sigma = c_0 j_1 + g_0 j_2 + m j_3$$

and we also will write j instead of j_4 from now on. The terms which belong to a given λ and σ are in A_λ :

$$\lambda(\zeta^\sigma \sum A_j \zeta^j + \zeta^{-\sigma} \sum B_j \zeta^{-j}).$$

Due to the symmetry of the equation for z the corresponding terms in C_λ are of the form:

$$\lambda(\zeta^\sigma \sum C_j \zeta^j - \zeta^{-\sigma} \sum C_j \zeta^{-j}).$$

Assuming terms in u_λ and z_λ of the same form, that is

$$\lambda(\zeta^\sigma \sum a_j \zeta^j + \zeta^{-\sigma} \sum b_j \zeta^{-j}) \qquad \text{and} \qquad \lambda(\zeta^\sigma \sum c_j \zeta^j - \zeta^{-\sigma} \sum c_j \zeta^{-j})$$

respectively, we compare coefficients and derive the following infinite set of linear equations

$$(5.2) \qquad (\sigma + j + m + 1)^2 a_j + \sum M_{j-k} a_k + \sum N_{j-k} b_k \;=\; A_k$$

$$(5.3) \qquad (\sigma + j - m - 1)^2 b_j + \sum N_{k-j} a_k + \sum M_{k-j} b_k \;=\; B_k$$

$$(5.4) \qquad\qquad (\sigma + j)^2 c_j \quad - 2\sum M_{j-k} c_k \;=\; C_k.$$

The equations have been written so that they hold for the case where j assumes all even integers and also for the case when j takes on all odd integers.

Obviously the first two equations belong together. With j_0 denoting an arbitrary index, we form the column vector of unknowns

$$\xi = (\dots, a_{j_0-2}, b_{j_0-2}, a_{j_0}, b_{j_0}, a_{j_0+2}, b_{j_0+2}, \dots)^T$$

and in the same way the vector of the right hand side

$$\eta = (\ldots, A_{j_0-2}, B_{j_0-2}, A_{j_0}, B_{j_0}, A_{j_0+2}, B_{j_0+2}, \ldots)^T.$$

Equations (5.2) and (5.3) represent an infinite system of linear equations, whose coefficient matrix will be displayed at the intersection of an arbitrary row and column j_0. We use the 2 by 2 submatrices

$$L_j = L_{-j}^T = \begin{pmatrix} M_{-j} & N_{-j} \\ N_j & M_j \end{pmatrix} \qquad j = 0, \pm 2, \pm 4, \ldots$$

and

$$\Delta_k = \begin{pmatrix} (k + \sigma + m + 1)^2 + M_0 & N_0 \\ N_0 & (k + \sigma - m - 1)^2 + M_0 \end{pmatrix}$$

in order to write the coefficient matrix in block form

(5.5) $\Theta(\sigma) =$

$$\begin{pmatrix}
\ddots & \vdots & \vdots & \vdots & \vdots & \vdots & \\
\cdots & \Delta_{j_0-4} & L_2 & L_4 & L_6 & L_8 & \cdots \\
\cdots & L_{-2} & \Delta_{j_0-2} & L_2 & L_4 & L_6 & \cdots \\
\cdots & L_{-4} & L_{-2} & \Delta_{j_0} & L_2 & L_4 & \cdots \\
\cdots & L_{-6} & L_{-4} & L_{-2} & \Delta_{j_0+2} & L_2 & \cdots \\
\cdots & L_{-8} & L_{-6} & L_{-4} & L_{-2} & \Delta_{j_0+4} & \cdots \\
& \vdots & \vdots & \vdots & \vdots & \vdots & \ddots
\end{pmatrix}$$

The coefficient matrix $\Theta(\sigma)$ is symmetric and has period 2 in σ. The matrix has a one dimensional null space when $\sigma = c_0$. As a matter of fact, one can determine c_0 from this condition.

The elements along the diagonal of $\Theta(\sigma)$ grow quadratically with j_0 and the terms in the vector η drop off quickly away from the index 0. Therefore there is no problem in truncating the infinite system $\Theta\xi = \eta$ to a finite one. The problem is the loss of significance in solving this system. The terms in η are affected by truncation errors. For some values of σ and j_0 these errors are amplified in the solution ξ.

Although all calculations are carried out with m replaced by its numerical value, $m \approx .08084 \cdots$, the loss of significance is best illustrated with the help of the formal expansion with respect to m. From (4.2) and (4.3) we find

$$\sigma = j_1 + j_2 + m(j_1 + j_2 + j_3) - \tfrac{3}{4}m^2(j_1 - j_2) - \tfrac{3}{32}m^3(67j_1 + 11j_2) \\ - \tfrac{3}{128}m^4(789j_1 + 35j_2) + \cdots.$$

Since M_k and N_k are at least of order $|k|$ in m, the system of equations can be solved accurately when the 2 by 2 matrices along the diagonal have determinants which do not include m as a factor. The factor m occurs when $j_0 + j_1 + j_2 = 1, 0$, or -1. Since j_0 is an arbitrary integer which is either even or odd, it will suffice to consider the first two cases.

When we solve the given system of equations, we can take into account that (5.5) is symmetric and that the matrices L_k have at least the factor m^k. We can use the Gaussian elimination method with the 2 by 2 submatrices of (5.5), and keeping all terms through m^4 we arrive at a system of linear equations for a_{j_0} and b_{j_0} with the coefficient matrix

$$(5.6) \qquad P_{j_0} - L_2^T P_{j_0-2}^{-1} L_2 - L_2 P_{j_0+2}^{-1} L_2^T + \mathcal{O}(m^5).$$

Here we have used the abbreviation $P_k = \Delta_k + L_0$. In deriving this coefficient matrix it suffices to look at the inner 3 by 3 block of (5.5), but a more careful analysis including all terms displayed in (5.5) shows that the above expression is correct even through higher orders of m as long as the determinant of P_k starts with a constant term for $k \neq j_0$.

In the case when $j_0 = 1 - j_1 - j_2$ the five 2 by 2 sub-matrices which are displayed along the diagonal of $\Theta(\sigma)$ have determinants which start with the following lowest order terms in m: $72 + \cdots$, $-2(1 + j_1 + j_2 + j_3)m + \cdots$, $2(-1 + j_1 + j_2 + j_3)m + \cdots$, $72 + \cdots$, and $600 + \cdots$. The third value vanishes for $j_3 = 1 - j_1 - j_2$ and the determinant starts then with $\frac{3}{2}(1 - j_1 + j_2)m^2 + \cdots$. Even this value can be zero when $j_1 = 1 - j_2$, so that the subdeterminant starts with $-(j_2 + 225)m^3/256 + \cdots$. Under these conditions the coefficient matrix (5.6) is

$$\begin{pmatrix} \frac{9}{2} + 9m + \frac{9}{4}m^2 - \frac{117}{2048}(512j_2 + 483)m^3 + \cdots & \frac{3}{2} + 3m + \frac{9}{4}m^2 - \frac{2277}{2048}m^3 + \cdots \\ \frac{3}{2} + 3m + \frac{9}{4}m^2 - \frac{2277}{2048}m^3 + \cdots & \frac{1}{2} + m + \frac{5}{4}m^2 + \frac{4761}{2048}m^3 \cdots \end{pmatrix}.$$

It is used to find a_{j_0} and b_{j_0} from $A_{j_0}^*$ and $B_{j_0}^*$, where the last two terms come from A_{j_0} and B_{j_0}, plus the contribution of the other unknowns in the linear system.

For convenience we will drop the index j_0. In order to simplify the system of equations, we divide by the off-diagonal elements and call the right hand sides again A and B. In essence we try to solve the following system of equations numerically:

$$(1 - m^2 - \tfrac{1}{64}(416j_2 + 217)m^3 + \cdots)3a + \qquad\qquad b \qquad\qquad = A$$
$$a \qquad\qquad + (1 + m^2 + \tfrac{217}{64}m^3 + \cdots)b/3 = B.$$

By solving this system one sees that a and b will have a singularity at $m = 0$, if $A - 3B$ does not have at least the factor m^3. It means also that any errors in A and B will be magnified by m^{-3} in a and b. Only $3a + b$ can be found with some accuracy remaining.

Since j_2 is even for terms in u the following makes up the list of critical arguments:

$$(5.7) \quad j_1 = 2k + 1, \quad j_2 = 2k, \quad j_3 = j_4 = -4k, \quad \text{with } k = \pm 1, \pm 2, \ldots.$$

In the case when $j_0 = -j_1 - j_2$ the values of the 2 by 2 sub-determinants along the diagonal of $\Theta(\sigma)$ are $240 + \cdots$, $12 + \cdots$, $-(j_1 + j_2 + j_3)^2 m^2 + \cdots$, $12 + \cdots$, and $240 + \cdots$. Only the sub-determinant in the middle can cause problems. For $j_3 = -j_2 - j_1$ its lowest order terms are $\frac{9}{128}(133 - 8(j_2 - j_1)^2)m^4 + \cdots$. It can not vanish for any values of j_1 and j_2, but the factor m^4 is an indication that terms in the solution can suffer a significant loss of accuracy. In order to see how accurately we can obtain a_{j_0} and b_{j_0} we again look at terms through order m^4 in (5.6). They are

$$
\begin{pmatrix} 3/2 & 3/2 \\ 3/2 & 3/2 \end{pmatrix}
$$

$$
+ \; m \begin{pmatrix} 3 & 3 \\ 3 & 3 \end{pmatrix}
$$

$$
+ \; \frac{3m^2}{2} \begin{pmatrix} j_2 - j_1 + 3/2 & 3/2 \\ 3/2 & j_1 - j_2 + 3/2 \end{pmatrix}
$$

$$
+ \; \frac{m^3}{16} \begin{pmatrix} -225j_1 - 9j_2 & 0 \\ 0 & 225j_1 + 9j_2 \end{pmatrix}
$$

$$
+ \; \frac{m^4}{16} \begin{pmatrix} 9(j_1 - j_2)^2 - \frac{3171}{4}j_1 - \frac{237}{4}j_2 - \frac{4209}{64} & -\frac{4209}{1024} \\ -\frac{4209}{1024} & 9(j_1 - j_2)^2 + \frac{3171}{4}j_1 + \frac{237}{4}j_2 - \frac{4209}{64} \end{pmatrix}
$$

$$
+ \; \cdots
$$

Again we will divide by the term off the diagonal and use the notation employed above. The system of equations we try to solve numerically is therefore of the following form

$$
\begin{aligned}
(1 + \epsilon_1 m^2 + \epsilon_2 m^4 + \cdots)a + b &= A \\
a + (1 - \epsilon_1 m^2 + \epsilon_2 m^4 + \cdots)b &= B
\end{aligned}
$$

where

$$
\epsilon_1 = j_1 - j_2 - \frac{m}{8}(59j_1 + 19j_2) - \frac{m^2}{32}(537j_1 - 25j_2) + \cdots
$$

$$
\epsilon_2 = \frac{3}{8}(j_2 - j_1)^2 + \cdots.
$$

By solving this system we get with

$$
\epsilon = 1/(2\epsilon_2 - \epsilon_1^2) = -4/(j_1 - j_2)^2 + \cdots
$$

$$
a = \epsilon(A - B)m^{-4} + A\epsilon\epsilon_1 m^{-2} + \cdots \qquad \text{and} \qquad b = \epsilon(B - A)m^{-4} - B\epsilon\epsilon_1 m^{-2} + \cdots
$$

so that

$$
a - b = 2\epsilon(B - A)m^{-4} + \cdots \qquad \text{and} \qquad a + b = \epsilon\epsilon_1(A - B)m^{-2} + \cdots.
$$

If the solution should be without a singularity at $m = 0$ we need $A - B = O(m^4)$. On the other hand if A and B already contain errors, the corresponding error in $a + b$ is magnified by m^{-2}, and in $a - b$ by m^{-4}.

The portion of the solution we are looking at is

$$a\zeta^{\sigma+j_0} + b\zeta^{-\sigma-j_0}$$

or in real and imaginary terms

$$(a + b)\cos(\sigma + j_0\tau) + i(a - b)\sin(\sigma + j_0\tau).$$

Therefore the larger errors can be found among the long periodic sine terms in y. They are long periodic since $\sigma + j_0 = \frac{3}{4}(j_2 - j_1)m^2 + \cdots$. Considering that a solution is valid only for a finite amount of time these errors behave more like secular terms.

The list of difficult arguments is in this case

(5.8) j_1, $j_2 = 2k$, $j_3 = j_4 = -2k - j_1$ with k and j_1 arbitrary.

Included in this list is the case $j_1 = j_2$, but it deserves special consideration as then ϵ_1 and ϵ_2 start out with higher order terms than indicated above. In this case we find

$$\epsilon_1 = -j_2 m^3 \left(\frac{39}{4} + 16m + \frac{39821}{768}m^2 + \frac{385015}{2304}m^3 + \cdots\right)$$

and

$$\epsilon_2 = \frac{4563}{128}j_2^2 m^6 + \cdots$$

and with it the solution

$$a = b = \frac{16(A - B)}{1521j_2^2 m^6} + \cdots.$$

Obviously, we have a singularity for $j_2 = 0$, but this is consistent with the fact that the matrix (5.5) is singular for $j_1 = j_2 = j_3 = 0$ because we have the freedom to select the constant term of the solution. Nevertheless, the special case of (5.8) $j_1 = j_2 = 2k$ and $j_3 = j_0 = -4k$ with $k \neq 0$ will lead to a serious loss of significance, once the right hand side is afflicted with errors. Fortunately, the first of these terms occurs only at order 12 and we did not have to include it in our solution.

Next we study the loss of accuracy among terms in the z-component which are found by solving (5.4). The column vector of unknowns is

$$\xi = (\ldots, c_{j_0-2}, c_{j_0}, c_{j_0+2}, \ldots)^T$$

and the right hand side is

$$\eta = (\ldots, C_{j_0-2}, C_{j_0}, C_{j_0+2}, \ldots)^T.$$

The coefficient matrix is this time

$$(5.9) \qquad \vartheta(\sigma) = \begin{pmatrix} \ddots & \vdots & \vdots & \vdots & \\ \cdots & \Delta_{j_0-2} & -2M_2 & -2M_4 & \cdots \\ \cdots & -2M_2 & \Delta_{j_0} & -2M_2 & \cdots \\ \cdots & -2M_4 & -2M_2 & \Delta_{j_0+2} & \cdots \\ & \vdots & \vdots & \vdots & \ddots \end{pmatrix}$$

where $\qquad \Delta_j = (j+\sigma)^2 - 2M_0$

The infinite system $\vartheta\xi = \eta$ can be truncated for the same reasons which were given earlier. The terms along the diagonal of $\vartheta(\sigma)$ start with a constant term unless $j_0 = 1 - j_1 - j_2$ or $j_0 = -1 - j_1 - j_2$. In the first case the 3 elements shown on the diagonal of $\vartheta(\sigma)$ have the following lowest order terms in m:

$$-2m(j_1 + j_2 + j_3 + 1) + \cdots, \quad 2m(j_1 + j_2 + j_3 - 1) + \cdots, \quad 8 + \cdots.$$

Either of the first two terms can be zero but not both together. When $j_3 = -1 - j_1 - j_2$ the first series starts with $\frac{3}{2}(j_1 - j_2 - 1)m^2 + \cdots$ and if $j_1 = 1 + j_2$ the series starts with $(225 + 234j_2)m^3/16 + \cdots$. The factor m^3 can lead to a loss of significance by amplifying any errors. Since j_2 is odd for terms of z this occurs when

$$(5.10) \quad j_1 = 2k, \quad j_2 = 2k - 1, \quad j_3 = -4k, \quad j_0 = 2 - 4k \quad \text{with } k = 1, 2, \ldots$$

When $j_3 = 1 - j_1 - j_2$ the second term on the diagonal starts with

$$\frac{3}{2}(-j_1 + j_2 - 1)m^2 + \cdots$$

and if $j_1 = j_2 - 1$ the series starts with

$$(225 - 234j_2)m^3/16 + \cdots.$$

A loss of significance can thus occur when

$$(5.11) \quad j_1 = 2k, \quad j_2 = 2k + 1, \quad j_3 = j_0 = -4k \quad \text{with } k = 1, 2, \ldots.$$

In the other case $j_0 = -1 - j_1 - j_2$ the same kind of arguments give the following two lists of difficult arguments

$$(5.12) \quad j_1 = 2k, \quad j_2 = 2k - 1, \quad j_3 = j_0 = -4k \quad \text{with } k = 1, 2, \ldots$$

and

$$(5.13) \quad j_1 = 2k, \quad j_2 = 2k + 1, \quad j_3 = -4k, \quad j_0 = -2 - 4k \quad \text{with } k = 1, 2, \ldots$$

In a more detailed analysis one has to consider the entire matrix $\vartheta(\sigma)$ and look at the 2 by submatrices along the diagonal which could be singular. In the case of (5.10) it is the upper left 2 by 2 submatrix which is displayed in (5.9). We then find out how accurately the pair (c_{j_0-2}, c_{j_0}) can be determined. The coefficient matrix for its system of linear equations is with the notation

$$P_k = \left(\begin{array}{cc} (k-2+\sigma)^2 - 2M_0 & -2M_2 \\ -2M_2 & (k+\sigma)^2 - 2M_0 \end{array} \right)$$

$$L_k = \left(\begin{array}{cc} -2M_{k+2} & -2M_{k+4} \\ -2M_k & -2M_{k+2} \end{array} \right)$$

given by

$$P_{j_0} - L_2^T P_{j_0-4}^{-1} L_2 - L_2 P_{j_0+4}^{-1} L_2^T$$

through order $\mathcal{O}(m^5)$. The inverse of this matrix has to be investigated.

6. Remarks. The fact that a formal solution to the equations of the motion of the moon will diverge is not surprising, as this is what we have to expect from the Kolmogorov-Arnold-Moser theorem. What should be more surprising is the fact that one is actually able to calculate a solution to a very high degree of accuracy. This is a consequence of the judicious choice of Hill's intermediate orbit as a starting point and the nature of the phase space in the vicinity of this orbit. It means that for the lunar theory one is away from any ergodic or quasi-ergodic domain of the phase space.

The numerical instability, that is the amplification of errors at long periodic terms which was discussed in the previous section, is of course an indication that the formal solution will not converge. Formulas (5.7) through (5.11) give the arguments where this numerical instability can be observed. The different arguments are given in the table below.

From (3.5) we see that (5.7), (5.10) and (5.11) produce terms which occur for the first time at order 7. On the other hand, the first term which is generated by (5.8) has the arguments $(1, 0, -1, -1)$ and appears already at order 3 in u with the characteristic $ee'\beta$. Considering that (5.8) is associated with a factor m^4 instead of m^3, this case is much more troublesome than all the others. Furthermore the same long periodic term will occur many times at higher order. The term just computed together with its error will contribute to the right hand sides, which are set up to determine the higher order terms with the same critical arguments.

At low orders it is possible to trace the contribution of each term to the different coefficients in A_λ and B_λ. Therefore, it is possible to calculate these terms separately with more precision, or conceivably in rational arithmetic. After a few orders this approach becomes impractical. It is simpler to compute all terms in extended precision. This approach

First occurrence	characteristic	j_1	j_2	j_3	j_4	see eqn
in u at order 3	$ee'\beta$	1	0	-1	-1	(5.8)
4	$k^2 e'^2$	0	2	-2	-2	(5.8)
4	$e^2 k^2$	2	-2	0	0	(5.8)
4	$e^2 e'^2$	2	0	-2	-2	(5.8)
5	$ek^2 e'\beta$	1	-2	1	1	(5.8)
7	$ek^2 e'^3\beta$	1	2	-3	-3	(5.8)
7	$e^3 k^2 e'\beta$	3	-2	-1	-1	(5.8)
7	$e^3 e'^3\beta$	3	0	-3	-3	(5.8)
7	$ek^2 e'^4$	-1	-2	4	4	(5.7)
in z at order 7	$e^2 k e'^4$	2	1	-4	-2	(5.10)
7	$e^2 k e'^4$	2	1	-4	-4	(5.12)
in u at order 8	$k^4 e'^4$	0	4	-4	-4	(5.8)
8	$e^2 k^4 e'^2$	2	-4	2	2	(5.8)
8	$e^2 k^2 e'^4$	2	2	-4	-4	(5.8)
8	$e^4 k^4$	4	-4	0	0	(5.8)
8	$e^4 k^2 e'^2$	4	-2	-2	-2	(5.8)
8	$e^4 k^4$	4	0	-4	-4	(5.8)
9	$ek^4 e'^3\beta$	1	-4	3	3	(5.8)
9	$e^3 k^4 e'\beta$	3	-4	1	1	(5.8)
9	$e^3 k^2 e'^4$	3	2	-4	-4	(5.7)
in z at order 9	$e^2 k^3 e'^4$	2	3	-4	-4	(5.11)
9	$e^2 k^3 e'^4$	2	3	-4	-6	(5.13)

has its limitations too. In order to get the higher accuracy, the threshold has to be set small but then many terms will be kept. At order 6 we reach the limit of our system which is 32767 terms for each individual series. It would have been possible to overcome this limitation but it seems pointless to manipulate series which become longer and longer, while in the end only a few terms will contribute to the final result.

Starting with order 6, we keep only those terms which are greater than 10^{-14} in absolute value after they have been evaluated at approximate values of the parameters. With this method we proceed to order 9. There the y-components of long periodic terms increase more than what seems acceptable. From this order on, most of the significant terms will have only the parameters e and k. Therefore we restrict ourselves to calculating terms with $j_3 = j_4 = 0$. This eliminates most of the troublesome long periodic terms, in particular $(1, 0, -1, -1)$ and $(1, 2, -3, -3)$. The latter argument appears for the first time at order 7 in u with the characteristic $ek^2 e'^3\beta$. Its coefficient and those of higher order terms with the same argument are especially difficult to determine. We terminate the calculations at order 12 because at this order very few terms remain significant.

In summary, the fact that the formal solution to the main problem of lunar theory will not converge manifests itself in the difficulties which one encounters in computing coefficients for long periodic terms accurately. The numerical instability is caused by the expediency of replacing a formal parameter by its numerical value from the beginning. This in turn leads to the use of floating point arithmetic and to the need to truncate infinite series at each order. Despite these difficulties, it is possible to compute a formal solution in the sense of an asymptotic approximation which matches the observational accuracy of the moon today.

REFERENCES

[1] E. Brown. Theory of the motion of the moon. *Memoirs of the Royal Astronomical Society of London*, 53:39–116, 1897; 54:1–64, 1900; 57:51–145, 1905; 59:1–103, 1908.

[2] M. Chapront-Touzé and J. Chapront. The lunar Ephemeris ELP 2000. *Astron. and Astrophys.*, 124:50–62, 1983.

[3] C. Delaunay. *Théorie du Mouvement de la Lune*. Volume 28 and 29, Mémoires de l'Académie des Sciences, Paris, 1860.

[4] A. Deprit. Lunar ephemeris: Delaunay's theory revisited. *Science*, 168:1569–70, 1970.

[5] M. Gutzwiller and D. Schmidt. The motion of the moon as computed by the method of Hill, Brown and Eckert. *Astronomical Papers, U.S. Naval Observatory*, 23:1–272, 1986.

[6] G. Hill. Researches in lunar theory. *Am. J. of Math*, 1:5–16, 129–147, 245–260, 1877.

[7] J. Kovalevsky. Modern lunar theory. In V. Szebehely, editor, Applications of Modern Dynamics to Celestial Mechanics and Astrodynamics. *D. Reidel Publ. Co., Dordrecht-Boston-London*, pages 59–76, 1982.

[8] D. S. Schmidt. Literal solution for Hill's lunar problem. *Celestial Mech.*, 19:279–289, 1979.

ON THE RAPIDLY FORCED PENDULUM

PREM KUMAR N. SWAMI*

Abstract. The rapidly forced pendulum equation $\ddot{x} + \sin x = \delta \sin t/\varepsilon$, with $\delta = \delta_0 \varepsilon^p$, $p \geq -2$ and δ_0, ε sufficiently small, is considered. We sketch our proof that the term proportional to δ^2 in the splitting distance $d(t_0)$ in the $(\dot{x}, t)$ plane has the form

$$\delta^2 \left(-\frac{\pi}{4} \varepsilon \, sech \, \frac{\pi}{2\varepsilon} + O(\varepsilon^2 \exp(-\pi/2\varepsilon)) \right) \sin \frac{2t_0}{\varepsilon}.$$

From this it follows that for $-2 \leq p < -1$, the Melnikov term $\pi \delta \sin \frac{1}{\varepsilon} t_0 \, sech \, \frac{1}{2\varepsilon} \pi$ is dominated by the $O(\delta^2)$ term as $\varepsilon \downarrow 0$.

1. Introduction. We consider the equation of the rapidly forced pendulum

$$(1.1) \qquad\qquad \ddot{x} + \sin x = \delta \sin t/\varepsilon,$$

where the amplitude of the forcing term has the form $\delta = \delta_0 \varepsilon^p$ and δ_0, $\varepsilon > 0$ are sufficiently small. For $\delta = 0$ Eq. (1.1) has a homoclinic orbit in the $(x, \dot{x} = y)$-phase plane connecting the equilibrium points $(\pm \pi, 0)$. As a solution of (1.1) it is given by means of the formulae:

$$(1.2) \quad x_0(s) = 2 \tan^{-1}(\sinh s), \quad y_0(s) = 2 \, sech \, s, \quad s = t - t_0, \quad -\infty < 1 < \infty.$$

For $\delta > 0$, but $\delta \varepsilon^2$ sufficiently small (i.e., $p \geq -2$) there exists a hyperbolic $2\pi\varepsilon$-periodic solution $(x_p(t), y_p(t))$ near $(\pi, 0)$. Also there is a parametrization of the stable manifold of $x_p(t)$ by solutions $x_s(t, t_0)$ of (1.1), with $t, t_0 \in \mathbb{R}$, $t \geq t_0$, such that $x_s(t_0, t_0) = 0$ for all t_0. These results can be obtained from the stable manifold theorem as stated in [8].

It is well known that in order to compute an expression for the splitting distance between the stable and the unstable manifold associated with this periodic solution in the present problem one cannot apply Melnikov's method directly, although the correctness of the term it yields can be justified under mild circumstances with alternative methods. The dominance of the Melnikov term was justified by different authors for different ranges of the value p: by Holmes, Marsden and Scheurle [2] for $p \geq 8$, by Gelfreich [1] for $p > 5$, by J.A. Ellison, M. Kummer and A.W. Saenz [7] for $p \geq 5$ and later [8] for $p \geq 3$, and finally by Delshams et al. [5] for $p > 0$.

The existence of the periodic solution together with the stable and unstable manifolds leads to the conjecture that the dominance of the Melnikov term in the splitting distance between the stable and unstable manifold associated with the periodic solution can be guaranteed for all values of δ, ε for which $\delta \varepsilon^2$ is sufficiently small, i.e., for $p \geq 2$. In this paper we sketch our

---

* Department of Mathematics, University of Toledo, Toledo, Ohio 43606.

proof that for $p < -1$ the terms of order two in δ dominate the Melnikov term.

Observe that "*if $x(t)$ is a solution of Eq. (1.1) then so is $x^*(t) := -x(-t)$.*" Hence letting $y_s(t, t_0) = D_1 x_s(t, t_0)$ it follows that $(x_u(t, t_0), y_u(t, t_0)) = (-x_s(-t, -t_0), y_s(-t, -t_0))$ characterizes the unstable manifold of the periodic solution $(x_p(t), y_p(t))$. Since $x_s(t_0, t_0) = 0 = x_u(t_0, t_0)$, the distance between the two manifolds (splitting distance) in the $(\dot{x}, t)$ plane at $t = t_0$ is:

$$(1.3) \qquad d(t_0) = y_s(-t_0, -t_0) - y_s(t_0, t_0).$$

It is proved in [7] and [8] that the termproportional to δ in $d(t_0)$ is $\delta\pi \sin \frac{t_0}{\varepsilon}$ sech $\frac{\pi}{2\varepsilon}$ which is the Melnikov term. The main result of this paper is contained in the following theorem:

Main Theorem: For $p \geq -2$, the term proportional to δ^2 in the splitting distance between the two manifolds has the following form:

$$\delta^2 \left(-\frac{\pi}{4} \varepsilon \text{ sech } \frac{\pi}{2\varepsilon} + O(\varepsilon^2 \exp(-\pi/2\varepsilon)) \right) \sin \frac{2t_0}{\varepsilon}.$$

In other words, the splitting distance is given by the formula:

$$d(t_0) = \pi\delta \left(\sin \frac{t_0}{\varepsilon} - \frac{\delta\varepsilon}{4}(1 + O(\varepsilon)) \sin \frac{2t_0}{\varepsilon} \right) \cdot \text{sech } \frac{\pi}{2\varepsilon} + O(\delta^3).$$

From this expression our contention that for $\delta = \delta_0 \varepsilon^p$ with $p < -1$ the Melnikov term does not give the correct approximation to the splitting distance follows.

The difficulty in getting exponentially small estimates for the splitting distance lies in extending the solutions $x_s(t, t_0)$ analytically into a complex domain that comes close to the singularities of the unperturbed separatrix $x_0(t)$. However, in order to prove our main theorem it is sufficient to construct such an extension for the terms in $x_s(t, t_0)$ up to order 2 in δ. Our proof is based on the previous work of Ellison, Kummer and Saenz ([6], [7], and [8]). In particular, the same Banach space X and the same linear operator $\mathcal{L} : X \to X$ which were used by these authors to prove the stable manifold theorem will also play a crucial role in our reasoning. We briefly describe these tools in Section 3. The reader may refer to [6], [7] or [8] for more details.

2. Preparations. Introducing the variables a, b through the relations:

(2.1)
$$\begin{aligned} x_s(t, t_0) &= x_0(s) + \delta a(s, t_0) \\ y_s(t, t_0) &= y_0(s) + \delta b(s, t_0), \end{aligned}$$

where $s = t - t_0$, $b(s, t_0) = D_1 a(s, t_0)$, Eq. (1.1) is transformed into:

(2.2)
$$a'' + g(s)a = F(a, s, \delta) + \sin\left(\frac{s + t_0}{\varepsilon}\right),$$

where:

(2.3)
$$F(a, s, \delta) := 2f(s)\frac{1 - \cos \delta a}{\delta} + g(s)\left(a - \frac{\sin \delta a}{\delta}\right)$$

with

(2.4)
$$f(s) : = \tanh s \, \mathrm{sech} \, s = \frac{1}{2}\sin x_0(s),$$

(2.5)
$$g(s) : = 2 \, \mathrm{sech}^2 \, s - 1 = \cos x_0(s).$$

Now, from (1.3) we obtain the following expression for the splitting distance.

(2.6)
$$d(t_0) = \delta(b(0, -t_0) - b(0, t_0)).$$

In particular the term proportional to δ^2 in $d(t_0)$ is:

(2.7)
$$d^{(2)}(t_0) = b^{(1)}(0, -t_0) - b^{(1)}(0, t_0),$$

where $b^{(1)}(s, t_0)$ is the term proportional to δ in $b(s, t_0)$ which is the derivative of the corresponding term $a^{(1)}$ in a

(2.8)
$$b^{(1)}(s, t_0) = D_1 a^{(1)}(s, t_0).$$

Putting $a = a^{(0)} + \delta a^{(1)} + O(\delta^2)$ in (2.2) and comparing terms of equal powers in δ we obtain:

(2.9)
$$\left(a^{(0)}(s, t_0)\right)'' + g(s)a^{(0)}(s, t_0) = \sin\frac{s + t_0}{\varepsilon},$$

and

(2.10)
$$\left(a^{(1)}(s, t_0)\right)'' + g(s)a^{(1)}(s, t_0) = f(s)\left(a^{(0)}(s, t_0)\right)^2.$$

The functions $a^{(0)}$ and $a^{(1)}$ satisfy the differential equations:

(2.11)
$$a(s, t_0)'' + g(s)a(s, t_0) = q(s, t_0)$$

with appropriate function $q(s, t_0)$ together with the initial conditions
$a(0, t_0) = 0$.

Now we separate the two time dependencies of the functions a and q by letting them depend on a "slow" time s and a fast "periodic" time $z := \frac{s}{\varepsilon}$ as indicated in the following formula:

(2.12) $a(s, \varepsilon\phi) = \alpha(s, z, \phi), \quad q(s, \varepsilon\phi) := q(s, z, \phi)$.

Here we assume that $\alpha(s, z, \phi)$ and $q(s, z, \phi)$ are periodic with period 2π in the variable z which hereafter will be considered to be independent of the variable s and since we have set $\phi := \frac{t_0}{\varepsilon}$, α and q also become periodic in ϕ with period 2π. Equs. (2.11) and (2.12) give rise to the following PDE for α:

(2.13) $$\left(\frac{\partial}{\partial s} + \frac{1}{\varepsilon}\frac{\partial}{\partial z}\right)^2 \alpha + g(s)\alpha = q(s, z, \phi)$$

with initial condition $\alpha(0, 0, \phi) = 0$.

Now we complexify s, z and ϕ and look for the solution of (2.13) in a certain domain $\mathcal{D} \times \mathcal{S} \times \mathcal{S}$ which will be defined in Section 3. We expand α into a Fourier Series w.r. to z:

(2.14) $$\alpha(s, z, \phi) = \sum_k \alpha_k(s, \phi)e^{ikz}.$$

The initial condition expressed in terms of the Fourier coefficients becomes:

(2.15) $$\alpha_0(0, \phi) = -\sum_{k \neq 0} \alpha_k(0, \phi).$$

Moreover, in view of (2.13) the Fourier coefficients must satisfy the ODE:

(2.16) $\left(e^{i(k/\varepsilon)s}\alpha_k(s, \phi)\right)'' + g(s)\alpha_k(s, \phi)e^{i(k/\varepsilon)s} = q_k(s, \phi)e^{i(k/\varepsilon)s}$,

where $q_k(s, \phi)$ is the kth Fourier coefficient of the function $q(s, z, \phi)$. Since

(2.17) $\ell(s) := \operatorname{sech} s, \quad h(s) := \sinh s + s\ell(s)$

is a fundamental set of solutions of the variational equation

(2.18) $$\xi'' + g(s)\xi = 0$$

along the homoclinic orbit, it follows that the bounded solutions of

$$\xi'' + g(s)\xi = q(s)$$

have the form

(2.19) $\xi(s) = x(0)\ell(s) - \frac{1}{2}\ell(s)\int_0^s h(u)q(u)du - \frac{1}{2}h(s)\int_s^\infty \ell(u)q(u)du$.

As in refs. [6]-[8], this result will be used to find solutions of the ODE (2.16) for the Fourier coefficients $\alpha_k(s, \phi)$ and in turn these solutions will be employed to construct a linear operator $\mathcal{L}$ acting in a suitable Banach space X. The next section is devoted to a review of these concepts which will play a crucial role in the proof of our main theorem.

3. The theoretical tools of our proof. Let $\mathcal{D}$ be the wedge shaped region in the right half plane of $\mathbb{C}$ (shown in Figure 3.1) and let $\mathcal{S} = \{\zeta \in \mathbb{C} : |Im\zeta| \leq w\}$.

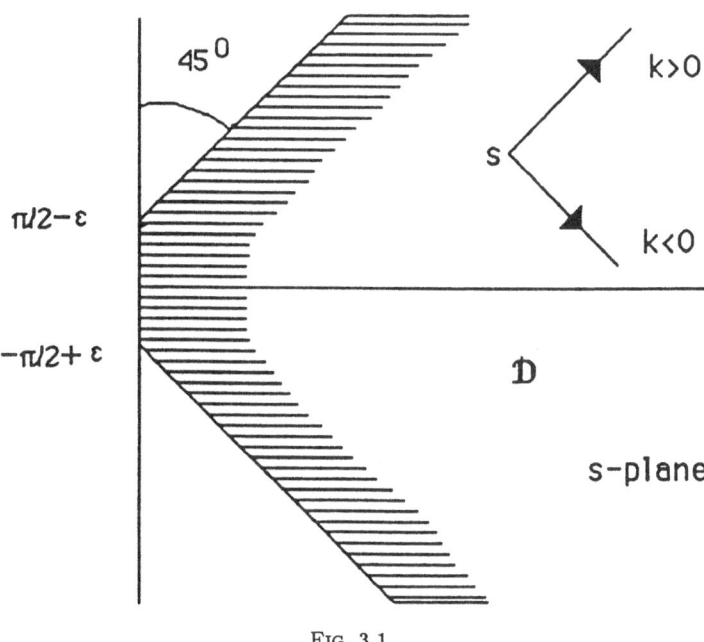

FIG. 3.1.

Let Y be the space of bounded complex valued functions $a(s, \phi)$ continuous on $\mathcal{D} \times \mathcal{S}$, real analytic on $Int\,\mathcal{D} \times \mathcal{S}$ and with period 2π in ϕ. Y is a Banach space under the sup norm:

$$(3.1) \qquad \qquad \|a\|_\infty := \sup_{(s,\phi)\in\mathcal{D}\times S} |a(s, \phi)|.$$

Now, let Y_0 be the closed subspace of Y consisting of $a \in Y$ such that $\lim_{s\to\infty} a(s, \phi) := a^\infty(\phi)$ exists uniformly in ϕ. Finally, we introduce the

Banach space:

$$X = \left\{ \alpha : \mathcal{D} \times S \times S \to \mathbb{C} : \alpha(s, z, \phi) = \sum_k \alpha_k(s, \phi) e^{ikz}, \right.$$

(3.2)

$$\left. \alpha_k(s, \phi) \in Y_0, \|\alpha\| := \sum_k \|\alpha_k\|_\infty e^{|k|w} < \infty \right\}.$$

It has been proved in [7] that X is a Banach algebra with respect to sum and product of functions. Acting in this algebra is a linear operator $\mathcal{L} : X \to X$ defined by means of the following prescription: For $q(s, z, \phi) = \sum_k q_k(s, \phi) e^{ikz} \in X$, set

$$(3.3) \quad (\mathcal{L}q)_k(s, \phi) = \frac{1}{2} \oint_s^\infty G(s, u) e^{\frac{ik}{\epsilon}(u-s)} q_k(u, \phi) du \quad (k \neq 0),$$

$$(3.4) \quad (\mathcal{L}q)_0(s, \phi) = -\frac{1}{2} \oint_0^s G(s, u) q_0(u, \phi) du + \frac{1}{2} h(s) \oint_0^\infty \ell(u) q_0(u, \phi) du +$$

$$\ell(u) \sum_{k \neq 0} \oint_0^\infty h(u) e^{\frac{iku}{\epsilon}} q_k(u, \phi) du,$$

where the function G is defined in terms of the functions introduced in (2.17) as follows:

$$(3.5) \qquad G(s, u) = \ell(s) h(u) - h(s) \ell(u).$$

In Formulae (3.3) and (3.4) integration, denoted by $\oint_s^\infty$, is supposed to proceed along a straight line through s with slope $+1$ if $k > 0$ and -1 if $k < 0$ (see Figure 3.1). It is shown in ref. [7] and [8] that $\mathcal{L} : X \to X$ is a bounded linear operator. According to our reasoning at the end of Section 2 (see (2.13)-(2.19) and also [7]) it is clear that $\mathcal{L}$ has been constructed in such a way that the image $\alpha(s, z, \phi) := (\mathcal{L}q)(s, z, \phi)$ under $\mathcal{L}$ satisfies the partial differential equation (2.13) together with the initial condition $\alpha(0, 0, \phi) = 0$.

4. Computation of term after Melnikov's. In this section we briefly describe the actual computation of the term after Melnikov's. Details will appear elsewhere [9]. First note that it follows from Eqs. (2.9)-(2.10) that:

(4.1)
$$\alpha^{(0)}(s, z, \phi) = \mathcal{L}(\sin(z + \phi)), \quad \text{and}$$
$$\alpha^{(1)}(s, z, \phi) = \mathcal{L}(f(s)(\alpha^{(0)}(s, z, \phi))^2).$$

Moreover, since $\sin(z + \phi) = \frac{e^{i\phi}}{2i} e^{iz} - \frac{e^{-i\phi}}{2i} e^{-iz}$ we deduce from Eqs. (3.3), (3.4) that:

$$\alpha^{(0)}(s, z, \phi) = \mathcal{L}(\sin(z + \phi))$$

$$= \frac{1}{4i} e^{i\phi} A_+(s) e^{iz} - \frac{1}{4i} e^{-i\phi} - \frac{1}{2}(M \sin \phi + \Delta A \cos \phi) \ell(s)$$

Here, the function $A_\pm(s)$ are analytic in $\mathcal{D}$ and are defined via the integrals:

$$(4.2) \qquad A_\pm(s) := \oint_s^\infty G(s,\sigma) e^{\pm i/\varepsilon(\sigma-s)} d\sigma,$$

The function G was defined in (3.5). Moreover M and ΔA are abbreviations of the following expressions

$$(4.3) \qquad M = \frac{1}{2}(A_+(0) + A_-(0)), \quad \Delta A = \frac{1}{2i}(A_+(0) - A_-(0)).$$

Expanding $\beta^{(1)}(s,z,\phi) := D_1\alpha^{(1)}(s,z,\phi) + \frac{1}{\varepsilon}D_2\alpha^{(1)}(s,z,\phi)$ into a Fourier series

$$\beta^{(1)}(s,z,\phi) = \sum_{k=-\infty}^{\infty} \beta_k^{(1)}(s,\phi)e^{ikz},$$

we verify that the Fourier coefficients of $\alpha^{(1)}$ and $\beta^{(1)}$ are related as follows:

$$\beta_k^{(1)}(s,\phi) = e^{iks/\varepsilon}\left(e^{iks/\varepsilon}\alpha_k^{(1)}(s,\phi)\right)'.$$

and we deduce from (3.3), (3.4) and (4.1) that

$$\beta_k^{(1)}(0,\phi) = -\oint_0^\infty \ell(u)f(u)(\alpha^{(0)}(u,z,\phi)^2)_k e^{iku/\varepsilon} du.$$

Finally, since $b^{(1)}(0,t_0) = D_1a^{(1)}(0,t_0) = \beta^{(1)}(0,0,\phi)$, where $\phi = t_0/\varepsilon$, we have:

$$b^{(1)}(0,-t_0) - b^{(1)}(0,t_0) = \beta^{(1)}(0,0,-\phi) - \beta^{(1)}(0,0,\phi)$$

$$= \sum_{k=-\infty}^{\infty} (\beta_k^{(1)}(0,-\phi) - \beta_k^{(1)}(0,\phi))$$

$$= \sum_{k=-\infty}^{\infty} \oint_0^\infty \ell(u)f(u)\left\{(\alpha^{(0)}(u,z,\phi)^2)_k e^{iku/\varepsilon} - (\alpha^{(0)}(u,-z,-\phi)^2)_k e^{-iku/\varepsilon}\right\} du.$$

As in [7] we introduce the function $Q(z,\phi)$ by means of the formulae:

$$Q(z,\phi) = \sum_k Q_k(\phi)e^{ikz}, \quad Q_k(\phi) = \beta_k(0,-\phi) - \beta_k(0,\phi).$$

For the term proportional to δ in Q we find the expression:

$$Q^{(1)}(z,\phi) = \oint_0^\infty \ell(u)f(u)\left\{\hat{\alpha}^{(0)}(u,z,\phi)^2 - \hat{\alpha}^{(0)}(u,-z,-\phi)^2\right\} du,$$

where we have set

$$\hat{\alpha}^{(0)}(s, z, \phi) := \alpha^{(0)}\left(s, z + \frac{s}{\varepsilon}, \phi\right).$$

The significance of the function Q lies in its close relationship to the distance function. Indeed, one checks that Eq. (2.7) can now be written in the form:

$$(4.4) \qquad\qquad d^{(2)}(t_0) = Q^{(1)}(0, t_0/\varepsilon).$$

The proof of our main theorem is based on the following:

PROPOSITION 4.1. *The term proportional to δ^2 in the splitting distance is given by*

$$d^{(2)}(t_0) = \left\{ \left(\frac{M}{\varepsilon} + \int_0^\infty \ell(t) \sin \frac{t}{\varepsilon} dt \right) \frac{\pi}{4} \operatorname{sech} \frac{\pi}{2\varepsilon} - \frac{1}{2\varepsilon} \left(\int_0^\infty \ell(t) \sin \frac{t}{\varepsilon} dt \right) \Delta A \right.$$

$$\left. - \frac{1}{4\varepsilon} \left(\oint_0^\infty \ell(s) H_+(s) e^{is/\varepsilon} ds + \oint_0^\infty \ell(s) H_-(s) e^{-is/\varepsilon} ds \right) \right\} \sin \frac{2t_0}{\varepsilon}.$$

where $M, \Delta A$ are as in (4.3), and $H_\pm(s)$ are defined as follows

$$H_\pm(s) := \oint_s^\infty h(u) e^{\pm iu/\varepsilon} du,$$

and they are analytic in the domain $\mathcal{D}$.

The proof of this proposition will be published elsewhere. Here we sketch the proof of our main theorem using this proposition.

Proof of the main theorem: First we make the following observations.

(i) $\Delta A = -\dfrac{\pi^2}{4} \tanh \dfrac{\pi}{2\varepsilon} \operatorname{sech} \dfrac{\pi}{2\varepsilon}.$

(ii) $M = \dfrac{2\varepsilon^2}{1+\varepsilon^2} \left\{ \displaystyle\int_0^\infty t \operatorname{sech}^3 t \cos \dfrac{t}{\varepsilon} dt - \dfrac{1}{\varepsilon} \int_0^\infty t \operatorname{sech} t \sin \dfrac{t}{\varepsilon} dt \right\} = -2\varepsilon^2 + O(\varepsilon^2).$

(iii) $\displaystyle\oint_0^\infty \ell(s) H_+(s) e^{is/\varepsilon} ds + \oint_0^\infty \ell(s) H_-(s) e^{-is/\varepsilon} ds = -2\Delta A \int_0^r \sec te^{-t/\varepsilon} dt$

$\qquad + O(e^{-2r/\varepsilon})$, where $r = \dfrac{\pi}{2} - \varepsilon.$

(iv) $\displaystyle\int_0^r \sec te^{-t/\varepsilon} dt - \int_0^\infty \ell(t) \sin \dfrac{t}{\varepsilon} dt = O(e^{-\pi/2\varepsilon}).$

(v) $\displaystyle\int_0^\infty \ell(t)\sin\frac{t}{\varepsilon}\,dt = \varepsilon + O(\varepsilon^3).$

The proofs of (i), (ii) and (v) are straightforward.

To prove (iii) we integrate along the boundary of the domain $\mathcal{D}$ (in the upper half plane for the first integral and in the lower half plane for the second integral) and obtain:

$$\oint_0^\infty \ell(s)H_+(s)e^{is/\varepsilon}\,ds + \oint_0^\infty \ell(s)H_-(s)e^{-is/\varepsilon}\,ds$$

$$= i\int_0^r \sec t\{H_+(it) - H_-(-it)\}e^{-t/\varepsilon}\,dt + O(e^{-2r/\varepsilon}).$$

Now we observe that:

$$H_+(it) - H_-(-it) = H_+(0) - H_-(0) = 2i\Delta A \text{ for all } t \in [0,r].$$

This proves (iii). In order to prove (iv) we observe that

$$\int_0^\infty \ell(t)\sin\frac{t}{\varepsilon}\,dt = \frac{1}{2i}\left\{\oint_0^\infty \ell(t)e^{it/\varepsilon}\,dt - \oint_0^\infty \ell(t)e^{-it/\varepsilon}\,dt\right\}.$$

Furthermore, changing the paths of integration as in (iii) we get:

$$\int_0^\infty \ell(t)\sin\frac{t}{\varepsilon}\,dt = \int_0^r \sec t e^{-t/\varepsilon}\,dt + O(e^{-\pi/2\varepsilon}).$$

Now the main theorem follows by substituting the expression (i)-(v) into the formula of the proposition. $\square$

5. Conclusion. By expanding the splitting distance between stable and unstable manifold for the periodic solution in the rapidly forced pendulum problem with respect to the amplitude $\delta = \delta_0\varepsilon^p$ of the forcing term, we have presented an expression for the next term after the Melnikov term. We have shown that for $p < -1$ the Melnikov term is dominated by the $O(\delta^2)$ term as $\varepsilon \downarrow 0$. Under the assumption that the ε-dependence of the higher order terms in δ is given by a convergent Laurent Series in ε (compare [4]), this implies that for $-2 \le p < -1$ this term no longer gives the correct approximation to the splitting distance between the two manifolds.

Following a suggestion made to us by Dr. Lochak at this conference an investigation into these matters may be simplified by considering a slightly modified problem in which the forcing term corresponds to a fast oscillating suspension of the pendulum. Indeed, this problem appears to be simpler than the original problem because the form of the periodic solution in the full problem remains the same as in the unperturbed problem. We intend to investigate this problem in a future publication.

Acknowledgements: I thank my thesis advisor Professor Martin Kummer for a ll the helpful discussions and suggestions.

REFERENCES

[1] V.G. GELFREICH, *Separatrices splitting for the rapidly forced pendulum*, preprint.

[2] P. HOLMES, J. MARSDEN AND J. SCHEURLE, *Exponentially small splittings of separatrices with applications to KAM theory and degenerate bifurcations*, Contemporary Mathematics 81 (1988), p. 213.

[3] J. SCHEURLE, *Chaos in a rapidly forced pendulum equation*, Contemporary Mathematics 97 (1989), p. 411.

[4] J. SCHEURLE, J. MARSDEN AND P. HOLMES, *Exponentially small estimates for separatrix splittings*, in Asymptotics beyond all orders, edited by H. Segur et al., Plenum Press, New York, NATO ASI Series. Series B: Physics Vol. 284 (1991), pp. 187-195.

[5] A. DELSHAMS AND T.M. SEARA, *An asymptotic expression for the splitting of separatrices of the rapidly forced pendulum*, preprint.

[6] M. KUMMER, J.A. ELLISON AND A.W. SAENZ, *Exponentially small phenomena in the rapidly forced pendulum*, in Asymptotics beyond all orders, edited by H. Segur et al., Plenum Press, New York, NATO ASI Series. Series B: Physics Vol. 284 (1991), pp. 197-211.

[7] J.A. ELLISON, M. KUMMER, AND A.W. SAENZ, *Transcendentally Small Transversality on the Rapidly Forced Pendulum*, accepted for Publication by Journal of Dynamics and Differential Equations.

[8] A paper with the same title and result but with the sharper condition $p > 3$ (instead of $p > 5$) is available in preprint form from the authors of [7].

[9] PREM KUMAR N. SWAMI, *On the term after Melnikov's in the Rapidly Forced Pendulum*, preprint.

EXISTENCE OF INVARIANT TORI FOR CERTAIN NON-SYMPLECTIC DIFFEOMORPHISMS

ZHIHONG XIA*

Abstract. In this paper, we consider certain diffeomorphisms on $I^m \times T^n$, where $I = [1,2] \subset \mathbb{R}$ and $T = S^1$. We show that under certain nondegeneracy conditions and intersection property, all of the maps sufficiently close to the integrable ones preserve a large set of m-dimensional invariant tori. Some applications of these results to forced oscillations with changing forcing frequencies are given.

1. Introduction and main results. Cheng & Sun [1] showed that for a three-dimensional volume preserving diffeomorphism sufficiently close to an integrable one, a large set of two-dimensional invariant tori is preserved, provided that a certain nondegeneracy condition is satisfied. Herman (cf. Yoccoz [8]) generalized this result to n-dimensional volume preserving maps with arbitrary n. Using a different approach, a similar result was obtained by Xia [7].

In this paper, we generalize the above results to more general diffeomorphisms on $I^m \times T^n$ that have a certain intersection property, where $I = [1,2] \subset \mathbb{R}$ and $T = S^1$. We show that, under certain nondegeneracy conditions, all the maps sufficiently close to integrable ones preserve a large set of m-dimensional invariant tori. Some applications to forced oscillations with changing forcing frequencies are also given at the end of the paper.

We consider the following map F defined on $I^m \times T^n$, where $I = [1,2] \subset \mathbb{R}$ is a closed interval and T^n is an n-dimensional torus, $T^n = S^1 \times S^1 \times \cdots \times S^1$:

$$F : (r_1, r_2, \ldots, r_m; \theta_1, \theta_2, \ldots, \theta_n) \to (r_1', r_2', \ldots, r_m'; \theta_1', \theta_2', \ldots, \theta_n')$$

with

$$
\begin{aligned}
r_i' &= r_i + \epsilon f_i(r_1, r_2, \ldots, r_m; \theta_1, \theta_2, \ldots, \theta_n) \\
\theta_j' &= \theta_j + g_j(r_1, r_2, \ldots, r_m) + \epsilon h_j(r_1, r_2, \ldots, r_m; \theta_1, \theta_2, \ldots, \theta_n)
\end{aligned}
$$

for $i = 1, 2, \ldots, m$ and $j = 1, 2, \ldots, n$, where $g_1, g_2, \ldots, g_n$, $h_1, h_2, \ldots, h_n$ and $f_1, f_2, \ldots, f_m$ are real analytic functions in all of their variables and ϵ is a small perturbation parameter. Assume $m \leq n$. We also make the following three assumptions:

(a) Intersection Property: for any homotopically nontrivial n-torus, T^n, its intersection with its image $F(T^n)$ is nonempty. i.e., $F(T^n) \cap T^n \neq \emptyset$;

* School of Mathematics, Georgia Institute of Technology, Atlanta, GA 30332. Research supported in part by the National Science Foundation.

(b) Nondegeneracy Condition: The $m \times n$ matrix

$$\left(\frac{\partial g_i(r_1, r_2, \ldots, r_m)}{\partial r_j} \right)$$

has the maximum rank m.

(c) Twist Condition: There exists an integer l such that the following matrix $A(\vec{r})$ has the maximum rank n, for all $\vec{r} \in I^m$, where $\vec{r} = (r_1, r_2, \ldots, r_m)$.

$$A(\vec{r}) = \begin{pmatrix} \frac{dg_1(\vec{r})}{d\vec{r}} & \frac{dg_2(\vec{r})}{d\vec{r}} & \cdots & \frac{dg_n(\vec{r})}{d\vec{r}} \\ \frac{d^2 g_1(\vec{r})}{d\vec{r}^2} & \frac{d^2 g_2(\vec{r})}{d\vec{r}^2} & \cdots & \frac{d^2 g_n(\vec{r})}{d\vec{r}^2} \\ \vdots & \vdots & \ddots & \vdots \\ \frac{d^l g_1(\vec{r})}{d\vec{r}^l} & \frac{d^l g_2(\vec{r})}{d\vec{r}^l} & \cdots & \frac{d^l g_n(\vec{r})}{d\vec{r}^l} \end{pmatrix}$$

where $d^j g_i(\vec{r})/d\vec{r}^j$ is a column vector of all mixed partial derivatives of g_i of order j.

With the nondegeneracy condition, we may assume, without loss of generality, that

$$g_i(\vec{r}) = r_i, \quad \text{for } i = 1, 2, \ldots, m,$$

and we may assume that all the functions involved are real analytic for all $r_i \in [1 - \delta, 1 + \delta]$ for some $\delta > 0$, $i = 1, 2, \ldots, m$. For simplicity, we also use the following vector notation:

$$\begin{aligned} \vec{\theta} &= (\theta_1, \theta_2, \ldots, \theta_n) \\ \vec{f}(\vec{r}, \vec{\theta}) &= (f_1(\vec{r}, \vec{\theta}), f_2(\vec{r}, \vec{\theta}), \ldots, f_m(\vec{r}, \vec{\theta})) \\ \vec{g}(\vec{r}) &= (g_1(\vec{r}), g_2(\vec{r}), \ldots, g_n(\vec{r})) \\ \vec{h}(\vec{r}, \vec{\theta}) &= (h_1(\vec{r}, \vec{\theta}), h_2(\vec{r}, \vec{\theta}), \ldots, h_n(\vec{r}, \vec{\theta})) \end{aligned}$$

Now, we can state our main theorem.

THEOREM 1.1. *There exists a positive number ϵ_0 such that if $0 < \epsilon \le \epsilon_0$, the map F admits a family of invariant tori of the form:*

$$\begin{aligned} r_i &= v_i(\xi_1, \xi_2, \ldots, \xi_n, \omega_1, \omega_2, \ldots, \omega_m), \\ \theta_j &= \xi_1 + u_j(\xi_1, \xi_2, \ldots, \xi_n, \omega_1, \omega_2, \ldots, \omega_m) \end{aligned}$$

for $i = 1, 2, \ldots, m$ and $j = 1, 2, \ldots, n$, with $u_0, u_1, \ldots, u_n$ real analytic functions of period 2π. Moreover, the map F induced on the torus is given by:

$$(1.1) \quad \begin{aligned} \xi_1' &= \xi_i + \omega_1, \\ \xi_2' &= \xi_2 + \omega_2 \\ &\cdots\cdots \\ \xi_m' &= \xi_m + \omega_m \\ \xi_{m+1}' &= \xi_{m+1} + g_{m+1}(\omega_1, \ldots, \omega_m) + g_{m+1}^*(\omega_1, \ldots, \omega_m, \epsilon). \\ &\cdots\cdots \\ \xi_n' &= \xi_n + g_n(\omega_1, \ldots, \omega_m) + g_n^*(\omega_1, \ldots, \omega_m, \epsilon). \end{aligned}$$

where $g_i^*(\omega_1, \ldots, \omega_m, \epsilon)$, $i = m+1, m+2, \ldots, n$ are functions depending on the perturbations $\vec{h}(\vec{r}, \vec{\theta})$ and $g_i^*(\omega_1, \omega_2, \ldots, \omega_m, 0) = 0$, for $i = m + 1, m + 2, \ldots, n$.

In fact, there exists a Cantor set $S(\epsilon) \subset I^m$ depending on the function $\vec{h}(\vec{r}, \vec{\theta})$ such that for each $(\omega_1, \omega_2, \ldots, \omega_m) \in S(\epsilon)$, there is a corresponding invariant torus of the form (1.1). Furthermore, the measure of the set $S(\epsilon)$ tends to 1 as $\epsilon \to 0$.

Before proving the theorem, we make the following remarks:

Remarks 1.1.

(1) When $m = n$ and F is exact symplectic, the theorem is the standard KAM theorem. On the other hand, if $m = 1$ and $n > 1$ and F is volume preserving, the theorem is the same as the one in Xia [7], except that the twist conditions here are a little weaker.

(2) The functions $g_i^*(\omega_1, \omega_2, \ldots, \omega_m, \epsilon)$, $i = m + 1$, $\ldots, n$ in the theorem are due to the frequency drift effect of the perturbation. For the symplectic maps in KAM theory, one has the same number of fast and slow variables (action-angle variables), hence each frequency component can be controlled by varying the corresponding slow variable. In this case, when $m < n$, we have less slow variables than angular variables. It is easy to see that any nonresonant (even with diophantine conditions) torus in the unperturbed system can be perturbed away. We emphasize that the theorem states only that the majority of the orbits lie on invariant tori and that these invariant tori need not come from original unperturbed invariant tori. Generally, any invariant torus of the map disintegrates as the perturbation changes (no matter how small the change is). However, the theorem states that, in the neighborhood of the disintegrated invariant torus, new invariant tori with new frequencies appear. In fact, given any number $(\omega_1, \ldots, \omega_m) \in I^m$, it cannot be predicted with the current method whether $(\omega_1, \ldots, \omega_m) \in S(\epsilon)$, i.e., whether there is an invariant torus of the map F corresponding to this number, even if the following diophantine condition is satisfied:

$$|k_0 + k_1 g_1(\omega_1, \ldots, \omega_m) + \ldots + k_n g_n(\omega_1, \ldots, \omega_m)| \geq c(|k_0| + \ldots + |k_n|)^{-\mu},$$

for all $(k_0, k_1, \ldots, k_n) \in \mathbb{Z}^{n+1} \backslash \mathbf{0}$ and for some $c > 0$ and $\mu > 0$. In fact, this presents the main difficulty of the proof.

(3) Because the proof of the Theorem is an application of KAM theory, smoothing techniques can be used [2]. Therefore, for the Theorem to be true, the perturbation need not to be analytic. It suffices to require that the functions $f_1, f_2, \cdots, f_m$ and $h_1, h_2, \cdots, h_n$ are C^r for sufficiently large r. In fact, $r > 2n + 1$ will suffice.

2. Proof of the main theorem. The proof of the theorem is similar to that of Xia [7] for volume preserving maps. The main idea is to suspend the system into a high-dimensional system so that KAM theory can be used and the drift effect of the perturbation can be controlled.

Let $\delta > 0$ be a positive real number and let $J \subset \mathbb{R}^n$ be the following set:

$$J = \{g_j(\vec{r}) - \delta \leq r_j \leq g_j(\vec{r}) + \delta, \ \text{for} \ j = m+1, m+2, \ldots, n, \vec{r} \in I^m\} \subset \mathbb{R}^n$$

We add $n - m$ "action" variables $r_{m+1}, \ldots, r_n$ to the system and consider the following augmented system $\tilde{F}: J \times T^n \to J \times T^n$:

$$
\begin{aligned}
r_i' &= r_i + \epsilon f_i(\vec{r}, \vec{\theta}), \ \text{for} \ i = 1, 2, \ldots, m \\
r_j' &= r_j + g_j(\vec{r}') - g_j(\vec{r}), \ \text{for} \ j = m+1, m+2 \ldots, n \\
\theta_i' &= \theta_i + r_i + \epsilon h_i(\vec{r}, \vec{\theta}) \ \text{for} \ i = 1, 2, \ldots, n
\end{aligned}
$$

For any ϵ, the diffeomorphism $\tilde{F}$ has the set $J_0 \times T^n$ as an invariant subset, where

$$J_0 = \{(r_1, r_2, \ldots, r_n) \in J \mid r_{m+1} = g_{m+1}(\vec{r}), \ldots, r_n = g_n(\vec{r})\}$$

and $\tilde{F}$ restricted to the set $J_0 \times T^n$ is exactly the map F. The map $\tilde{F}$ defined above also preserves the following quantities: $r_i - g_i(\vec{r})$, for $i = m+1, m+2, \ldots, n$. We have the following theorem:

THEOREM 2.1. *There exists a positive number ϵ_0 such that if $0 < \epsilon \leq \epsilon_0$, the map $\tilde{F}$ admits a family of invariant tori of the form:*

$$
\begin{aligned}
r_i &= v_i(\xi_1, \xi_2, \ldots, \xi_n, \omega), \\
\theta_i &= \xi_i + u_i(\xi_1, \xi_2, \ldots, \xi_n, \omega),
\end{aligned}
$$

for all $i = 1, 2, \ldots, n$ and where u_i, v_i for $i = 1, 2, \ldots, n$ are real analytic functions of period 2π in $\xi_i, i = 1, \ldots, n$ and $\omega = (\omega_1, \omega_2, \ldots, \omega_n)$ is a rotation vector. Moreover, the map $\tilde{F}$ induced on the torus is given by:

$$\xi_i' = \xi_i + \omega_i, \ \text{for} \ i = 1, 2, \ldots, n$$

Furthermore, there is an invariant torus of the above form for any $\omega = (\omega_1, \omega_2, \ldots, \omega_n)$ satisfying the diophantine condition:

$$(2.1) \qquad |k_0 + \mathbf{k} \cdot \omega| \geq c(|k_0| + |\mathbf{k}|)^{-\mu} \ \text{for some} \ c > 0, \mu > n + 1$$

where $\mathbf{k} = (k_1, k_2, \ldots, k_n)$ and $|\mathbf{k}| = |k_1| + |k_2| + \ldots + |k_n|$.

Outline of the Proof: The proof of this theorem follows precisely the method of KAM (Kolmogorov-Arnold-Moser) theory. In KAM theory, the map considered there either is exact symplectic, or it has some integral invariant [2,6,5]. In fact, the only place where the exact symplecticity is used in the proof is in the estimation of the averaged perturbation in the direction of the action variables. For the exact symplectic map, any torus close to an unperturbed invariant torus has to intersect the image of itself under the map. This and the smoothness of the map that the averaged perturbation in the direction perpendicular to the torus to be small. Our map $\tilde{F}$ does not have exactly the same intersection property as the exact symplectic map. However, if one follows the rapid convergence scheme used in KAM theory, in each coordinate transformation for the map $\tilde{F}$, the higher order terms of the transformation do not depend on the variables $r_{m+1}, r_{m+2}, \ldots, r_n$. Therefore, the resulting map after each coordinate transformation takes the same form as $\tilde{F}$, i.e., the perturbation terms of the map depend only on the variables $r_1, r_2, \ldots, r_m, \theta_1, \ldots, \theta_n$.

Therefore, at each step of iteration of the transformation, the intersection property is preserved. i.e., at each iteration, there exists $(\theta_1, \theta_2, \ldots, \theta_n)$ such that $r_i'(\vec{r}, \vec{\theta}) = r_i$ for $i = 1, 2, \ldots, m$. This is exactly the property needed to obtain the proper estimate and to carry on the iteration procedure. We do not give a complete proof here. For details, see Moser [2], Siegel & Moser [5], and Svanidze [6].

Without confusion, we now identify the set J with the set

$$\{(r_1, \ldots, r_n, \theta_1, \ldots, \theta_n) \in J \times T^n \mid (r_1, \ldots, r_n) \in J; \vec{\theta} = (0, 0, \ldots, 0)\}$$

and similarly identify the subset J_0 of J. The invariant tori of Theorem 2.1 are graphs over T^n and therefore each torus intersects the set J once and only once. The intersection of J with the set of all invariant tori with rotation vector satisfying diophantine condition (2.1) with diophantine constant (c, μ) forms a Cantor set. Now fix the constants $c > 0$ and $\mu > n + 1$, and let $\Omega(c, \mu)$ denote this set. For each point $p \in \Omega(c, \mu)$, we let $\omega(p) = (\omega_1, \omega_2, \ldots, \omega_n)$ be the rotation vector of the invariant torus. To prove our main theorem, we need the following important lemma:

LEMMA 2.2. *Let* $\omega : \Omega(c, \mu) \to \mathbb{R}^n$ *be the function defined above. then for any* $r \geq 0$, ω *is an injective differentiable* C^r-*map at every point of* $\Omega(c, \mu)$. *Moreover, the above map can be smoothly extended to* J *and* $\|\omega - Id\|_{C^r} \to 0$ *as* $\epsilon \to 0$, *where* Id *is the identity map.*

The proof of this lemma follows from the details of the proof of Theorem 2.1 with some careful estimates. In fact, it can be shown that the map ω defined above is C^∞. For details, see Pöschel [4] for the proof in the case of Hamiltonian systems and Svanidze [6] in the case of maps.

Now we come to the proof of our main theorem. From the definition of the map $\tilde{F}$, the restriction of $\tilde{F}$ to the invariant subset $J_0 \times T^n$ is exactly the original map F. Also, it follows from Theorem 2.1 that any invariant torus with rotation number ω satisfying diophantine condition (2.1) persists under small perturbations. However, because of the frequency drift, the invariant torus originally on the subset J_0 may drift away and the perturbed invariant torus with the same rotation number need not stay on this subset. In fact, it may happen that every invariant torus of the unperturbed system on the subset J_0 moves out of the set J_0 under certain small perturbations. However, we show that the set J_0 is still mostly occupied by invariant tori, although they may not be the original ones; i.e., some invariant tori originally not on the subset J_0 may enter the subset under certain small perturbations. This is because the smoothness of the map ω and the fact that the majority of the invariant tori of $\tilde{F}$ persist under small perturbation.

Given a rotation vector $\omega = (\omega_1, \ldots, \omega_n)$ satisfying the diophantine condition (2.1), to show that we have an invariant torus for the map F with rotation number ω, we need to show that the corresponding invariant torus lands in the subset $J_0 \times T^n$. Let $D(c, \mu) \subset \mathbb{R}^n$ be the set of rotation vectors satisfying the diophantine condition (2.1). To prove Theorem 1.1, we need to show that

$$D(c, \mu) \cap \omega(J_0) \neq \varnothing$$

and that the measure of the set $D(c, \mu) \cap \omega(J_0)$ approaches the full measure, as $\epsilon \to 0$.

Recall that J_0 is defined by

$$J_0 = \{(r_1, r_2, \ldots, r_n) \in J \mid r_{m+1} = g_{m+1}(\vec{r}), \ldots, r_n = g_n(\vec{r})\}$$

By lemma 2.2, $\|\omega - Id\|_{C^r} \to 0$ as $\epsilon \to 0$, therefore, $\omega(J_0)$ is given by

$$
\begin{aligned}
\omega_{m+1} &= g_{m+1}(\omega_1, \omega_2, \ldots, \omega_m) + g_{m+1}^*(\omega_1, \omega_2, \ldots, \omega_n, \epsilon) \\
\omega_{m+2} &= g_{m+2}(\omega_1, \omega_2, \ldots, \omega_m) + g_{m+2}^*(\omega_1, \omega_2, \ldots, \omega_n, \epsilon) \\
&\qquad \cdots \cdots \\
\omega_n &= g_n(\omega_1, \omega_2, \ldots, \omega_m) + g_n^*(\omega_1, \omega_2, \ldots, \omega_n, \epsilon)
\end{aligned}
$$
(2.2)

where the functions $g_i^*(\omega_1, \omega_2, \ldots, \omega_n, \epsilon)$, $i = m+1, m+2, \ldots, n$ are C^r-differentiable functions and $\|g_i^*(\omega_1, \omega_2, \ldots, \omega_n, \epsilon)\|_{C^r} \to 0$ as $\epsilon \to 0$.

The problem now becomes whether the set of equations (2.2) has any solutions at all on $D(c, \mu)$, for given $(\omega_1, \omega_2, \ldots, \omega_m) \in \mathbb{R}^m$. We need the following lemma, whose proof is postponed to the next section:

LEMMA 2.3. *Let* $S(c, \mu)$ *be the solutions of the following set of equations and inequalities:*

$$
\begin{aligned}
&\omega_i = g_i(\omega_1, \omega_2, \ldots, \omega_m) \quad \text{for } i = m+1, m+2, \ldots, n \\
&|k_0 + \mathbf{k} \cdot (\omega_1, \omega_2, \ldots, \omega_n)| \geq c|\mathbf{k}|^{-\mu} \quad \text{for all } (k_0, k_1, \ldots, k_n) \\
&\in \mathbb{Z}^{n+1} \backslash 0
\end{aligned}
$$
(2.3)

where $\mathbf{k} = (k_1, k_2, \ldots, k_n)$, $|\mathbf{k}| = |k_1| + |k_2| + \ldots + |k_n|$. *Assume that the functions* $g_1, g_2, \ldots, g_n$ *are* C^{l+1}, *they satisfy the twist condition, and that* $\mu > l^2 + l - 1$. *There exists a constant* $c_0 > 0$ *such that for all* $c \leq c_0$, $S(c, \mu)$ *is a nonempty Cantor set where the measure of* $S(c, \mu)$ *approaches to 1 as* $c \to 0$.

Now consider the equations (2.2). By the implicit function theorem, we may assume that g_i^*, $i = m+1, m+2, \ldots, n$ is independent of $\omega_{m+1}, \omega_{m+2}, \ldots, \omega_n$. The functions $g_i + g_i^*$, $i = m+1, m+2, \ldots, n$ are smooth and for $\epsilon > 0$ small, they satisfy the twist condition. Now lemma 2.3 applies to our function $g_i + g_i^*$ and we obtain a set $S(c, \mu, \epsilon) \in I^m$ such that for any

$(\omega_1, \omega_2, \ldots, \omega_m) \in I^m$, there exists a rotation vector $\omega = (\omega_1, \omega_2, \ldots, \omega_n)$ satisfying both the diophantine condition with diophantine constants (c, μ) and equations (2.2). Therefore, the invariant torus of $\tilde{F}$ with rotation vector $(\omega_1, \omega_2, \ldots, \omega_n)$ is in the subset $J_0 \times T^n$ and hence it is an invariant torus of F, our original map.

This completes the proof of the theorem.   □

3. Proof of lemma 2. In this section, we give the proof of lemma 2 of the last section. We prove it by induction on m. As a first step, we assume $m = 1$.

First, fix c and μ, for $\{k_1, \ldots, k_n\} \neq 0$ and $k_0 \in \mathbb{Z}$, let $Z(k_0, \mathbf{k}) \subset [1, 2]$ be the following set:

$$Z(k_0, \mathbf{k}) = \{\omega_1 \in [1, 2] \mid |k_0 + k_1\omega_1 + k_2 g_2(\omega_1) + \ldots + k_n g_n(\omega_1)| < c|\mathbf{k}|^{-\mu}\}$$

Let $\omega_1 = \omega^* \in [1, 2]$ be a solution of the following equations:

(3.1) $k_0 + k_1\omega_1 + k_2 g_2(\omega_1) + \ldots + k_n g_n(\omega_1) + c^* = 0$

for some $c^* \in \mathbb{R}$. Then nearby ω^*, we have the following:

$$\begin{aligned}
&k_0 + k_1\omega_1 + k_2 g_2(\omega_1) + \ldots + k_n g_n(\omega_1) + c^* \\
&= k_1(\omega_1 - \omega^*) + k_2 g_2'(\omega^*)(\omega_1 - \omega^*) + \ldots + k_n g_n'(\omega^*)(\omega_1 - \omega^*) \\
&\quad + k_2 g_2''(\omega^*)\frac{(\omega_1 - \omega^*)^2}{2} + \ldots + k_n g_n''(\omega^*)\frac{(\omega_1 - \omega^*)^2}{2} \\
&\quad + \cdots \cdots \\
&\quad + k_2 g_2^{(l)}(\omega^*)\frac{(\omega_1 - \omega^*)^l}{l!} + \ldots + k_n g_n^{(l)}(\omega^*)\frac{(\omega_1 - \omega^*)^l}{l!} \\
&\quad + k_2 g_2^{(l+1)}(\tilde{\omega})\frac{(\omega_1 - \omega^*)^{l+1}}{(l+1)!} + \ldots + k_n g_n^{(l+1)}(\tilde{\omega})\frac{(\omega_1 - \omega^*)^{l+1}}{(l+1)!}
\end{aligned}$$

where by Taylor's theorem, $\tilde{\omega}$ is a real number in (ω^*, ω_1). Let $A(\omega^*)$ be the following matrix:

$$(3.2) \quad A(\omega^*) = \begin{pmatrix} g_1'(\omega^*) & g_2'(\omega^*) & \cdots & g_n'(\omega^*) \\ g_1''(\omega^*)/2 & g_2''(\omega^*)/2 & \cdots & g_n''(\omega^*)/2 \\ \vdots & \vdots & \ddots & \vdots \\ g_1^{(l)}(\omega^*)/l! & g_2^{(l)}(\omega^*))/l! & \cdots & g_n^{(l)}(\omega^*)/l! \end{pmatrix}$$

and let $\mathbf{x}^t = ((\omega_1 - \omega^*), (\omega_1 - \omega^*)^2, \ldots, (\omega_1 - \omega^*)^l)$, then

$$k_0 + k_1\omega_1 + k_2 g_2(\omega_1) + \ldots + k_n g_n(\omega_1) = \mathbf{x}^t A\mathbf{k} + (\omega_1 - \omega^*)^{l+1}\mathbf{h}(\tilde{\omega}) \cdot \mathbf{k}$$

Where $\mathbf{h}(\tilde{\omega})$ is a uniformly bounded function in $\tilde{\omega} \in [1, 2]$,

$$\mathbf{h}(\tilde{\omega}) = \left(\frac{g_1^{(l+1)}(\tilde{\omega})}{(l+1)!}, \frac{g_2^{(l+1)}(\tilde{\omega})}{(l+1)!}, \ldots, \frac{g_n^{(l+1)}(\tilde{\omega})}{(l+1)!} \right)$$

By the twist condition, A has the maximum rank n, for all $\omega^* \in [1,2]$. Thus, $|A\mathbf{k}| \geq c_1 |\lambda| |\mathbf{k}|$ for some $c_1 > 0$. Hence, we have the following lemma:

LEMMA 3.1. *Let* $u(\omega_1, k_0, \mathbf{k}, c^*) = k_0 + k_1 \omega_1 + k_2 g_2(\omega_1) + \ldots, k_n g_n(\omega_1) + c^*$ *with* $\mathbf{k} \neq 0$ *and* $\omega_1 \in [1,2]$, *then, the order of all the zeros and critical points of* $u(\omega_1, k_0, \mathbf{k}, c^*)$ *in* $[1,2]$ *is at most* l, *for any* $c^* \in \mathbb{R}$, $k_0 \in \mathbb{Z}$ *and* $\mathbf{k} \in \mathbb{Z}^n \backslash 0$.

From this, we can prove the following lemma:

LEMMA 3.2. *Let* $N(Z(k_0, \mathbf{k}))$ *be the number of connected components of the open set* $Z(k_0, \mathbf{k})$ *with* $\mathbf{k} \neq 0$. *Then there exist positive integers* N_1 *and* N_2, *depending only on* $g_1, g_2, \ldots, g_n$ *such that*

  1. $N(Z(k_0, \mathbf{k})) = 0$, *if* $|k_0| > N_1 |\mathbf{k}|$;

  2. $N(Z(k_0, \mathbf{k})) \leq N_2$, *if* $|k_0| \leq N_1 |\mathbf{k}|$.

Proof. For $\mathbf{k} \neq 0$, consider the following equivalent definition of $Z(k_0, \mathbf{k})$:

$$(3.3)\quad Z(k_0, \mathbf{k}) = \left\{ \begin{array}{c} \omega_1 \in [1,2] \, | \, |\frac{k_0}{|\mathbf{k}|} + \frac{k_1}{|\mathbf{k}|}\omega_1 + \frac{k_2}{|\mathbf{k}|}g_2(\omega_1) \\ + \ldots, \frac{k_n}{|\mathbf{k}|}g_n(\omega_1)| < c|\mathbf{k}|^{-(\mu+1)} \end{array} \right\}$$

Since $\frac{k_1}{|\mathbf{k}|}\omega_1 + \frac{k_2}{|\mathbf{k}|}g_2(\omega_1) + \ldots, \frac{k_n}{|\mathbf{k}|}g_n(\omega_1)$ and $c|\mathbf{k}|^{-(\mu+1)}$ are bounded, it is obvious that there exists a positive integer N_1 such that if $|k_0| > N_1 |\mathbf{k}|$, the above equation has no solution and therefore, $N(Z(k_0, \mathbf{k})) = 0$. Now, let us restrict ourselves to the case $|k_0| \leq N_1 |\mathbf{k}|$. Let $Z_1(k_0, \mathbf{k}) \subset Z$ be a connected component of Z, it is easy to see that if the closure of Z_1 does not contain both boundaries of $[1,2]$, there must be a point ω^* in the closure of Z_1 such that ω^* is a zero of the following equation:

$$\frac{k_0}{|\mathbf{k}|} + \frac{k_1}{|\mathbf{k}|}\omega_1 + \frac{k_2}{|\mathbf{k}|}g_2(\omega_1) + \ldots, \frac{k_n}{|\mathbf{k}|}g_n(\omega_1) = \pm c|\mathbf{k}|^{-(\mu+1)}$$

By Lemma 3.1, the order of each zero of the above equation is at most n and thus it has only finitely many zeros. Moreover, in a small neighborhood of the parameters k_0, $\mathbf{k}/|\mathbf{k}|$ and $c|\mathbf{k}|^{-(\mu+1)}$ in the parameter space, the above equations have only finitely many zeros, again by Lemma 3.1. With our parameters restricted to a compact set, we see that there is a uniform bound on the number of roots of the above equation for any $k_0/|\mathbf{k}|, \mathbf{k}/|\mathbf{k}|$ in the domain $[1,2]$. Denoting this bound by $N_2 - 2$, the lemma follows.

Let ω^* be a point in $Z(k_0, \mathbf{k})$, and let $Z_1(k_0, \mathbf{k}, \omega^*) \subset Z$ be a connected component of Z containing ω^* in $[1,2]$. $Z_1(k_o, \mathbf{k}, \omega^*)$ is an open subset of $[1, 2]$. We would like to estimate the size of Z_1. Suppose $A(\omega^*)\mathbf{k}/|\mathbf{k}| = (a_1, a_2, \cdots, a_l)$, where $A(\omega^*)$ is the matrix defined in equation (3.2) and let $j, 1 \leq j \leq l$ be an integer such that $a_j \geq c_1/l$. Such j always exists, by the twist condition. We write

$$a_1(\omega_1 - \omega^*) + a_2(\omega_1 - \omega^*)^2 + \cdots + a_j(\omega_1 - \omega^*)^j$$
$$= a_j(\omega_1 - \omega^*)(\omega_1 - b_2) \cdots (\omega_1 - b_j)$$

where $b_2, \cdots, b_j$ are complex numbers. Now, $Z(k_0, \mathbf{k})$ can be rewritten as the following set:

$$
\begin{aligned}
\{\omega_1 \in [1,2] \quad | \quad &|(k_0 + k_1\omega^* + \ldots + k_n g_n(\omega^*))/|\mathbf{k}| + \\
&a_j(\omega_1 - \omega^*)(\omega_1 - b_2) \cdots (\omega_1 - b_j) \\
&+a_{j+1}(\omega_1 - \omega^*)^{j+1} + \ldots + \\
&+a_n(\omega_1 - \omega^*)^n + (\omega_1 - \omega^*)^{n+1}\mathbf{h}(\tilde{\omega}) \cdot \mathbf{k}/|\mathbf{k}| \,| \\
&< c|\mathbf{k}|^{-(\mu+1)}\}
\end{aligned}
$$

Let c_2 be a positive number such that, for all $\omega_1 \in [1,2]$,

$$
\begin{aligned}
|a_{j+1}(\omega_1 - \omega^*)^{j+1} + \ldots + a_n(\omega_1 - \omega^*)^n &+ (\omega_1 - \omega^*)^{n+1}\mathbf{h}(\tilde{\omega}) \cdot \tfrac{\mathbf{k}}{|\mathbf{k}|}| \\
&\leq c_2|\omega_1 - \omega^*|^{j+1}
\end{aligned}
$$

Since $\omega^* \in Z(k_0, \mathbf{k})$, we have

$$
\frac{|(k_0 + k_1\omega^* + \ldots + k_n g_n(\omega^*))|}{|\mathbf{k}|} < \frac{c}{|\mathbf{k}|^{\mu+1}}
$$

Therefore, $Z(k_0, \mathbf{k})$ is a subset of the following set:

$$
\left\{\omega_1 \in [1,2] \,|\, |(\omega_1 - \omega^*)(\omega_1 - b_2) \cdots (\omega_1 - b_j)| < \frac{c_3 c}{|\mathbf{k}|^{\mu+1}} + c_2|\omega_1 - \omega^*|^{j+1}\right\}
$$

Where $c_3 = 1 + \frac{1}{c_1 l|\lambda|}$.

Let Z_0 be the set of the points in $[1, 2]$ such that one of the following inequalities is satisfied:

$$
|\omega_1 - \omega^*| < \left(\frac{c_3 c}{|\mathbf{k}|^{\mu+1}} + c_2|\omega_1 - \omega^*|^{j+1}\right)^{1/j}
$$

$$
|\omega_1 - b_2| < \left(\frac{c_3 c}{|\mathbf{k}|^{\mu+1}} + c_2|\omega_1 - \omega^*|^{j+1}\right)^{1/j}
$$

$$
\cdots\cdots
$$

$$
|\omega_1 - b_j| < \left(\frac{c_3 c}{|\mathbf{k}|^{\mu+1}} + c_2|\omega_1 - \omega^*|^{j+1}\right)^{1/j}
$$

It is obvious that $Z_1 \subset Z \subset Z_0$. Now considering the intersection of the set Z_0 with the following interval I' in $[1, 2]$:

$$
|\omega_1 - \omega^*|^{j+1} < \frac{c_3 c}{|\mathbf{k}|^{\mu+1}}
$$

choose c small enough such that

$$
\left(\frac{c_3 c(1 + c_2)}{|\mathbf{k}|^{\mu+1}}\right)^{1/j} \leq \frac{1}{4l}\left(\frac{c_3 c}{|\mathbf{k}|^{\mu+1}}\right)^{1/(j+1)}
$$

for all $\mathbf{k} \neq 0$. We see that if $b_i \notin I'$ for some $2 \leq i \leq j$, then the following set

$$|\omega_1 - b_i| < \left(\frac{c_3 c}{|\mathbf{k}|^{\mu+1}} + c_2 |\omega_1 - \omega^*|^{j+1} \right)^{1/j}$$

does not intersect the interval defined by:

$$|\omega_1 - \omega^*| < 3l \left(\frac{c_3 c(1 + c_2)}{|\mathbf{k}|^{\mu+1}} \right)^{1/j}$$

Since there are at most $j \leq l$ b_i's in I', this implies that Z_1 is a subset of the following set:

$$|\omega_1 - \omega^*| < 2l \left(\frac{c_3 c(1 + c_2)}{|\mathbf{k}|^{\mu+1}} \right)^{1/j}$$

It follows from above that

$$mes(Z_1(k_0, \mathbf{k}, \omega^*)) \leq c_4 c^{1/j} |\mathbf{k}|^{-(\mu+1)/j} \leq c_5 c^{1/l} |\mathbf{k}|^{-(\mu+1)/l}$$

for some $c_4 > 0$ and $c_5 > 0$, where $mes(S)$ is the Lebegue measure of the set S in the real line. It follows from Lemma 3.2 that

$$\sum_{k_0 \in \mathbb{Z}} mes(Z(k_0, \mathbf{k})) \leq N_2(1 + 2N_1) c_5 c^{1/l} |\mathbf{k}|^{1-(\mu+1)/l}$$

Now, the proof of lemma 2.3 follows easily. By definition,

$$mes(S(c, \mu)) \geq 1 - \sum_{|\mathbf{k}|=1}^{\infty} \sum_{k_0=-\infty}^{\infty} mes(Z(k_0, \mathbf{k}))$$
$$\geq 1 - \sum_{|\mathbf{k}|=1}^{\infty} N_2(1 + 2N_1) c_5 c^{1/l} |\mathbf{k}|^{1-(\mu+1)/l}$$

the above series converges when $1-(\mu+1)/l < -l$ therefore for $\mu > l^2+l-1$, there exists a positive number c_6 such that $mes(S(c, \mu)) \geq 1 - c_6 c^{1/l}$. This proves lemma 2.3 for the case $m = 1$.

Next, we proceed with $m = 2$, i.e., we assume that there are two independent frequencies ω_1 and ω_2. Fix $\omega_2 \in [1, 2]$ and follow the above proof with $m = 1$. Now, $A(\omega^*)$ is dependent of ω_2 and therefore the coefficients $(a_1, a_2, \ldots, a_l)$ are functions of ω_2. Define $Z_1(k_0, \mathbf{k}, \omega^*)$ and $Z(k_0, \mathbf{k})$ the same way as above; similar argument shows that, for any $j > 0$,

$$mes(Z_1(k_0, \mathbf{k}, \omega^*)) \leq c_4 c^{1/j} |\mathbf{k}|^{-(\mu+1)/j} a_j^{-1/j}$$

for some $c_4 > 0$.

Now, we need to estimate $a_j(\omega_2)$ for all $\omega_2 \in [1, 2]$. We can expand $a_j(\omega_2)$ in a Taylor series at any point $\omega_2 \in [1, 2]$. Let $a_{j,i}$ be the ith coefficient of the expansion. By the twist condition, there exists a constant

$c_8 > 0$, such that for any point $(\omega^*, \omega_2) \in I^2$, there exists positive integers j and j' with $j + j' \leq l$ such that $a_{j,j'} > c_8$. By compactness, we may assume that j and j' are fixed for all $(\omega^*, \omega_2) \in I^2$. Applying the same technique as above, we obtain the following estimate:

$$mes(\{\omega_2 \in [1,2] \mid |a_j(\omega_2)| \leq \epsilon'\}) \leq c_9 \epsilon'^{1/j'}$$

for any $\epsilon' > 0$ and for some $c_9 > 0$. The integer j' here is from the twist condition. Therefore, we have

$$mes(\{\omega_2 \in [1,2] \mid |a_j(\omega_2)| \leq (c|\mathbf{k}|^{-(\mu+1)})^{j'/(j+j')}\}) \leq c_9(c|\mathbf{k}|^{-(\mu+1)})^{1/(j+j')}$$

We divide the set of $\omega_2 \in [1,2]$ into two sets. Let

$$S_1 = \{\omega_2 \in [1,2] \mid |a_j(\omega_2)| > (c|\mathbf{k}|^{-(\mu+1)})^{j'/(j+j')}\}.$$

For any $\omega_2 \in S_1$, we have

$$mes(Z_1(k_0, \mathbf{k}, \omega^*)) \leq c_4(c|\mathbf{k}|^{-(\mu+1)})^{1/(j+j')}$$

Also, it is easy to see that for $\omega_2 \in S_1$, Lemma 4 is still true; therefore, we have:

$$\sum_{k_0 \in \mathbb{Z}} mes(Z(k_0, \mathbf{k})) \leq c_{10}|\mathbf{k}|(c|\mathbf{k}|^{-(\mu+1)})^{1/(j+j')}$$

for some c_{10} and for all $\omega_2 \in S_1$. Let $S_2 = [1,2]\backslash S_1$. Then for any $\omega_2 \in S_2$, $mes(Z(k_0, \mathbf{k})) \leq 1$ and $mes(S_2) \leq c_9(c|\mathbf{k}|^{-(\mu+1)})^{1/(j+j')}$. Using an argument similar to that of Lemma 4, we see that

$$\sum_{k_0 \in \mathbb{Z}} mes(S_2) \leq c_{11}|\mathbf{k}|(c|\mathbf{k}|^{-(\mu+1)})^{1/(j+j')}$$

for some $c_{11} > 0$. Let $MES(S)$ be the Lebegue measure of a two-dimensional set $S \in \mathbb{R}^2$, then

$$MES(I^2 \backslash S(c, \mu)) \leq \sum_{|\mathbf{k}|=1}^{\infty} \left(\frac{c_{11}|\mathbf{k}|(c|\mathbf{k}|^{-(\mu+1)})^{1/(j+j')} \cdot 1 + 1 \cdot c_{10}|\mathbf{k}|}{(c|\mathbf{k}|^{-(\mu+1)})^{1/(j+j')}} \right).$$

The above series converges for $1 - (\mu+1)/l < l$; therefore, for $\mu > l^2 + l - 1$, we obtain the desired estimate: $MES(S(c, \mu)) \geq 1 - c_{12}c^{1/l}$ for some $c_{12} > 0$.

The case with $m \geq 3$ can be proved similarly. This completes the proof of Lemma 2. $\square$

4. A forced oscillation problem. In this section, we apply our theorem to a specially forced oscillation problem. The notation in this section is independent of that of previous sections.

Consider the following forced oscillation problem with an "almost" periodic forcing. Let

$$H(I, \theta) = H^0(I) + \epsilon H^1(I, \theta, \omega, t)$$

where $(I, \theta, \omega, t) \in \mathbb{R} \times S^1 \times T^n \times S^1$. We assume that the function H is periodic in t and in $\omega = (\omega_1, \omega_2, \ldots, \omega_n)$ with period 1. The equation of motion we shall consider is as follows:

$$
\begin{aligned}
\dot{I} &= -H_\theta(I, \theta, \omega, t) \\
\dot{\theta} &= H_I(I, \theta, \omega, t) \\
\dot{\omega} &= F(I, \theta, \omega, t) = F^0(I) + \epsilon F^1(I, \theta, \omega, t)
\end{aligned}
$$

where $F(I, \theta, \omega, t) = (F_1, F_2, \ldots, F_n)$ is a smooth vector function and is periodic in t and ω with period 1. One may consider F as the frequency function of ω, which, instead of being fixed, may change with respect to time and position. If F_i, $i = 1, \ldots, n$ are constants, i.e., if the forcing frequencies are fixed, and further if they satisfy diophantine conditions, then KAM theory asserts that, under the weaker non-degeneracy condition, the system is stable and most of the orbits will lie on some invariant tori, for sufficiently small ϵ.

Here we consider the case where forcing frequency changes with respect to time and the position of the motion. We must point out that the resulting system is no longer Hamiltonian. Because the functions F and H are periodic in t with period 1, we may study the time-1 map of the system. To apply the main theorem of this paper, we impose the following conditions on F_0 and H_0:

(1) $\sum_{i=1}^{n} \partial F_i^0 / \partial \omega_i = 0$. This implies the intersection property.
(2) $\partial H^0 / \partial I \neq 0$ and there exist an integer l such that the following matrix has the maximum rank $n + 1$:

$$
\begin{pmatrix}
(H^0)'(I) & (F^0)'_1(I) & \cdots & (F^0)'_n(I) \\
(H^0)''(I) & (F^0)''_1(I) & \cdots & (F^0)''_n(I) \\
\vdots & \vdots & \ddots & \vdots \\
(H^0)^{(n)}(I) & (F^0)^{(n)}_1(I) & \cdots & (F^0)^{(n)}_n(I)
\end{pmatrix}
$$

Under the above two conditions, Theorem 1 asserts that most of the orbits of the system lie on invariant tori, and because the invariant tori have co-dimension one, the system is topologically stable.

Similar results may be obtained for higher degree of freedom systems. However, it seems to be difficult to formulate the intersection property in terms of the perturbations.

REFERENCES

[1] C.Q. CHENG, Y.S. SUN, *Existence of invariant tori in three-dimensional measure preserving mappings*, Celestial Mechanics 47 (1990) 275–292.

[2] J. MOSER, *On invariant curves of area preserving mappings of an annulus*, Nachr. Akad. Wiss. Göttingen Math.-Phys. KI. II (1962) 1–20.

[3] J. MOSER, *Stable and Random Motions in Dynamical Systems*, Annals of Mathematics Studies No. 77 (1973) Princeton University Press.

[4] J. PÖSCHEL, *Integrability of hamiltonian systems on cantor sets*, Comm. Pure Appl. Math. 35 (1982) 653–695.

[5] C.L. SIEGEL, J.K. MOSER, *Lectures on celestial mechanics* (1971) Springer-Verlag, Berlin, Heidelberg, New York.

[6] N.V. SVANIDZE, *Small perturbations of an integrable dynamical system with an integral invariant*, Proc. Steklov Inst. Math. 147 (1980).

[7] Z. XIA, *The existence of the invariant tori in the volume preserving diffeomorphisms*, Erg. Th. & Dyn. Sys. 12 (1992) 621–631.

[8] J.-C. YOCCOZ, *Travaux De Herman Sur Les Tores Invariants*, Séminaire Bourbaki 754, (1992).